Martín Lara

Hamiltonian Perturbation Solutions for Spacecraft Orbit Prediction

De Gruyter Studies in Mathematical Physics

Volume 54

Martín Lara

Hamiltonian Perturbation Solutions for Spacecraft Orbit Prediction

The Method of Lie Transforms

DE GRUYTER

Physics and Astronomy Classification 2010
45.10.Hj, 45.20.Jj, 45.50.Pk, 07.87.+v, 05.45.-a

Author
Dr. Martín Lara
GRUCACI - University of La Rioja
CCT building
C/ Madre de Dios, 53
26006 Logroño
Spain
malarac@unirioja.es

ISBN 978-3-11-066722-6
e-ISBN (PDF) 978-3-11-066851-3
e-ISBN (EPUB) 978-3-11-066732-5
ISSN 2194-3532

Library of Congress Control Number: 2021932014

Bibliographic information published by the Deutsche Nationalbibliothek
The Deutsche Nationalbibliothek lists this publication in the Deutsche Nationalbibliografie; detailed bibliographic data are available on the Internet at http://dnb.dnb.de.

Cover image: cemagraphics / iStock / Getty Images Plus
Typesetting: VTeX UAB, Lithuania
Printing and binding: CPI books GmbH, Leck

www.degruyter.com

Preface

By the beginning of the 1990s I moved to Zaragoza to complete my technical education in the University. There, I followed the course of celestial mechanics taught by S. Ferrer, whose lectures were based on a manuscript he was compiling under the close supervision of A. Deprit. Regrettably, the manuscript was never concluded and sent to print, but some parts of it were distributed to the course students like handouts. That was my first contact with perturbation methods.

At those times Deprit visited regularly the University of Zaragoza, and I had the opportunity of attending to different lectures and seminars given by him. In particular, his course on symbolic manipulation with algebra systems was unique. The implementation of consistent simplification rules as well as the precise definition of the mathematical properties to be assigned to the variety of symbols that supplement the standard set of Keplerian elements, and infest common expressions of celestial mechanics, was central to the course. At the end it was the construction of the tables of partial derivatives that were essential in the assembly of perturbation solutions when expansions in the eccentricity are to be avoided.

But it was not until several years later that I paid true attention to perturbation methods. In my periodic attendances to the AAS meetings I soon realized that approximate mean-element solutions smartly obtained by astrodynamicists could be easily extended and refined with the help of standard perturbation methods. Moreover, the fact that Deprit's perturbation algorithm by Lie transforms is readily implemented in commercial computer algebra systems paved the way for achieving higher orders in a perturbation approach. Most of my research since then has focused on astrodynamics applications of perturbation methods, and this monograph is mainly the result of that work.

An introductory chapter provides some context on perturbation methods, whose use in the solution of astrodynamics problems is as old as the space era, yet reading it is not essential. Part I deals with the fundamentals of the Lie transforms method in the form in which it was lately explained by Deprit. My experience is that application of the Lie transforms method to integrable cases is a good exercise that helps newcomers in grasping the essence of what is being carried out. That is why two sample applications to classical integrable problems are also provided in this part.

Part II is devoted to the Earth–artificial satellite problem, which, as customary, is approached like a perturbed two-body problem. In the first chapter, the integration of the Kepler problem follows the usual Hamiltonian reduction to action-angle variables, which are the natural variables in which perturbed Keplerian motion is approached in subsequent chapters. The important contribution of the second zonal harmonic to the dynamics of close-Earth orbits is profusely discussed in the third chapter. The rest of the chapters of Part II describe how gravitational and non-gravitational effects acting on Earth orbiting satellites are treated by perturbations.

https://doi.org/10.1515/9783110668513-201

Part III deals with applications of perturbation methods to specific cases of the restricted three-body problem that are of interest in astrodynamics. After an initial chapter recalling basic facts of this non-reducible model, three distinct applications of the perturbation approach are proposed. First, the investigation of the dynamics about planetary satellites is approached like a perturbed Keplerian problem in a rotating frame. Due to the unstable dynamics that commonly affect science orbits about planetary satellites, this application shows that, beyond the insights provided by the mean-element dynamics, the process of mission design gets a significant benefit from the use of the mean to osculating element transformation. The suitability of the Lie transforms method for the automatic computation of higher orders of the perturbation solution is clearly illustrated in the next chapter with the analytical computation of halo orbits. Finally, the last chapter shows the difficulties that may arise in a perturbation approach when elliptic functions are involved in the process. While this is not a flaw of the method in itself, it shows the convenience of finding efficient procedures to automatically process special functions in a calculus of perturbations.

Different persons and circumstances kept my interest in the perturbation approach alive along the years. My acquaintance with the Lie transforms method grew thanks to many conversations with J. F. Palacián during his short summer visits to Real Observatorio de la Armada the years 2008 through 2010, in which our technical discussions were usually ended in a *chiringuito* after a refreshing bath at one of the splendid beaches of Cádiz or Chiclana de la Frontera. Later, the interest of P. Gurfil in onboard orbit propagation under limited resources reawakened my own interest in intermediary solutions of the artificial satellite problem, giving rise to three fruitful visits to Technion's Asher Space Research Institute, in Haifa, in the course of the years 2011–2014. Frequent and vivid discussions with S. Ferrer about Hamiltonian simplification procedures (and more), and with J. F. San-Juan on the technicalities involved in the computation of higher orders of the artificial satellite problem in closed form were always important stimuli in my research. The latter was also of invaluable help in finding some of the few old references that are not yet available in the collections so kindly compiled by the SAO/NASA Astrophysics Data System.

Attention paid by the new generations of astrodynamicists to the potential of semi-analytical propagation gave me the needed impetus to write these notes—whose original motivation could be traced back to early conversations on the topic with R. P. Russell. Interaction with R. Armellin and D. Hauteserres while developing software for ESA and CNES, respectively, for the efficient semi-analytical propagation of highly elliptical orbits, deserves particular mention. Also, the contact with young (and not so young!) colleagues during the KePASSA meetings has been a source of inspiration.

Lastly, the two PhD courses given in the Department of Aerospace Science and Technology of Politecnico di Milano the years 2018 and 2020 by the kind invitation of C. Colombo (the last one via telecon due to the pandemic), together with the encouragement of M. Efroimsky, editor of the series De Gruyter Studies in Mathematical Physics, were crucial to the completion of the product that is now in your hands, and I hope that it will meet your expectations.

Contents

To my lovely wife Chon. Who else!

1 Introduction

Before the advent of electronic computing machines, analytical solutions to the orbital motion of celestial objects were the common source for computing the ephemerides that are needed for scheduling astronomical observations. These approximate solutions were generally computed by calculating the variations or *perturbations* of the orbital motion with respect to a Keplerian ellipse or to other more sophisticated "intermediary" orbit. With the beginning of the space era, the same methods were imported to the realm of astrodynamics for predicting the motion of artificial satellites. However, the increasing accuracy of observations of both natural and artificial celestial objects soon made analytical solutions become unpractical. Indeed, including more and more perturbation effects in the theory in order to reach the precision required by observations augmented severely the difficulties in obtaining the solution, on the one hand, and made the analytical series representing the orbit expand unwieldy, on the other, with the consequent growth of the computational burden needed to evaluate the analytical solution.

With the irruption of electronic computers, numerical "special perturbations" methods took clear advantage over the "general perturbations" provided by the analytical approach. Still, the latter survived thanks to software progress in automatic symbolic manipulation. In the end, the much faster rate of hardware advances with respect to software development made the general perturbation methods to be displaced from the original purpose of making accurate predictions. But in no way the analytical approach became obsolete. New problems originating from the increasing saturation of the Earth's close space, as, for instance, the maintenance of space catalogs comprising thousands or millions of objects for collision avoidance purposes, can be efficiently carried out with analytical perturbation solutions. Their lower accuracy yet much faster computation plays a complementary role to the high-fidelity numerical integration that would be mandatory when a hazardous conjunction is detected under the accuracy provided by the analytical prediction. In addition, the analytical approach is useful in simplifying the dynamical model by removing non-essential short-period effects, in this way easing the design and optimization of the end-of-life disposal maneuvers that are required for compliance with current space law. Moreover, efficient guidance and control algorithms for relative motion, which are advantageously designed in mean elements, take great benefit of the accurate analytical conversion between mean and osculating elements that can be achieved with general perturbations.

Among the different general perturbation methods that make the computation of (approximate) analytical solutions feasible, the advantages provided by the Lie transforms method turn it into the standard of these days. It is systematic, specifically designed for automatic computation by machine, and versatile enough to deal with different simplification procedures. In addition, this method avoids the drawback

https://doi.org/10.1515/9783110668513-001

of other perturbation methods of relying on series reversion procedures. The method of Lie transforms applies generally to perturbations of vectorial flows, but it is particularly efficient in the treatment of Hamiltonian perturbations, which are the most common case in orbit propagation problems. Non-conservative effects, like the atmospheric drag which drives the dynamics of the lower altitude satellites orbiting the Earth, can always be incorporated into the Hamilton equations like generalized forces in a final step of the simplification procedure.

For these reasons, the most common orbital perturbations affecting artificial satellite missions are discussed in this monograph under the light of the method of Lie transforms. In particular, the treatment of perturbations of the Keplerian motion originating from the non-homogeneous distribution of the mass of the Earth, as well as lunisolar perturbations, is profusely discussed. Besides, solutions to the approximation of the circular restricted three-body problem provided by the Hill problem are also discussed with full detail in three particular cases that are achievable by perturbations and clearly illustrate the power of the Lie transforms method—as well as the difficulties that may limit its application. Namely, the motion of a space probe about a natural satellite perturbed by the gravitational pull of the mother planet, the motion about the libration points, and the coorbital motion of the (massless) orbiter and the smaller-mass primary around the primary of bigger mass are analytically solved by perturbations.

1.1 Perturbed integrable problems

There are just a few dynamical models that can be solved analytically. Real-world problems include a variety of effects that normally prevent, or at least complicate to a significant extent, the achievement of analytical solutions. However, in many cases the effects that frustrate integrability are small and one may reasonably expect that actual solutions behave like basic integrable models that were slowly distorting with time. Even though the time evolution of non-integrable problems may be unpredictable (chaotic), the time scale in which chaos manifests is very long in perturbation problems, at least in some regions of phase space, and hence, computing approximate analytical solutions makes full sense.

Original efforts in the computation of perturbation solutions to the planetary motion fructified with Lagrange's successful method of *variation of parameters* (see [42], for instance) in which the solution to the Newtonian equations of motion can be approached by Picard iterations. However, expansions carried out in the iterative process usually make the time explicit, producing the consequent deterioration of the solution as time grows. The appearance of secular or mixed secular-periodic terms in the expansion of the variation equations may be avoided with the use of Lindstedt series [459], in which case the method of undetermined coefficients is used to split the variation equations into a chain of differential equations that are then solved sequentially [516, 526].

Later efforts in solving perturbation problems resorted to Hamilton's ideas of finding integrals of a differential system by applying canonical transformations of variables, and gave rise to canonical perturbation methods [55, 144]. The perturbation approach is now applied to the scalar Hamiltonian function, contrary to the variation equations, and the solution is found by the stepwise construction of such canonical transformation that, up to some truncation order of the perturbation series, completely reduces the Hamiltonian to a function that only depends on the momenta conjugate to the canonical coordinates [30]. Finding the solution to the equations of motion (Hamilton equations) is trivial in the new variables, and the problem becomes solved in the original variables when the transformation constructed in the procedure is applied to this solution.[1]

The subsequent appearance of Poincaré's *méthodes nouvelles* [558] produces a breakthrough in the development of perturbation methods. In the Hamilton–Jacobi style, Poincaré derives the canonical transformation that solves the perturbation problem from a generating function in mixed variables. Poincaré's method is not directly applicable to degenerate perturbed Hamiltonians, a case in which Poincaré himself suggested to add to the generating function arbitrary functions of the angle variables [198, 407, 558]. In the latter case, variations of Poincaré's perturbation method introduced by von Zeipel [698] in his studies of the motion of minor planets, and later applied by Brouwer [75] to the solution to the artificial satellite problem, achieved such a great success that the modifications to Poincaré's method for dealing with perturbed degenerate Hamiltonian problems are these days customarily known as either the Brouwer–von Zeipel [667] or von Zeipel–Brouwer method [198].

The fact that Poincaré's generating function is in mixed variables makes the use of series reversion procedures necessary, and commonly complicates obtaining the solution to higher than the first order of the perturbation approach. On the other hand, infinitesimal contact transformations defined by Lie series [254, 458] can be derived from a generator in an explicit manner, thus paving the way for the eventual appearance of the method of Lie transforms. In the latter case, the canonical transformation is obtained explicitly from a generator in non-mixed variables, in this way avoiding the need of series reversion. Beyond the basic ideas in a seminal paper by Hori [303], the Lie transform method was thoroughly developed by Deprit [151] as a technique specifically devised for computer implementation.[2] In Deprit's conception, the method of Lie

---

1 Integrals found with this method do not exist in general, yet may survive in some non-resonant regions [29, 353].

2 The independent algorithms of Hori and Deprit are, in fact, equivalent, as discussed by different authors [91, 288, 486]. On the other hand, efficient alternatives to Deprit's algorithm are commonly used in other fields than astrodynamics and celestial mechanics, like plasma physics, optics, and molecular dynamics [177, 178, 619]. Still, the different approaches are essentially equivalent [356].

transforms enjoys a great generality encompassing a variety of applications [490]. In particular, different Hamiltonian simplification procedures [166, 167] have been devised based on Deprit's perturbation algorithm [14, 133, 154, 169, 402, 602], which furnish the method of Lie transforms with a much wider scope than the traditional normalization of a Hamiltonian by the introduction of formal integrals [238]. Moreover, the original application of the method of Lie transforms to Hamiltonian perturbations was very soon extended to generally deal with perturbations of vectorial flows [282, 304, 333], or with mixed perturbation models [39].

Even though the latter invention of two new operations over power series—the skew composition and the skew reversion of series—extended the functionalities of Poincaré's perturbation method to the same level of the method of Lie transforms [174], the latter is generally accepted as the standard perturbation method of today due to its great generality and the simplicity derived from a recursive algorithmic definition [491]. Because of that, the perturbation solutions in this monograph deal exclusively with the method of Lie transforms. The mathematical foundations of the method are these days rigorously described in different reports and textbooks [54, 93, 94, 491, 510, 627], and hence we only deal with the practical implementation of the method, to which we devote the first part of the monograph. Thus, Chapter 2 presents the description of the fundamental algorithm and its application to conservative Hamiltonian perturbations. The case in which the perturbation Hamiltonian depends on time can be equally approached by the simple expedient of moving to the extended phase space [15, 558, 621], a case that is illustrated later in Chapter 8 with the treatment of third-body perturbations. The method deals with formal series in the sense that convergence issues are not discussed [239, 595], yet the appearance of small divisors in the perturbation series is tackled in Chapter 7 in reference to tesseral resonances.

The essence of the method of Lie transforms is illustrated in Chapter 3, where it is applied to the computation of approximate solutions of two integrable problems, namely the mathematical pendulum and the triaxial free rigid body. These integrable problems are approached like perturbations of simpler integrable problems: the spherical rotor in the case of the pendulum (§3.1.2), and the uniaxial free rigid body (§3.2.6) or the harmonic oscillator for the free rigid body (§3.2.7). The free rigid body application serves also to show that, in spite of the reduction of a Hamiltonian perturbation problem to its secular frequencies being naturally approached in the action-angle variables of the integrable part, which is taken as the zeroth order of the perturbation Hamiltonian, the procedure is expedited in some cases with the use of other sets of canonical variables. Indeed, while Hamiltonian perturbation methods are naturally implemented within an action-angle variable framework, in which averaging operations are customary [594], a good knowledge of the particular algebra of the functions that comprise the perturbation Hamiltonian when formulated in other set of canonical variables is of great help in the computation of the perturbation solution. In particular, the use of complex variables is recognized to be advantageous when dealing with perturbed harmonic motion [373, 399]. At the end, since the Hamiltonian reduction is unique [30] one can always recover the action-angle variable formulation if found convenient.

Hamiltonian simplification algorithms based on Lie transforms are not discussed in this first part. The point of view of these days is that they are simple particularizations of the general procedure based on the solution of the so-called homological equation. Indeed, at each order of the Lie transform approach the user is endowed with an ample freedom to choose the form of the new Hamiltonian term, as well as the arbitrary integration "constant" that leaves undetermined the generating function term [402, 408, 438, 440–442, 511, 593]. On the contrary, details of the implementation of the different simplification algorithms are supplied in the chapters where they are first applied. In particular, the *elimination of the parallax* and the *elimination of the perigee* are discussed in Chapter 6, while the *relegation algorithm* is introduced in Chapter 7.

On the other hand, it is worth recalling that physical phenomena are measured in observable variables (for instance, spherical coordinates), contrary to action-angle variables. Therefore, having a detailed description of the transformation from the usual observables to the action-angle variables in which the zeroth-order Hamiltonian term is solved becomes essential.[3] Because of that, the reduction of the integrable Hamiltonian that defines the zeroth-order term of the perturbation Hamiltonian via the Hamilton–Jacobi equation [199, 243] is thoroughly discussed in different chapters of the monograph.

1.2 Artificial satellite theory

Departure from Keplerian motion of close-Earth satellites is mostly driven by the noncentralities of the Earth's gravitational potential, the main part of which is due to the Earth's oblateness. Disturbances produced by the second zonal harmonic, whose nondimensional coefficient is customarily noted J_2, cause the precession of the orbital plane on the equator, and induce a secular trend on the motion of the perigee, which advances or regresses in the orbital plane depending on orbital inclination and gets fixed at the *critical inclination* of 63.4 degrees. In consequence, Keplerian solutions only are useful in forecasting short arcs of Earth's satellite orbits under low accuracy requirements. Even in the short times spent in the free-flight trajectory of a ballistic missile, the Earth's oblateness needs to be taken into account [678]. Therefore, early efforts in making accurate predictions of the motion of Earth's artificial satellites focused on the search for analytical solutions to the "main problem" of artificial satellite theory, also called the J_2-problem. Even though the J_2-problem is not integrable [102, 312], the time scale in which chaos manifests is only relevant for large values

3 This rule has exceptions and there are cases in which the observables are precisely the good solution variables [644].

of J_2 [70, 138]. On the contrary, in view of the smallness of the Earth's J_2 coefficient, which is of the order of one thousandth, the search for integrable approximations of the main problem of artificial satellite theory makes full sense.

A wealth of analytical approximations to the main problem dynamics have been proposed since the beginning of the space era. These solutions are grouped under the general name of *intermediary orbits*, and commonly based on the separability of the Hamilton–Jacobi equation of some modified version of the main problem Hamiltonian [6, 111, 154, 216, 217, 219, 225, 532, 620], yet non-canonical solutions also exist [33, 326, 606]. While most intermediaries were constrained to mimic the main problem dynamics up to first-order effects of J_2, remarkable exceptions were successful in handling the whole effect of J_2 as well as some second-order effects of the geopotential [5, 37, 38, 315, 466, 662–666]. Still, intermediary solutions fail in predicting the libration dynamics of the perigee in the vicinity of the critical inclination [218, 261, 301, 316, 555, 625], a fact that is sometimes attributed precisely to their separability [119, 168, 406]. Because of that, and due to the increase of computational power, classical intermediary solutions were soon abandoned in favor of the higher-accuracy solutions obtained with perturbation theory [299]. On the other hand, the interest in intermediary orbits is experiencing some revival these days for their application to onboard short-term, orbit propagation under limited computational resources [51, 259, 268, 269, 393, 475, 603]. Computing perturbations of a more meaningful intermediary than the Kepler problem should provide more compact equations and seems a desirable aim [7, 633, 689]. In this sense, the obtention of the action-angle variables of Vinti's oblate spheroidal intermediary seems a promising result [681, 688], yet the higher complexity of the functions involved in the solution with respect to the simpler Keplerian orbit might counterbalance the expected improvements to some extent.

Dealing with perturbed Keplerian motion is conveniently approached in orbital elements, which disclose the existence of short- and long-period terms, as well as secular perturbations (see [341], for instance). Then, in the general framework provided by perturbation methods [526], solutions of the problem of artificial satellite theory are commonly decomposed into the mean-elements equations, which provide the orbit evolution, and the short-period corrections that are needed for ephemeris computation. The transformation from mean to osculating variables given by the short-period corrections is obtained analytically, whereas the mean-elements equations remain in the form of a reduced differential system. The latter is numerically integrated with very long step sizes because it only depends on the lower frequencies of the motion [354, 453]. This semi-analytical integration scheme can cope with many different perturbations of the pure Keplerian motion, and it is implemented in different orbit integration packages [81, 123, 355, 435, 478, 674, 692]. If, besides, the long-period effects are removed from the mean-elements equations by means of a new transformation of variables, which is also computed analytically, one obtains the secular frequencies of the motion. The latter are trivially integrated, thus providing the secular terms that complete the solution in a pure analytical form [58, 75, 171, 360, 461, 599].

Alternatively, both the short- and long-period terms can be removed at once, with a sole transformation, yielding a single set of analytical corrections that may allow for better code optimization [407]. Evaluation of analytical solutions computed in either of these manners is straightforward and permits the implementation of very fast an efficient orbit propagators [114, 121, 298, 380, 590, 641]. However, it must be noted that the validity of the analytical approach is limited to the case of motion far away enough from inclination-related resonances, which require specific treatment [120, 222, 316, 330, 394].

A single general perturbation solution to the artificial satellite problem makes no sense if higher-order effects need to be taken into account. On the contrary, specific perturbation solutions must be computed depending on particular orbital regimes in which different disturbing effects may dominate the dynamics. Indeed, below the geosynchronous distance the non-centralities of the geopotential have the most important effect, which is clearly dominated by the J_2 harmonic. Conversely, the gravitational pull of the moon is the most important perturbation above the geosynchronous distance, followed by that of the sun. The solar radiation pressure yields higher-order effects for usual satellites, yet they may become comparable to the J_2 effect at high altitudes.

Alternatively to the reduction of the differential system by normalization, Picard's iterative approach is sometimes used to solve perturbed Keplerian motion in what is customarily known as a numerical–analytical approach [4]. Namely, the variation of parameters equations are solved *analytically* under the usual averaging assumption that the parameters (orbital elements) remain constant in the right side of the variation equations. The validity of this analytical solution, which strongly relays on the use of special functions, is limited to one anomalistic period, and hence initial conditions must be *numerically* updated in steps of this size or shorter. Improvements are found when solving the perturbation equations with respect to an intermediary orbit in Eulerian elements, as opposite to the variations of Keplerian elements [244].

Perturbed Keplerian motion is efficiently approached in Delaunay action-angle variables. Therefore, the second part of this monograph starts in Chapter 4 with the solution of the Kepler problem via the Hamilton–Jacobi reduction to action-angle variables, yet constrained to the case of bounded, elliptic motion. Besides, the main problem dynamics is discussed to a considerably extent in Chapter 5. This chapter is not only useful in understanding intermediary solutions, but it serves to prepare the reader for the Hamiltonian simplification methods that will be thoroughly discussed in following chapters of this part.

1.2.1 Significance of Lyapunov instability

Two particular application of the perturbation approach that are conveniently approached in the mean elements setting are the analysis of end-of-life disposal strategies [27, 28, 124] and the design of formation flying reconfiguration and station-keeping maneuvers [16, 211]. The need of having available efficient equations for carrying out the accurate transformation from osculating to mean elements and vice versa [256], arises in both kinds of problems. Traditional methods for the computation of the initialization constants of the perturbation solution base either on series reversion or the more common root-finding procedures [90, 675]. On the contrary, the Lie transforms method provides explicit expressions for the evaluation of both transformations, direct and inverse, in this way allowing for the instant accurate conversion from osculating to mean or secular elements [212, 213].

Because of the Lyapunov instability that is inherent in Keplerian motion, it happens that the most sensitive element to be transformed is the osculating orbit semimajor axis, a fact that was clearly observed when checking the accuracy of Brouwer's [75] seminal solution against numerical integration [469]. Indeed, an error in the osculating to mean transformation of the semimajor axis directly translates into a comparable error in the computation of the mean motion in mean variables, with the consequent impact on the secular frequencies of the motion [63, 181, 393]. This fact makes it quite desirable to achieve the transformation from osculating to mean elements, at least for the semimajor axis, up to the same truncation order as the secular terms. On the contrary, an error in the opposite transformation, from mean to osculating elements, has only periodic implications in the computation of ephemerides. Therefore, truncating this last transformation to a lower order than that of the secular terms usually provides the required accuracy for most applications, and makes the evaluation of the analytical solution much more efficient [269].

The way in which the accurate conversions from mean to osculating elements, and vice versa, affect the precision of an analytical perturbation solution is illustrated in §6.5 for a higher-order perturbation solution of the main problem.

1.2.2 Geopotential long-period effects in closed form

Traditional efforts in extending perturbation solutions to higher orders resorted to the use of expansions of the elliptic motion in powers of the eccentricity [78, 171, 346, 374]. The literal calculation of these kinds of expansions motivated the development of the first computerized procedures [172, 173, 289], which paved the way for the later appearance of the so-called Poisson series processors [69, 83, 140, 236, 286, 323, 570]. However, these kinds of expansions cast the analytical solution in the form of long multivariate Fourier series, whose laborious evaluation discouraged practitioners from the use perturbation solutions. Alternatively, the computation of the perturbation solution in closed form yields a notable reduction in the size of the series to be evaluated, on the one hand, and widens applicability of the solution to the case of high-eccentricity orbits, on the other. However, reaching higher orders than that achieved by Brouwer in his closed-form solution [75] finds real (yet solvable) difficulties in the indefinite integration of the equation of the center—a periodic but non-trigonometric function—and related functions [8, 156, 363, 487, 536].

In the usual approach, in which the short-period terms are removed first, in the style of Brouwer, these difficulties can be delayed to higher orders after an ad hoc preprocessing of the original Hamiltonian using the elimination of the parallax simplification technique [154]. Indeed, the elimination of the parallax strips the zonal potential of non-essential effects and performs wonders in a higher-order normalization in closed form [118, 119, 155]. The main features of the elimination of the parallax canonical simplification are described in §6.2.1.

Alternatively, carrying out the normalization in the reverse order, by first eliminating the long-period terms through the standard normalization of the total angular momentum, notably simplifies the subsequent removal of short-period terms [408]. This latter approach does not deny the use of the elimination of the parallax as a preliminary simplification, after which the elimination of the perigee [14, 440] is carried out to reduce the Hamiltonian to a one-degree-of-freedom system depending only on short-period terms. The result of both eliminations, which can be combined into a single Lie transformation [593], is what is sometimes called the *elimination of the latitude* [133, 166]. The consequent removal of the remaining short-period terms with the standard Delaunay normalization [156] turns the conjugate momentum to the mean anomaly into the third (formal) integral of the satellite problem, thus achieving the complete Hamiltonian reduction. Full details of the reverse normalization are given in §6.4.

Regrettably, the equations of the canonical transformation that eliminates the perigee are flawed with divisors involving the difference between $\frac{4}{5}$ and the square of the sine of the inclination. Therefore, it becomes singular at the *critical* direct (resp. inverse) inclination of ~ 63.4 deg (resp. 116.6 deg), and prevents convergence of the perturbation solution for inclinations close to that value. This singularity is essential to the artificial satellite problem [120, 135] due to a resonance between the anomalistic and draconitic frequencies of the orbiter [216, 394], and hence cannot be avoided with a different choice of variables.

It deserves to be mentioned that, when just dealing with the solution in mean elements, the removal of short-period terms in closed form commonly deprives the mean-elements equations from some long-period terms, which, on the contrary, remain hidden in the transformation equations of the averaging. This fact may force the orbit in mean elements to depart from the average evolution of the true orbit. While this issue was clearly identified in classical closed-form averaging procedures [363], it may remain hidden when the short-period averaging is carried out after the elimination of the parallax. In both cases, the situation is partially amended by the addition of an adequate integration "constant" to the corresponding term of the generating function of the transformation [197, 437, 442, 488, 618]. This problem does not happen in the computation of the secular terms, because all the periodic effects are removed in that case. Issues related to getting the mean elements close to the average value of the osculating elements are further discussed in §6.7. We remark that this kind of problem is only relevant for the propagation of the mean-elements orbit, and it has no effect in a semi-analytical integration scheme, in which the periodic terms are recovered analytically to get osculating elements.

1.2.3 Tesseral effects

The longitude-dependent terms of the gravitational potential expansion, tesseral and sectoral harmonics, which are encompassed hereof under the name of tesseral harmonics, in general just produce short-period variations of small amplitude on the motion of an artificial satellite. Among them, the more relevant effects are related to those tesseral terms that are free from the argument of the latitude. These terms are sometimes denoted "m-daily" terms because their trigonometric arguments depend only on integer values of the longitude of the node, and, therefore, repeat their values m times a day [335].

Interactions between zonal and tesseral harmonics only produce short-period variations that, besides, remain at higher orders of the perturbation solution [687]. On the contrary, interactions between different tesseral harmonics may produce secular and long-period variations in addition to the short-period variations. Nevertheless, their amplitudes are very small and only modify the solution to a very minor extent [489, 686].

Taking the effect of the tesseral harmonics into account notably complicates the computation of the analytical perturbation solution in closed form. For this reason tesseral harmonics perturbations are customarily approached after expanding the gravitational potential in powers of the eccentricity [82, 341, 348, 563, 638, 685]. Still, part of these tesseral terms can be previously simplified in closed form with the elimination of the parallax preprocessing [117]. A notable exception is found in the case of low Earth orbits, where the fact that the mean orbital motion and the Earth's rotation rate are of different orders of magnitude allows for the closed-form removal of tesseral terms related to the mean anomaly in the usual form [220, 221, 439, 539, 541]. This approach is illustrated in §7.2.

Efforts in developing algorithms for the closed-form elimination of tesseral effects in the general case led to the invention of the relegation algorithm [169, 538]. Application of this algorithm for the closed-form elimination of tesseral terms in the artificial satellite problem, the so-called "relegation of the node", found some success in the case of super-synchronous orbits [97, 591, 602]. However, the efficiency of the relegation algorithm is more questionable in the case of sub-synchronous orbits. Indeed, due to a hidden dependence on the eccentricity in the algorithm for the sub-synchronous relegation [602], the relegation of tesseral effects rarely offers clear advantages when compared with classical procedures for their elimination based on the traditional expansions in powers of the eccentricity [438]. Despite this shortcoming, it has been suggested that the dependence on the eccentricity of the sub-synchronous relegation can be used to modify the classical relegation algorithm in order to obtain enhanced results in the elimination of tesseral terms of low-eccentricity orbits close to tesseral resonances [441]. The classical relegation algorithm for the removal tesseral effects in closed form is discussed in §7.4, where details on the particular version of the relegation for low-eccentricity orbits are provided, and its efficiency is illustrated with sample applications.

On the other hand, the short-period elimination of tesseral harmonics in closed form can be exactly reduced to quadratures [471, 540]. However, the integrals to be solved involve non-integer combinations of the true and mean anomalies and their closed-form solution is not known. In consequence, one must resort to series representations that involve either the eccentricity or the ratio of the Earth's rotation rate to the satellite's mean motion in order to solve these integrals analytically, thus making the procedure analogous to the relegation algorithm. Alternatively, these quadratures can be solved numerically, yet they still rely in the series representation to compute the integration constant of the generating function, which cannot be left arbitrary when the integration is approached numerically. The exact reduction to quadratures of the short-period elimination of tesseral effects is outlined in §7.3.

In addition to the short-period tesseral effects of small amplitude, which slightly affect the precision of the ephemeris provided by the perturbation solution, orbital configurations leading to tesseral resonance produce noticeable long-period effects on the semimajor axis. Resonant effects due to tesseral harmonics become apparent in orbital regimes in which the satellite advances Q nodal periods in the time in which the Earth rotates P times relative to the precessing orbital plane, P and Q being mutually prime integers. That is, the tesseral resonance happens when the combined rate of variation of the argument of the perigee and the mean anomaly is commensurable with the combined rate of variation of the longitude of the node and the Earth's rotation rate. Therefore, tracing tesseral-resonant terms of the geopotential for particular orbit regimes requires the explicit appearance of the mean anomaly, thus making it unavoidable to resort to the usual expansions of elliptic motion.

Tesseral resonances bring significant perturbations on time scales of interest for mission planning without constraint to the case of deep resonances [180]. After removing short-period terms from the tesseral Hamiltonian by perturbations, the resonant combination is replaced by a single variable, the so-called longitude of the stroboscopic mean node, which defines the geographic position of the subsatellite point at the time of intersection of the Earth's equator by the satellite orbit.[4] Station-keeping control strategies are properly developed in the pair given by the stroboscopic mean node and the semimajor axis [183, 228, 334], the latter being the orbital element that is primarily affected by the resonance phenomenon.

4 The nomenclature "stroboscopic mean node" has been attributed to Garfinkel [228].

The analytical approach has been quite successful in the relevant problem of geosynchronous satellites [18, 52, 53, 341, 509]. In that case, the resonant dynamics is mainly driven by the $J_{2,2}$ term, which is the longitude-dependent term associated with the ellipticity of the Earth's equator. Isolated resonances enjoy the simple pendulum dynamics, which is an ideal case that can be integrated analytically [18, 19, 222, 228, 331, 381]. However, when different resonances overlap, the problem generally remains of two degrees of freedom and, except for special configurations [434, 443, 612], it is customarily studied by the numerical integration of the long-term equations, showing that chaos may arise [142, 145, 149, 186–188]. Still, attempts to deal analytically with the case of non-isolated resonances using Bohlin's theory [55, 198] have also been carried out [507, 508, 572, 573].

The peculiarities of the perturbation approach when applied to resonant orbital regimes is illustrated in §7.5. Because the important case of geosynchronous satellites, which undergo the effects of the 1:1 tesseral resonance, is profusely discussed in the literature [45, 509, 522, 531, 613], we rather deal with the case of orbits of traditional constellations providing global navigation services, which are also significantly affected by different tesseral resonances [108, 146, 580]. Thus, the orbital configuration of Galileo operational satellites repeats after 17 orbits of the satellite, which is commensurate with 10 rotations of the Earth. While the effects of this shallow 10 to 17 tesseral resonance are not too relevant due to the high altitude of Galileo satellites, the operational orbits are also slightly affected by the 3 to 5 tesseral resonance, whose effect can be much more important in the case of Galileo disposal orbits. Besides, GLONASS satellites are in shallow 8 to 17 resonance, and the GPS constellation is in deep 1 to 2 resonance. Thus, GPS satellites complete two nodal orbits while the Earth completes one rotation. Due to this strong resonance all GPS satellites exhibit evident long-period variations in the semimajor axis [601], with different resonant geopotential terms having observable effects [309]. In particular, the harmonic coefficients of degree three and order two produce the main effects on GPS orbits. Still, perturbations due to this tesseral resonance cannot be considered isolated, and the resonance overlapping of other tesseral harmonics must be taken into account. On the other hand, due to the high altitudes in which global navigation satellite systems reside, low-order truncations of the geopotential normally suffice for different applications [204]. In consequence, the sample applications provided in §7.5 are limited to a geopotential truncated to degree and order five.

1.2.4 Lunisolar perturbations

Depending on the orbital regime, lunisolar perturbations caused by the gravitational pull of both the moon and the sun may also show important effects in the long-term dynamics of Earth orbits. When dealing analytically with gravitational perturbations of a distant body, the third-body disturbing function is customarily expanded in the ratio of satellite's radius and third body's radius as an infinite series in Legendre polynomials. For low Earth orbits lunisolar perturbations are small when compared with the Earth's oblateness disturbing effect, and can be modeled as second order of J_2 effects in a perturbation approach. In that case, neglecting parallactic terms of the third-body perturbation provides an acceptable modeling of the dynamics [128, 359, 450, 519]. However, parallactic effects due to the lunar attraction are clearly observable in high Earth orbits, a fact that motivated the early incorporation of these kinds of terms into the lunar disturbing potential [521].

Initial formulations of the lunar disturbing function referred the angular elements of the moon to the equatorial plane of the Earth [340], with respect to which the inclination and longitude of the node of the moon orbital plane experience important variations. Alternatively, the inclination of the moon orbit takes an almost constant value of about 5° when referred to the ecliptic, whereas the longitude of the ascending node in that plane can be approximated by a linear function of time [80, 481]. Therefore, it is advantageous in some cases to refer the moon's orbital elements to the ecliptic [366, 367, 581].

Truncations of the lunar disturbing function up to the fourth degree in the parallactic ratio are considered acceptable up to the geosynchronous region [121, 640], yet the fifth degree is also added when a more accurate modeling of the dynamics is required [437, 488]. Still, higher-degree truncations may be needed when modeling high Earth orbits [337, 338, 436]. Available recurrence relations allow for the extension of the series expansion of the third-body disturbing function to any degree [98, 230, 233, 348]. Typos and errors in these references that were pointed out in the literature [382] included in turn additional typos and errors, which only were amended recently for the more relevant formulas [101]. Analogous general derivations are presented in §8.1, in which we depart from tradition and rather rely on the vectorial approach in the apsidal frame [425].

In the presence of third-body perturbations, the removal by perturbation methods of short-period effects related to the mean anomaly of the satellite provides an efficient way of solving the problem semi-analytically. Formulas for the mean-elements Hamiltonian as well as the generating function from which short-period corrections are derived, are discussed in §8.2 for an arbitrary truncation of the third-body disturbing function. Mean motion resonances between the artificial satellite and the moon are of concern only in cislunar space, and they do not cause trouble in the case of close Earth orbits. Hence, removing monthly effects, related to the mean anomaly of the moon, and, in some cases, annual effects related to that of the sun, in addition to short-period effects associated to the satellite's mean anomaly, is common practice to improve performance of the numerical propagations used in the investigation of the long-term dynamics [122, 382, 437, 443, 617, 683]. The double-averaging procedure that removes the mean anomaly of the third body is discussed in §8.3 based on the particular characteristics of the orbits of the Galileo constellation.

On the contrary, resonances between the mean motion of either the sun or the moon and the rate of variation of other angular elements of the artificial satellite's orbit are common in orbital regimes below the geostationary region [128, 305, 307, 308]. In particular, *apsidal* resonances occur when the secular motion of the line of apsides is commensurable with the mean motion of the third body, and *nodal* resonances happen when the critical argument is the right ascension of the ascending node [64–66]. These resonances may induce important variations in the eccentricity of the satellite's orbit and are these days scrutinized as possible natural ways of speeding the satellite de-orbiting up at the end of life [12, 108, 240, 580]. On the other hand, special configurations have been detected in which secular variations in the eccentricity and inclination cancel out giving rise to the so-called balanced Earth satellite orbits [371, 372]. The latter are discussed in §8.5 from the perspective of the ecliptic frame formulation.

1.2.5 Non-conservative perturbations

Perturbed two-body problems in which the perturbations involve the velocity appeared first in connection to the solar system dynamics [556, 561, 568, 690]. Non-conservative perturbations may have appreciable effects on Earth satellite orbits [43, 344, 445, 496, 659, 677], but also in other different kinds of orbits [49, 86, 262, 383, 493, 596, 597]. In particular, the force exerted by solar radiation has great importance in the (passive) dynamical evolution of objects with high area-to-mass ratio, like fragments of thermal blankets that may detach from the satellite's body or other similar objects [25, 95, 107, 210, 460, 576, 639]. But it also provides an inexhaustible source of power for the propulsion of space vehicles within the solar system that can be effectively exploited with solar sails [226, 463, 470, 479, 634, 648] and allows for the operation of novel satellite concepts like "smart dust" devices [32, 126, 484, 528, 699].

Effects of solar radiation pressure (SRP) were clearly identified like the origin of discrepancies between observed perigee heights of the orbits of some of the first artificial satellites and corresponding predictions based on conservative force fields models [523, 545]. This fact led to subsequent theoretical efforts to understand and estimate the long- and short-period effects induced by SRP on the dynamics of artificial satellites [128, 520], which, under general assumptions, can be modeled like a potential function [306, 340], in this way making the SRP effect amenable to Hamiltonian treatment. The modifications of the SRP acceleration resulting from the intermittent eclipsing of satellites by the Earth were taken into account since the very beginning [10, 85, 195, 361, 375, 605]. However, the effects of the Earth shadow, which repeat with the orbital period, are customarily neglected in studies of the long-term behavior, where long-period terms make the principal contribution to the orbit dynamics. Disturbances produced by subtle phenomena such as penumbra transitions and albedo effects have also received important attention in the literature [59, 584, 668–671].

While the disturbing effects of SRP commonly average out in the long-term, some orbital configurations lead to resonances between SRP and the motion of the line of apsides, or the combined motion of this line and the line of nodes [76, 128, 306, 520]. These kinds of resonances produce long-period oscillations of the eccentricity that may have significant amplitudes, and, similarly to the case of third-body resonances, are explored these days as a possibility for accelerating de-orbiting of low Earth orbit satellites at the end of their lifes [13].

The perturbed Keplerian dynamics induced by SRP can be represented by a time-dependent three degrees of freedom Hamiltonian. However, the convenient formulation in a rotating frame may be used to remove the time dependency from the Hamiltonian. Except for particular integrable cases, the problem is intractable in its generality. Still, an important insight into the long-term behavior can be obtained through the perturbation approach. After the usual removal of short-period terms from the Hamiltonian, the reduced two-degrees-of-freedom problem becomes integrable [105, 158, 159, 493, 495]. This fact furnishes the reduced model with an analogous category to the zonal intermediaries of the main problem [224]. On the other hand, simple solutions taking the oblateness disturbing effects into account in addition to SRP [494, 520] show that Earth's artificial satellites can undergo dramatic orbital changes under small variations of the initial conditions as well as the force parameters [125, 369, 370]. The perturbative treatment of SRP disturbances is presented in §9.1.

Another important non-conservative effect is the atmospheric drag, which may cause important perturbations on the lower Earth orbits. Indeed, the dissipation of orbital energy caused by the interaction of the upper atmosphere with the satellite, yields a reduction of the semimajor axis and induces a circularization trend in the orbit. These dissipative effects are noticeable on altitudes, say, below 2000 km and become the dominant perturbation in reentry orbits. While atmospheric drag perturbations are not derived from a disturbing function, their effects can still be added like generalized forces to the Hamilton equations. They are also amenable to perturbation treatment due to the previously mentioned extensions of the Lie transforms method to vectorial flows [282, 304, 333].

The drag force is usually modeled as the product of the atmospheric density, the square of the satellite's velocity relative to the atmosphere, the reference surface area of the satellite scaled by the satellite mass, and the "drag coefficient" C_D. The latter depends on gas-surface interactions, which involve a variety of facets like the atmospheric temperature and composition, or the satellite's shape [501]. While accurate determinations of the drag coefficient are important for precise orbit predictions [483], taking a constant value $C_D = 2.2$ is still customary in long-term orbit prediction of standard satellites in low Earth orbits [129]. Indeed, due to the difficulties in modeling the upper atmosphere accurately, the atmospheric density is a major source of error in predicting the drag force, a fact that may make efforts in improving the drag coefficient superfluous [513].

Pioneering perturbation solutions described the contraction of almost circular satellite orbits under the disturbing effect of the atmospheric drag alone [131], but following approaches soon included the coupling between the drag and oblateness perturbations [79, 378], and removed previous limitations of the perturbation solution to the case of small eccentricities [379].[5] Since then, a variety of analytical solutions of perturbed Keplerian motion that include atmospheric drag effects have been proposed [40, 148, 285, 475, 476].

On the other hand, it is often enough just to deal with averaged drag effects, which are obtained after removing the mean anomaly from the drag equations [462]. This is the common case of orbit propagators in which complex force models are integrated semi-analytically. Besides, in view of the increasing complexity of atmospheric models, the averaging is most efficiently carried out by numerical quadrature. This last case is the only one that is tackled in this monograph, and it is discussed in §9.2.

1.2.6 Action-angle and non-singular variables

The reduction of perturbed Keplerian motion is commonly carried out in Delaunay canonical variables, which are the action-angle variables of the Kepler problem [144, 243, 395]. Still, it is well known that Delaunay variables suffer the same deficiencies as their non-canonical counterpart: the Keplerian orbital elements. Indeed, Delaunay variables are singular for circular orbits, a case in which the argument of the perigee is not defined, and also in the case of equatorial orbits, where the longitude of the node is not defined. However, these deficiencies of the Delaunay variables are easily avoided by reformulating both the secular frequencies and the periodic corrections of the perturbation solution in non-singular variables. Poincaré canonical variables [468] or the set of equinoctial orbital elements [31, 72] are popular sets of non-singular variables used in perturbation theory, yet other different options have been proposed in the literature [294, 296, 647].

When low-inclination orbits are not of concern, a popular set of non-singular variables for zero eccentricity orbits based on the traditional set of Keplerian elements includes the mean distance to the node (or mean argument of the latitude), and the semi-equinoctial elements that materialize the eccentricity vector in the nodal frame [130]. These non-singular variables are usually complemented with the longitude of the node, the inclination, and the semimajor axis [132, 181]. In the Hamiltonian setting, the two last are commonly replaced by the third component of the angular momentum vector and the Delaunay action, respectively. Still, this last set of variables lacks canonical structure [171].

5 The solution of [79] was later amended in [658] using a technique developed in [196] for removing spurious secular terms in dissipative systems.

Another set of variables that is free from singularities in the case of circular orbits is the canonical set of polar-nodal variables, yet the singularity remains for equatorial orbits. Still, the trouble with small divisors only arises when the perturbation model takes odd zonal harmonics into account. Alternative non-singular variables based on the polar-nodal set have been proposed to avoid the poor convergence of the short-period corrections for almost equatorial inclinations, either in the non-canonical [9] or canonical modalities [121]. Furthermore, the use of polar-nodal variables reduces dramatically the size of the perturbation series needed in the analytical approach, in this way improving evaluation of the solution [317, 363]. Their use is then encouraged, and it has been suggested that the solution of almost equatorial orbits should be treated separately. In this last case non-singular variables based on polar-nodal variables admit radical simplifications [392, 395, 444].

It deserves to be mentioned, however, that polar-nodal variables are not very useful in the integration of the mean-element equations, and their utility in dealing with analytical or semi-analytical solutions of perturbed Keplerian problems is usually constrained to the efficient evaluation of the periodic corrections. Moreover, polar-nodal variables facilitate the reduction of perturbed Keplerian motion to quasi-Keplerian systems [154]. This approach provides an efficient alternative to usual orbit prediction procedures [259, 269, 393] that can also be used in the case of relative motion applications [416].

In spite of the compact formulation and wider applicability of closed-form solutions, the perturbation solution can be notably simplified in the case of orbits with specific characteristics. This is, in particular, the case of low-eccentricity orbits, where the closed form is efficiently replaced by short truncations of the usual expansions of the elliptic motion to the lower orders of the eccentricity. In this way the evaluation of the perturbation solution is made notably easier [181, 389, 637].

On the other hand, the use of vectorial elements in orbital problems, which can be traced back to the early works of Strömgren [623] and Milankovitch [497], provides a compact and elegant representation of the equations of motion, on the one hand, and releases them from the dependence on a particular reference system, on the other. The advantages of using Milankovitch's selection of the eccentricity vector and angular momentum vector like the vectorial elements in the treatment of very low-eccentricity orbits were soon recognized [291], and attaching the name of Milankovitch to this set of orbital parameters seems to have been popularized among some authors after [21]. While vectorial formulations increase the dimension of the differential system to integrate, this redundancy can be used to evaluate the quality of the numerical integration, in which, besides, the evaluation of trigonometric functions is reduced to a minimum [290, 291]. The latter fact, together with the more symmetric form taken by the equations of motion, makes the differential system simpler to program and faster to evaluate, thus making the vectorial formulations quite appealing in the formulation of efficient special perturbation schemes [152, 397, 518, 583, 636].

The vectorial approach also shows real merits in the formulation of analytical perturbations. Indeed, after the seminal application by Musen [520] and subsequent applications by Allan [17, 20], the use of the vectorial approach in the propagation of mean elements is experiencing a revival these days [112, 577, 676]. In particular, the vectorial formulation has demonstrated high efficiency in the integration of the mean vectorial elements of highly eccentric orbits under third-body perturbations [425].

Vectorial elements have been also useful to show that Keplerian Hamiltonians are invariant with respect to the group of rotations in a four-dimensional Euclidean space [512]. This fact becomes evident when the eccentricity and angular momentum vectors are replaced by its sum and difference vectors (see §45 of [621]), which are customarily known as Moser elements or Cartan coordinates [92, 202].[6] Since these two vectors enjoy the same constant modulus, which is proportional to the orbit's semimajor axis, an orbit can be represented by a pair of points on the surface of a sphere, as illustrated in §8.2.3. This characterization of Keplerian motion provides another utility of the vectorial elements, like the visualization of a high number of orbits by a cloud of points. In this way, for instance, the structure of subpopulations of debris orbits is readily made apparent [629].

1.3 Non-Earth orbits and perturbed non-Keplerian orbits

The first artificial satellites missions were shortly followed by the launching of unmanned lunar probes, as well as other interplanetary missions. While low Earth orbits are accurately represented by the perturbed Keplerian dynamics, the trajectories of space probes go beyond the distance in which the Earth's gravitational attraction is dominant. Then, the orbits do no longer resemble distorted Keplerian ellipses about the central body, and they need to be computed as particular solutions of a different problem. More specifically, in different instances the orbit dynamics of artificial satellites is better described in the more general frame of the three-body problem, in which three point masses evolve only under the influence of their mutual gravitational attraction (see [646], for instance).

1.3.1 The restricted three-body problem

Contrary to the Kepler problem, the three-body problem cannot be completely reduced by integrals [558], thus lacking a closed-form solution. Conversely, analytical solutions to the three-body problem can be computed in the form of power series [626], yet

6 Analogous elements seem to have been customarily used by quantum physicists, and they are sometimes attributed to Jauch and Hill; cf. [120] p. 374.

the poor convergence of these series makes them useless for computational purposes [44]. Therefore, it is customary to explore the qualitative aspects of the three-body problem dynamics by searching for particular solutions of the system, like singular points, periodic orbits, and other invariant manifolds. In particular, when referring the motion to the center of mass, the nine degrees of freedom of the three-body problem can be reduced to just three due to the existence of the energy integral as well as the preservation of the linear and angular momentum vectors, which, in addition, are combined with Jacobi's elimination of the nodes [157, 321, 684].

On the other hand, the model provided by the three-body problem can be notably simplified when dealing with spacecraft orbits. In that case the mass of the orbiter is negligible when compared to the mass of most natural celestial bodies of interest. Therefore, it does not produce observable effects in the motion of the two massive bodies, the primaries, which then are assumed to evolve about their mutual center of mass in the two-body problem approximation. When the motion of the primaries is circular, the simplified problem is called the *circular restricted* three-body problem, or CRTBP in short, which is conveniently formulated in a rotating frame with the rotation rate of the primaries.

In spite of the radical simplifications introduced by the CRTBP approximation, a general closed-form analytical solution is not known either. Well-known facts of the dynamics of the CRTBP are that it accepts the Jacobi integral and enjoys five equilibrium solutions or *libration* points, three of them collinear with the primaries and the other two equilateral to them. Besides, a wealth of periodic solutions are customarily computed by numerical integration [84, 176, 278, 628]. In particular, periodic orbits stemming from the libration points are shown to exist after linearization of the flow. In the case of the collinear points, these kinds of solutions give rise to the planar and vertical Lyapunov orbits. The analytical continuation of these infinitesimal orbits into families of periodic orbits of the CRTBP for variations of the Jacobi constant reveals the existence of new families of periodic orbits that bifurcate from them. Of special relevance for their application to space mission design are the so-called halo orbits [62, 191], which bifurcate out of the plane of the primaries from the family of planar Lyapunov orbits.

In spite of the fact that the applications of perturbation theory in this monograph do not deal with the CRTBP, but with the simpler Hill problem, we adhere to the tradition of presenting the latter as a limiting case of the former. Hence, the standard derivation of the CRTBP is first provided in §10.1.

1.3.2 Hill problem simplifications

When the massless body evolves closer to the central body than the distance between the two point masses, the ratio of the two distances is less than one and the third-body potential can be replaced by some truncation of its series expansion in Legendre polynomials. As long as this ratio is small the first terms of the expansion can be representative of the dynamics. If, besides, the mass of the central body is much smaller than the mass of the other primary, additional simplifications can be carried out that give rise to the so-called Hill problem [292].

The simplifications of the CRTBP that lead to the Hill problem are described in §10.2. It must be said, however, that Hill's equations have greater generality than being a refinement of the restricted three-body problem. Indeed, they can be independently derived from the general three-body problem when two of the three involved masses are much smaller than the mass of the heavier body [281, 616]. In that case, the two lighter masses evolve about the heavier body basically in terms of two separate two-body problems except when they are close enough to each other, so that their mutual attraction becomes comparable to the differential attraction of the heavier body.

A remarkable feature of the Hill problem is that it does not depend on physical parameters, a fact that is shown after scaling the equations of motion using suitable units of length and time [628]. This useful characteristic provides a wide generality to the Hill problem, which can be used to represent the dynamics of a small body under the gravitational action of different pairs of bodies of the solar system. Specific aspects of a particular binary system are then manifested when recovering the physical units. Hence the Hill problem is well suited to the study of a variety of astronomical and astrodynamics problems, like, for instance, the investigation of satellite encounters [554], cf. [534, 642, 643, 645], the dynamics of coorbital motion [47, 524], or the dynamics of relative spacecraft motion [71, 113, 336, 446]. It is also useful in the description of the spacecraft motion about planetary satellites [419, 430], as well as the dynamics about asteroids [597]. On the other hand, this simple model may need to be supplemented with other effects in addition to the mass-point attraction in order to provide a more accurate model of the dynamics of particular problems. Thus, the non-centralities of the gravitational potential of the smaller primary, and, in particular, those related to the ellipsoidal figure, are commonly taken into account in studies of orbital dynamics close to planetary satellites [432, 585], while solar radiation pressure effects may play an important role in the dynamics about asteroids [215]. Also, the dynamical coupling produced by large structures, like space tethers, may have relevant implications on the Hill problem dynamics [550, 551].

The global dynamics of the Hill problem must necessarily be investigated numerically. In particular, thorough computations of periodic and quasi-periodic orbits have been carried out by various authors [275–277, 279, 492]. Besides, due to its interest for space mission design, investigation of low-energy transit orbits [127] has received particular attention, and comprehensive studies of the stable and unstable manifolds associated to libration point orbits have been carried out in the investigation of the center manifold of the collinear libration points [250, 477]. In addition, the construction of periapsis Poincaré maps has been revealed to be a fruitful procedure in the characterization of escaping and capture trajectories [579, 660]. Other detailed accounts on the Hill problem dynamics are given in different studies [611], in which the planar case is described globally for energy values yielding bounded motion.

Alternatively, purely analytical approaches may provide useful information on the Hill problem dynamics in some regions of phase space. In particular, three different instances in which the Hill problem can be approached by perturbations are discussed in this monograph; namely, the perturbed Keplerian motion about planetary satellites (Chapter 11), the perturbed harmonic motion about the libration points (Chapter 12), and the coorbital motion of the smaller-mass primary and the spacecraft (Chapter 13).

1.3.3 Motion about planetary satellites

Most natural satellites have masses that are notably smaller than the mass of their mother planets, and, therefore, are well suited to the small-mass assumption of the Hill problem, cf. Table 1 of [598]—the Sun–Mercury system being also naturally included in this class of models. On the other hand, prospective mapping missions about planetary satellites require low-altitude orbiters, whose semimajor axis is usually much shorter than the semimajor axis of the orbit of the natural satellite about the mother planet, in this way fulfilling also the negligible parallax assumption of the Hill problem. Then, the Hill problem dynamics fits well to the description of the motion about planetary satellites. Besides, in those cases in which the orbiter evolves clearly inside the Hill sphere, which is not always the case [694, 695], the Hill problem Hamiltonian is naturally arranged in the form of perturbed Keplerian motion, and hence it is a good candidate for the analytical approach [182, 343].

A low-order perturbation approach suffices for describing the major effects of the long-term dynamics about planetary satellites. After removing first the mean anomaly of the orbiter in §11.1.1, and then the argument of the node in the rotating frame in §11.1.2, the reduced problem becomes integrable. The reduced dynamics is discussed in §11.1.3 and shows the unstable character of high inclination circular orbits. It also reveals the existence of librational motion of the eccentricity vector about stable eccentric orbits in what is known as the Lidov–Kozai resonance [48, 71, 364, 450].[7] Beyond the qualitative aspects of the planetary satellite dynamics, the accurate description of the solution may require one to achieve much higher orders of the perturbation approach [357, 418, 419, 535, 652, 656], a procedure that is carried out in §11.2.

On the other hand, because of the usual proximity of a mapping orbiter to the planetary satellite, the non-centralities of its gravitational field may cause clearly observable effects in the long-term dynamics. Therefore, these effects must be included

7 A recent proposal suggests to enlarge the prefix to von Zeipel–Lidov–Kozai to credit the much earlier findings of Hugo von Zeipel on this topic [313, 697].

as additional perturbations [365, 432, 452, 457, 592, 598, 650]. In particular, the oblateness and dynamical ellipticity of the planetary satellite are commonly incorporated into the perturbation model, yet the usual synchronous rotation of the satellite with its orbital motion may allow for some simplifications. Still, because higher-order gravity terms may have possible implications in the long-term dynamics [3], the latitudinal asymmetry of the central body is sometimes included in the perturbation model [389, 421, 547]. The modifications of the dynamics introduced by the ellipsoidal figure of the natural satellite are discussed in §11.3.

A relevant application of the analytical solution is found in the computation of long-lifetime science orbits, which is approached in §11.4. Indeed, due to the mentioned instability of almost circular high-inclination orbits induced by the planetary perturbations, the eccentricity of the spacecraft orbit grows exponentially until the orbiter impacts on the planetary satellite's surface, if uncontrolled, in relatively short times. Strategies for maximizing the lifetime of science orbits based on time-to-impact analytical predictions [598] or the use of the stable/unstable manifold dynamics of the averaged problem [547] give promising results. Moreover, beyond common linearization procedures [548], the use of higher orders of the mean to osculating elements transformation provided by the Lie transform perturbation approach may help in improving lifetimes in the mission design procedure [410, 411]. In addition, as illustrated in §11.4.2, for the low eccentricities and high inclinations required for science orbits, the whole sequence of transformations from mean to osculating elements may be replaced by a single set of simplified non-singular transformation equations [389].

1.3.4 Libration points orbits

Another region in which the Hill problem is amenable to perturbative treatment is in the vicinities of the libration points, whose specific characteristics make them attractive for different space mission applications [11, 190, 249]. In particular, the normalization approach is customarily used in removing the hyperbolic instability associated with the saddle × center × center character of the libration points equilibria [246]. This procedure reduces the restricted three-body problem dynamics to its *center manifold*, which is of just two degrees of freedom and, therefore, can be explored with the usual tools on non-linear dynamics, as the computation of Poincaré surfaces of section or the numerical computation of invariant manifolds [250]. On the other hand, particular analytical solutions are known to exist either in the linearized or non-linearized dynamics [515], and analytical approximations of the main existing periodic orbits about the libration points have been computed eventually, based on Lindstedt series or analogous expansions [191, 566, 693].

For energy values close enough to the energy of the libration points, the dynamics of the center manifold can also be approached analytically by perturbations, as an alternative to the use of Poincaré surfaces of section, yet the dynamics is no longer that of perturbed Keplerian motion. Indeed, the reduced Hamiltonian of the center manifold takes the form of a planar perturbed harmonic oscillator that is in the quasi-resonance condition. The following introduction of a detuning parameter [283] casts the problem in the form of a perturbed *elliptic* oscillator. That is, we take an oscillator in the 1-1 resonance condition for the unperturbed term, whereas the detuned terms are moved to the disturbing function. That this class of resonant systems can be efficiently approached using perturbation theory is a very well-known fact [198, 472]. In particular, the short-period terms of the center manifold of the Hill problem Hamiltonian are conveniently disclosed when reformulated in Deprit's Lissajous variables [160]. Then the resonant normal form Hamiltonian is constructed by standard averaging over the elliptic anomaly.

Splitting the reduction process into two different canonical transformations, the reduction to the center manifold and the short-period elimination, with focus explicitly on the separation of long- and short-period effects of the perturbed motion, provides the insight usually wanted by astrodynamicists [396]. However, when higher orders of the perturbation approach are required, the reduction of the problem to a one-degree-of-freedom Hamiltonian by means of a single Lie transformation is computationally more efficient [103, 238, 564]. Furthermore, when this single transformation is carried out in complex variables the normalization becomes a simple exercise of polynomial algebra. The resulting expressions in the normalization process only involve arithmetic operations, in this way making the evaluation of higher orders of the analytical, perturbation solution very efficient [424]. This approach clearly illustrates the versatility of the Lie transforms method, and is the one that we choose for the construction of the reduced Hamiltonian in §12.1.

The normalized Hamiltonian with the short-period effects removed is of one degree of freedom. The reduced phase space, which is the sphere [134, 373], is conveniently described in Hopf coordinates [300]. Most strikingly, the equations of the reduced dynamics turn out to be a particular case of the equations of motion of a free gyrostat,[8] whose analytical integration has been investigated since the pioneering work of Zhukovski [700] and Volterra [672] (see also [41]). This fact provides one more instance of the analogies between orbital and rotational motions that have been recurrently mentioned in the literature [345].

On the other hand, while the complete description of the reduced dynamics on the sphere is properly done in Hopf variables [163, 498], the use of Deprit's Lissajous variables provides immediate insight into the nature of particular solutions. Moreover, the simple evaluation of the equilibria solutions in these variables at each value of the elliptic anomaly between 0 and 2π allows for the straightforward reconstruction

8 This fact has been pointed out to the author by Alexander Burov, Dorodnitsyn Computing Centre of the Russian Academy of Sciences, in private communication of November 2018.

of their partner periodic orbits, whose period is easily computed from the Hamilton equation for the variation of the elliptic anomaly [396]. These aspects of the reduced dynamics are discussed in §12.2.

When higher orders of the perturbation approach are required, their computation is customarily approached using floating-point arithmetic because of simplicity and efficiency [171]. Still, this procedure may increase non-negligibly the numerical errors due to the number representation in the computer. The propagation of the truncation errors can be studied with the help of interval arithmetic [100, 328]. Alternatively, the use of complex variables provides a simple way of estimating the accumulation of truncation errors at each order of the perturbation theory by tracking the residual complex terms that remain in the normalized Hamiltonian after recovering the real variables [424].

Alternatively, to mitigate the growth of the truncation errors associated with the floating-point arithmetic, the perturbation solution is approached exactly using integer arithmetic. However, the increase in the size of the rational coefficients handled with the perturbation order, which may become enormous at relatively moderate orders [162], makes the computation of very high orders with integer arithmetics unpractical. Still, the integer-arithmetic solution can be used as a benchmark with which to compare the lower orders of the floating-point arithmetic solution, and to estimate the highest order that may make sense for a floating-point solution [424]. To expedite evaluation of the perturbation solution, the higher orders of the Lie transform method computed in §12.3 to improve the accuracy of the analytical solution, have been approached in floating-point arithmetics.

1.3.5 Coorbital motion with low eccentricity

The Hill problem is also useful in describing the interaction of two small bodies orbiting a massive one. Particular instances of the case in which both small bodies evolve in almost circular orbits with the same semimajor axis give rise to the so-called quasi-satellite orbits, also named distant retrograde orbits, in which the massless body seems to evolve in orbit about the smaller primary. As far as this apparent motion occurs out of the Hill sphere, it cannot be approached as a case of perturbed Keplerian motion about the smaller primary.

Discussions of the possibility of this kind of relative motion in the solar system can be traced back to the beginning of the twentieth century [319]. Since then, families of retrograde periodic orbits have been computed numerically by different authors, either for the Copenhagen problem (primaries with equal masses) [87, 273, 274, 624], the case of Earth–Moon masses [68], or the pure Hill problem [275]. Brief reviews on the topic can be consulted in [560, 608, 673], whereas different natural objects in quasi-satellite orbits that have been identified in the solar system can be found in [205] and the references therein.

The strong stability characteristics of quasi-satellite orbits make them appealing for different astrodynamics applications, like quarantine orbits [50, 376], surveillance missions related to the protection of the Earth [622], or for "orbiting" about bodies of non-negligible dimensions and very low mass, as is the case of Phobos [35, 352, 530].

Quasi-satellite orbits are customarily computed numerically like periodic solutions of a restricted three-body problem [46, 235, 429, 456, 499]. Approximate analytical solutions to the coorbital motion have also been reported that are useful in the qualitative description of the main facts of the quasi-satellite orbits' dynamics [47, 524]. While these rough analytical solutions can be improved when the problem is approached by Hamiltonian perturbations [454], difficulties arising in the practical implementation of the perturbation approach derived from the essential dependence of the solution on special functions constrained the perturbation solution to the lower orders.[9] This fact led to the conclusion that the perturbation approach is just partially successful [455], with a limited application to the description of the long-term orbital behavior obtained from the numerical integration of the evolution equations [456]. It was, perhaps, the untimely demise of Lidov which made the perturbation approach be abandoned until recently. Still, new applications of the perturbation approach were, again, limited to the orbit evolution, whose equations are obtained by numerically averaging the higher frequencies of the motion [608]. Numerical averaging techniques [184, 467, 635] present an alternative to effectively deal with the complications in the analytical integration of special functions, but they also hinder the computation of higher orders of the perturbation.

The arrangement of the Hill problem Hamiltonian in a form useful for the perturbation approach to the quasi-satellite orbits problem is discussed in §13.1, yet limited to the planar case. The Hamiltonian is split into the quadratic part, which is integrable, and the non-linear term containing the interaction of the orbiter and the primary. Reformulation in epicyclic variables turns the quadratic terms into a completely reduced Hamiltonian, which is immediately integrated to show that the orbiter moves in an ellipse whose center may evolve with linear motion along the direction orthogonal to the line of the primaries. This linear trend can be removed by the proper selection of initial conditions, thereby making place to purely periodic motion. When the interaction of the smaller primary with the orbiter becomes significant, the linear motion of the center of the ellipse may change into a slow oscillatory motion that gives rise to the quasi-satellite configuration.

The perturbation approach proceeds by removing the phase of the orbiter in its epicycle, an operation that is efficiently achieved by the method of Lie transforms. Details of this procedure, including the computation of the short-period corrections,

9 The efficient treatment of special functions in a calculus of perturbation is a problem that seems to have remained open since many years ago [36, 82, 83, 153].

are given in §13.2. In this way, the planar case of the Hill problem is reduced to an integrable one-degree-of-freedom Hamiltonian describing the motion of the center of the reference ellipse (the "deferent" of the epicyclical motion). The nature of this solution is discussed in §13.3, where, for the lower orders of the perturbation approach, it is shown that the motion is made of harmonic oscillations and can be nicely expressed in closed form [401]. Still, the validity of the closed-form solution is constrained to the case of small amplitude librations of the deferent motion. Higher orders of the single-averaged Hamiltonian prevent the closed-form solution, yet the long-term motion is efficiently integrated semi-analytically with very long steps. Alternatively, analytical solutions in the form of Lindstedt series extend the validity of the purely analytical approach [398, 403].

As an option to the Lindstedt series approach, a new Lie transformation is carried out in §13.4 to remove the phase of the deferent. In this way, the planar Hill problem Hamiltonian is completely reduced, up to the truncation order of the perturbation approach, thus yielding a trivially integrable Hamiltonian. The evaluation of the secular terms of the double-averaged solution is computationally undemanding, and it provides an instant way of exploring quasi-satellite orbit evolution. In addition, when required, the computation of an ephemeris is straightforward by simply adding the periodic corrections to the secular solution. These corrections are made of both long- and short-period effects, and they are obtained analytically as a result of the perturbation procedure. Moreover, the formal integrals of the secular solution can be used as design parameters in the implementation of quasi-satellite orbits with specific characteristics [404]. In particular, it is shown in §13.5 how commensurability between the (secular) periods of the orbital and libration motions results in quasi-satellite orbits of the planar Hill problem that are periodic on average, and approximately periodic in the non-averaged dynamics. As customary in preliminary mission design procedures [386, 387, 390], differential corrections can then be used to slightly modify the initial conditions provided by the analytical solution in order to get an exact periodic orbit of the non-averaged problem with the required characteristics.

Part I: Hamiltonian perturbations by Lie transforms

2 The method of Lie transforms

The solution to different problems of mechanics is made notably easier when discovering a convenient set of variables that simplifies the mathematical formulation of the problem. In Hamiltonian mechanics, this is the particular case in which we find a transformation of canonical variables such that, in the new variables, the Hamiltonian is reduced to a *normal* form that is function of only the momenta. Being cyclic all the coordinates, the problem is trivially integrated. In the new variables, the momenta are integrals of the motion whereas their conjugate coordinates evolve linearly with frequencies that formally depend on the momenta. The transformation is commonly found by the Hamilton–Jacobi method through a generating function in mixed variables, old coordinates and new momenta [30]. This is in the same spirit as Hamiltonian perturbation methods: to find an infinitesimal contact transformation such that, in the new variables, the perturbed integrable Hamiltonian takes a simpler form that makes the integration of the Hamiltonian flow easier.

Beyond traditional perturbation approaches to particular problems, Poincaré proposed a general algorithm that, in the style of the Hamilton–Jacobi approach, relies on a generating function in mixed variables [558]. The construction of the perturbation solution with Poincaré's method is straightforward up to the first order of the perturbation approach. But to overcome the difficulties that appear in the computation of higher orders, which stem from the implicit form of the transformation, one must deal with sophisticated operations on power series [174]. Alternatively, perturbation methods based on Lie transforms [151, 303] provide systematic algorithms to extend the perturbation solution to higher orders, which, besides, are easily implemented in modern commercially available symbolic algebra systems. We only deal with the practical aspects that lead to the implementation of the latter, which is generally recognized as the standard perturbation method of these days. Moreover, the following descriptions are constrained to the case of conservative transformations and functions. The case in which the time appears explicitly is approached analogously by the simple expedient of turning to the extended phase space.

2.1 Lie transformation of a function

Let $\boldsymbol{q}$ be a vector of canonical coordinates $\boldsymbol{q} = (q_1, q_2, \dots, q_l)$, and let $\boldsymbol{Q} = (Q_1, Q_2, \dots, Q_l)$ be their conjugate momenta. Let $\mathcal{W} \equiv \mathcal{W}(\varepsilon; \boldsymbol{q}, \boldsymbol{Q})$ be an analytical function of ε given by the Taylor series expansion

$$\mathcal{W} = \sum_{n\geq 0} \frac{\varepsilon^n}{n!} \mathcal{W}_{n+1}(\boldsymbol{q}, \boldsymbol{Q}). \tag{2.1}$$

https://doi.org/10.1515/9783110668513-002

Then the solution to the differential system

$$\frac{\mathrm{d}q_k}{\mathrm{d}\varepsilon} = \frac{\partial \mathcal{W}}{\partial Q_k}, \quad \frac{\mathrm{d}Q_k}{\mathrm{d}\varepsilon} = -\frac{\partial \mathcal{W}}{\partial q_k}, \quad k = 1, 2, \ldots, l, \tag{2.2}$$

for the initial conditions

$$\boldsymbol{q}(\boldsymbol{p}, \boldsymbol{P}, 0) = \boldsymbol{p}, \quad \boldsymbol{Q}(\boldsymbol{p}, \boldsymbol{P}, 0) = \boldsymbol{P}, \tag{2.3}$$

defines an infinitesimal mapping $\varphi : (\boldsymbol{p}, \boldsymbol{P}, \varepsilon) \mapsto (\boldsymbol{q}, \boldsymbol{Q})$ given by

$$\boldsymbol{q} = \boldsymbol{q}(\boldsymbol{p}, \boldsymbol{P}, \varepsilon), \quad \boldsymbol{Q} = \boldsymbol{Q}(\boldsymbol{p}, \boldsymbol{P}, \varepsilon), \tag{2.4}$$

where $\boldsymbol{p} = (p_1, p_2, \ldots, p_l)$ are canonical coordinates and $\boldsymbol{P} = (P_1, P_2, \ldots, P_l)$ their conjugate momenta, which is known as the *Lie transformation* with generating function $\mathcal{W}$. Lie transformations are *completely* canonical transformations [151, 491] and their existence in a neighborhood of $\varepsilon = 0$ is guaranteed by the basic theory of differential equations.

Let $F \equiv F(\boldsymbol{q}, \boldsymbol{Q}, \varepsilon)$ be an analytical function depending on the canonical set $(\boldsymbol{q}, \boldsymbol{Q})$, as well as on the scalar small parameter ε. F can be expanded as the Taylor series

$$F = \sum_{n\geq 0} \frac{\varepsilon^n}{n!} F_{n,0}(\boldsymbol{q}, \boldsymbol{Q}), \quad F_{n,0} = \left.\frac{\mathrm{d}^n F}{\mathrm{d}\varepsilon^n}\right|_{\varepsilon=0}, \tag{2.5}$$

where the convenience of using the double subindex notation $F_{n,0}$ for the total derivatives of F with respect to the small parameter will become apparent soon.

The function F is reformulated in the $(\boldsymbol{p}, \boldsymbol{P})$ variables by replacing Eq. (2.4) into Eq. (2.5). By direct replacement we obtain $G = G(\boldsymbol{p}, \boldsymbol{P}, \varepsilon) \equiv \varphi F$ in the form

$$G = F(\boldsymbol{q}(\boldsymbol{p}, \boldsymbol{P}, \varepsilon), \boldsymbol{Q}(\boldsymbol{p}, \boldsymbol{P}, \varepsilon), \varepsilon) = \sum_{n\geq 0} \frac{\varepsilon^n}{n!} F_{n,0}(\boldsymbol{q}(\boldsymbol{p}, \boldsymbol{P}, \varepsilon), \boldsymbol{Q}(\boldsymbol{p}, \boldsymbol{P}, \varepsilon)), \tag{2.6}$$

which is not a Taylor series. To obtain G in the form of a Taylor series, we would need to expand and rearrange the right side of Eq. (2.6).

Rather than solving Eq. (2.2) to get the transformation (2.4), replace it in Eq. (2.6) and carry out the consequent rearrangement, we will rely on basic operations on power series to obtain directly G like the Taylor series

$$G = \sum_{n\geq 0} \frac{\varepsilon^n}{n!} G_n(\boldsymbol{p}, \boldsymbol{P}), \quad G_n = \left.\frac{\mathrm{d}^n G}{\mathrm{d}\varepsilon^n}\right|_{\varepsilon=0}. \tag{2.7}$$

The direct computation of the coefficients G_n in Eq. (2.7) is done by taking into account that $G = \varphi F \equiv F(\boldsymbol{q}(\boldsymbol{p}, \boldsymbol{P}, \varepsilon), \boldsymbol{Q}(\boldsymbol{p}, \boldsymbol{P}, \varepsilon), \varepsilon)$ is a composite function, to which we apply the chain rule. Thus, working in the $(\boldsymbol{q}, \boldsymbol{Q})$ variables,

$$G_1 = \left.\frac{\mathrm{d}G}{\mathrm{d}\varepsilon}\right|_{\varepsilon=0}, \quad \text{where} \quad \frac{\mathrm{d}G}{\mathrm{d}\varepsilon} = \frac{\partial F}{\partial \varepsilon} + \sum_{k=1}^{l} \left(\frac{\partial F}{\partial q_k} \frac{\mathrm{d}q_k}{\mathrm{d}\varepsilon} + \frac{\partial F}{\partial Q_k} \frac{\mathrm{d}Q_k}{\mathrm{d}\varepsilon} \right),$$

in which, since we are dealing with a Lie transformation, the total derivatives of q_k and Q_k with respect to the small parameter are replaced using Eq. (2.2). Hence,

$$\frac{\mathrm{d}G}{\mathrm{d}\varepsilon} = \frac{\partial F}{\partial \varepsilon} + \{F; \mathcal{W}\}, \tag{2.8}$$

where

$$\{F; \mathcal{W}\} = \sum_{k=1}^{l} \left(\frac{\partial F}{\partial q_k} \frac{\partial \mathcal{W}}{\partial Q_k} - \frac{\partial F}{\partial Q_k} \frac{\partial \mathcal{W}}{\partial q_k} \right) \tag{2.9}$$

stands for the Poisson bracket of F and $\mathcal{W}$.

2.1.1 The fundamental recursion

The partial derivative of F with respect to the small parameter in Eq. (2.8) is readily computed from (2.5),

$$\frac{\partial F}{\partial \varepsilon} = \sum_{n\geq 0} \frac{\varepsilon^n}{n!} F_{n+1,0}. \tag{2.10}$$

On the contrary, the computation of the Poisson bracket of F and $\mathcal{W}$ is a little bit more involved. First, we plug Eqs. (2.1) and (2.5) into Eq. (2.9), to obtain

$$\{F; \mathcal{W}\} = \sum_{k=1}^{l} \left(\sum_{j\geq 0} \frac{\varepsilon^j}{j!} \frac{\partial F_{j,0}}{\partial q_k} \sum_{i\geq 0} \frac{\varepsilon^i}{i!} \frac{\partial \mathcal{W}_{i+1}}{\partial Q_k} - \sum_{j\geq 0} \frac{\varepsilon^j}{j!} \frac{\partial F_{j,0}}{\partial Q_k} \sum_{i\geq 0} \frac{\varepsilon^i}{i!} \frac{\partial \mathcal{W}_{i+1}}{\partial q_k} \right). \tag{2.11}$$

At this point, we recall that the product of the power series $s_1 = \sum_{j\geq 0} a_j \varepsilon^j$ and $s_2 = \sum_{i\geq 0} b_i \varepsilon^i$, is another power series $p = s_1 s_2 = \sum_{n\geq 0} c_n \varepsilon^n$, whose coefficients are given by the Cauchy product $c_n = \sum_{m=0}^{n} a_{n-m} b_m$. Since we are dealing with Taylor series, the coefficients of the series s_1, s_2 and p are reorganized in the form $a_j = A_j/j!$, $b_i = B_i/i!$, $c_n = C_n/n!$. Therefore, the coefficient $C_n = n! c_n$ of the Cauchy product reads

$$C_n = n! \sum_{m=0}^{n} \frac{A_{n-m}}{(n-m)!} \frac{B_m}{m!} = \sum_{m=0}^{n} \binom{n}{m} A_{n-m} B_m.$$

Computing the series products in Eq. (2.11) in this way, we readily obtain

$$\{F; \mathcal{W}\} = \sum_{k=1}^{l} \sum_{n\geq 0} \frac{\varepsilon^n}{n!} \sum_{m=0}^{n} \binom{n}{m} \left(\frac{\partial F_{n-m,0}}{\partial q_k} \frac{\partial \mathcal{W}_{m+1}}{\partial Q_k} - \frac{\partial F_{n-m,0}}{\partial Q_k} \frac{\partial \mathcal{W}_{m+1}}{\partial q_k} \right),$$

where the term inside the parentheses is easily identified with a Poisson bracket after exchanging the order of the summations. That is,

$$\{F; \mathcal{W}\} = \sum_{n\geq 0} \frac{\varepsilon^n}{n!} \sum_{m=0}^{n} \binom{n}{m} \{F_{n-m,0}; \mathcal{W}_{m+1}\}. \tag{2.12}$$

Next, Eqs. (2.10) and (2.12) are plugged into Eq. (2.8) to yield

$$\frac{\mathrm{d}G}{\mathrm{d}\varepsilon} = \sum_{n\geq 0} \frac{\varepsilon^n}{n!} F_{n,1}, \tag{2.13}$$

where

$$F_{n,1} = F_{n+1,0} + \sum_{m=0}^{n} \binom{n}{m} \{F_{n-m,0}; \mathcal{W}_{m+1}\}. \tag{2.14}$$

So far, all the computations have been carried out in the original $(\boldsymbol{q}, \boldsymbol{Q})$ variables. To compute the term $G_1 \equiv G_1(\boldsymbol{p}, \boldsymbol{P})$ it only remains to use Eq. (2.3) in order to evaluate Eq. (2.13) in $\varepsilon = 0$. That is,

$$G_1 = \left.\frac{\mathrm{d}G}{\mathrm{d}\varepsilon}\right|_{\varepsilon=0} = \sum_{n\geq 0} \frac{\varepsilon^n}{n!} F_{n,1}(\boldsymbol{q}, \boldsymbol{Q})|_{\varepsilon=0} = F_{0,1}(\boldsymbol{q}|_{\varepsilon=0}, \boldsymbol{Q}|_{\varepsilon=0}) = F_{0,1}(\boldsymbol{p}, \boldsymbol{P}),$$

where $F_{0,1} = F_{1,0} + \{F_{0,0}; \mathcal{W}_1\}$ from Eq. (2.14).

In summary: we started from Eq. (2.5) and arrived at Eq. (2.13), which equations are formally equal except for the coefficients $F_{n,0}$ being replaced by $F_{n,1}$. That is,

$$\frac{\mathrm{d}G}{\mathrm{d}\varepsilon} = \frac{\mathrm{d}}{\mathrm{d}\varepsilon} \sum_{n\geq 0} \frac{\varepsilon^n}{n!} F_{n,0}(\boldsymbol{q}, \boldsymbol{Q}) = \sum_{n\geq 0} \frac{\varepsilon^n}{n!} F_{n,1}(\boldsymbol{q}, \boldsymbol{Q}).$$

Proceeding likewise with $\mathrm{d}G/\mathrm{d}\varepsilon$, which is again handled as a composite function, straightforward computations show that the second derivative of G with respect to ε is

$$\frac{\mathrm{d}^2G}{\mathrm{d}\varepsilon^2} = \frac{\mathrm{d}}{\mathrm{d}\varepsilon} \sum_{n\geq 0} \frac{\varepsilon^n}{n!} F_{n,1}(\boldsymbol{q}, \boldsymbol{Q}) = \sum_{n\geq 0} \frac{\varepsilon^n}{n!} F_{n,2}(\boldsymbol{q}, \boldsymbol{Q}),$$

in which

$$F_{n,2} = F_{n+1,1} + \sum_{m=0}^{n} \binom{n}{m} \{F_{n-m,1}; \mathcal{W}_{m+1}\}.$$

Then an arbitrary derivative $\mathrm{d}^nG/\mathrm{d}\varepsilon^n$ is computed by induction showing that, starting from the qth derivative of G,

$$\frac{\mathrm{d}^qG}{\mathrm{d}\varepsilon^q} = \sum_{n\geq 0} \frac{\varepsilon^n}{n!} F_{n,q}(\boldsymbol{q}, \boldsymbol{Q}),$$

we obtain the $(q+1)$th derivative,

$$\frac{\mathrm{d}^{q+1}G}{\mathrm{d}\varepsilon^{q+1}} = \sum_{n\geq 0} \frac{\varepsilon^n}{n!} F_{n,q+1}(\boldsymbol{q}, \boldsymbol{Q}),$$

in which

$$F_{n,q+1} = F_{n+1,q} + \sum_{m=0}^{n} \binom{n}{m} \{F_{n-m,q}; \mathcal{W}_{m+1}\}. \tag{2.15}$$

Once the total derivatives have been computed up to the desired order, it only remains to evaluate them at $\varepsilon = 0$, namely,

$$\left.\frac{\mathrm{d}^q G}{\mathrm{d}\varepsilon^q}\right|_{\varepsilon=0} = \sum_{n\geq 0} \frac{\varepsilon^n}{n!} F_{n,q}(\boldsymbol{q}, \boldsymbol{Q})|_{\varepsilon=0} = F_{0,q}(\boldsymbol{q}|_{\varepsilon=0}, \boldsymbol{Q}|_{\varepsilon=0}) = F_{0,q}(\boldsymbol{p}, \boldsymbol{P}),$$

to get the Lie transformation of Eq. (2.5) in the form of the genuine Taylor series given in Eq. (2.7).

2.1.2 Deprit's triangle

Deprit's fundamental recursion (2.15) is visualized in the form of a triangular table

$$\begin{array}{cccccc} F_{0,0} & F_{0,1} & F_{0,2} & F_{0,3} & F_{0,4} & \cdots \\ F_{1,0} & F_{1,1} & F_{1,2} & F_{1,3} & \cdots & \\ F_{2,0} & F_{2,1} & F_{2,2} & \cdots & & \\ F_{3,0} & F_{3,1} & \cdots & & & \\ F_{4,0} & \cdots & & & & \\ \cdots & & & & & \end{array} \tag{2.16}$$

in which the computation of a given term only involves the preceding terms in the diagonal passing through it, and those in the preceding columns above that diagonal.

For instance, the computation of $F_{0,2}$ only involves the diagonal terms $F_{1,1}$ and $F_{2,0}$, those above $F_{1,1}$, namely $F_{0,1}$, and those above $F_{2,0}$, namely $F_{1,0}$ and $F_{0,0}$. Indeed, making $n = 0$, $q = 1$ in Eq. (2.15) yields $F_{0,2} = F_{1,1} + \{F_{0,1}; \mathcal{W}_1\}$, where $\mathcal{W}_1$ is known, and Eq. (2.15) is used again for computing $F_{1,1}$ (with $n = 1$ and $q = 0$) and $F_{0,1}$ (with $n = q = 0$). That is

$$\begin{aligned} F_{1,1} &= F_{2,0} + \binom{1}{0}\{F_{1,0}; \mathcal{W}_1\} + \binom{1}{1}\{F_{0,0}; \mathcal{W}_2\}, \\ F_{0,1} &= F_{1,0} + \{F_{0,0}; \mathcal{W}_1\}. \end{aligned}$$

A final remark is in order. While the transformation of the analytical function F given by Eq. (2.5) into the new variables has been achieved without solving the differential system (2.2) that defines the Lie transformation, it obviously plays a fundamental role in the computation of the Taylor series expansion (2.7) since Eq. (2.2) has certainly been used in the construction of Deprit's recursion (2.15).

2.1.3 The direct transformation

Note that given the generating function $\mathcal{W}$, the transformation (2.4) can be computed without need of solving the differential system (2.2). Indeed, the components of a vectorial function are scalar functions, to each of which the fundamental recursion (2.15) can be applied independently.

Thus, denote by $x = q_1$ one of the canonical coordinates of $\boldsymbol{q}$. It can be assumed to be given formally by the Taylor series

$$x = \sum_{n\geq 0} \frac{\varepsilon^n}{n!} x_{n,0}(\boldsymbol{q}, \boldsymbol{Q}),$$

where $x_{0,0} = x$, and $x_{n,0}$ vanish for $n > 0$. Then Eq. (2.15), in which the symbol F is replaced by the symbol x of the coordinate, is repeatedly applied to compute the terms

$$\begin{aligned}
x_{0,1} &= \{x; \mathcal{W}_1\} \\
&\quad x_{1,1} = \{x; \mathcal{W}_2\} \\
x_{0,2} &= x_{1,1} + \{x_{0,1}; \mathcal{W}_1\} \\
&\qquad x_{2,1} = \{x; \mathcal{W}_3\} \\
&\quad x_{1,2} = x_{2,1} + \{x_{1,1}; \mathcal{W}_1\} + \{x_{0,1}; \mathcal{W}_2\} \\
x_{0,3} &= x_{1,2} + \{x_{0,2}; \mathcal{W}_1\} \\
&\dots
\end{aligned} \tag{2.17}$$

and so on. Recall that this sequence is computed in the $(\boldsymbol{q}, \boldsymbol{Q})$ variables, but, in the end, the terms $x_{0,n}$ must be evaluated at $\varepsilon = 0$, in which case $\boldsymbol{q}(\boldsymbol{p}, \boldsymbol{P}, 0) = \boldsymbol{p}$, $\boldsymbol{Q}(\boldsymbol{p}, \boldsymbol{P}, 0) = \boldsymbol{P}$. That is, in the end, $\boldsymbol{q}$ is simply replaced by $\boldsymbol{p}$, and $\boldsymbol{Q}$ by $\boldsymbol{P}$, to give the transformation equation

$$x = \sum_{n\geq 0} \frac{\varepsilon^n}{n!} x_{0,n}(\boldsymbol{p}, \boldsymbol{P}),$$

where $x_{0,0}(\boldsymbol{p}, \boldsymbol{P}) = p_1$ is the corresponding coordinate to $x = q_1$ in the new variables.

Obviously, the same procedure can be applied to the other coordinates as well as their conjugate momenta, in this way obtaining explicitly the *direct* Lie transformation (2.4) in the form of the Taylor series

$$\boldsymbol{q} = \sum_{n\geq 0} \frac{\varepsilon^n}{n!} \boldsymbol{q}_{0,n}(\boldsymbol{p}, \boldsymbol{P}), \quad \boldsymbol{Q} = \sum_{n\geq 0} \frac{\varepsilon^n}{n!} \boldsymbol{Q}_{0,n}(\boldsymbol{p}, \boldsymbol{P}), \tag{2.18}$$

where $\boldsymbol{q}_{0,0} = \boldsymbol{p}$, and $\boldsymbol{Q}_{0,0} = \boldsymbol{P}$.

2.1.4 Composition of Lie transformations

Let $\psi : (\boldsymbol{s}, \boldsymbol{S}, \varepsilon) \mapsto (\boldsymbol{p}, \boldsymbol{P})$ be a new Lie transformation from the $(\boldsymbol{p}, \boldsymbol{P})$ variables to the $(\boldsymbol{s}, \boldsymbol{S})$ ones. If

$$\mathcal{V} = \sum_{n\geq 0} \frac{\varepsilon^n}{n!} \mathcal{V}_{n+1}(\boldsymbol{p}, \boldsymbol{P}) \tag{2.19}$$

is the generating function of ψ, the transformation equations $\boldsymbol{p} = \boldsymbol{p}(\boldsymbol{s}, \boldsymbol{S}, \varepsilon)$, $\boldsymbol{P} = \boldsymbol{P}(\boldsymbol{s}, \boldsymbol{S}, \varepsilon)$ are now obtained from the solution of the new differential system

$$\frac{\mathrm{d}p_k}{\mathrm{d}\varepsilon} = \frac{\partial \mathcal{V}}{\partial P_k}, \quad \frac{\mathrm{d}P_k}{\mathrm{d}\varepsilon} = -\frac{\partial \mathcal{V}}{\partial p_k}, \quad k = 1, 2, \ldots, l, \tag{2.20}$$

for the initial conditions $\boldsymbol{p}(\boldsymbol{s}, \boldsymbol{S}, 0) = \boldsymbol{s}$, $\boldsymbol{S}(\boldsymbol{s}, \boldsymbol{S}, 0) = \boldsymbol{S}$.

We will check that the composition

$$\begin{aligned} \boldsymbol{q} &= \boldsymbol{q}(\boldsymbol{p}(\boldsymbol{s}, \boldsymbol{S}, \varepsilon), \boldsymbol{P}(\boldsymbol{s}, \boldsymbol{S}, \varepsilon), \varepsilon) = \boldsymbol{q}(\boldsymbol{s}, \boldsymbol{S}, \varepsilon), \\ \boldsymbol{Q} &= \boldsymbol{Q}(\boldsymbol{p}(\boldsymbol{s}, \boldsymbol{S}, \varepsilon), \boldsymbol{P}(\boldsymbol{s}, \boldsymbol{S}, \varepsilon), \varepsilon) = \boldsymbol{Q}(\boldsymbol{s}, \boldsymbol{S}, \varepsilon), \end{aligned}$$

of the Lie transformations φ and ψ is a new Lie transformation.

To do that we compute the total derivatives of $q_k = \varphi q_k \equiv q_k(\boldsymbol{p}, \boldsymbol{P}, \varepsilon)$, and $Q_k = \varphi Q_k \equiv Q_k(\boldsymbol{p}, \boldsymbol{P}, \varepsilon)$ with respect to ε. By direct application of the chain rule we obtain

$$\frac{\mathrm{d}q_k}{\mathrm{d}\varepsilon} = \sum_{j=1}^{l} \left(\frac{\partial \varphi q_k}{\partial p_j} \frac{\mathrm{d}p_j}{\mathrm{d}\varepsilon} + \frac{\partial \varphi q_k}{\partial P_j} \frac{\mathrm{d}P_j}{\mathrm{d}\varepsilon} \right) + \frac{\partial \varphi q_k}{\partial \varepsilon}, \tag{2.21}$$

where the total derivatives of the $(\boldsymbol{p}, \boldsymbol{P})$ variables are taken from Eq. (2.20). Besides, since the transformation $q_k = \varphi q_k$ is the solution of Eqs. (2.2)–(2.3), it must yield the identity when plugged into Eq. (2.2). That is,

$$\frac{\mathrm{d}\varphi q_k}{\mathrm{d}\varepsilon} = \frac{\partial \varphi q_k}{\partial \varepsilon} \equiv \varphi \frac{\partial \mathcal{W}}{\partial Q_k}.$$

Therefore, Eq. (2.21) turns into

$$\frac{\mathrm{d}q_k}{\mathrm{d}\varepsilon} = \sum_{j=1}^{l} \left(\frac{\partial \varphi q_k}{\partial p_j} \frac{\partial \mathcal{V}}{\partial Q_j} - \frac{\partial \varphi q_k}{\partial P_j} \frac{\partial \mathcal{V}}{\partial q_j} \right) + \varphi \frac{\partial \mathcal{W}}{\partial Q_k} = \{\varphi q_k, \mathcal{V}\} + \varphi \frac{\partial \mathcal{W}}{\partial Q_k}, \tag{2.22}$$

in which the terms in the right side are functions of the $(\boldsymbol{p}, \boldsymbol{P})$ variables.

Moreover, due to the invariance of Poisson brackets with respect to completely canonical transformations, we can evaluate the Poisson bracket in Eq. (2.22) either in the $(\boldsymbol{p}, \boldsymbol{P})$ or in the $(\boldsymbol{q}, \boldsymbol{Q})$ variables; namely, $\{\varphi q_k, \mathcal{V}\} = \{q_k, \varphi^{-1}\mathcal{V}\}$. Therefore, we can rewrite the whole rightmost side of Eq. (2.22) in terms of the $(\boldsymbol{q}, \boldsymbol{Q})$ variables. Namely,

$$\frac{\mathrm{d}q_k}{\mathrm{d}\varepsilon} = \{q_k, \varphi^{-1}\mathcal{V}\} + \frac{\partial \mathcal{W}}{\partial Q_k} = \frac{\partial \varphi^{-1}\mathcal{V}}{\partial Q_k} + \frac{\partial \mathcal{W}}{\partial Q_k} = \frac{\partial \mathcal{U}}{\partial Q_k}, \tag{2.23}$$

where

$$\mathcal{U} = \mathcal{U}(\boldsymbol{q}, \boldsymbol{Q}, \varepsilon) \equiv \mathcal{W} + \varphi^{-1}\mathcal{V}. \tag{2.24}$$

Proceeding analogously with the momenta Q_k we readily obtain

$$\frac{\mathrm{d}Q_k}{\mathrm{d}\varepsilon} = -\frac{\partial \mathcal{U}}{\partial q_k}. \tag{2.25}$$

In consequence, the solution of Eqs. (2.23)–(2.25) for the initial conditions $\boldsymbol{q}(\boldsymbol{s}, \boldsymbol{S}, 0) = \boldsymbol{s}$, $\boldsymbol{Q}(\boldsymbol{s}, \boldsymbol{S}, 0) = \boldsymbol{S}$, defines a Lie transformation $\phi : (\boldsymbol{s}, \boldsymbol{S}, \varepsilon) \mapsto (\boldsymbol{q}, \boldsymbol{Q})$, given by the composition of φ and ψ, which is obtained from the generating function $\mathcal{U}$ defined in Eq. (2.24).

The formulation of the generating function $\mathcal{V} = \mathcal{V}(\boldsymbol{p}, \boldsymbol{P}, \varepsilon)$ in the $(\boldsymbol{q}, \boldsymbol{Q})$ variables $\varphi^{-1}\mathcal{V} = \mathcal{V}(\boldsymbol{p}(\boldsymbol{q}, \boldsymbol{Q}, \varepsilon), \boldsymbol{P}(\boldsymbol{q}, \boldsymbol{Q}, \varepsilon), \varepsilon)$, is readily obtained by a new application of the Lie transforms method, and follows analogous steps as those shown in the next section.

2.1.5 The inverse transformation

The composition of the direct transformation $\varphi : (\boldsymbol{p}, \boldsymbol{P}, \varepsilon) \mapsto (\boldsymbol{q}, \boldsymbol{Q})$ and the inverse one $\varphi^{-1} : (\boldsymbol{q}, \boldsymbol{Q}, \varepsilon) \mapsto (\boldsymbol{p}, \boldsymbol{P})$ obviously yields the identity. In this particular case, Eq. (2.24) turns into $\mathcal{W}+\varphi^{-1}\mathcal{V} = 0$ and the inverse transformation is obtained from the generating function $\mathcal{V} = -\varphi\mathcal{W}$. That is,

$$\mathcal{V} = -\mathcal{W}(\boldsymbol{q}(\boldsymbol{p}, \boldsymbol{P}, \varepsilon), \boldsymbol{Q}(\boldsymbol{p}, \boldsymbol{P}, \varepsilon), \varepsilon). \tag{2.26}$$

Therefore, to compute the inverse transformation of Eq. (2.18), the first step is to write Eq. (2.26) as an explicit function of $(\boldsymbol{p}, \boldsymbol{P})$ in the form of a Taylor series. This is done by standard application of the fundamental recursion (2.15) to the function $R = -\mathcal{W}(\boldsymbol{q}, \boldsymbol{Q}, \varepsilon)$. That is, starting from

$$R(\boldsymbol{q}, \boldsymbol{Q}, \varepsilon) = \sum_{n\geq 0} \frac{\varepsilon^n}{n!} R_{n,0},$$

in which, by comparison with Eq. (2.1), $R_{n,0} \equiv -\mathcal{W}_{n+1}(\boldsymbol{q}, \boldsymbol{Q})$, we construct

$$R(\boldsymbol{p}, \boldsymbol{P}, \varepsilon) = \sum_{n\geq 0} \frac{\varepsilon^n}{n!} R_{0,n}.$$

Straightforward computations show that

$$\begin{aligned}
R_{0,0} &= -\mathcal{W}_1, \\
R_{0,1} &= -\mathcal{W}_2, \\
R_{0,2} &= -\mathcal{W}_3 - \{\mathcal{W}_2; \mathcal{W}_1\}, \\
R_{0,3} &= -\mathcal{W}_4 - 2\{\mathcal{W}_3; \mathcal{W}_1\} - \{\{\mathcal{W}_2; \mathcal{W}_1\}; \mathcal{W}_1\}, \\
R_{0,4} &= \dots.
\end{aligned} \tag{2.27}$$

Finally, we make $\mathcal{V}_{n+1} = R_{0,n}(\boldsymbol{p}, \boldsymbol{P})$ to get the generator of the inverse transformation in the form of the Taylor series

$$\mathcal{V} = \sum_{n\geq 0} \frac{\varepsilon^n}{n!} \mathcal{V}_{n+1}(\boldsymbol{p}, \boldsymbol{P}).$$

Once $\mathcal{V} \equiv \mathcal{V}(\boldsymbol{p}, \boldsymbol{P})$ has been obtained, the computation of the inverse transformation

$$\boldsymbol{p} \equiv \sum_{n\geq 0} \frac{\varepsilon^n}{n!} \boldsymbol{p}_{0,n}(\boldsymbol{q}, \boldsymbol{Q}), \quad \boldsymbol{P} \equiv \sum_{n\geq 0} \frac{\varepsilon^n}{n!} \boldsymbol{P}_{0,n}(\boldsymbol{q}, \boldsymbol{Q}), \tag{2.28}$$

follows exactly the same steps as in the direct case §2.1.3 simply replacing $\mathcal{W}$ by $\mathcal{V}$.

2.2 Deprit's perturbations approach

Hamiltonian perturbation problems are generally stated in the form

$$\mathcal{H} = \mathcal{H}_0 + \mathcal{D}, \tag{2.29}$$

where $\mathcal{H}_0$ is an integrable Hamiltonian and $\mathcal{D}$ is a disturbing function that only modifies slightly the integrable Hamiltonian flow, that is $|\mathcal{D}| \ll |\mathcal{H}_0|$ at any time. The general goal of the perturbation approach is to find a canonical transformation such that the transformed Hamiltonian is simpler in the new variables than in the original form. The simplification will depend on the problem at hand, yet the more common case consists of making cyclic one of the variables which the new Hamiltonian depends upon.

2.2.1 Hamiltonian simplification by Lie transforms

The Lie transforms method provides a general frame for the computation of Hamiltonian perturbation solutions. Indeed, Eq. (2.29) can be written in the form

$$\mathcal{H} = \sum_{m\geq 0} \frac{\varepsilon^m}{m!} \mathcal{H}_{m,0}(\boldsymbol{q}, \boldsymbol{Q}), \tag{2.30}$$

where $\mathcal{H}_{0,0} \equiv \mathcal{H}_0$ and ε is a small parameter that represents the smallness of the disturbing function. In the more favorable case ε would be a physical parameter, but in general it will be a formal small parameter ($\varepsilon \equiv 1$) used to visualize some dynamical conditions which would apply to the solutions of interest.

In particular Deprit's perturbation approach provides the means for computing the completely canonical Lie transformation

$$\mathcal{T} : (\boldsymbol{q}, \boldsymbol{Q}) \mapsto (\boldsymbol{p}, \boldsymbol{P}; \varepsilon), \tag{2.31}$$

leading to the desired simplification in the Hamiltonian (2.30). That is,

$$\mathcal{T}:\mathcal{H}=\mathcal{H}(\boldsymbol{q}(\boldsymbol{p},\boldsymbol{P},\varepsilon),\boldsymbol{Q}(\boldsymbol{p},\boldsymbol{P},\varepsilon),\varepsilon)\equiv\sum_{m\geq 0}\frac{\varepsilon^m}{m!}\mathcal{H}_{0,m}(\boldsymbol{p},\boldsymbol{P}) \tag{2.32}$$

is "simpler" than Eq. (2.30) up to some truncation order $\mathcal{O}(\varepsilon^k)$.

The Lie transformation (2.31) is derived from a generating function of the form of Eq. (2.1), which is no longer assumed to be known in advance. On the contrary, the coefficients $\mathcal{W}_{n+1}$ of the generating function must now be determined stepwise from the simplification criterion used when approaching each particular problem.

Thus, replacing the generic function F by $\mathcal{H}$ in Deprit's recursion (2.15), in which we make $n = q = 0$, we obtain $\mathcal{H}_{0,1} = \{\mathcal{H}_{0,0}; \mathcal{W}_1\} + \mathcal{H}_{1,0}$. This equation is rearranged in the form

$$\{\mathcal{W}_1;\mathcal{H}_{0,0}\}=\widetilde{\mathcal{H}}_{0,1}-\mathcal{H}_{0,1}, \tag{2.33}$$

where $\mathcal{H}_{0,0}$ and $\widetilde{\mathcal{H}}_{0,1} = \mathcal{H}_{1,0}$ are known functions of the $(\boldsymbol{q},\boldsymbol{Q})$ variables, as given in Eq. (2.30), but both $\mathcal{H}_{0,1}$ and $\mathcal{W}_1$ are undetermined by now. In the perturbation approach $\mathcal{H}_{0,1}$ is *chosen* according to the simplification criterion. Finally, $\mathcal{W}_1$ must be solved from the partial differential equation that is obtained after the evaluation of the Poisson bracket on the left side of Eq. (2.33).

Once $\mathcal{W}_1$ has been computed, the second order of Deprit's recursion (2.15) yields

$$\mathcal{H}_{0,2}=\{\mathcal{H}_{0,1};\mathcal{W}_1\}+\mathcal{H}_{1,1} \tag{2.34}$$

where, again from Eq. (2.15),

$$\mathcal{H}_{1,1}=\{\mathcal{H}_{0,0};\mathcal{W}_2\}+\{\mathcal{H}_{1,0};\mathcal{W}_1\}+\mathcal{H}_{2,0}. \tag{2.35}$$

As before, after plugging Eq. (2.35) into Eq. (2.34), the latter is rearranged in the form

$$\{\mathcal{W}_2;\mathcal{H}_{0,0}\}=\widetilde{\mathcal{H}}_{0,2}-\mathcal{H}_{0,2}, \tag{2.36}$$

in which $\widetilde{\mathcal{H}}_{0,2}$ comprises all the terms that have become known hitherto. Namely,

$$\widetilde{\mathcal{H}}_{0,2}=\mathcal{H}_{2,0}+\{\mathcal{H}_{1,0};\mathcal{W}_1\}+\{\mathcal{H}_{0,1};\mathcal{W}_1\}. \tag{2.37}$$

Again, the new Hamiltonian term $\mathcal{H}_{0,2}$ is chosen according to the simplification criterion, and $\mathcal{W}_2$ must be solved from the partial differential equation stemming from Eq. (2.36).

Once the procedure has been extended up to the desired truncation order of the perturbation approach, Eq. (2.32) is obtained in the new variables by simply replacing $\boldsymbol{q}$ by $\boldsymbol{p}$ and $\boldsymbol{Q}$ by $\boldsymbol{P}$ in the terms of $\mathcal{H}_{0,m}$.

2.2.2 Example: Small oscillations of the simple pendulum

The simple pendulum is discussed in more detail in §3.1, from which we take the pendulum Hamiltonian

$$\mathcal{H} = \frac{1}{2}\Theta^2 + 2\omega^2 \sin^2 \frac{1}{2}\theta, \tag{2.38}$$

where Θ is the conjugate momentum to the angle θ and ω is a (dispensable) parameter. In the case of small oscillations the angle with which the pendulum departs from the vertical is $\theta \ll 1$, and hence $\sin^2 \frac{1}{2}\theta = \frac{1}{4}\theta^2 - \frac{1}{48}\theta^4 + \frac{1}{1440}\theta^6 + \cdots$ is replaced into Eq. (2.38). After rearrangement, Eq. (2.38) is written in the form

$$\mathcal{H} = \frac{1}{2}(\Theta^2 + \omega^2\theta^2) - \frac{1}{4!}\omega^2\theta^4 + \frac{1}{6!}\omega^2\theta^6 + \cdots \tag{2.39}$$

of the typical perturbation Hamiltonian (2.30) in which the small parameter is formal ($\varepsilon \equiv 1$), the zeroth-order term $\mathcal{H}_{0,0} = \frac{1}{2}(\Theta^2 + \omega^2\theta^2)$ is a harmonic oscillator of frequency ω in the "Cartesian" coordinate θ and conjugate momentum Θ, and the perturbation terms are $\mathcal{H}_{1,0} = -\frac{1}{4!}\omega^2\theta^4$, $\mathcal{H}_{2,0} = \frac{1}{6!}\omega^2\theta^6, \dots$.

First of all, we apply the harmonic transformation $(\phi, \Phi; \omega) \mapsto (\theta, \Theta)$ given by

$$\Theta = \sqrt{2\omega\Phi}\cos\phi, \quad \theta = \sqrt{2\Phi/\omega}\sin\phi, \tag{2.40}$$

which is canonical and converts Cartesian variables into harmonic variables. We obtain

$$\mathcal{H}_{0,0} = \omega\Phi, \quad \mathcal{H}_{1,0} = -\frac{\Phi^2}{6}\sin^4\phi, \quad \mathcal{H}_{2,0} = \frac{\Phi^3}{45\omega}\sin^6\phi, \dots. \tag{2.41}$$

That is, the action Φ is an integral of the unperturbed problem. If we extend this integral to the perturbed problem, then it will be trivially solved.

To do that, we apply the procedure described in §2.2.1 to find a canonical transformation $(\phi', \Phi'; \epsilon) \mapsto (\phi, \Phi)$ such that it transforms the Hamiltonian in harmonic variables (ϕ, Φ) into the Hamiltonian

$$\mathcal{H}(\phi(\phi', \Phi'; \epsilon), \Phi(\phi', \Phi'; \epsilon); \epsilon) = \sum_{m=0}^{n} \frac{\epsilon^m}{m!}\mathcal{H}_{0,m}(-, \Phi') + \sum_{j>n} \frac{\epsilon^j}{j!}\mathcal{H}_{0,j}(\phi', \Phi'),$$

in the new variables (ϕ', Φ'). That is, after truncation to order n, the new Hamiltonian will only depend on the momentum Φ', which, therefore, becomes a formal integral of the perturbed problem.

The first step is to solve Eq. (2.33), in which $\widetilde{\mathcal{H}}_{0,1} = \mathcal{H}_{1,0}$ and

$$\{\mathcal{W}_1; \mathcal{H}_{0,0}\} = \{\mathcal{W}_1; \omega\Phi\} = \omega\frac{\partial \mathcal{W}_1}{\partial \phi}.$$

Therefore, $\mathcal{W}_1$ is solved from Eq. (2.33) by indefinite integration. That is,

$$\mathcal{W}_1 = \frac{1}{\omega} \int \left[-\frac{1}{6} \Phi^2 \sin^4 \phi - \mathcal{H}_{0,1}(\phi, \Phi) \right] \mathrm{d}\phi, \tag{2.42}$$

where $\sin^4 \phi \equiv \frac{1}{8}(3 - 4\cos 2\phi + \cos 4\phi)$. Choosing

$$\mathcal{H}_{0,1}(\phi, \Phi) = -\frac{1}{16}\Phi^2,$$

we remove the angle ϕ from the first-order term of the new Hamiltonian. In addition, this choice converts the integrand of Eq. (2.42) in a purely periodic trigonometric function of ϕ, which is trivially solved to give

$$\mathcal{W}_1 = \frac{1}{192} \frac{\Phi^2}{\omega} (8 \sin 2\phi - \sin 4\phi) + C_1(\Phi).$$

The role of the arbitrary functions of the type of C_1 is not discussed here, and we simply make $C_1 = 0$.

Note that $\mathcal{H}_{0,1}$ is the average of $\widetilde{\mathcal{H}}_{0,1} = -\frac{1}{6}\Phi^2 \sin^4 \phi$ in the time in which the angle ϕ advances one period. That is, the new Hamiltonian term $\mathcal{H}_{0,1} = \langle \widetilde{\mathcal{H}}_{0,1} \rangle_\phi$ has been chosen by "averaging".

Once $\mathcal{W}_1$ is obtained we can proceed to the second order of the perturbation approach, in which, after evaluation of the Poisson bracket, Eq. (2.36) is also solved by indefinite integration. Namely,

$$\mathcal{W}_2 = \frac{1}{\omega} \int [\widetilde{\mathcal{H}}_{0,2}(\phi, \Phi) - \mathcal{H}_{0,2}] \mathrm{d}\phi, \tag{2.43}$$

where the computation of $\widetilde{\mathcal{H}}_{0,2}$ from Eq. (2.37) only involves the evaluation of Poisson brackets. Straightforward computations yield

$$\widetilde{\mathcal{H}}_{0,2} = -\frac{\Phi^3}{1920\omega} (15 - 35 \cos 2\phi + 2 \cos 4\phi + 3 \cos 6\phi).$$

Then the choice

$$\mathcal{H}_{0,2} = \langle \mathcal{H}_{0,2} \rangle_\phi \equiv -\frac{\Phi^3}{128\omega}$$

removes the angle from the new Hamiltonian, as desired, and leaves the integrand of Eq. (2.43) in the form of a purely periodic function of ϕ, which is trivially integrated to give

$$\mathcal{W}_2 = \frac{1}{3840} \frac{\Phi^3}{\omega^2} (35 \sin 2\phi - \sin 4\phi - \sin 6\phi) + C_2(\Phi).$$

Once more, we make the trivial choice $C_2 = 0$.

Because we neglected higher orders of the perturbation in Eq. (2.41), we stop the computations here. The transformed Hamiltonian is obtained by simply changing old by new (prime) variables in the terms $\mathcal{H}_{0,m}$, $(m = 0, 1, 2)$, yielding

$$\mathcal{H}' = \omega\Phi'\left[1 - \frac{1}{16}(\Phi'/\omega) - \frac{1}{256}(\Phi'/\omega)^2 + \mathcal{O}(\Phi'/\omega)^3\right],$$

whose Hamilton equations are trivially solved to give the *secular* terms of the solution

$$\Phi' = \Phi'_0 - \frac{\partial\mathcal{H}'}{\partial\phi'}\tau = \Phi'_0, \tag{2.44}$$

$$\phi' = \phi'_0 + \frac{\partial\mathcal{H}'}{\partial\Phi'}\tau = \phi'_0 + \omega\left(1 - \frac{1}{8}\frac{\Phi'_0}{\omega} - \frac{3}{256}\frac{\Phi'^2_0}{\omega^2}\right)\tau, \tag{2.45}$$

where τ denotes the time. That is, in the case of small oscillations, the simple pendulum behaves, on average, like a harmonic oscillator that evolves with constant, perturbed frequency $\tilde{\omega} = \omega[1 - \frac{1}{8}(\Phi'_0/\omega) - \frac{3}{256}(\Phi'_0/\omega)^2]$.

In order to obtain the solution in the original variables we need the direct transformation, which is readily obtained from Eq. (2.17). For the angle ϕ we compute the sequence

$$\begin{aligned}
\phi_{0,1} &= \{\phi; \mathcal{W}_1\} = \frac{\partial\mathcal{W}_1}{\partial\Phi} = \frac{1}{96}\frac{\Phi}{\omega}(8\sin 2\phi - \sin 4\phi),\\
\phi_{1,1} &= \{\phi; \mathcal{W}_2\} = \frac{\partial\mathcal{W}_2}{\partial\Phi} = \frac{1}{1280}\frac{\Phi^2}{\omega^2}(35\sin 2\phi - \sin 4\phi - \sin 6\phi),\\
\phi_{0,2} &= \phi_{1,1} + \{\phi_{0,1}; \mathcal{W}_1\} = \phi_{1,1} + \frac{\partial\phi_{0,1}}{\partial\phi}\frac{\partial\mathcal{W}_1}{\partial\Phi} - \frac{\partial\phi_{0,1}}{\partial\Phi}\frac{\partial\mathcal{W}_1}{\partial\phi}\\
&= \frac{1}{46080}\frac{\Phi^2}{\omega^2}(1280\sin 2\phi + 124\sin 4\phi - 96\sin 6\phi + 5\sin 8\phi).
\end{aligned}$$

After replacing the original by prime variables, and up to the truncation order of the perturbation approach, we obtain the direct transformation of the angle ϕ:

$$\begin{aligned}
\phi = \phi' &+ \frac{1}{96}\frac{\Phi'}{\omega}(8\sin 2\phi' - \sin 4\phi') + \frac{1}{2!}\frac{1}{46080}\frac{\Phi'^2}{\omega^2}\\
&\times (1280\sin 2\phi' + 124\sin 4\phi' - 96\sin 6\phi' + 5\sin 8\phi').
\end{aligned} \tag{2.46}$$

Analogously,

$$\begin{aligned}
\Phi_{0,1} &= \{\Phi; \mathcal{W}_1\} = -\frac{\Phi^2}{\omega}\frac{4\cos 2\phi - \cos 4\phi}{48},\\
\Phi_{1,1} &= \{\Phi; \mathcal{W}_2\} = -\frac{\Phi^3}{\omega^2}\frac{35\cos 2\phi - 2\cos 4\phi - 3\cos 6\phi}{1920},\\
\Phi_{0,2} &= \Phi_{1,1} + \{\Phi_{0,1}; \mathcal{W}_1\} = \frac{\Phi^3}{\omega^2}\frac{85 - 150\cos 2\phi + 6\cos 4\phi + 14\cos 6\phi}{5760},
\end{aligned}$$

which, after replacing the original by prime variables, lead to the direct transformation of the action Φ:

$$\Phi = \Phi'\left[1 - \frac{1}{48}\frac{\Phi'}{\omega}(4\cos 2\phi' - \cos 4\phi') + \frac{1}{2!}\frac{1}{5760}\frac{\Phi'^2}{\omega^2}(85 - 150\cos 2\phi' + 6\cos 4\phi' + 14\cos 6\phi')\right]. \tag{2.47}$$

The appearance of Φ' and ϕ' on the right sides of Eqs. (2.46) and (2.47) must be replaced by corresponding expressions in Eqs. (2.44) and (2.45), respectively, to obtain the perturbation solution as a function of time and initial conditions. Furthermore, the inverse transformation is needed for the initialization of the constants $\phi'_0 = \phi'(\phi_0, \Phi_0)$, $\Phi'_0 = \Phi'(\phi_0, \Phi_0)$, which the secular solution depends upon. We recall that the first- and second-order terms of the generating function of the inverse transformation are just the opposites of the corresponding terms of the direct generating function when it is rewritten in the prime variables—a fact that is no longer true when higher orders are taken into account, as checked in Eq. (2.27). Straightforward computations yield the inverse transformation

$$\begin{aligned}\phi' &= \phi - \frac{1}{96}\frac{\Phi}{\omega}(8\sin 2\phi - \sin 4\phi) \\ &\quad - \frac{1}{2!}\frac{1}{46080}\frac{\Phi^2}{\omega^2}(1240\sin 2\phi - 196\sin 4\phi + 24\sin 6\phi - 5\sin 8\phi), \\ \Phi' &= \Phi\left[1 + \frac{1}{48}\frac{\Phi}{\omega}(4\cos 2\phi - \cos 4\phi) \right.\\ &\quad \left. + \frac{1}{2!}\frac{1}{5760}\frac{\Phi^2}{\omega^2}(85 + 60\cos 2\phi - 6\cos 4\phi - 4\cos 6\phi)\right],\end{aligned}$$

in which ϕ and Φ are replaced by the initial conditions ϕ_0 and Φ_0 in order to obtain the corresponding initial conditions in the prime variables that feed the secular solution (2.44)–(2.45).

2.2.3 The homological equation

The procedure that led to Eqs. (2.33) and (2.36) can be repeated to extend the perturbation approach to any order. It is summarized in the so-called *homological* equation

$$\mathcal{L}_0(\mathcal{W}_m) = \widetilde{\mathcal{H}}_{0,m} - \mathcal{H}_{0,m}, \tag{2.48}$$

in which the operator

$$\mathcal{L}_0 \equiv \{\ \ ; \mathcal{H}_{0,0}\} \tag{2.49}$$

is customarily known as the *Lie derivative* in the Hamiltonian flow $\mathcal{H}_{0,0}$. Remark that the Lie derivative provides the time variation of a function along the Hamiltonian flow stemming from $\mathcal{H}_{0,0}$. This is the only case in which the time derivatives of the coordinates and their conjugate momenta can be replaced by corresponding Hamilton equations of the integrable part $\mathcal{H}_{0,0}$.

Terms $\widetilde{\mathcal{H}}_{0,m}$ in Eq. (2.48) are known from previous computations obtained after successive evaluations of Deprit's recursion (2.15), whereas particular choices of the terms $\mathcal{H}_{0,m}$ depend on the aim of the perturbation approach. Finally, $\mathcal{W}_m$ is obtained as a particular solution of the partial differential equation (2.48), contrary to the general or complete solution.

The selection of the terms $\mathcal{H}_{0,m}$ is arbitrary but only to some extent. Obviously, to make the perturbation approach feasible, the homological equation must be solvable for $\mathcal{W}_m$. The conditions that make Eq. (2.48) solvable depend on the algebraic structure of the functions $\mathcal{H}_{m,0}$ that are comprised by the original Hamiltonian (2.30). For instance, because the terms $\mathcal{H}_{m,0}$ in the example §2.2.2 are trigonometric polynomials in ϕ, the choice of the new Hamiltonian terms $\mathcal{H}_{0,m} = \langle\widetilde{\mathcal{H}}_{0,m}\rangle_\phi$, $m = 1, 2$, made in Eqs. (2.42) and (2.43), guaranteed that, in addition to obtaining a new Hamiltonian that only depends on the action, the terms $\mathcal{W}_m$ of the generating function are solvable and only depend on trigonometric terms. In this way, the solution of the homological equation is obtained at any order within the algebra of trigonometric functions.

In more abstract terms, the Lie derivative in the Hamiltonian flow $\mathcal{H}_{0,0} = \omega\Phi$ maps Fourier series of the form

$$F = \sum_{j\geq 0}[A_j(\Phi)\cos j\phi + B_j(\Phi)\sin j\phi], \tag{2.50}$$

into elements

$$\mathcal{L}_0 : F(\phi,\Phi) \mapsto \{F;\omega\Phi\} = \omega\frac{\partial F}{\partial\phi},$$

pertaining either to the *kernel* of the Lie derivative, which comprises such functions $F = A_0(\Phi)$ that $\mathcal{L}_0(F) = 0$, or to the *image* of the Lie derivative, which is made of such functions $F = \sum_{j\geq 1}[A_j(\Phi)\cos j\phi + B_j(\Phi)\sin j\phi]$ that $\mathcal{L}_0(F) = \sum_{j\geq 1}[A_j'(\Phi)\cos j\phi + B_j'(\Phi)\sin j\phi]$.

Applying this decomposition to terms $\widetilde{\mathcal{H}}_{0,m}$ entering the right side of the homological equation, then selecting $\mathcal{H}_{0,m}$ in such a way that it cancels out all the terms of $\widetilde{\mathcal{H}}_{0,m}$ pertaining to the kernel, is what makes $\mathcal{W}_m$ to pertain to the image, thus being solvable—and this is exactly what we did in §2.2.2.

In general, it is a good strategy to identify the algebra of functions to which the disturbing function of a particular problem pertains, and to choose the new Hamiltonian term in such a way that it cancels out all the terms on the right side of Eq. (2.48) pertaining to the kernel of the Lie derivative of the given problem. This strategy, when feasible, guarantees that the right side of the homological equation pertains to the image of the Lie derivative and, in consequence, the terms of the generating function can be solved up to arbitrary order of the perturbation approach.

Finally, it must be noted that, in general, the solution of the homological equation is not unique. Indeed, if the function C_0 pertains to the kernel of the Lie derivative, then replacing $\mathcal{W}_m$ by $\mathcal{W}_m + C_0$ in Eq. (2.48) also satisfies the homological equation. This fact was already illustrated in the example of §2.2.2 with the introduction of the arbitrary integration constants C_1 and C_2.

3 Application to integrable problems

Solutions of integrable systems are fruitfully approached in action-angle variables by solving the Hamilton–Jacobi equation [30, 199, 243]. Because this method involves the determination of a generating function in mixed variables, the solution is commonly obtained in the form of a mixed transformation. When this solution is achieved in terms of elementary functions, the transformation to action-angle variables can be obtained explicitly in closed form, as, for instance, in the case of the harmonic oscillator [627]. However, when the solution relies on special functions, whose evaluation depends on one or more parameters in addition to the function's argument, the action-angle variables approach may provide the closed-form solution in implicit form. While this is not troublesome in the evaluation of the solution, which is readily done with the help of root-finding procedures, the implicit form as well as the dependence on special functions may deprive the analytical solution of physical clarity. On the other hand, when dealing with a perturbed integrable motion the disturbing function is customarily expressed in the action-angle variables of the integrable problem. This process makes expanding the (implicit) transformation to action-angle variables as a Fourier series in the argument of the special functions necessary. These kinds of expansions are not at all trivial—yet these days one finds enormous assistance in computer algebra systems—and obtaining them was regarded as a notable achievement [345, 587, 588].

When the closed-form solution involves elliptic functions, the normal way of proceeding is to replace them by their definitions in terms of Jacobi theta functions, which in turn are replaced by their usual Fourier series expansion in trigonometric functions of the elliptic argument, whose coefficients are powers of the elliptic nome [89, 175, 447, 465]. This laborious procedure is further complicated when the modulus of the elliptic function remains as an implicit function of the action-angle variables, a case that requires its additional expansion followed by the series reversion, as it happens with the simple pendulum case [413]. Needless to say that carrying out expansions only makes sense when the closed-form solution depends on something that is small, either a physical parameter or the maximum value achieved by some variables, in this way making the power series to converge. But then the cumbersome procedure of making the expansions and their subsequent reversions can be completely avoided by directly approaching the solution of the integrable problem by perturbations.

In this chapter we provide two examples that illustrate the use of the Lie transforms method in the direct computation of the explicit, expanded solution of integrable problems whose closed-form solutions in action-angle variables depend on special functions and remain implicit. The first case is the simple mathematical pendulum, which is a problem of one degree of freedom, and is free from essential physical parameters. Since the traditional approach for the oscillatory regimen has already been outlined in §2.2.2, we only discuss the rotation regime following the descriptions in [413, 414]. The second example is the free rigid body, which is a system of two de-

https://doi.org/10.1515/9783110668513-003

grees of freedom and depends essentially on physical parameters. This fact serves to illustrate the perturbation approach in the case in which the small parameter is physical [200], and also to discuss the limitations of perturbation solutions when it relies on a dynamical (formal) small parameter [391, 399].

3.1 The simple gravity pendulum

The simple gravity pendulum consists of a bob of mass m that is attached to one end of a rod of length l and negligible mass, whose other end is fixed. It evolves under the action of the local gravity acceleration g without friction, and it is one of the simplest integrable models. The dynamical system is only of one degree of freedom, but the nonlinear motion may evolve in different regimes, and one must resort to the use of special functions to express its general solution in closed form [533]. In particular, the solution involves the use of Jacobi elliptic functions, and the action-angle variables can be obtained either by complete Hamiltonian reduction or directly by a canonical transformation of the traditional solution [67]. A good brief account on the topic from the point of view of Hamiltonian mechanics can be found in Appendix B of [198].

However elegant the closed-form solution in action-angle variables may be, it is not practical in common applications because, in addition to depending on special functions, it is obtained in implicit form. The expansion of the solution in action-angle variables, while feasible, is not trivial at all [413]. On the contrary, when approached by perturbations, the computation of the expansion of the explicit solution of the pendulum in action-angle variables is straightforward and systematic [414].

The case of small oscillations about the stable equilibrium position is customarily studied with linearized dynamics. Extending the solution further than the linear terms by perturbations is well documented in the literature [209, 449, 552] and was already briefly discussed in §2.2.2.

3.1.1 Hamiltonian reduction

The pendulum Hamiltonian represents the total energy $\mathcal{H} = T+V$. The potential energy is $V = mgh$, where the height $h = l(1 - \cos\theta)$ is measured with respect to the reference level in which the pendulum reaches the vertical direction, and θ is the angle with respect to that direction. The kinetic energy is $T = \frac{1}{2}\Theta^2/I$, where $I = ml^2$ denotes the moment of inertia, and the angular momentum Θ is the conjugate momentum to the generalized coordinate θ.

A time scaling $\tau = It$ yields

$$\mathcal{K} = I\mathcal{H} \equiv \frac{1}{2}\Theta^2 + \omega^2(1 - \cos\theta), \tag{3.1}$$

where $\omega = I\sqrt{g/l}$. Now, from Hamilton equations

$$\frac{\mathrm{d}\theta}{\mathrm{d}\tau} = \frac{\partial \mathcal{K}}{\partial \Theta} = \Theta, \quad \frac{\mathrm{d}\Theta}{\mathrm{d}\tau} = -\frac{\partial \mathcal{K}}{\partial \theta} = -\omega^2 \sin\theta,$$

it is immediately derived the usual equation of the simple pendulum

$$\frac{\mathrm{d}^2\theta}{\mathrm{d}\tau^2} + \omega^2 \sin\theta = 0, \tag{3.2}$$

where the parameter ω can be further eliminated by a convenient choice of units.

Alternatively, for a given value of the energy $\mathcal{K}(\theta_0, \Theta_0) = E$, the trajectories in phase space are directly obtained from Eq. (3.1) like

$$\Theta = \pm\omega\sqrt{2}\sqrt{\cos\theta - 1 + E/\omega^2}, \tag{3.3}$$

from which the phase space, the cylinder (θ, Θ), is readily represented without need of integration. This is illustrated in Fig. 3.1, where the trajectories are traveled from left to right for positive heights on the cylinder ($\Theta > 0$) and from right to left for negative heights ($\Theta < 0$), and discloses the two different regimes that may exist depending on the energy value. Namely, the rotation regime $E > 2\omega^2$, in which Eq. (3.3) always take real values, and the oscillation regime $0 \le E < 2\omega^2$, where the motion of θ is constrained to the interval in which Eq. (3.3) takes real values. In this last region, Fig. 3.1 shows the existence of a fixed point of the elliptic type at $E = 0$, corresponding to the stable equilibrium of the pendulum in the downward position ($\Theta = 0$, $\theta = 0$).

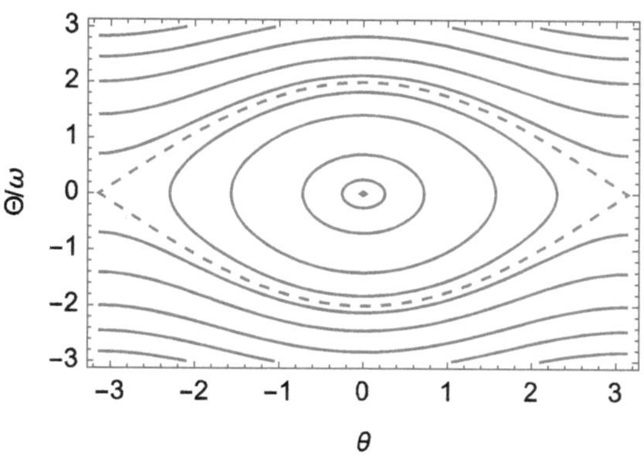

Figure 3.1: Phase space of the simple pendulum.

Oscillations and rotations are separated by the trajectory $\Theta = \pm 2\omega\cos\frac{1}{2}\theta$ corresponding to the energy $E = 2\omega^2$; the dashed line in Fig. 3.1 tends asymptotically to $\Theta = 0$, $\theta = \pm\pi$, which is a fixed point of the hyperbolic type on the surface of the cylinder that corresponds to the unstable equilibrium of the pendulum in the upward position. The branches of this trajectory that depart from (resp. arrive at) the hyperbolic fixed point are known as their unstable (resp. stable) manifolds.

As an alternative to the classical integration of Eq. (3.2), the solution of the simple pendulum can be computed by Hamiltonian reduction using the Hamilton–Jacobi equation [413]. Thus, we look for a canonical transformation $(\theta,\Theta) \mapsto (\theta',\Theta')$ that transforms the pendulum Hamiltonian (3.1) on a new Hamiltonian $\mathcal{K}(\theta(\theta',\Theta'),\Theta(\theta',\Theta')) \equiv \Psi(-,\Theta')$ that is cyclic in the new coordinate θ', whose integration is trivial: $\Theta' = \Theta'_0, \theta' = \theta'_0 + (\partial\Psi/\partial\Theta')\tau$.

The required transformation,

$$\theta' = \frac{\partial S}{\partial \Theta'}, \qquad \Theta = \frac{\partial S}{\partial \theta}, \tag{3.4}$$

is derived from a generating function in mixed variables $S \equiv S(\theta,\Theta')$, from which the second equation is replaced into Eq. (3.1) to arrange the Hamilton–Jacobi equation $\mathcal{K}(\theta,\partial S/\partial\theta) = \Psi(\Theta')$ from which S must be solved [30]. In particular, replacing $\Theta = \partial S/\partial\theta$ in Eq. (3.1), S is solved by indefinite integration,

$$S = \sqrt{2}\int \left[\Psi(\Theta') - 2\omega^2\sin^2(\theta/2)\right]^{1/2}\mathrm{d}\theta. \tag{3.5}$$

Plugging Eq. (3.5) into Eq. (3.4) yields

$$\theta' = \sqrt{2}\frac{\partial\Psi}{\partial\Theta'}\int \frac{\mathrm{d}(\theta/2)}{\sqrt{\Psi(\Theta') - 2\omega^2\sin^2(\theta/2)}}, \tag{3.6}$$

$$\Theta = \sqrt{2}\left[\Psi(\Theta') - 2\omega^2\sin^2(\theta/2)\right]^{1/2}, \tag{3.7}$$

in which the form of the new Hamiltonian $\Psi(\Theta')$ remains undetermined, thus giving rise to a whole family of transformations parameterized by Ψ [199].

3.1.2 Rotation regime. Solution in action-angle variables

The form of the solution of Eq. (3.6) depends on the dynamical regime in which the pendulum evolves. In the rotation regime $\Psi(-,\Theta') = E > 2\omega^2$, and the family of transformations given by Eqs. (3.7) and (3.6) is expressed in the form

$$\Theta = 2\frac{\omega}{k}\sqrt{1 - k^2\sin^2\psi}, \tag{3.8}$$

$$\theta' = \frac{k}{\omega}\frac{\partial\Psi}{\partial\Theta'}\mathrm{F}(\psi,k^2), \tag{3.9}$$

in which

$$\psi = \frac{1}{2}\theta, \quad k = \omega\sqrt{2/\Psi(\Theta')} < 1, \tag{3.10}$$

and $\mathrm{F}(\psi, k^2) \equiv \int_0^{\psi}(1-k^2\sin^2\alpha)^{-1/2}\mathrm{d}\alpha$ is the incomplete elliptic integral of the first kind of amplitude ψ and elliptic modulus k.

From the definition of the k in Eq. (3.10), the reduced Hamiltonian of the simple pendulum is written in the standard form

$$\Psi = \frac{2\omega^2}{k(\Theta')^2}, \tag{3.11}$$

which allows one to parameterize the family of transformations by k rather than Ψ. Indeed, from Eq. (3.11) we compute $\partial\Psi/\partial\Theta' = -4(\omega^2/k^3)\mathrm{d}k/\mathrm{d}\Theta'$, which is placed into Eq. (3.9), to obtain

$$\theta' = -4\frac{\omega}{k^2}\frac{\mathrm{d}k}{\mathrm{d}\Theta'}\mathrm{F}(\psi, k^2). \tag{3.12}$$

While any selection $k = k(\Theta')$ will produce the desired Hamiltonian reduction, it is common to choose k in such a way that the new, prime variables remain of the same nature as the original ones. In particular, the condition $\oint \mathrm{d}\theta' = 2\pi$, where the integral is computed along a closed curve in θ, will turn $\theta' = \theta'(\theta, \Theta)$ into an angle [30]. That is, Eq. (3.12) must fulfill the condition $\theta'(2\pi, \Theta') - \theta'(0, \Theta') = 2\pi$. Hence,

$$-4\frac{\omega}{k^2}\frac{\mathrm{d}k}{\mathrm{d}\Theta'}[\mathrm{F}(\pi, k^2) - \mathrm{F}(0, k^2)] = 2\pi.$$

From the properties of the elliptic functions, $\mathrm{F}(0, k^2) = 0$ and $\mathrm{F}(\pi, k^2) = 2\mathrm{K}(k^2)$, where $\mathrm{K}(k^2)$ denotes the complete elliptic integral of the first kind. Therefore,

$$\frac{\mathrm{d}k}{\mathrm{d}\Theta'} = -\frac{\pi}{4\omega}\frac{k^2}{\mathrm{K}(k^2)}, \tag{3.13}$$

which is plugged into Eq. (3.12) to obtain

$$\theta' = \frac{\pi}{\mathrm{K}(k^2)}\mathrm{F}(\psi, k^2). \tag{3.14}$$

On the other hand, because Eq. (3.13) is in separate variables we readily solve it to obtain

$$\Theta' = \frac{4\omega}{\pi k}\mathrm{E}(k^2), \tag{3.15}$$

where $\mathrm{E}(k^2)$ is the complete elliptic integral of the second kind.

For given (θ, Θ), k is obtained from Eq. (3.10) where $\Psi = \mathcal{H}(\theta, \Theta) = E$. Therefore, Eqs. (3.14)–(3.15) provide explicitly the transformation to the action-angle variables in which the pendulum Hamiltonian is completely reduced.

To obtain the transformation from action-angle variables to original variables we must solve $\theta = 2\psi$ from Eq. (3.14), which, jointly with Eq. (3.8), yields

$$\theta = 2\,\mathrm{am}(u, k^2), \quad \Theta = 2(\omega/k)\,\mathrm{dn}(u, k^2), \tag{3.16}$$

where $u = \mathrm{K}(k^2)\theta'/\pi$, am denotes the Jacobi amplitude function, and dn stands for the Jacobi delta amplitude. Evaluation of Eq. (3.16) requires the previous computation of $k = k(\Theta')$ from the Eq. (3.15). For numerical evaluation purposes it is readily done with the usual root-finding procedures. On the contrary, the implicit character of k in Eq. (3.15) prevents the explicit representation of the completely reduced Hamiltonian (3.11) in closed form as a function of the new momentum Θ'.

The formal inversion of Eq. (3.15) to get k as an explicit function of Θ' is needed in the solution of perturbed pendular motion. After expanding the right side of Eq. (3.15) in powers of $k < 1$, we find

$$2\frac{\omega}{\Theta'} = k\left(1 + \frac{1}{4}k^2 + \frac{7}{64}k^4 + \frac{15}{256}k^6 + \cdots\right),$$

which, setting $\epsilon = (\omega/\Theta)^2 < 1$, is followed by a series reversion procedure, to obtain

$$k = 2\sqrt{\epsilon}\left(1 - \epsilon + \frac{5}{4}\epsilon^2 - \frac{7}{4}\epsilon^3 + \frac{161}{64}\epsilon^4 + \cdots\right). \tag{3.17}$$

Now, the expanded value $k = k(\Theta')$ can be replaced into both the standard Hamiltonian (3.11) and the transformation (3.16)—the latter having been preprocessed using standard expansions of the Jacobi elliptic functions [175].

3.1.3 Expanded solution by Lie transforms

This involved procedure above is completely avoided by standard application of the Lie transforms method. Thus, after neglecting the constant term ω^2, the Hamiltonian (3.1) is rearranged in the form

$$\mathcal{K} = \frac{1}{2}\Theta^2[1 - 2(\omega/\Theta)^2\cos\theta], \tag{3.18}$$

which for values $(\omega/\Theta)^2 \ll \frac{1}{2}$ takes the form of a perturbation Hamiltonian. That is, the simple pendulum in the rotation regime can be viewed as the spherical rotor $\mathcal{K}_0 = \frac{1}{2}\Theta^2$ perturbed by the local gravity [627]. In that case, we can use the Lie transforms method to compute directly the explicit canonical transformation $(\theta, \Theta) \mapsto (\theta', \Theta'; \varepsilon)$ that, up to some truncation order ε^m, converts Eq. (3.18) into a new Hamiltonian depending only on Θ'.

To this aim, Eq. (3.18) is arranged in the form of the perturbation Hamiltonian (2.30) with $\mathcal{K}$ replacing $\mathcal{H}$. Namely, $\mathcal{K}_{0,0} = \frac{1}{2}\Theta^2$, $\mathcal{K}_{1,0} = -\cos\theta$, $\mathcal{K}_{m,0} = 0$ for $m \geq 2$, and the small parameter $\varepsilon = \omega^2$ has dimensions of angular momentum.

The Lie derivative (2.49) is $\mathcal{L}_0 = \Theta\partial/\partial\theta$. It immediately shows that its kernel is made of functions that do not depend on θ, whereas the image of the Lie derivative is made of trigonometric polynomials of θ whose coefficients are functions of the kernel. Then the homological equation (2.48) is solved by indefinite integration,

$$\mathcal{W}_m = \frac{1}{\Theta}\int(\widetilde{\mathcal{K}}_{0,m} - \mathcal{K}_{0,m})\,\mathrm{d}\theta. \tag{3.19}$$

At first order, $\widetilde{\mathcal{K}}_{0,1} = \mathcal{K}_{1,0} = -\cos\theta$, which is purely periodic. Therefore, we choose $\mathcal{K}_{0,1} = 0$ and trivially integrate Eq. (3.19) to get $\mathcal{W}_1 = -\Theta^{-1}\sin\theta$, where we ignored the arbitrary integration constant. At second order, Eq. (2.37) yields $\widetilde{\mathcal{K}}_{0,2} = \{\mathcal{K}_{1,0}; W_1\} = \Theta^{-2}\sin^2\theta$, where the terms of the kernel are easily identified by recalling that $\sin^2\theta = \frac{1}{2} - \frac{1}{2}\cos 2\theta$. The choice $\mathcal{K}_{0,2} = \frac{1}{2}\Theta^{-2}$ cancels the terms of the kernel in the integrand of Eq. (3.19) out, which then becomes a function of the image. Then Eq. (3.19) is solved to give $\mathcal{W}_2 = -\frac{1}{4}\Theta^{-3}\sin 2\theta$. Finally, we fill Deprit's triangle (2.16) by computing the intermediate term $\mathcal{K}_{1,1} = \frac{1}{2}\Theta^{-2}$ using Deprit's recursion (2.15).

At third order, after successive applications of Eq. (2.15), we obtain

$$\widetilde{\mathcal{K}}_{0,3} = -\frac{1}{4}\Theta^{-4}(5\cos\theta + 3\cos 3\theta),$$

which pertains to the image of the Lie derivative. In consequence, we choose $\mathcal{K}_{0,3} = 0$, compute $\mathcal{W}_3 = -\frac{1}{4}\Theta^{-5}(5\sin\theta + 3\sin 3\theta)$, from Eq. (3.19), and fill Deprit's triangle with the terms $\mathcal{K}_{2,1} = 2\Theta^{-4}\cos\theta$ and $\mathcal{K}_{1,2} = \Theta^{-4}\cos\theta$, which will be needed in following orders.

Analogous computations yield

$$\mathcal{K}_{0,4} = \frac{15}{8}\Theta^{-6}, \quad \mathcal{W}_4 = -\frac{3}{32}\Theta^{-7}(72\sin 2\theta + 5\sin 4\theta),$$

and so on. After reaching the desired order, the procedure ends writing the terms $\mathcal{K}_{0,m}$ in the new variables. In this way, we obtain the new, completely reduced Hamiltonian $\Psi = \sum_{m\geq 0}(\varepsilon^m/m!)\mathcal{K}_{0,m}(\Theta')$ given by

$$\Psi = \frac{1}{2}\Theta'^2\left[1 + \frac{1}{2}\epsilon^2 + \frac{5}{32}\epsilon^4 + \frac{9}{64}\epsilon^6 + \mathcal{O}(\epsilon^8)\right], \tag{3.20}$$

where the non-dimensional small parameter $\epsilon = (\omega/\Theta')^2 < 1$, is used for brevity instead of the dimensional one $\varepsilon = \omega^2$ of the perturbation approach. Comparison of Eq. (3.20) with the one obtained from the direct expansion of the standard Hamiltonian (3.11), after replacing k by the right side of Eq. (3.17), will show that the two expansions match term by term.

The direct transformation from prime to original variables is computed by standard application of Deprit's recursion (2.15). Indeed, since the generating function is known, replacing x by θ in Eq. (2.17) and evaluating the resulting Poisson brackets, we find

$$\begin{aligned}
\theta_{0,1} &= \Theta^{-2}\sin\theta,\\
\theta_{1,1} &= \frac{3}{4}\Theta^{-4}\sin 2\theta,\\
\theta_{0,2} &= -\frac{1}{2}\Theta^{-4}\sin 2\theta,\\
\theta_{1,2} &= \frac{3}{8}\Theta^{-6}(13\sin\theta + \sin 3\theta),\\
\theta_{2,1} &= \frac{5}{4}\Theta^{-6}(5\sin\theta + \sin 3\theta),
\end{aligned}$$

and so on. Then, after replacing original by prime variables, the transformation is obtained in the form of a Taylor series $\theta = \sum_{m\geq 0}(\varepsilon^m/m!)\theta_{0,m}(\theta',\Theta')$.

We carry out analogous computations for Θ, and reorganize both transformations in the form of Fourier series whose coefficients are truncated series in the small parameter. Using again the non-dimensional abbreviation ϵ instead of the small parameter $\varepsilon = \omega^2$, we obtain

$$\theta = \theta' + \sum_{j\geq 1} z_j(\epsilon)\epsilon^j \sin j\theta', \quad \Theta = \Theta' \sum_{j\geq 0} Z_j(\epsilon)\epsilon^j \cos j\theta', \tag{3.21}$$

where the coefficients z_j and Z_j are given in Table 3.1. Like before, it can be checked that the traditional expansion Eq. (3.16) matches Eq. (3.21) term by term.

Table 3.1: Coefficients in Eq. (3.21) up to $\mathcal{O}(\epsilon^6)$; $Z_0 = 1 - \frac{1}{2}\epsilon^2 - \frac{15}{32}\epsilon^4 - \frac{45}{64}\epsilon^6$.

j:	1	2	3	4	5	6
z_j	$1 + \frac{11}{16}\epsilon^2 + \frac{247}{256}\epsilon^4$	$\frac{1}{8} + \frac{3}{16}\epsilon^2 + \frac{707}{2048}\epsilon^4$	$\frac{1}{48} + \frac{9}{192}\epsilon^2$	$\frac{1}{256} + \frac{3}{256}\epsilon^2$	$\frac{1}{1280}$	$\frac{1}{6144}$
Z_j	$1 + \frac{3}{16}\epsilon^2 + \frac{39}{256}\epsilon^4$	$\frac{1}{4} + \frac{1}{4}\epsilon^2 + \frac{395}{1024}\epsilon^4$	$\frac{1}{16} + \frac{7}{64}\epsilon^2$	$\frac{1}{64} + \frac{5}{128}\epsilon^2$	$\frac{1}{256}$	$\frac{1}{1024}$

Computation of the expansion of Eqs. (3.14)–(3.15) by Lie transforms requires the preliminary computation of the generating function of the inverse transformation $\mathcal{V} = -\mathcal{W}(\theta(\theta',\Theta'),\Theta(\theta',\Theta'))$, which is also obtained by Lie transforms in the form of a Taylor series as described in §2.1.3.

3.2 The free rigid body

Another example of the suitability of the Lie transforms for approaching integrable problems is the free rotation of a rigid body about a fixed point. External forces being absent, the linear momentum is conserved and the motion is conveniently described by the rotation of the rigid body with respect to its center of mass O. Traditionally, the motion is decomposed into the rotation referred to the body (rotating) frame $(O, \boldsymbol{b}_1, \boldsymbol{b}_2, \boldsymbol{b}_3)$—which is attached to the rigid body's center of mass and defined by the principal axis of the body—after which the attitude of the body in the space (inertial) frame $(O, \boldsymbol{s}_1, \boldsymbol{s}_2, \boldsymbol{s}_3)$ is materialized by the Euler angles [243].

Due to the lack of external forces or torques the free rigid body conserves both the angular momentum vector $\boldsymbol{G} = \mathbb{I}\,\boldsymbol{\omega}$, where $\boldsymbol{\omega}$ is the angular velocity and $\mathbb{I}$ is the inertia tensor, and the energy, which is limited to the kinetic one,

$$T = \frac{1}{2}\boldsymbol{\omega} \cdot \mathbb{I}\,\boldsymbol{\omega} = \frac{1}{2}\boldsymbol{G} \cdot \mathbb{I}^{-1}\,\boldsymbol{G}. \tag{3.22}$$

Since they are independent, the free rigid body motion accepts one more integral than the number of degrees of freedom, thus constraining the possible trajectories on the surface of a torus and making the problem superintegrable [192].

3.2.1 Rotation in the body frame

In the body frame $\mathbb{I}$ is constant. If, besides, the body axes are chosen parallel to the axes of principal inertia, then the products of inertia vanish and

$$\mathbb{I} = \begin{pmatrix} A & 0 & 0 \\ 0 & B & 0 \\ 0 & 0 & C \end{pmatrix}$$

where $0 < A \leq B \leq C$ are the principal moments of inertia, and $A + B \geq C$ from the definition of the inertia tensor.

For a given energy $T = E > 0$ constant, Eq. (3.22) constrains the possible solutions to the surface

$$\frac{1}{2}A\omega_1^2 + \frac{1}{2}B\omega_2^2 + \frac{1}{2}C\omega_3^2 = \frac{1}{2A}g_1^2 + \frac{1}{2B}g_2^2 + \frac{1}{2C}g_3^2 = E,$$

where $(\omega_1, \omega_2, \omega_3)$ and (g_1, g_2, g_3) are the components of $\boldsymbol{\omega}$ and $\boldsymbol{G}$, respectively, in the body frame. This constraint takes the form of an ellipsoid of axes $2AE$, $2BE$, $2CE$. Because $\|\boldsymbol{G}\| = G$ is also constant, the motion is constrained to the sphere $g_1^2+g_2^2+g_3^2 = G^2$, of radius G. The possible trajectories of the angular momentum vector in the body frame are, therefore, given by the intersection of both surfaces [474], which is illustrated in Fig. 3.2 for a Moon-type body.

The instantaneous rotation in the body frame is obtained from Euler equations for the variation of the angular momentum $(\mathrm{d}\mathbb{I}\boldsymbol{\omega}/\mathrm{d}t)_s = (\mathrm{d}\mathbb{I}\boldsymbol{\omega}/\mathrm{d}t)_b + \boldsymbol{\omega} \times \mathbb{I}\,\boldsymbol{\omega}$. For the free rigid body, we obtain

$$\dot{\omega}_1 = -\frac{C-B}{A}\omega_2\omega_3, \quad \dot{\omega}_2 = \frac{C-A}{B}\omega_3\omega_1, \quad \dot{\omega}_3 = -\frac{B-A}{C}\omega_1\omega_2. \tag{3.23}$$

In the particular case of mass distribution with spherical symmetry $A = B = C$, all the solutions are equilibria. In the case $A = B \neq C$ of axial symmetry with respect

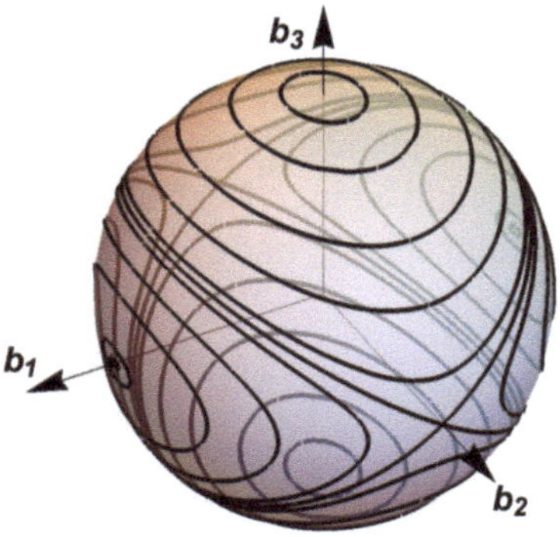

Figure 3.2: Trajectories of the angular momentum vector of a Moon-like free rigid body in the body frame ($A/C = 0.9994$, $B/C = 0.9996$ [208]).

to the axis of maximum inertia (or short axis), ω_3 is constant and ω_1, ω_2 evolve with harmonic motion. The case $A \neq B = C$ of axial symmetry with respect to the axis of minimum inertia (or long axis), is analogous to the previous one.

When $A < B < C$, simple inspection of Eq. (3.23) shows that free rotations around the principal axes correspond to dynamical equilibria, whose stability character can be guessed from Fig. 3.2. Thus, stable rotations around the axis of maximum inertia $\boldsymbol{b}_3$ ($\omega_1 = \omega_2 = 0$) occur for the minimum energy

$$E_3 = \frac{1}{2}C\omega^2 = \frac{1}{2C}G^2. \tag{3.24}$$

Unstable rotations around the axis of intermediate inertia $\boldsymbol{b}_2$ ($\omega_1 = \omega_3 = 0$) occur for the intermediate energy

$$E_2 = \frac{1}{2}B\omega^2 = \frac{1}{2B}G^2, \tag{3.25}$$

and the range $E_3 \leq E < E_2$ defines a regime in which the angular momentum vector rotates about the axis of maximum inertia. Finally, stable rotations around the axis of minimum inertia $\boldsymbol{b}_1$ ($\omega_2 = \omega_3 = 0$) correspond to the maximum energy

$$E_1 = \frac{1}{2}A\omega^2 = \frac{1}{2A}G^2, \tag{3.26}$$

between which value and E_2 the angular momentum vector rotates about the axis of minimum inertia. The different energy regimes are separated by the trajectory with $E = E_2$ that links the unstable equilibria.

The general solution of Eq. (3.23) is obtained for the different energy regimes in closed form in terms of Jacobi's elliptic functions, and in terms of hyperbolic functions in the case of the separatrix [377, 473].

3.2.2 Attitude in the space frame

Once the rotation is solved in the body frame, it remains to relate the body and space frames through the Euler angles (ψ, θ, φ) for precession, inclination and rotation, respectively. They are illustrated in Fig. 3.3, in which the unit vector $\boldsymbol{n} = \boldsymbol{s}_3 \times \boldsymbol{b}_3/\|\boldsymbol{s}_3 \times \boldsymbol{b}_3\|$ defines the node of the body's equatorial plane $(\boldsymbol{b}_1, \boldsymbol{b}_2)$ over the space plane $(\boldsymbol{s}_1, \boldsymbol{s}_2)$, and $\boldsymbol{m} = \boldsymbol{s}_3 \times \boldsymbol{n}$.

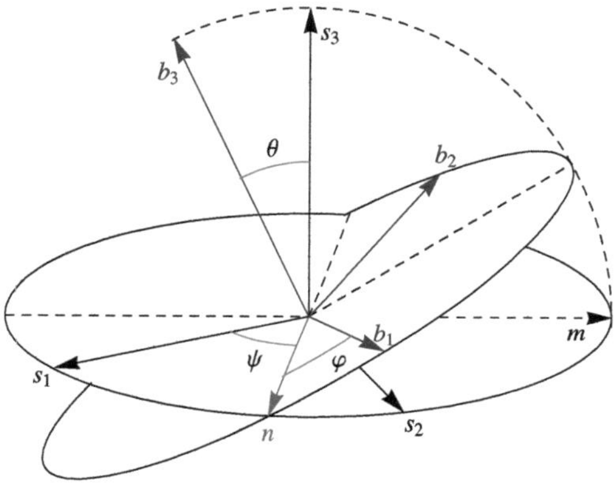

Figure 3.3: Euler angles relating the body and space frames.

The desired relation between the components of $\boldsymbol{\omega}$ in the body (rotating) frame and the derivatives of the Euler angles is obtained following the derivations in [682] (see also [544]). Thus, let $\boldsymbol{v}$ be a vector whose components in the space (inertial) frame $\boldsymbol{v}_s = (\xi, \eta, \zeta)$ are obtained from the corresponding ones in the body frame $\boldsymbol{v}_b = (x, y, z)$ by a rotation. Using matrix notation, $(\xi, \eta, \zeta)^\tau = R\,(x, y, z)^\tau$, where τ denotes transposition, and $R \equiv R_3(-\psi)\,R_1(-\theta)\,R_3(-\phi)$, where

$$R_3 = \begin{pmatrix} \cos & \sin & 0 \\ -\sin & \cos & 0 \\ 0 & 0 & 1 \end{pmatrix}, \quad R_1 = \begin{pmatrix} 1 & 0 & 0 \\ 0 & \cos & \sin \\ 0 & -\sin & \cos \end{pmatrix}, \tag{3.27}$$

are the usual rotation matrices. Differentiation of the rotation yields $(\dot{\xi}, \dot{\eta}, \dot{\zeta})^\tau = R\,(\dot{x}, \dot{y}, \dot{z})^\tau + \dot{R}\,(x, y, z)^\tau$, where $(\dot{\xi}, \dot{\eta}, \dot{\zeta})$ are the components of the velocity vector in the space frame, and $(\dot{x}, \dot{y}, \dot{z})$ are those in the body frame. Hence, on account of R being an orthogonal matrix,

$$\begin{pmatrix} \dot{\xi} \\ \dot{\eta} \\ \dot{\zeta} \end{pmatrix} = R \left[\begin{pmatrix} \dot{x} \\ \dot{y} \\ \dot{z} \end{pmatrix} + R^\tau \dot{R} \begin{pmatrix} x \\ y \\ z \end{pmatrix} \right]. \tag{3.28}$$

Alternatively, from the rule for differentiation of a vector in a moving frame,

$$\frac{\mathrm{d}\boldsymbol{x}}{\mathrm{d}t} = \dot{\boldsymbol{x}} + \boldsymbol{\omega} \times \boldsymbol{x}, \tag{3.29}$$

where the dot over a vector means here derivation of the vector in the rotating frame, we get the components in the space frame,

$$\begin{pmatrix} \dot{\xi} \\ \dot{\eta} \\ \dot{\zeta} \end{pmatrix} = R\left[\begin{pmatrix} \dot{x} \\ \dot{y} \\ \dot{z} \end{pmatrix} + \Omega\begin{pmatrix} x \\ y \\ z \end{pmatrix}\right], \quad \Omega = \begin{pmatrix} 0 & -\omega_3 & \omega_2 \\ \omega_3 & 0 & -\omega_1 \\ -\omega_2 & \omega_1 & 0 \end{pmatrix}. \tag{3.30}$$

From Eqs. (3.28) and (3.30) we obtain $\Omega = R^T\dot{R}$, from which[1]

$$\begin{aligned} \omega_1(t) &= \dot{\psi}\sin\theta\sin\varphi + \dot{\theta}\cos\varphi, \\ \omega_2(t) &= \dot{\psi}\sin\theta\cos\varphi - \dot{\theta}\sin\varphi, \\ \omega_3(t) &= \dot{\psi}\cos\theta + \dot{\varphi}, \end{aligned} \tag{3.31}$$

which is readily solved in the derivatives of the angles to obtain

$$\begin{aligned} \dot{\psi} &= \left[\omega_1(t)\sin\varphi + \omega_2(t)\cos\varphi\right]\csc\theta, \\ \dot{\theta} &= \omega_1(t)\cos\varphi - \omega_2(t)\sin\varphi, \\ \dot{\varphi} &= \omega_3(t) - \dot{\psi}\cos\theta. \end{aligned} \tag{3.32}$$

The time solution of Eq. (3.32) involves the use of Jacobi theta functions [322, 571].

Note, however, that an interesting simplification arises when taking the plane perpendicular to the angular momentum vector as the inertial plane. Then $\boldsymbol{G} = G\boldsymbol{s}_3$ and the components of the angular momentum vector in the body frame are simply

$$(g_1, g_2, g_3)^T = (A\omega_1, B\omega_2, C\omega_3)^T = R_3(\varphi)\,R_1(\theta)\,(0, 0, G)^T,$$

from which φ and θ are solved without need of integration, whereas the following integration of $\dot{\psi}$ from Eq. (3.32) provides ψ in closed form as a function of the elliptic integrals of the first and the third kinds [245]. Then, referring the solution to a different fixed plane only involves additional rotations of fixed Euler angles, say $(\psi_1, \theta_1, \varphi_1)$.

3.2.3 The invariable plane

The plane perpendicular to the angular momentum vector is customarily called the *invariable* plane, yet it will not remain fixed in the presence of torques. It provides a natural link between the body frame and any fixed plane.

1 Customary arguments to justify the vectorial decomposition $\boldsymbol{\omega} = \dot{\psi}\,\boldsymbol{s}_3 + \dot{\theta}\,\boldsymbol{n} + \dot{\varphi}\,\boldsymbol{b}_3$ have been pointed out as erroneous in [448].

Indeed, let $\boldsymbol{m}$ be a unit vector in the direction of the angular momentum vector, $\boldsymbol{G} = G\boldsymbol{m}$, and let $\boldsymbol{l}$ be a unit vector in the direction defined by the intersection of the space plane and the invariable plane. Then (see Fig. 3.4 for reference), $\boldsymbol{s}_3 \cdot \boldsymbol{m} = \cos I$, where $0 \le I < \pi$ is the angle encompassed by $\boldsymbol{s}_3$ and $\boldsymbol{m}$ reckoned counterclockwise from $\boldsymbol{s}_3$ (the inclination between the invariable and space planes). Besides, we have $\boldsymbol{s}_3 \times \boldsymbol{m} = \boldsymbol{l} \sin I$, where $\boldsymbol{l} = \boldsymbol{s}_1 \cos\lambda + \boldsymbol{s}_2 \sin\lambda$ and $0 \le \lambda < 2\pi$ is the angle encompassed by $\boldsymbol{s}_1$ and $\boldsymbol{l}$ reckoned counterclockwise from $\boldsymbol{s}_1$—the precession angle of the invariable plane on the space plane.

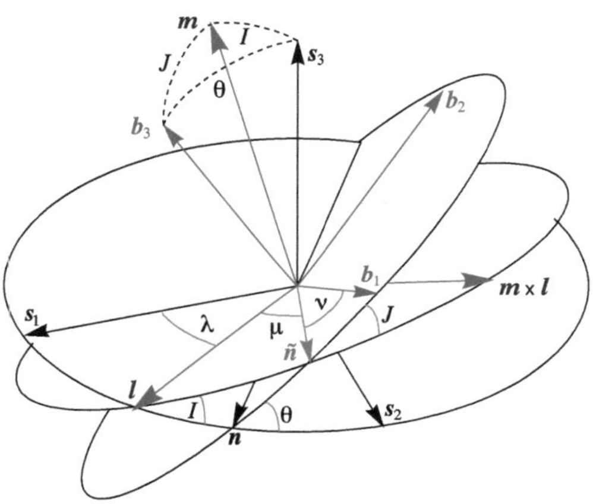

Figure 3.4: The invariable plane $\boldsymbol{m}$ as an intermediate reference.

Analogously, let $\tilde{\boldsymbol{n}}$ be a unit vector in the direction defined by the intersection of the invariable plane and the equatorial plane of the body. Then $\boldsymbol{m} \cdot \boldsymbol{b}_3 = \cos J$, where $0 \le J < \pi$ is the angle encompassed by $\boldsymbol{m}$ and $\boldsymbol{b}_3$ reckoned counterclockwise from $\boldsymbol{m}$—the inclination between the invariable plane and the body's equatorial plane. Besides, $\boldsymbol{m} \times \boldsymbol{b}_3 = \tilde{\boldsymbol{n}} \sin J$, where $\tilde{\boldsymbol{n}} = \boldsymbol{b}_1 \cos\nu + \boldsymbol{b}_2 \sin\nu$ and $0 \le \nu < 2\pi$ is the angle encompassed by $\tilde{\boldsymbol{n}}$ and $\boldsymbol{b}_1$ reckoned counterclockwise from $\tilde{\boldsymbol{n}}$—the rotation angle of the body frame. Finally, the precession angle of the equatorial plane on the invariable plane μ is defined from $\boldsymbol{l} \cdot \tilde{\boldsymbol{n}} = \cos\mu$, $(\boldsymbol{m} \times \boldsymbol{l}) \cdot \tilde{\boldsymbol{n}} = \sin\mu$, where $0 \le \mu < 2\pi$ is the angle encompassed by $\boldsymbol{l}$ and $\tilde{\boldsymbol{n}}$ reckoned counterclockwise from $\boldsymbol{l}$.

Then the components of the vector $\boldsymbol{v}$ in the body frame $\boldsymbol{v}_b$ and space frame $\boldsymbol{v}_s$ are related by the sequence of rotations $\boldsymbol{v}_b = R_3(\nu)\, R_1(J)\, R_3(\mu)\, R_1(I)\, R_3(\lambda)\, \boldsymbol{v}_s$. Alternatively, $\boldsymbol{v}_s = R_3(-\lambda)\, R_2(-I)\, R_3(-\mu)\, R_2(-J)\, R_3(-\nu)\, \boldsymbol{v}_b$. That is, the body and space frames can be related by five rotations based on two sets of Euler angles, say $(\psi_1, \theta_1, \phi_1, \psi_2, \theta_2, \phi_2)$, such that $\lambda = \psi_1$, $I = \theta_1$, $\mu = \phi_1 + \psi_2$, $J = \theta_2$, and $\nu = \phi_2$. Needless to say that the construction described above is purely geometric, and no assumption has been made on the conservation of $\boldsymbol{G}$.

3.2.4 Hamiltonian formulation

Like in the case of the simple pendulum, an alternative to the classical integration is given by the complete Hamiltonian reduction of the free rigid body Hamiltonian.

The first step is to compute the Lagrangian $\mathfrak{L} \equiv \mathfrak{L}(\varphi, \theta, \psi, \dot\varphi, \dot\theta, \dot\psi)$. On account of there not being potential for the torque-free rotation, the total energy is limited to the kinetic component in Eq. (3.22). Hence, $\mathfrak{L} = \frac{1}{2}(A\,\omega_1^2 + B\,\omega_2^2 + C\,\omega_3^2)$, where the components of the angular velocity in the body frame are those in Eq. (3.31). Because the Lagrangian only involves homogeneous terms of the second degree in the generalized velocities the Hamiltonian is the total energy [243]. Therefore, $\mathcal{H} = T = \mathfrak{L}$.

Using Eqs. (3.32) we find that the conjugate momenta to the Euler angles can be written in the form

$$\Phi = \frac{\partial \mathfrak{L}}{\partial \dot\varphi} = \boldsymbol{G}\cdot\boldsymbol{b}_3, \qquad \Theta = \frac{\partial \mathfrak{L}}{\partial \dot\theta} = \boldsymbol{G}\cdot\boldsymbol{n}, \qquad \Psi = \frac{\partial \mathfrak{L}}{\partial \dot\psi} = \boldsymbol{G}\cdot\boldsymbol{s}_3. \tag{3.33}$$

However, rather than constructing the Hamiltonian in Euler variables, it is advisable to resort to the canonical set of Andoyer variables $(\lambda, \mu, \nu, \Lambda, M, N)$ given by the angles λ, μ, and ν, used in the description of the invariable plane, and the actions $\Lambda = G\cos I$, $M = G$, and $N = G\cos J$ [24]. The canonical transformation from Euler variables to Andoyer variables can be found, for instance, in [56]. Singularities of Andoyer variables when I or J vanishes can be avoided using, for instance, the alternative sets proposed in [206, 207, 287].

From Eq. (3.22) and on account of $\mathcal{H} = T$ for the free rigid body, we write

$$\mathcal{H} = \frac{1}{2}\left(\frac{1}{A}g_1^2 + \frac{1}{B}g_2^2 + \frac{1}{C}g_3^2\right), \tag{3.34}$$

where

$$\begin{pmatrix} g_1 \\ g_2 \\ g_3 \end{pmatrix} = R_3(\nu)\,R_1(J)\begin{pmatrix} 0 \\ 0 \\ G \end{pmatrix} = G\begin{pmatrix} \sin J\sin\nu \\ \sin J\cos\nu \\ \cos J \end{pmatrix}.$$

Then, replacing $G = M$ and $J = \arccos(N/M)$, the free rigid body Hamiltonian in Andoyer variables is

$$\mathcal{H} = \frac{1}{2}\left(\frac{1}{A}\sin^2\nu + \frac{1}{B}\cos^2\nu\right)(M^2 - N^2) + \frac{1}{2C}N^2, \tag{3.35}$$

where Λ, λ, and μ are ignorable variables. Therefore, M, Λ, and λ are integrals of the torque-free motion [150]. These integrals decouple the flow of the free rigid body, whose reduced dynamics is obtained from the integration of the Hamilton equations

$$\frac{\mathrm{d}\nu}{\mathrm{d}t} = \frac{\partial \mathcal{H}}{\partial N} = -\left(\frac{1}{A}\sin^2\nu + \frac{1}{B}\cos^2\nu - \frac{1}{C}\right)N, \tag{3.36}$$

$$\frac{dN}{dt} = -\frac{\partial \mathcal{H}}{\partial \nu} = \left(\frac{1}{A} - \frac{1}{B}\right)(M^2 - N^2) \sin \nu \cos \nu. \tag{3.37}$$

The time history of μ is then obtained by indefinite integration.

The Hamiltonian (3.35) provides an elegant alternative to the geometric representation of the flow in Fig. 3.2. Thus, for a given energy manifold, say $\mathcal{H}(\nu, N; M) = E$, Eq. (3.35) can be solved for N to give

$$N = \pm\sqrt{Q}M, \tag{3.38}$$

where

$$Q = \frac{(1/A)\sin^2 \nu + (1/B)\cos^2 \nu - 2E/M^2}{(1/A)\sin^2 \nu + (1/B)\cos^2 \nu - 1/C}. \tag{3.39}$$

Phase curves (ν, N) are then obtained by evaluation of Eq. (3.38) for different values of E, as illustrated in Fig. 3.5 for the same Moon-type body used in Fig. 3.2. Curves in Fig. 3.5 correspond to scaled energy levels $\mathcal{E} = 2CE/G^2$, and are traversed right to left for $N > 0$ and left to right for $N < 0$.

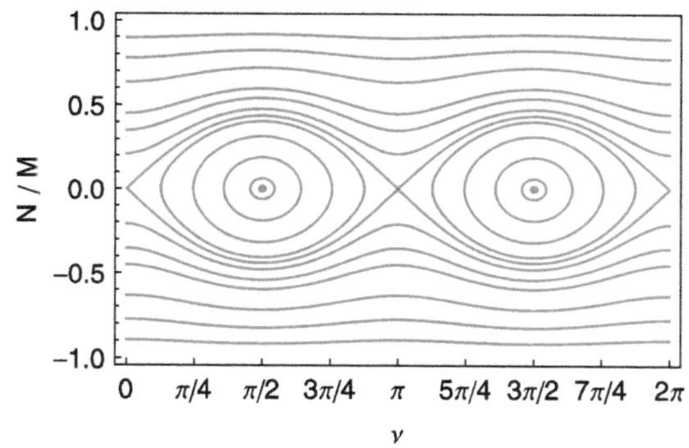

Figure 3.5: Phase space of the free rotation of a Moon-like body (after [150]).

When $N = \pm M$ the variation of N in Eq. (3.37) vanishes, and the phase lines $N/M = \pm 1$ of Fig. 3.2 correspond to permanent rotations about the axis of maximum inertia. Besides, the configurations $\nu = 0$, $N = 0$, and $\nu = \frac{\pi}{2}$, $N = 0$, are equilibria of Eqs. (3.36)–(3.37). The former corresponds to a hyperbolic fixed point of Fig. 3.2 that yields permanent rotations about the axis of intermediate inertia, whereas the latter corresponds to an elliptic fixed point of the same figure that yields permanent rotations about the axis of minimum inertia.

3.2.5 Closed-form solution by complete reduction

Like in the case of the simple pendulum, the integration of the torque-free motion can be achieved by finding a canonical transformation $(\lambda, \mu, \nu, \Lambda, M, N) \mapsto (\ell, g, h, L, G, H)$

that completely reduces Hamiltonian (3.35) to a function that, in the new variables, depends only on momenta. In what follows, we closely adhere to the descriptions in [412].

Because the conjugate pair (λ, Λ) is ignorable in Eq. (3.35), we trivially take $h = \lambda$, $H = \Lambda$, and look for a transformation

$$\ell = \frac{\partial S}{\partial L}, \quad g = \frac{\partial S}{\partial G}, \quad M = \frac{\partial S}{\partial \mu}, \quad N = \frac{\partial S}{\partial \nu}, \tag{3.40}$$

derived from the generating function in mixed variables $S = S(\mu, \nu, L, G)$, such that $\mathcal{H}(-, \nu, M, N) = \Upsilon(-, -, L, G)$.

Because μ is a cyclic variable, we choose the generating function in separate variables $S = G\mu + W(\nu, L, G)$, from which $M = G$ in Eq. (3.40). Then the Hamilton–Jacobi equation

$$\frac{1}{2}\left(\frac{\sin^2 \nu}{A} + \frac{\cos^2 \nu}{B}\right)\left[G^2 - \left(\frac{\partial W}{\partial \nu}\right)^2\right] + \frac{1}{2C}\left(\frac{\partial W}{\partial \nu}\right)^2 = \Upsilon(L, G) \tag{3.41}$$

is solved for W, yielding

$$W = G \int \sqrt{Q(\nu, L, G)}\, \mathrm{d}\nu, \tag{3.42}$$

where Q is the same as in Eq. (3.39), but now the total energy E is replaced by the formal Hamiltonian $\Upsilon(L, G)$. Namely

$$Q = \frac{(1/A)\sin^2 \nu + (1/B)\cos^2 \nu - 1/\Delta}{(1/A)\sin^2 \nu + (1/B)\cos^2 \nu - 1/C}, \tag{3.43}$$

in which we introduced the auxiliary variable of the moment-of-inertia type

$$\frac{1}{\Delta} = \frac{2}{G^2}\Upsilon(L, G). \tag{3.44}$$

Note that $A < \Delta < C$, as follows from Eqs. (3.24) and (3.26).

Then, by applying the chain rule, the transformation in Eq. (3.40) is

$$\ell = \frac{1}{G}\frac{\partial \Upsilon}{\partial L}\mathcal{I}, \quad g = \mu + \frac{1}{G}W + \left(\frac{1}{G}\frac{\partial \Upsilon}{\partial G} - \frac{1}{\Delta}\right)\mathcal{I}, \quad M = G, \quad N = G\sqrt{Q}, \tag{3.45}$$

in which

$$\mathcal{I} = C \int \frac{R(\nu)}{\sqrt{Q(\nu, L, G)}}\, \mathrm{d}\nu, \tag{3.46}$$

with the abbreviation

$$R = \frac{1}{C}\frac{\partial Q}{\partial(1/\Delta)} = \frac{1}{1 - (C/A)\sin^2 \nu - (C/B)\cos^2 \nu}. \tag{3.47}$$

The solutions of integrals (3.42) and (3.46) depend on the energy regime in which the free rigid body evolves. We focus on the case of rotation around the axis of maximum inertia in which case ν can take any value. Then the condition $0 < Q < 1$ translates into $B < \Delta < C$, as follows from Eq. (3.25) and the definition of Δ in Eq. (3.44). The case of rotation around the axis of minimum inertia can be approached analogously using Fukushima's alternative to Andoyer variables [164, 206, 207].

The integration of Eqs. (3.42) and (3.46) is made easier with the introduction of the following auxiliary quantities: the non-dimensional parameter

$$f = \frac{1/A - 1/B}{1/B - 1/C} > 0, \tag{3.48}$$

the non-dimensional function of the new momenta

$$m = pf, \quad p = \frac{1/\Delta - 1/C}{1/A - 1/\Delta} > 0, \quad 0 < m < 1, \tag{3.49}$$

and the angle ψ, defined unambiguously from

$$\cos\nu = \frac{\sqrt{1+f}\sin\psi}{\sqrt{1+f\sin^2\psi}}, \qquad \sin\nu = \frac{\cos\psi}{\sqrt{1+f\sin^2\psi}}, \tag{3.50}$$

from which

$$\mathrm{d}\nu = -\frac{\sqrt{1+f}}{1+f\sin^2\psi}\,\mathrm{d}\psi. \tag{3.51}$$

Then, replacing Eq. (3.50) into Eqs. (3.43) and (3.47), after some rearrangement we find

$$Q = \frac{1}{1+p}(1 - m\sin^2\psi), \quad R = -\frac{1}{C}\frac{1}{(1/A - 1/C)}(1+f\sin^2\psi),$$

which, jointly with Eq. (3.51), are plugged into Eq. (3.46) to give

$$\mathcal{I} = \frac{AC}{C-A}\sqrt{(1+f)(1+m/f)}\,\mathrm{F}(\psi|m). \tag{3.52}$$

The integration of W is a little bit more involved. Thus, after replacing the auxiliary variables into Eq. (3.42), we get

$$W = G\sqrt{(1+f)(1+m/f)}\int \frac{-f + mf\sin^2\psi}{(f+m)\sqrt{1-m\sin^2\psi}\,(1+f\sin^2\psi)}\,\mathrm{d}\psi,$$

which is rearranged by adding and subtracting m to the numerator of the integrand, to give

$$W = G\sqrt{(1+f)(1+m/f)}\left[\frac{m}{f+m}\,\mathrm{F}(\psi|m) - \Pi(-f,\psi|m)\right], \tag{3.53}$$

where $\Pi(-f,\psi|m)$ is the elliptic integral of the third kind of parameter m, amplitude ψ, and characteristic $-f$.

The computation of the mixed canonical transformation given in Eq. (3.45) is then completed. In fact, in view of $\mathcal{I}$ and W depending on m, which in turn depends on $\Delta = \Delta(\Upsilon)$, it is a family of transformations parameterized by Υ.

On the other hand, Υ can be solved from Eq. (3.44) as a function of Δ, which in turn is solved from Eq. (3.49) as a function of m. In this way, we get the standard Hamiltonian [587]

$$\Upsilon = \frac{G^2}{2A}\left(1 - \frac{C-A}{C}\frac{f}{f+m}\right), \tag{3.54}$$

which shows that the family of canonical transformations that achieve the complete reduction of the Hamiltonian in Andoyer variables is, in fact, parameterized by m.

Moreover, to reflect that $m = m(L,G)$ is non-dimensional, we make $m = m(\rho)$, with $\rho = L/G$. Hence, by application of the chain rule,

$$\frac{\partial\Upsilon}{\partial L} = \frac{\partial\Upsilon}{\partial m}\frac{\partial m}{\partial\rho}\frac{\partial\rho}{\partial L} = \frac{G}{2A}\frac{C-A}{C}\frac{f}{(f+m)^2}\frac{\partial m}{\partial\rho}, \tag{3.55}$$

$$\frac{\partial\Upsilon}{\partial G} = \frac{G}{A}\left(1 - \frac{C-A}{C}\frac{f}{f+m}\right) + \frac{\partial\Upsilon}{\partial m}\frac{\partial m}{\partial\rho}\frac{\partial\rho}{\partial G} = \frac{2\Upsilon}{G} - \rho\frac{\partial\Upsilon}{\partial L}, \tag{3.56}$$

which are plugged into Eq. (3.45) to give the family of transformations parameterized by $m = m(\rho)$ [199]. That is,

$$\ell = \frac{f}{2(f+m)^2}\sqrt{(1+f)(1+m/f)}\,\mathrm{F}(\psi|m)\frac{\partial m}{\partial\rho}, \tag{3.57}$$

$$g = \mu + \sqrt{(1+f)(1+m/f)} \times\left[\left(m - \frac{\rho}{2}\frac{f}{f+m}\frac{\partial m}{\partial\rho}\right)\frac{\mathrm{F}(\psi|m)}{f+m} - \Pi(-f,\psi|m)\right], \tag{3.58}$$

$$M = G, \tag{3.59}$$

$$N = G\frac{1}{\sqrt{1+m/f}}\sqrt{1-m\sin^2\psi}. \tag{3.60}$$

Now, we particularize the transformation by requiring that both $\ell = \ell(\nu,m)$ and $g = g(\mu,\nu,m)$ be angles. That is, $\oint \mathrm{d}\ell = 2\pi$ when ν advances by 2π or, equivalently, ψ recedes by 2π. Hence, $\ell(\psi=0) - \ell(\psi=2\pi) = 2\pi$, from which

$$\frac{\partial m}{\partial\rho} = -\frac{(f+m)^{3/2}\pi}{\mathrm{K}(m)\sqrt{f(1+f)}}. \tag{3.61}$$

On the other hand, the angle condition for g requires that it increases by 2π when its partner angle μ advances by 2π, while remaining unaffected by a 2π increase of ν [30]. When this condition is applied to Eq. (3.58), we get

$$\left(m - \frac{\rho}{2}\frac{f}{f+m}\frac{\partial m}{\partial\rho}\right)\frac{\mathrm{K}(m)}{f+m} - \Pi(-f|m) = 0. \tag{3.62}$$

The auxiliary variable $\rho = L/G$ is solved after eliminating $\partial m/\partial\rho$ between Eqs. (3.61) and (3.62). We get

$$\rho = \frac{2}{\pi}\sqrt{(1+f)(1+m/f)}\left[\Pi(-f|m) - \frac{m}{f+m}\mathrm{K}(m)\right], \tag{3.63}$$

which leaves m as *implicit* function of ρ.[2]

The transformation between Andoyer variables (μ, ν, M, N) and action-angle variables (ℓ, g, G, L) is thus completed. However, the reduced Hamiltonian Υ must remain implicit, in the standard form of Eq. (3.54) as far as m is an implicit function of L and G. Still, the Hamilton equations are obtained by replacing Eq. (3.61) into Eqs. (3.55) and (3.56), which provide the numeric values of the constant, secular frequencies once m has been computed.

In summary, given the Andoyer variables, the action-angle variables are explicitly obtained from the following sequence. First, make $G = M$ from Eq. (3.59), and compute $\Upsilon = \mathcal{H}(\nu, N, M)$ from Eq. (3.35), Δ from Eq. (3.44), and m from Eq. (3.49). Then we solve the inverse of Eq. (3.50) for ψ, namely

$$\cos\psi = \frac{\sqrt{1+f}\sin\nu}{\sqrt{1+f\sin^2\nu}}, \qquad \sin\psi = \frac{\cos\nu}{\sqrt{1+f\sin^2\nu}}. \tag{3.64}$$

Next, plug Eq. (3.61) into Eq. (3.57) to compute

$$\ell = -\frac{\pi}{2\mathrm{K}(m)}\,\mathrm{F}(\psi|m). \tag{3.65}$$

Plug Eqs. (3.61) and (3.63) into Eq. (3.58) to compute

$$g = \mu + \sqrt{(1+f)(1+m/f)}\left[\frac{\Pi(-f|m)}{\mathrm{K}(m)}\mathrm{F}(\psi|m) - \Pi(-f,\psi|m)\right], \tag{3.66}$$

and replace $\rho = L/G$ into Eq. (3.63) to obtain

$$L = \frac{2M}{\pi}\sqrt{(1+f)(1+m/f)}\left[\Pi(-f|m) - \frac{m}{f+m}\mathrm{K}(m)\right]. \tag{3.67}$$

The mapping $(\ell, g, G, L) \mapsto (\mu, \nu, M, N)$ is given by an analogous sequence which starts with $M = G$ from Eq. (3.59). Then m is computed from Eq. (3.67) using a root-finding procedure. Next, Eq. (3.65) is inverted to obtain

$$\psi = \mathrm{am}\left[-(2/\pi)\mathrm{K}(m)\ell|m\right]. \tag{3.68}$$

---

2 It can be checked that ρ is solution of both Eq. (3.61) and Eq. (3.62), so in this way we have obtained ρ as an anti-derivative of integrals for which we still lack rules to solve.

Finally, ν is computed from Eq. (3.50), μ is solved form Eq. (3.66), and N is obtained from Eq. (3.60).

In order to apply the free rigid body solution to perturbed rigid body rotation, Andoyer variables must be explicitly obtained in terms of the action-angle variables. This requires, as a first step, one to solve formally m from the implicit equation (3.67). Like in the case of the simple pendulum, solving m explicitly needs to make use of series expansion and reversion procedures. One must note, however, that while in the case of the pendulum expansion of Eq. (3.15) converges, as shown in Eq. (3.17), more care must be taken when solving formally Eq. (3.67).

Indeed, recalling that $p = m/f$, from Eq. (3.49), standard expansions of the elliptic integrals in Eq. (3.67) produce

$$\frac{L}{G} = 1 - \frac{1}{2}\sqrt{1+f}\,p\Big[1 - \frac{1}{8}(6-f)p + \frac{1}{64}(40 - 8f + 3f^2)p^2 - \frac{1}{1024}(560 - 120f + 54f^2 - 25f^3)p^3 + \mathcal{O}(p^4)\Big],$$

whose convergence may be compromised for small f. The convergence limit $p < 1$ means that

$$\frac{A}{C} < \frac{\Delta/C}{2 - \Delta/C}, \tag{3.69}$$

which establishes a relation between physical and dynamical features of the torque-free motion. In particular, the expansions will converge only in the gray region of Fig. 3.6, defined by Eq. (3.69), whereas convergence fails between the border of that region and the dashed diagonal that marks the limit $\Delta = A$.

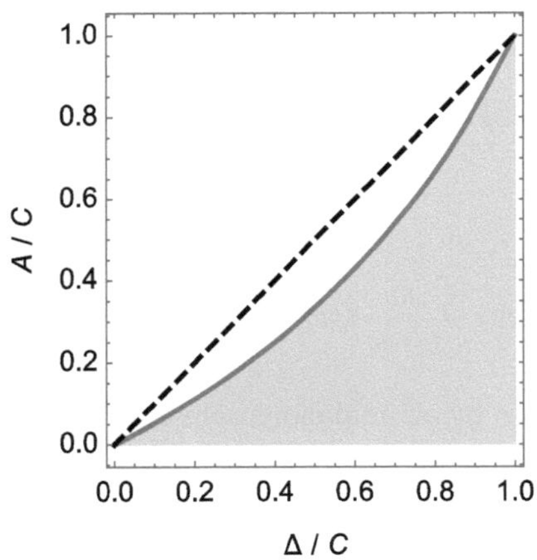

Figure 3.6: Convergence domain (gray region) of the expansion of Eq. (3.67).

The computation of the explicit transformation based on expansions of the closed-form solution is awkward yet feasible [345, 347, 614]. Alternatively, we will show in

§3.2.6 and §3.2.7 how the Lie transforms approach provides both the expanded Hamiltonian and the explicit transformation in a systematic way [200, 391, 399].

3.2.6 The case of low triaxiality

A simple rearrangement leaves Hamiltonian (3.35) in the form

$$\mathcal{H} = \frac{M^2}{2C}[1 + \alpha \sin^2 J(1 - \beta \cos 2\nu)], \tag{3.70}$$

where we recall that $J = J(N, M) \equiv \arccos(N/M)$, and the new inertia parameters $0 \leq \alpha$ and $0 \leq \beta \leq 1$, which are solved from $\alpha\,(1 + \beta) = C/A - 1$, $\alpha\,(1 - \beta) = C/B - 1$, have clear physical meaning. The limit case $\alpha = 0$ corresponds to a rigid body with spherical mass distribution ($A = B = C$), whereas the extreme values $\beta = 0$ and $\beta = 1$ correspond to axisymmetric oblate ($B = A$) and prolate mass distribution ($B = C$), respectively. Therefore, the *triaxiality coefficient*,

$$\beta = \frac{1}{\alpha}\left(\frac{C}{A} - \frac{C}{B}\right) = \frac{B - A}{B - A + 2A(1 - B/C)}, \tag{3.71}$$

provides a measure of how much the rigid body departs from axisymmetrical mass distribution.

In those cases in which β is small, Eq. (3.70) can be viewed like the perturbation Hamiltonian $\mathcal{H} = \mathcal{H}_0 + \beta\,\mathcal{H}_1$, in which

$$\mathcal{H}_0 = \frac{M^2}{2C}\left[1 + \alpha\left(1 - \frac{N^2}{M^2}\right)\right] = \frac{1}{2B^*}M^2 - \frac{1}{2}\left(\frac{1}{B^*} - \frac{1}{C}\right)N^2 \tag{3.72}$$

is formally the same as an axisymmetric oblate body with intermediate moment of inertia $B^* = C/(1 + \alpha)$, and

$$\mathcal{H}_1 = -\frac{M^2}{2C}\alpha\left(1 - \frac{N^2}{M^2}\right)\cos 2\nu \tag{3.73}$$

is a perturbation. Therefore, the expanded solution of the Hamiltonian flow stemming from Eq. (3.70) can be approached directly by perturbations based on Lie transforms.

We start from the usual perturbation Hamiltonian (2.30), in which $\varepsilon = \beta$ is a physical small parameter, $\mathcal{H}_{0,0}$ is given by Eq. (3.72), $\mathcal{H}_{1,0}$ by Eq. (3.73), $\mathcal{H}_{m,0} = 0$ for $m \geq 2$, α is a physical parameter, and M is an integral. Then we look for a Lie transformation $(\nu, N; \beta) \mapsto (\nu', N')$ derived from the generating function $\mathcal{W} = \sum_{m\geq 0}(\beta^m/m!)\mathcal{W}_{m+1}(\nu, N)$, such that, up to some truncation order β^m, it converts the Hamiltonian into a function of only the new momenta N'.

Due to the particular form of the zeroth order Hamiltonian (3.72), the Lie derivative (2.49) is $\mathcal{L}_0 = -(\alpha/C)N\partial/\partial\nu$. Therefore, the homological equation (2.48) is solved by indefinite integration,

$$\mathcal{W}_m = -\frac{1}{(\alpha/C)N}\int(\widetilde{\mathcal{H}}_{0,m} - \mathcal{H}_{0,m})\,\mathrm{d}v. \tag{3.74}$$

At first order $\widetilde{\mathcal{H}}_{0,1} = \mathcal{H}_{1,0}$, which is purely periodic in v, and, therefore, pertains to the image of the Lie derivative. Hence, we choose $\mathcal{H}_{0,1} = 0$ and compute

$$\mathcal{W}_1 = -\frac{1}{4}Nt^2\sin 2v,$$

where we abbreviated $t \equiv \tan J = \sqrt{M^2/N^2 - 1}$. It is worth noting that, because t can grow without bound, the perturbation approach will fail in such dynamical configurations where the rotation may depart notably from the axis of maximum inertia.

At second order, the known terms given in Eq. (2.37) result in

$$\widetilde{\mathcal{H}}_{0,2} = \{\mathcal{H}_{1,0}, \mathcal{W}_1\} = (\alpha/C)\frac{1}{8}N^2t^2(4 + t^2 - t^2\cos 4v).$$

To cancel the terms in the integrand of Eq. (3.74) pertaining to the kernel of the Lie derivative, we choose

$$\mathcal{H}_{0,2} = \langle\widetilde{\mathcal{H}}_{0,2}\rangle_v = (\alpha/C)\frac{1}{8}N^2t^2(4 + t^2),$$

and we solve the homological equation (3.74) to obtain

$$\mathcal{W}_2 = -\frac{1}{32}Nt^4\sin 4v.$$

Before going to the third order, we fill the first diagonal of Deprit's triangle (2.16) with the term $\mathcal{H}_{1,1}$, which is solved from Eq. (2.34) and happens to be the same as $\mathcal{H}_{0,2}$. The homological equation of the third order is then obtained by computing the term $\mathcal{H}_{0,3}$ form Deprit's recursion (2.15). After evaluation of the computable Poisson brackets at this step, the known terms as a result are found to be

$$\widetilde{\mathcal{H}}_{0,3} = -(\alpha/C)\frac{1}{32}N^2t^2[(32 + 8t^2 + 5t^4)\cos 2v + 3t^4\cos 6v)].$$

Because $\widetilde{\mathcal{H}}_{0,3}$ is purely periodic in v, we choose $\mathcal{H}_{0,3} = 0$ and compute

$$\mathcal{W}_3 = -\frac{1}{64}Nt^2[(32 + 8t^2 + 5t^4)\sin 2v + t^4\sin 6v],$$

from Eq. (3.74). Once $\mathcal{W}_3$ is known, the terms $\mathcal{H}_{1,2}$ and $\mathcal{H}_{2,1}$ are computed, using Deprit's recursion (2.15), in order to complete the second diagonal of Deprit's triangle (2.16).

One further iteration of the perturbation approach yields

$$\begin{aligned}\widetilde{\mathcal{H}}_{0,4} = (\alpha/C)\frac{3}{128}N^2t^2[&64 + 16t^2 + 8t^4 + 5t^6\\ &- 4(16 + 16t^2 + 9t^4)t^2\cos 4v - 5t^6\cos 8v],\end{aligned}$$

from which we choose

$$\mathcal{H}_{0,4} = (\alpha/C)\frac{3}{128}N^2t^2(64 + 16t^2 + 8t^4 + 5t^6)$$

and compute

$$\mathcal{W}_4 = -\frac{3}{128}Nt^4\left[(16 + 16t^2 + 9t^4)\sin 4\nu + \frac{5}{8}t^4 \sin 8\nu\right],$$

from Eq. (3.74). We stop here, which is the same order provided in the seminal paper by Kinoshita [345], but the procedure is easily extended to the computation of higher orders [200].

After truncation to $\mathcal{O}(\beta^4)$, the new, completely reduced Hamiltonian is

$$\mathcal{K} = \mathcal{H}(\nu(\nu', N'), N(\nu', N')) \equiv \sum_{m=0}^{4} \frac{\beta^m}{m!}\mathcal{H}_{0,m},$$

which, after changing original by prime variables in the computed $\mathcal{H}_{0,m}$ terms, reads

$$\mathcal{K} = \frac{M^2}{2B^*} - \frac{N'^2}{2C}\alpha\left[1 - \frac{\beta^2}{2!}\frac{1}{4}t^2(1 + t^2) - \frac{\beta^4}{4!}\frac{3}{64}t^2(64 + 16t^2 + 8t^4 + 5t^6)\right],$$

where now $t = \sqrt{(N'/M)^2 - 1}$.

The direct transformation is computed from Eq. (2.17), changing x by ν or N, and using the computed terms of the generating function. Finally, after changing primes by original variables in the terms $\nu_{0,m}$, $N_{0,m}$, the direct transformation is

$$\nu = \nu' + \sum_{m=1}^{4} \frac{\beta^m}{m!}\nu_{0,m}(\nu', N'), \quad N = N' + N' \sum_{m=1}^{4} \frac{\beta^m}{m!}N_{0,m}(\nu', N'), \tag{3.75}$$

whose first coefficients are given in Table 3.2. The inverse transformation is analogously computed using the generating function $\mathcal{V} = -\mathcal{W}(\nu(\nu', N'), N(\nu', N'))$. The first four coefficients of Eq. (2.27) are listed in Table 3.3.

The method is obviously valid when additional perturbations are added to the original Hamiltonian (3.70). The additional terms only need to be included in the perturbation arrangement in the place corresponding to the magnitude of its disturbing effect [415].

3.2.7 Short-axis-mode rotation

When the free rigid body rotation happens in such a dynamical regime that the rotation axis keeps close to the body's short axis—the axis of maximum inertia—then $N \approx M$ and, therefore, $\sin^2 J \ll 1$. However, it must be noted that both the physical

Table 3.2: Coefficients $v_{0,m}$ (top) and $N_{0,m}$ (bottom) in Eq. (3.75).

0,1 :	$\frac{1}{4}(2+t^2)\sin 2v$
0,2 :	$\frac{1}{32}(8+8t^2+t^4)\sin 4v$
0,3 :	$\frac{3}{128}(32+16t^2+14t^4+11t^6)\sin 2v+\frac{1}{128}(2+t^2)(16+16t^2+t^4)\sin 6v$
0,4 :	$\frac{3}{64}(32+32t^2+16t^4+14t^6+3t^8)\sin 4v+\frac{3}{1024}(128+256t^2+160t^4+32t^6+t^8)\sin 8v$
0,1 :	$\frac{1}{2}t^2\cos 2v$
0,2 :	$-\frac{1}{8}t^2(4+2t^2-t^2\cos 4v)$
0,3 :	$\frac{3}{64}t^2[(32+8t^2+3t^4)\cos 2v+t^4\cos 6v]$
0,4 :	$-\frac{3}{128}t^2[6(32+24t^2+12t^4+5t^6)-16t^2(4+2t^2+t^4)\cos 4v-t^6\cos 8v]$

Table 3.3: Coefficients of the generating function of the inverse transformation.

$\mathcal{V}_1=\frac{1}{4}N't^2\sin 2v'$
$\mathcal{V}_2=\frac{1}{32}N't^4\sin 4v'$
$\mathcal{V}_3=\frac{1}{128}N't^2[(64+24t^2+15t^4)\sin 2v'+3t^4\sin 6v']$
$\mathcal{V}_4=\frac{1}{1024}N't^4[8(64+96t^2+55t^4)\sin 4v'+31t^4\sin 8v']$

parameter α and the dynamical quantity $\sin^2 J$ just scale Eq. (3.70). Indeed, because M is constant, the Hamiltonian flow stemming from Eq. (3.70) is the same as the one stemming from $\mathcal{H}^*=\sin^2 J(1-\beta\cos 2v)$, yet in a different time scale and with respect to a different energy level. Hence, neither α nor $\sin^2 J$ are useful in providing a perturbation arrangement of the free rigid body Hamiltonian. Still, by simply rewriting $\sin J$ in terms of the half angle, and recalling that $\sin^2\frac{1}{2}J=\frac{1}{2}(1-\cos J)=\frac{1}{2}(1-N/M)$, we can write the free rigid body Hamiltonian (3.70) like the perturbation Hamiltonian $\mathcal{H}=\mathcal{H}_0(N)+\varepsilon\,\mathcal{H}_1(v,N)$, in which the small parameter ε is now formal ($\varepsilon\equiv 1$), the zeroth order term is

$$\mathcal{H}_0=\frac{M^2}{2C}\left[1+(4\alpha)\sin^2\!\left(\frac{1}{2}J\right)(1-\beta\cos 2v)\right], \tag{3.76}$$

and the perturbation is

$$\mathcal{H}_1=-\frac{M^2}{2C}(4\alpha)\sin^4\!\left(\frac{1}{2}J\right)(1-\beta\cos 2v). \tag{3.77}$$

The fourth power to which $\sin\frac{1}{2}J$ is raised in Eq. (3.77) shows the smallness of $\mathcal{H}_1$ compared to $\mathcal{H}_0$.

While the term $\mathcal{H}_0$, the *main problem* of short-axis-mode (SAM) rotation [391], remains integrable, at variance with the case of small triaxiality discussed in §3.2.6 the intermediary Hamiltonian $\mathcal{H}_0$ is not reduced. Therefore, the customary complete reduction of the zeroth order term is carried out before approaching the solution by perturbations.

We note that Eq. (3.76) is formally analogous to Eq. (3.70), in which J is replaced by $J^* = \frac{1}{2}J$ and α by $\alpha^* = 4\alpha$. But this subtle change from J to $\frac{1}{2}J$ has removed the exponent 2 from the variable N, thus preventing the appearance of square roots in the solution of the Hamilton–Jacobi equation and the consequent appearance of elliptic functions. While the computation of the solution of the Hamilton–Jacobi equation is now notably simpler than in the full problem [391], a more direct approach is as follows.

First of all, we recall that Andoyer variables are singular when $J = 0$, a case that is close to the usual SAM rotation. Hence, it is customary to reformulate the problem in nonsingular variables [287]. Thus, we start applying the canonical transformation $(\mu, \nu, M, N) \mapsto (g, \theta, G, \Theta)$, given by

$$g = \mu + \nu, \quad G = M, \quad \theta = -\sqrt{2(M-N)}\sin\nu, \quad \Theta = \sqrt{2(M-N)}\cos\nu, \tag{3.78}$$

which formally converts Eq. (3.76) into the Hamiltonian of a harmonic oscillator,

$$\mathcal{H}_0 = \frac{G^2}{2C}\left[1 + \frac{1-\beta}{G}\alpha\frac{1}{2}(\Theta^2 + \omega^2\theta^2)\right], \tag{3.79}$$

of frequency $\omega = \sqrt{(1+\beta)/(1-\beta)}$. Next, the standard transformation to harmonic variables (2.40), which we now write

$$\Theta = \sqrt{2\omega L}\cos\ell, \quad \theta = \sqrt{2L/\omega}\sin\ell, \tag{3.80}$$

is applied to Eq. (3.79) to give

$$\mathcal{H}_0 = \frac{G^2}{2C} + \alpha(1-\beta)\frac{G}{C}\omega L, \tag{3.81}$$

which completely reduces the Hamiltonian of the main problem of the SAM rotation in the action-angle variables (ℓ, g, L, G); cf. [391].

By further applying the transformations (3.78) and (3.80) to the perturbation $\mathcal{H}_1$ in Eq. (3.77), we obtain

$$\mathcal{H}_1 = -\left[\alpha(1-\beta)\frac{G}{C}\omega L\right]\omega\frac{L}{G}\left(\frac{1}{2} - \frac{\beta}{1+\beta}\sin^2\ell\right),$$

in which the coefficient L/G manifests the smallness of $\mathcal{H}_1$ with respect to $\mathcal{H}_0$. The Hamiltonian $\mathcal{H} = \mathcal{H}_0(L) + \epsilon\,\mathcal{H}_1(\ell, L)$ takes the form of a perturbed harmonic oscillator, whose solution can be approached by Lie transforms following analogous steps to those carried out in the example in §2.2.2.

However, to further illustrate the application of the Lie transform method rather than the standard case in which the disturbing function is expanded as a Fourier series, we go back to Eq. (3.79) and, instead of using the harmonic transformation (3.80), we rewrite it in real (g, G) and complex variables (u, U) [399]. To do that, we apply the canonical transformation

$$\theta = (u - \mathbf{i}U)\sqrt{2\omega}, \quad \Theta = (U - \mathbf{i}u)\sqrt{\omega/2}, \tag{3.82}$$

where $\mathbf{i} \equiv \sqrt{-1}$ denotes the imaginary unit. It yields

$$\mathcal{H}_0 = \frac{G^2}{2C}\left(1 - \frac{2\kappa}{G}\mathbf{i}uU\right), \tag{3.83}$$

in which we abbreviated $\kappa = \alpha(1+\beta)/\omega = \alpha\sqrt{1-\beta^2}$.

On the other hand, consecutive application of the transformations (3.78) and (3.82) to the perturbation term (3.77) yields

$$\mathcal{H}_1 = \frac{\alpha}{4C}[2u^2U^2 - \mathbf{i}\beta(u^3U - uU^3)], \tag{3.84}$$

which is a homogeneous polynomial of degree 4 in u^jU^k. Because $\mathcal{H}_1$ is free from g and G, the integral G plays the role of a dynamical parameter, and the Lie derivative (2.49) of the Hamiltonian flow (3.83) takes the form

$$\mathcal{L}_0 = G\frac{\kappa}{C}\mathbf{i}\left(U\frac{\partial}{\partial U} - u\frac{\partial}{\partial u}\right). \tag{3.85}$$

For a generic monomial u^jU^k, $\mathcal{L}_0(u^jU^k) = G(\kappa/C)(k-j)\mathbf{i}u^jU^k$, which vanishes when $j = k$. Therefore, the elements of the kernel of the Lie derivative (3.85) are the monomials u^jU^k with $j = k$, whereas the image of the Lie derivative is made of monomials u^jU^k with $k \neq j$.

The solution of the homological equation is then trivial in complex variables by simply noting that each term of the image $q_{j,k}u^jU^k$, where $q_{j,k}$ denotes some numeric coefficient, contributes a term $p_{j,k}q_{j,k}u^jU^k$ to the generating function, where

$$p_{j,k} = \frac{C}{G\kappa}\frac{\mathbf{i}}{j-k}. \tag{3.86}$$

In consequence, the construction of the perturbation solution of a free rigid body in SAM rotation becomes a simple exercise of polynomial algebra. Let us check this.

The perturbation Hamiltonian is $\mathcal{H} = \sum_{m\geq 0}(\varepsilon^m/m!)\mathcal{H}_{m,0}(u, U)$, with a formal small parameter $\varepsilon \equiv 1$, $\mathcal{H}_{0,0} = \mathcal{H}_0$ from Eq. (3.83), $\mathcal{H}_{1,0} = \mathcal{H}_1$ from Eq. (3.84), and $\mathcal{H}_{m,0} = 0$ for $m \geq 2$. At first order

$$\widetilde{\mathcal{H}}_{0,1} = \mathcal{H}_{1,0} = \frac{\alpha}{2C}u^2U^2 - \frac{\alpha\beta}{4C}\mathbf{i}u^3U + \frac{\alpha\beta}{4C}\mathbf{i}uU^3, \tag{3.87}$$

where the first summand on the right side pertains to the kernel of the Lie derivative (3.85), whereas the other two summands are members of the image. Then we select

$$\mathcal{H}_{0,1} = \frac{\alpha}{2C}u^2U^2.$$

Next, the second summand on the right side of Eq. (3.87)—whose monomial $u^j U^k$ has exponents $j = 3$ and $k = 1$—is multiplied by $p_{3,1} = \mathbf{i}C/(2G\kappa)$, as follows from Eq. (3.86), whereas the last summand is multiplied by $p_{1,3} = -\mathbf{i}C/(2G\kappa)$, to give

$$\mathcal{W}_1 = \frac{\alpha\beta}{8G\kappa}(u^3U + uU^3).$$

At second order, we find

$$\widetilde{\mathcal{H}}_{0,2} = \frac{\alpha^2\beta}{2GC\kappa}u^2U^4 - \frac{\alpha^2\beta}{2GC\kappa}u^4U^2 - \frac{\alpha^2\beta^2}{2GC\kappa}\mathbf{i}u^3U^3,$$

where the last summand pertains to the kernel and the other two are elements of the image. Then we choose

$$\mathcal{H}_{0,2} = -\frac{\alpha^2\beta^2}{2GC\kappa}\mathbf{i}u^3U^3$$

and solve the homological equation by adding the result of multiplying the first summand of $\widetilde{\mathcal{H}}_{0,2}$ by $p_{2,4}$, and the second summand by $p_{4,2}$. Using Eq. (3.86), we get

$$\mathcal{W}_2 = -\mathbf{i}\frac{\alpha^2\beta}{4G^2\kappa^2}(u^4U^2 + u^2U^4).$$

It follows the computation of $\mathcal{H}_{1,1}$ to complete the corresponding diagonal of Deprit's triangle (2.16) before proceeding to the next order.

Straightforward computations, and the final replacement of the original variables by the new ones, yield the completely reduced Hamiltonian

$$\mathcal{K} = \frac{G^2}{2C}\left[1 + \kappa\mathbf{i}\frac{u'U'}{G}\sum_{m\geq 0} b_m(\beta)\left(-\frac{\alpha}{\kappa}\mathbf{i}\frac{u'U'}{G}\right)^m\right], \tag{3.88}$$

where $b_0 = -2$, $b_1 = 1$, and the remaining b_m denote polynomials in the triaxiality coefficient β, the first of which are given in Table 3.4.

Table 3.4: Triaxiality polynomials in Eq. (3.88) [399].

$b_2 = \beta^2$	$b_6 = \frac{1}{128}\beta^2(45\beta^4 + 354\beta^2 + 128)$
$b_3 = \frac{5}{8}\beta^2$	$b_7 = \frac{9}{1024}\beta^2(265\beta^4 + 650\beta^2 + 128)$
$b_4 = \frac{3}{32}\beta^2(3\beta^2 + 8)$	$b_8 = \frac{5}{8192}\beta^2(953\beta^6 + 14888\beta^4 + 17120\beta^2 + 2048)$
$b_5 = \frac{7}{32}\beta^2(5\beta^2 + 4)$	$b_9 = \frac{11}{8192}\beta^2(4075\beta^6 + 20212\beta^4 + 13104\beta^2 + 1024)$

Needles to say that the desired Hamiltonian reduction has been effectively achieved by the Lie transform process. Indeed, in spite of the new Hamiltonian (3.88) depending on both complex variables, in addition to the integral G, it must be noted that $\mathcal{K} = \mathcal{K}(G, -\mathbf{i}u'U')$, where $-\mathbf{i}uU = L$, as readily checked by applying to the complex variables the inverse transformation of Eq. (3.82), namely

$$u = (\mathbf{i}\Theta + \omega\theta)/\sqrt{2\omega}, \quad U = (\Theta + \mathbf{i}\omega\theta)/\sqrt{2\omega},$$

followed by Eq. (3.80).

A caveat is in order in reference to the coefficient $\alpha/\kappa = (1-\beta^2)^{-1/2}$ in Eq. (3.88), which can make the Hamiltonian in new variables converge slowly in the case of almost prolate bodies ($\beta \approx 1$). In such a critical case, carrying out a perturbation approach based on a physical parameter $\tilde{\beta}(\beta) \ll 1$ should be a better option. The choice $\tilde{\beta} = (1-\beta)/(1+3\beta)$, which plays the symmetric role of β in the case of rotations about the axis of minimum inertia [295], may apply.

As regards the generating function, it is written in the form

$$\mathcal{W} = \beta(u^2 + U^2)uU \sum_{m\geq 0} \frac{1}{m!} \frac{\alpha^m}{\kappa^m G^m} (uU)^{\lfloor m/2 \rfloor} w_m, \tag{3.89}$$

where $\lfloor m/n \rfloor$ denotes the integer division of the integers m and n, $w_1 = \frac{1}{8}$, $w_2 = -\frac{1}{4}\mathbf{i}$ and the first few terms w_n are given in Table 3.5; cf. [399]. The relations $U^2 + u^2 = 2\mathbf{i}L \sin 2\ell$, $U^2 - u^2 = 2L\cos 2\ell$, $U^4 + u^4 = 2L^2 \cos 4\ell$, and $U^6 - u^6 = 2L^3 \cos 6\ell$ are useful in the translation of the generating function from complex to action-angle variables.

Table 3.5: Coefficients w_m in Eq. (3.89), cf. [399].

$w_3 =$	$-\frac{3}{8}(\beta^2+2)uU + \frac{5}{64}\beta\mathbf{i}(U^2-u^2)$
$w_4 =$	$\frac{3}{32}(57\beta^2+32)\mathbf{i}uU + \frac{3}{64}\beta(9\beta^2+20)(U^2-u^2)$
$w_5 =$	$\frac{1}{32}(343\beta^4+2024\beta^2+480)u^2U^2 - \frac{11}{64}\beta^2(\beta^2+2)(u^4+U^4) - \frac{3}{32}\beta(147\beta^2+100)\mathbf{i}uU(U^2-u^2)$
$w_6 =$	$-\frac{15}{32}(909\beta^4+1588\beta^2+192)\mathbf{i}u^2U^2 + \frac{15}{64}\beta^2(63\beta^2+44)\mathbf{i}(u^4+U^4)$
	$-\frac{15}{32}\beta(91\beta^4+653\beta^2+200)uU(U^2-u^2)$
$w_7 =$	$-\frac{45}{2048}\beta^3(283\beta^2+186)\mathbf{i}(U^6-u^6) + \frac{45}{64}\beta^2(116\beta^4+917\beta^2+308)uU(u^4+U^4)$
	$+\frac{45}{2048}\beta(124067\beta^4+272282\beta^2+44800)\mathbf{i}u^2U^2(U^2-u^2)$
	$-\frac{45}{32}(720\beta^6+8345\beta^4+6496\beta^2+448)u^3U^3$

The computation of the direct and inverse transformation equations is standard and only requires the evaluation of Poisson brackets, as described in §2.1.3.

Finally, in the case in which both β and $\sigma = 2\sin^2\frac{1}{2}J$ are small, the question of which perturbation approach would be the most convenient emerges naturally. The answer will obviously depend on the particular values taken by the triaxiality parameter and the inclination on the equatorial plane of the body with respect to the invariable plane. However, in common cases, the SAM perturbation approach provides a much more efficient procedure than the one based on a small triaxiality. This fact is illustrated in Table 3.6, in which the physical small parameter β is computed from Eq. (3.71), and J_0 is a bound for J [347].

Table 3.6: Inertia parameters and inclination angle for different solar system bodies. Adapted by permission from Springer: [391].

Body [Ref.]	A/C	B/C	β	J_0	σ
Mars [143, 614]	0.994292	0.994981	0.0646	0.1″	$\mathcal{O}(10^{-13})$
Earth [208]	0.996720	0.996722	0.0003	1″	$\mathcal{O}(10^{-11})$
Moon [208, 527]	0.999368	0.999601	0.2261	6.2″	$\mathcal{O}(10^{-10})$
Eros [615]	0.229427	0.963754	0.9779	55″	$\mathcal{O}(10^{-8})$

Part II: Perturbed elliptic motion: Artificial satellite theory

4 The Kepler problem

The Kepler problem describes the relative motion of two point masses under their mutual gravitational attraction. The three degrees of freedom of the Kepler problem are reduced by the integrals derived from the conservation of the angular momentum vector, which defines the orbital plane and makes the problem integrable. Moreover, the Kepler problem is a superintegrable Hamiltonian system [192] due to the additional integrals provided by the eccentricity vector.[1] Because of that, it can be reduced to a set of five constant elements, which determine the nature of the orbit, and a single variable that reckons the relative motion of one of the particles with respect to the other from some initial epoch. Obviously, the orbital elements in any form are unavoidably tied to the two fundamental vectors of the Kepler problem. In addition, the angular momentum vector and the eccentricity vector are the basis of the apsidal frame, which is fundamental in the vectorial formulation of perturbed Keplerian motion.

The elliptic case of the Keplerian motion is the basic integrable model in which bounded orbital motion hinges on. When the reduction to elements is carried out by the Hamilton–Jacobi method, it provides the action-angle variables in which perturbed Keplerian motion is naturally approached by perturbation methods [61, 358]. Therefore, the solution of the Hamilton–Jacobi equation of the Kepler Hamiltonian and the following particularization of the transformation to the case of action-angle variables are discussed in some detail, mostly following analogous descriptions in [395] (see also [154, 171]). Alternative derivations of these variables rely on the use of Lagrange brackets [1, 78] or on purely geometric considerations [189].

4.1 The orbital frame

Let $(O, \boldsymbol{i}, \boldsymbol{j}, \boldsymbol{k})$ be an inertial orthonormal frame, and let $\boldsymbol{x}$ be the position vector of a particle with respect to the origin O, and $\dot{\boldsymbol{x}} = \mathrm{d}\boldsymbol{x}/\mathrm{d}t$ its velocity. The angular momentum (per unit of mass) $\boldsymbol{G} = \boldsymbol{x} \times \dot{\boldsymbol{x}}$ defines the *orbital plane*, which is the instantaneous plane orthogonal to $\boldsymbol{G}$ where the motion takes place.

Let $G = \|\boldsymbol{G}\|$ and define the unit vector $\boldsymbol{n} = \boldsymbol{G}/G$ in the normal direction to the orbital plane. Then $\boldsymbol{n} \cdot \boldsymbol{k} = \cos I$, where the angle $0 \le I \le \pi$ is the inclination of the orbital plane with respect to the inertial $(\boldsymbol{i}, \boldsymbol{j})$ plane. The product $\boldsymbol{k} \times \boldsymbol{n} = \boldsymbol{\ell} \sin I$ defines the unit vector $\boldsymbol{\ell} = \boldsymbol{i} \cos \nu + \boldsymbol{j} \sin \nu$ that materializes the direction of the ascending node $0 \le \nu < 2\pi$ of the orbital plane on the inertial plane. The *nodal frame* is then defined by the orthonormal frame $(O, \boldsymbol{\ell}, \boldsymbol{n} \times \boldsymbol{\ell}, \boldsymbol{n})$.

Finally, let $r = \|\boldsymbol{x}\|$, and let $\boldsymbol{u} = \boldsymbol{x}/r$ be a unit vector in the radial direction. Then the *orbital frame* is defined by the radial, transversal and normal directions $(O, \boldsymbol{u}, \boldsymbol{n} \times \boldsymbol{u}, \boldsymbol{n})$.

1 For historical considerations on the discovery of this vector and the different names attached to it, the interested reader is referred to the erudite discussions in [241, 242].

https://doi.org/10.1515/9783110668513-004

The components of the radial direction in the nodal frame $\boldsymbol{u} = \boldsymbol{\ell}\cos\theta + (\boldsymbol{n}\times\boldsymbol{\ell})\sin\theta$ define the argument of the latitude $0 \le \theta < 2\pi$, which is the polar angle in the orbital plane.

The components of the position and velocity vectors in the orbital frame are

$$\boldsymbol{x} = (r, 0, 0), \quad \dot{\boldsymbol{x}} = (\dot{r}, r\dot{\theta}, 0). \tag{4.1}$$

Hence,

$$\boldsymbol{G} = r^2\dot{\theta}\boldsymbol{n}, \tag{4.2}$$

from which, replacing $\boldsymbol{G} = G\boldsymbol{n}$, we obtain the fundamental relation

$$r^2 \mathrm{d}\theta = G\,\mathrm{d}t. \tag{4.3}$$

On the other hand, if (x, y, z) and $(\dot{x}, \dot{y}, \dot{z})$ are the components of the position and velocity vectors in the inertial frame, respectively, we obtain

$$\begin{pmatrix} x & \dot{x} \\ y & \dot{y} \\ z & \dot{z} \end{pmatrix} = R_3(-\nu)\,R_1(-I)\,R_3(-\theta)\begin{pmatrix} r & \dot{r} \\ 0 & r\dot{\theta} \\ 0 & 0 \end{pmatrix}. \tag{4.4}$$

4.2 Kepler Hamiltonian

Let us consider a system of two points of masses m_1 and m_2, respectively, under the only action of their mutual gravitational attraction. The conservation of linear momentum constrains the center of mass of the system to evolve with linear motion. Let $\boldsymbol{\xi} = \boldsymbol{\xi}_0 + \boldsymbol{c}t$, with $\boldsymbol{\xi}_0$ and $\boldsymbol{c}$ constant, be the position of the center of mass in the inertial frame, and let $\boldsymbol{\xi}_j = \boldsymbol{\xi} + \boldsymbol{x}_j$, $j = 1, 2$, be the position of the particle of mass m_j. Then, from Newton's gravitational law,

$$m_j\frac{\mathrm{d}^2\boldsymbol{\xi}_j}{\mathrm{d}t^2} = m_j\frac{\mathrm{d}^2\boldsymbol{x}_j}{\mathrm{d}t^2} = (-1)^j\frac{\mathcal{G}m_1m_2}{\|\boldsymbol{x}_1 - \boldsymbol{x}_2\|^3}(\boldsymbol{x}_1 - \boldsymbol{x}_2), \quad j = 1, 2,$$

where $\mathcal{G}$ denotes the gravitational constant. Then the equation of motion of the relative motion $\boldsymbol{x} = \boldsymbol{x}_2 - \boldsymbol{x}_1$ is

$$\frac{\mathrm{d}^2\boldsymbol{x}}{\mathrm{d}t^2} = -\frac{\mu}{\|\boldsymbol{x}\|^3}\boldsymbol{x}, \tag{4.5}$$

where $\mu = \mathcal{G}(m_1 + m_2)$ is the gravitational parameter. Note that Eq. (4.5) is singular for $\|\boldsymbol{x}\| = 0$.

The right side of Eq. (4.5) is the opposite of the gradient of the potential energy $V = -\mu/(\boldsymbol{x}\cdot\boldsymbol{x})^{1/2}$, which, jointly with the kinetic energy (per unit of mass) $T = \frac{1}{2}(\dot{\boldsymbol{x}}\cdot\dot{\boldsymbol{x}})$, defines the Lagrangian of the Kepler problem,

$$\mathfrak{L} = T - V = \frac{1}{2}(\dot{\boldsymbol{x}}\cdot\dot{\boldsymbol{x}}) + \frac{\mu}{\sqrt{\boldsymbol{x}\cdot\boldsymbol{x}}}. \tag{4.6}$$

The conjugate momentum per unit of mass to the position vector is

$$\boldsymbol{X} = \frac{\partial\mathfrak{L}}{\partial\dot{\boldsymbol{x}}} = \dot{\boldsymbol{x}}, \tag{4.7}$$

from which

$$\mathcal{H} = \frac{1}{2}\boldsymbol{X}\cdot\boldsymbol{X} - \frac{\mu}{\sqrt{\boldsymbol{x}\cdot\boldsymbol{x}}}, \tag{4.8}$$

which is a Hamiltonian of three degrees of freedom when the vectors are expressed by their Cartesian coordinates $\boldsymbol{x} = (x, y, z)$, $\boldsymbol{X} = (X, Y, Z)$.

The Lagrangian (4.6) is written in polar coordinates using Eq. (4.1). We obtain $\mathfrak{L} = \frac{1}{2}(\dot{r}^2 + r^2\dot{\theta}^2) + \mu/r$, from which the conjugate momenta to r and θ are, respectively,

$$R = \frac{\partial\mathfrak{L}}{\partial\dot{r}} = \dot{r}, \quad \Theta = \frac{\partial\mathfrak{L}}{\partial\dot{\theta}} = r^2\dot{\theta}. \tag{4.9}$$

That is, for the Kepler problem the conjugate momentum to r is the radial velocity, and the conjugate momentum to θ is the total angular momentum, as follows from Eq. (4.2). Hence,

$$\mathcal{H} = \frac{1}{2}\left(R^2 + \frac{\Theta^2}{r^2}\right) - \frac{\mu}{r}, \tag{4.10}$$

where θ is a cyclic variable and, in consequence, Θ is an integral of the Kepler problem. That is, by the simple expedient of transforming Cartesian into polar variables one has carried out a double reduction of the Kepler Hamiltonian, which in polar variables is a Hamiltonian of one degree of freedom, $\mathcal{H} = \mathcal{H}(r, R)$, thus showing the integrability of the Kepler problem.

If we now use Eqs. (4.7) and (4.9) to replace velocities by momenta in Eq. (4.4), we obtain

$$\begin{pmatrix} x & X \\ y & Y \\ z & Z \end{pmatrix} = R_3(-\nu)\,R_1(-I)\,R_3(-\theta)\begin{pmatrix} r & R \\ 0 & \Theta/r \\ 0 & 0 \end{pmatrix}, \tag{4.11}$$

from which it is readily checked that $\boldsymbol{X}\cdot\mathrm{d}\boldsymbol{x} = R\,\mathrm{d}r + \Theta\,\mathrm{d}\theta + \Theta\cos I\,\mathrm{d}\nu$, where

$$N = \Theta\cos I, \tag{4.12}$$

is the third component of the angular momentum $\boldsymbol{G} \cdot \boldsymbol{k}$ in the particular case of Keplerian motion. Therefore, the mapping $(\boldsymbol{x}, \boldsymbol{X}) \mapsto (r, \theta, \nu, R, \Theta, N)$ from Cartesian to *polar-nodal* variables, sometimes called Hill [293] or Whittaker variables [679], defines a canonical transformation of the Mathieu type [680].

We remark that the definition of the canonical transformation from Cartesian to polar variables given by Eqs. (4.11)–(4.12) does not involve the Kepler problem, and hence is a purely geometrical definition. To the contrary, Eq. (4.9) is a particularization for the Kepler problem. Therefore, the exact physical meaning of R and Θ will depend on the specific problem in which these variables are used.

4.3 Hamilton–Jacobi reduction

The simple formulation of the Kepler Hamiltonian in polar variables has disclosed three integrals of the problem: the total angular momentum $\Theta = G$, the third component of the angular momentum vector in the inertial frame $N = \Theta \cos I$, and the argument of the node ν. That is, because

$$(\boldsymbol{G} \cdot \boldsymbol{i}, \boldsymbol{G} \cdot \boldsymbol{j}, \boldsymbol{G} \cdot \boldsymbol{k})^{\tau} = R_3(-\nu)\, R_1(-I)(0, 0, \Theta)^{\tau}, \tag{4.13}$$

these three integrals represent the conservation of the angular momentum vector $\boldsymbol{G}$—a case, in which Eq. (4.3) reduces to Kepler's law of areas. Still, additional integrals will be disclosed in a following integration of Eq. (4.10).

Like in the examples of Chapter 3, the Kepler problem is solved by complete Hamiltonian reduction of Eq. (4.10) using the Hamilton–Jacobi equation. That is, we look for a canonical transformation $(r, \theta, \nu, R, \Theta, N) \mapsto (\ell, g, h, L, G, H)$ such that it transforms Eq. (4.10) into a function of only the new momenta. Because N, ν, and θ are ignorable variables in Eq. (4.10), we choose the generating function of the transformation in separate variables $S = \nu H + \theta G + W(r, L, G)$.

Then the transformation equations

$$N = \frac{\partial S}{\partial \nu} = H, \quad \Theta = \frac{\partial S}{\partial \theta} = G, \quad h = \frac{\partial S}{\partial H} = \nu, \tag{4.14}$$

and

$$\ell = \frac{\partial W}{\partial L}, \quad g = \theta + \frac{\partial W}{\partial G}, \quad R = \frac{\partial W}{\partial r}, \tag{4.15}$$

are plugged into Eq. (4.10) to form the Hamilton–Jacobi equation

$$\frac{1}{2}\left[\left(\frac{\partial W}{\partial r}\right)^2 + \frac{G^2}{r^2}\right] - \frac{\mu}{r} = \Phi(L, G),$$

which is solved for W to give $W = G \int \sqrt{Q(r; \Phi(L, G), G)}\, \mathrm{d}r$, where the radicand $Q \geq 0$ is the quadratic form

$$Q = -\frac{1}{r^2} + 2\frac{\mu}{G^2}\frac{1}{r} + 2\frac{\Phi}{G^2}. \tag{4.16}$$

The nontrivial part of the transformation is found by replacing W into Eq. (4.15). We obtain

$$R = G\sqrt{Q}, \tag{4.17}$$

and, after straightforward computations,

$$\ell = \frac{\partial \Phi}{\partial L}\mathcal{I}_1, \quad g = \theta + \frac{\partial \Phi}{\partial G}\mathcal{I}_1 + G\mathcal{I}_2, \tag{4.18}$$

where the integrals

$$\mathcal{I}_1 = \int_{r_0}^{r} \frac{1}{R(r)}\,\mathrm{d}r, \quad \mathcal{I}_2 = \int_{s_0}^{s} \frac{1}{R(s)}\,\mathrm{d}s, \quad s = \frac{1}{r}, \tag{4.19}$$

will be solved for the case of bounded motion $0 < r_\mathrm{P} \le r \le r_\mathrm{A} < \infty$, in which we choose the lower integration limit $r_0 = 1/s_0 = r_\mathrm{P}$.

First of all, we rearrange Eq. (4.16) in the form

$$Q = \left(\frac{1}{r} - \frac{1}{r_\mathrm{A}}\right)\left(\frac{1}{r_\mathrm{P}} - \frac{1}{r}\right), \tag{4.20}$$

where, calling

$$p = G^2/\mu, \quad a = \mu/(-2\Phi), \tag{4.21}$$

from the properties of the roots of a quadratic equation we obtain

$$\frac{1}{r_\mathrm{A}} + \frac{1}{r_\mathrm{P}} = \frac{2}{p}, \quad \frac{1}{r_\mathrm{A}}\frac{1}{r_\mathrm{P}} = \frac{1}{ap}. \tag{4.22}$$

That is, $r_\mathrm{A} = a(1 + \sqrt{1-p/a}) \ge r_\mathrm{P} = a(1 - \sqrt{1-p/a})$, from which $0 < p \le a$. Besides, Eq. (4.21) shows that bounded Keplerian motion is constrained to negative energies ($\Phi < 0$).

If we further define

$$e = \sqrt{1 - p/a}, \quad 0 \le e < 1, \tag{4.23}$$

from which $p/a = 1 - e^2$, the roots of Eq. (4.20) are written either in the form

$$r_\mathrm{A} = a(1+e), \quad r_\mathrm{P} = a(1-e), \tag{4.24}$$

or

$$\frac{1}{r_\mathrm{A}} = \frac{1-e}{p}, \quad \frac{1}{r_\mathrm{P}} = \frac{1+e}{p}. \tag{4.25}$$

Now, to solve $\mathcal{I}_1$ we make the change of variable

$$r = a(1 - e\cos u), \tag{4.26}$$

which must be inserted into Eq. (4.19) to obtain both $R(r)$ and $\mathrm{d}r$ in terms of u. We check in Eq. (4.24) that $r(u = 0) = r_\mathrm{P}$ and $r(u = \pi) = r_\mathrm{A}$. Then plugging Eqs. (4.26) and (4.24) into (4.20), which in turn is put into Eq. (4.17), we obtain

$$\frac{1}{R} = \sqrt{\frac{a}{\mu}}\frac{1 - e\cos u}{e\sin u}. \tag{4.27}$$

Replacing $\mathrm{d}r = ae\sin u\,\mathrm{d}u$ and Eq. (4.27) in $\mathcal{I}_1$ in Eq. (4.19) with the lower integration limit $u = 0$, we readily obtain

$$\mathcal{I}_1 = \sqrt{\frac{a^3}{\mu}}(u - e\sin u). \tag{4.28}$$

Analogous computations making the change of variable

$$\frac{1}{r} = \frac{1 + e\cos f}{p} \tag{4.29}$$

and using Eq. (4.25), that is, $r(f = 0) = r_\mathrm{P}$ and $r(f = \pi) = r_\mathrm{A}$, yield

$$R = \frac{G}{p}e\sin f. \tag{4.30}$$

Then $\mathrm{d}s = \mathrm{d}(1/r) = -(e/p)\sin f\,\mathrm{d}f$ is computed from Eq. (4.29) and replaced, together with Eq. (4.30), into Eq. (4.19), to obtain

$$\mathcal{I}_2 = -\frac{1}{G}f \Rightarrow g = \theta - f + \frac{\partial\Phi}{\partial G}\sqrt{\frac{a^3}{\mu}}(u - e\sin u). \tag{4.31}$$

That is, Eq. (4.18) turns into

$$\ell = \frac{\partial\Phi}{\partial L}\sqrt{\frac{a^3}{\mu}}(u - e\sin u), \quad g = \theta - f + \frac{\partial\Phi}{\partial G}\sqrt{\frac{a^3}{\mu}}(u - e\sin u), \tag{4.32}$$

where we recall that $a = a(\Phi)$ and $e = e(G, \Phi)$, as given by their definitions in Eqs. (4.21) and (4.23), respectively. Therefore, Eqs. (4.14), (4.17), and (4.32) define a family of canonical transformations, parameterized by $\Phi = \Phi(L, G)$.

On the other hand, using the definitions of a, p, and e in Eqs. (4.21)–(4.23), the totally reduced Hamiltonian can be written in the standard form

$$\Phi = -\frac{\mu^2}{2G^2}(1 - e^2), \tag{4.33}$$

where, because $e = e(L, G)$ is nondimensional, we take $e = e(\eta)$ with

$$\eta = G/L. \tag{4.34}$$

Then, by the chain rule,

$$\frac{\partial\Phi}{\partial L} = \frac{\partial\Phi}{\partial e}\frac{\partial e}{\partial\eta}\frac{\partial\eta}{\partial L} = -\frac{\mu^2}{G^3}e\eta^2\frac{\mathrm{d}e}{\mathrm{d}\eta},$$

$$\frac{\partial\Phi}{\partial G} = \frac{\mu^2}{G^3}(1-e^2) + \frac{\partial\Phi}{\partial e}\frac{\partial e}{\partial\eta}\frac{\partial\eta}{\partial G} = \frac{\mu^2}{G^3}\left(1-e^2+e\eta\frac{\mathrm{d}e}{\mathrm{d}\eta}\right),$$

that are replaced in turn into Eq. (4.32), to give

$$\ell = -\frac{e}{(1-e^2)^{3/2}}\eta^2\frac{\mathrm{d}e}{\mathrm{d}\eta}(u - e\sin u), \tag{4.35}$$

$$g = \theta - f + \frac{1}{(1-e^2)^{3/2}}\left(1-e^2+e\eta\frac{\mathrm{d}e}{\mathrm{d}\eta}\right)(u - e\sin u), \tag{4.36}$$

which, jointly with Eqs. (4.14) and (4.30), now parameterize by $e = e(\eta)$ the family of transformations that reduce the Kepler Hamiltonian to the standard form (4.33).

4.4 Solution in Delaunay variables

Now, we impose the requirement that $\ell = \ell(r)$ be an angle, where we recall that $r = r(u)$ and the auxiliary variable u is an angle. Therefore, the angle condition $\oint \mathrm{d}\ell = 2\pi$ implies that ℓ must increase by 2π when r varies from $r(u = 0)$ to $r(u = 2\pi)$. When the angle condition is applied to Eq. (4.35), we obtain

$$\frac{\mathrm{d}e}{\mathrm{d}\eta} = -\frac{(1-e^2)^{3/2}}{e\eta^2}, \tag{4.37}$$

which is in separate variables and, hence, is solved by indefinite integration to give

$$e = \sqrt{1-\eta^2} = \sqrt{1-(G/L)^2}. \tag{4.38}$$

Alternatively, we could have applied the angle condition also to Eq. (4.36), and replace the derivative $\mathrm{d}e/\mathrm{d}\eta$ by Eq. (4.37) to obtain the solution (4.38) without need of integration—like we did in §3.2.5 to compute Eq. (3.63).

After plugging Eq. (4.38) into Eqs. (4.35) and (4.36), we get

$$\ell = u - e\sin u, \tag{4.39}$$

$$g = \theta - f, \tag{4.40}$$

the first of which is the famous Kepler equation.

Therefore, the transformation to action-angle variables given by Eqs. (4.39), (4.40), (4.14), and (4.30), completely reduces the Kepler Hamiltonian in polar-nodal variables, given in Eq. (4.10), to

$$\Phi = \Phi(-,-,-,L,-,-) \equiv -\frac{1}{2}\mu^2/L^2, \tag{4.41}$$

whose Hamilton equations show that the only non-null frequency of the Kepler problem is

$$\frac{d\ell}{dt} = \frac{\partial\Phi}{\partial L} = \frac{\mu^2}{L^3}. \tag{4.42}$$

Then the solution of the Kepler problem in action-angle variables is simply

$$\ell = \ell_0 + nt, \quad g = g_0, \quad h = h_0, \quad L = L_0, \quad G = G_0, \quad H = H_0, \tag{4.43}$$

where

$$n = \mu^2/L^3 \tag{4.44}$$

is the usual abbreviation for the Keplerian mean motion. Note that $a = L^2/\mu$, as follows from the definition of a in Eq. (4.21) and Eq. (4.41). Then, from Eq. (4.44),

$$L = na^2. \tag{4.45}$$

The solution (4.43) shows that both g and L are integrals of the Kepler problem in addition to the previously computed G, H, and h.

The set (ℓ, g, h, L, G, H) of action-angle canonical variables is traditionally known as Delaunay variables. Note the dynamical constraint $|H| \leq G \leq L$.

In spite of the Delaunay variables having been derived in the traditional way from the Hamiltonian reduction of the Kepler problem (see Chapter VII of [631], for instance), alternative derivations can be made based on the properties of Lagrange brackets (cf. [78] Chapter XI, §4–§9), or directly from a generating function [154]. In particular, the latter derivation unambiguously shows that the transformation from polar variables to Delaunay canonical variables is a purely geometric operation. Thus, for instance, while $d\ell/dt = n = \mu^2/L^3$ in the case of Keplerian motion, this is no longer true for perturbed Keplerian motion, in which the variation of ℓ must be obtained from the corresponding Hamilton equation of the perturbed Keplerian Hamiltonian.

In the Keplerian case, the meaning of the Delaunay variables and the auxiliary variables introduced in their computation is easily identified. Thus, besides the trivial equivalence $G = \Theta$, for the total angular momentum, $H = N$, for the third component of the angular momentum vector, and $h = \nu$ for the argument of the node, the known relations on the ellipse show that ℓ, u, and f, are the mean, eccentric, and true anomalies, respectively, and g is the argument of the periapsis [42, 582]. Hence, n is the mean motion, and Eqs. (4.26) and (4.29) identify a with the semi-major axis of the ellipse, e with the eccentricity, and p with the conic parameter. Finally, L is called the Delaunay action.

Delaunay variables are the canonical counterpart of the classic Keplerian elements

$$a = \frac{L^2}{\mu}, \quad e = \sqrt{1 - \frac{G^2}{L^2}}, \quad I = \arccos\frac{H}{G}, \quad \Omega = h, \quad \omega = g, \quad M = \ell, \tag{4.46}$$

where Ω stands for the longitude of the ascending node, ω for the argument of the periapsis, and M for the mean anomaly. As such, they are flawed with the same singularities. Thus, the eccentricity vanishes when $G = L$, a case in which the periapsis is not defined. Besides, the case $H = \pm G$ implies that either $I = 0$ or $I = \pi$ (equatorial orbits), and the node is not defined.

The complete Hamiltonian reduction process obviously provides the same results as the classical integration of the Newtonian equations (4.5); cf. [42, 139]. Indeed, starting from the Newtonian equations of motion, classical results demonstrate the conservation of the angular momentum vector

$$\boldsymbol{G} = \boldsymbol{x} \times \dot{\boldsymbol{x}}, \tag{4.47}$$

of the eccentricity vector

$$\boldsymbol{e} = \frac{1}{\mu}\dot{\boldsymbol{x}} \times \boldsymbol{G} - \frac{\boldsymbol{x}}{r}, \tag{4.48}$$

and of the energy

$$E = -\frac{\mu}{2a}. \tag{4.49}$$

The first provides three integrals Ω, I and Θ, which are equivalent to those (Θ, $\nu = \Omega$, and $H = \Theta\cos I$) provided by the first Hamiltonian reduction from Cartesian to polar-nodal variables carried out in §4.2, and which were explicitly written in Eq. (4.13). On the other hand, in view of $\boldsymbol{G}\cdot\boldsymbol{e} = 0$, the eccentricity vector integral found in the classical approach only provides two new independent integrals, which are given by its components in the nodal frame $(e\cos\omega, e\sin,\omega, 0)$. That is, e and ω are constant. The second Hamiltonian reduction, from polar-nodal to Delaunay variables, also shows that $g = \omega$ is an integral, but, instead of e, it introduces the integral L, which essentially shows the conservation of energy, as shown in Eq. (4.41). Conversely, the energy is a derived integral in the Newtonian approach $E = E(\boldsymbol{e}, \boldsymbol{G}) \equiv \frac{1}{2}\mu(1 - e^2)/G^2$, whereas in the Hamiltonian reduction process the derived integral is the eccentricity $e = e(G, L) \equiv \sqrt{1 - G^2/L^2}$.

Finally, it is worth noting that Kepler laws are easily identified in the Hamilton–Jacobi reduction process. The first one comes from the fact that the change of variable in Eq. (4.29) is the equation of a conic. The law of areas is given by the differential relation (4.3) and the fact that G is constant in the Kepler problem. The third Kepler law $T^2 \propto a^3$ is derived by computing the orbital period $T = 2\pi/n$ from Eq. (4.44), in which L is replaced by $\sqrt{\mu a}$, as follows from the elimination of Φ between Eqs. (4.21) and (4.41).

4.5 Useful relations for perturbed Keplerian motion

A series of relations that are of general application to perturbed Keplerian motion are readily derived from the integrable case of Keplerian motion. In particular, the redundant vectorial formulation of the perturbed Keplerian motion is of interest in the case of third-body perturbations, both of gravitational and of nongravitational origin, as will be shown in §8.2 and §9.1. Besides, since the Lie derivative constrains the variation of a given function to some integrable flow, we will show how standard differential relations between the different anomalies of the Keplerian motion help in solving the homological equation of perturbed Keplerian motion in closed form.

4.5.1 The apsidal frame. Fundamental vectors

The angular momentum vector and the eccentricity vector are fundamental vectors of the Keplerian motion that are used to define the *apsidal frame* $(O, \hat{\boldsymbol{e}}, \hat{\boldsymbol{b}}, \boldsymbol{n})$, in which the unit vector $\hat{\boldsymbol{e}} = \boldsymbol{e}/e$ has the direction of the eccentricity vector (4.48), the unit vector $\boldsymbol{n} = \boldsymbol{G}/G$ has the direction of the angular momentum vector (4.47), and the unit vector $\hat{\boldsymbol{b}} = \boldsymbol{n} \times \hat{\boldsymbol{e}}$ completes an orthonormal frame. This frame is fixed, contrary to the orbital frame introduced in §4.1, and hence is quite useful in isolating the contribution of short-period effects. In particular,

$$\boldsymbol{r} = (\hat{\boldsymbol{e}} \cdot \boldsymbol{r})\hat{\boldsymbol{e}} + (\hat{\boldsymbol{b}} \cdot \boldsymbol{r})\hat{\boldsymbol{b}} = (r\cos f)\hat{\boldsymbol{e}} + (r\sin f)\hat{\boldsymbol{b}}. \tag{4.50}$$

The apsidal frame is also useful in the description of perturbed Keplerian motion, a case in which it is no longer a fixed frame, but evolves slowly. Therefore, it is worth to recall some basic properties of the Poisson brackets of the fundamental vectors $\boldsymbol{G}$ and $\boldsymbol{A} = L\boldsymbol{e}$, both having dimensions of angular momentum, which the apsidal frame is based upon. Namely, for $1 \le i, j \le 3$,

$$\{G_i; G_j\} = \{A_i; A_j\} = \sum_{k=1}^{3} \varepsilon_{i,j,k} G_k, \quad \{G_i; A_j\} = \{A_i; G_j\} = \sum_{k=1}^{3} \varepsilon_{i,j,k} A_k, \tag{4.51}$$

where the Levi-Civita symbol $\varepsilon_{i,j,k}$ is the signature of the permutation of the set $(1, 2, 3)$. That is, $\varepsilon_{i,j,k} = 1$ if the permutation is even, $\varepsilon_{i,j,k} = -1$ if the permutation is odd, and $\varepsilon_{i,j,k} = 0$ otherwise.

These properties are easily checked by direct computation of the Poisson brackets in Delaunay variables using Eq. (4.13). That is,

$$(G_1, G_2, G_3)^\tau = R_3(-h)\, R_1(-I)(0, 0, G)^\tau. \tag{4.52}$$

Analogously, the components of the eccentricity vector are

$$(A_1, A_2, A_3)^\tau = R_3(-h)\, R_1(-I)\, R_3(-g)(Le, 0, 0)^\tau. \tag{4.53}$$

Linear combinations of these vectors are also useful in the description of perturbed Keplerian motion. In particular, calling

$$\boldsymbol{S} = \frac{1}{2}(\boldsymbol{G} + \boldsymbol{A}), \quad \boldsymbol{D} = \frac{1}{2}(\boldsymbol{G} - \boldsymbol{A}), \tag{4.54}$$

which have dimensions of angular momentum, we readily obtain from Eq. (4.51)

$$\{S_i; S_j\} = \sum_{k=1}^{3} \varepsilon_{i,j,k} S_k, \quad \{D_i; D_j\} = \sum_{k=1}^{3} \varepsilon_{i,j,k} D_k, \quad \{S_i; D_j\} = 0. \tag{4.55}$$

The vectors $\boldsymbol{S}$ and $\boldsymbol{D}$, or scaled versions of them, are sometimes known as Moser elements, and enjoy the property $\boldsymbol{S} \cdot \boldsymbol{S} = \boldsymbol{D} \cdot \boldsymbol{D} = L^2$, which allows one to represent a Keplerian orbit by a point on the surface of two spheres of radius L. Particular applications of these vectors will be presented in §8.2.3 and §9.1.5.

4.5.2 Variation equations in vectorial elements

Let us assume that the disturbing function of perturbed Keplerian motion is written in terms of the fundamental vectors

$$\mathcal{H} = -\frac{\mu}{2a(L)} + \mathcal{P}(\boldsymbol{G}, \boldsymbol{A}, \ell, L),$$

where $\boldsymbol{G} = \boldsymbol{G}(g, h, G, H)$ and $\boldsymbol{A} = \boldsymbol{A}(g, h, L, G, H)$, from Eqs. (4.52) and (4.53), respectively. Then, from the chain rule,

$$\frac{\partial \mathcal{H}}{\partial \xi} = \frac{\partial \mathcal{P}}{\partial \xi} = \sum_{j=1}^{3} \left(\frac{\partial \mathcal{P}}{\partial G_j} \frac{\partial G_j}{\partial \xi} + \frac{\partial \mathcal{P}}{\partial A_j} \frac{\partial A_j}{\partial \xi} \right), \quad \xi \in (g, h, G, H),$$

from which we readily obtain, $i = 1, 2, 3$,

$$\frac{\mathrm{d}A_i}{\mathrm{d}t} = \{A_i; \mathcal{P}\} = \sum_{j=1}^{3} \left(\{A_i, G_j\} \frac{\partial \mathcal{P}}{\partial G_j} + \{A_i, A_j\} \frac{\partial \mathcal{P}}{\partial A_j} \right) - \frac{\partial A_i}{\partial L} \frac{\partial \mathcal{P}}{\partial \ell},$$

$$\frac{\mathrm{d}G_i}{\mathrm{d}t} = \{G_i; \mathcal{P}\} = \sum_{j=1}^{3} \left(\{G_i, G_j\} \frac{\partial \mathcal{P}}{\partial G_j} + \{G_i, A_j\} \frac{\partial \mathcal{P}}{\partial A_j} \right),$$

$$\frac{\mathrm{d}\ell}{\mathrm{d}t} = \frac{\partial \mathcal{H}}{\partial L} + \sum_{j=1}^{3} \frac{\partial \mathcal{P}}{\partial A_i}\frac{\partial A_i}{\partial L} = n + \frac{\partial \mathcal{P}}{\partial L} + \frac{1}{Le^2}\sum_{j=1}^{3} A_i \frac{\partial \mathcal{P}}{\partial A_i},$$

whereas the variation of L does not need to be integrated because $L^2 = \boldsymbol{A}\cdot\boldsymbol{A} + \boldsymbol{G}\cdot\boldsymbol{G}$. Now, using Eq. (4.51),

$$\frac{\mathrm{d}A_i}{\mathrm{d}t} = \sum_{k=1}^{3} G_k \sum_{j=1}^{3} \varepsilon_{i,j,k}\frac{\partial \mathcal{P}}{\partial A_j} - \sum_{k=1}^{3} A_k \sum_{j=1}^{3} \varepsilon_{i,j,k}\frac{\partial \mathcal{P}}{\partial G_j} - \frac{A_i}{Le^2}\frac{\partial \mathcal{P}}{\partial \ell},$$

$$\frac{\mathrm{d}G_i}{\mathrm{d}t} = \sum_{k=1}^{3} G_k \sum_{j=1}^{3} \varepsilon_{i,j,k}\frac{\partial \mathcal{P}}{\partial G_j} + \sum_{k=1}^{3} A_k \sum_{j=1}^{3} \varepsilon_{i,j,k}\frac{\partial \mathcal{P}}{\partial A_j},$$

and hence the variation equations are written in vectorial notation like

$$\frac{\mathrm{d}\boldsymbol{G}}{\mathrm{d}t} = -\boldsymbol{G}\times\nabla_{\boldsymbol{G}}\mathcal{P} - \boldsymbol{A}\times\nabla_{\boldsymbol{A}}\mathcal{P}, \tag{4.56}$$

$$\frac{\mathrm{d}\boldsymbol{A}}{\mathrm{d}t} = -\boldsymbol{G}\times\nabla_{\boldsymbol{A}}\mathcal{P} - \boldsymbol{A}\times\nabla_{\boldsymbol{G}}\mathcal{P} - \frac{1}{Le^2}\frac{\partial \mathcal{P}}{\partial \ell}\boldsymbol{A}, \tag{4.57}$$

$$\frac{\mathrm{d}\ell}{\mathrm{d}t} = n + \frac{\partial \mathcal{P}}{\partial L} + \frac{1}{Le^2}\boldsymbol{A}\cdot\nabla_{\boldsymbol{A}}\mathcal{P}. \tag{4.58}$$

An alternative derivation is provided in [21].

When the long-term dynamics is investigated after removing the short-period terms from the disturbing function $\mathcal{P}^* = \mathcal{P}^*(\boldsymbol{G},\boldsymbol{A},-,L)$, and L becomes a formal integral. Then it is common to write the evolution equations in terms of the nondimensional angular momentum

$$\boldsymbol{\eta} = \eta\boldsymbol{n} = \boldsymbol{G}/L, \tag{4.59}$$

and the (nondimensional) eccentricity vector $\boldsymbol{e} = \boldsymbol{A}/L$. Thus,

$$\frac{\mathrm{d}\boldsymbol{\eta}}{\mathrm{d}t} = \boldsymbol{\eta}\times\nabla_{\boldsymbol{\eta}}\mathcal{Q} + \boldsymbol{e}\times\nabla_{\boldsymbol{e}}\mathcal{Q}, \tag{4.60}$$

$$\frac{\mathrm{d}\boldsymbol{e}}{\mathrm{d}t} = \boldsymbol{e}\times\nabla_{\boldsymbol{\eta}}\mathcal{Q} + \boldsymbol{\eta}\times\nabla_{\boldsymbol{e}}\mathcal{Q}, \tag{4.61}$$

in which we replaced $\mathcal{Q} = -(1/L)\mathcal{P}^*(\boldsymbol{\eta},\boldsymbol{e},-,L)$.

4.5.3 Differential relations and closed-form integration

By differentiation of both sides of Eq. (4.39), and using Eq. (4.26) we obtain the useful differential relation

$$\mathrm{d}\ell = (1-\cos u)\,\mathrm{d}u = (r/a)\,\mathrm{d}u \tag{4.62}$$

between the mean and eccentric anomalies. Analogously, computing $\mathrm{d}\theta = \mathrm{d}f$ from Eq. (4.40), $\mathrm{d}t = (1/n)\,\mathrm{d}\ell$ from the first of Eq. (4.43), plugging these values into Eq. (4.3), and taking into account that

$$G = na^2\eta, \tag{4.63}$$

as follows from Eqs. (4.34) and (4.45), we obtain the differential relation between the mean and true anomalies,

$$\mathrm{d}\ell = \frac{n}{G}r^2\,\mathrm{d}f = \frac{r^2}{a^2\eta}\,\mathrm{d}f. \tag{4.64}$$

The differential relations (4.62) and (4.64) are crucial in the closed-form integration of the perturbed Keplerian problems that will be discussed in the following chapters. Besides, the difference between the true and mean anomalies,

$$\phi = f - \ell, \tag{4.65}$$

which is customary known as the *equation of the center*, is a purely periodic function of the mean anomaly that plays a fundamental role in the closed-form integration of perturbed Keplerian motion, as will be shown in the following chapters.

Finally, in view of Eq. (2.49), the Lie derivative when dealing with perturbed Keplerian motion, that is, the perturbation Hamiltonian (2.30) in which $\mathcal{H}_{0,0} = \Phi(L)$ is the Keplerian in Eq. (4.41), is simply

$$\mathcal{L}_0 = \frac{\partial\mathcal{H}_{0,0}}{\partial L}\frac{\partial}{\partial\ell} = \frac{\mu^2}{L^3}\frac{\partial}{\partial\ell} = n\frac{\partial}{\partial\ell}. \tag{4.66}$$

Therefore, the homological equation (2.48) turns into

$$n\frac{\partial\mathcal{W}_m}{\partial\ell} = \widetilde{\mathcal{H}}_{0,m} - \mathcal{H}_{0,m}, \tag{4.67}$$

which is solved by indefinite integration,

$$\mathcal{W}_m = \frac{1}{n}\int(\widetilde{\mathcal{H}}_{0,m} - \mathcal{H}_{0,m})\mathrm{d}\ell. \tag{4.68}$$

For perturbed Keplerian motion the disturbing function is commonly expressed in terms of the true, rather than the mean anomaly. The latter can always be made explicit from a usual expansion of the elliptic motion [78]. Nevertheless, the closed-form integration of (4.68) is many times feasible in the perturbation approach, based on the differential relations (4.62) and (4.64).

Alternative forms of the homological equation may be useful when the perturbation approach is based on a normalization process in which the mean anomaly is removed from the new Hamiltonian. That is, $\partial\mathcal{H}_{0,m}/\partial\ell = 0$. Thus, when the disturbing function is given in terms of the true anomaly, as will be the case of zonal perturbations in Chapter 6, we use Eq. (4.64) to rewrite Eq. (4.68) in the form

$$\mathcal{W}_m = \frac{1}{n}\left[-\mathcal{H}_{0,m}\ell + \int \widetilde{\mathcal{H}}_{0,m}\frac{n}{G}r^2\mathrm{d}f\right],$$

which, after adding and subtracting the term $\mathcal{H}_{0,m}$ to the integrand, is rearranged like

$$\mathcal{W}_m = \frac{1}{n}\mathcal{H}_{0,m}\phi + \frac{1}{n}\int\left(\widetilde{\mathcal{H}}_{0,m}\frac{n}{G}r^2 - \mathcal{H}_{0,m}\right)\mathrm{d}f, \tag{4.69}$$

where the integrand is purely periodic in f. If, on the contrary, the perturbation is naturally expressed in terms of the eccentric anomaly, as is the case of third-body perturbations in Chapter 8, or radiation effects in §9.1, Eq. (4.62) is used to rewrite Eq. (4.68) in the form

$$\mathcal{W}_m = \frac{1}{n}\left[-\mathcal{H}_{0,m}\ell + \int \widetilde{\mathcal{H}}_{0,m}\frac{r}{a}\,\mathrm{d}u\right],$$

which, after adding and subtracting the term $\mathcal{H}_{0,m}$ to the integrand, and replacing $u - \ell = e\sin u$ from the Kepler equation (4.39), is rearranged like

$$\mathcal{W}_m = \frac{1}{n}\mathcal{H}_{0,m}e\sin u + \frac{1}{n}\int\left(\widetilde{\mathcal{H}}_{0,m}\frac{r}{a} - \mathcal{H}_{0,m}\right)\mathrm{d}u, \tag{4.70}$$

in which the integrand is purely periodic in u.

Note that using the differential relations of the Keplerian motion, or any other relation, when dealing with perturbed Keplerian motion by Lie transforms is in no way an approximation; it is exact. This is because, different from a total derivative, the Lie derivative is constrained to the integrable flow, which in the case of perturbed Keplerian motion is the Kepler problem.

4.5.4 Principal relations of the ellipse

Useful relations between the true anomaly f and the eccentric anomaly u come from standard relations on the ellipse that are derived in textbooks (see [582] Section 4.5.4, or [42] Section 4.3, for instance). In particular, we borrow the following directly from Gauss' *principal relations* i–xii in [227]. Thus (viii and ix),

$$r\sin f = a\eta\sin u, \quad r\cos f = a\cos u - ae, \tag{4.71}$$

with r taken from Eq. (4.26) in the computation of f, and from Eq. (4.29) in the computation of u. Besides, we have (v and vi)

$$\sqrt{1+e}\sin\frac{1}{2}u = \sqrt{\frac{r}{a}}\sin\frac{1}{2}f, \quad \sqrt{1-e}\cos\frac{1}{2}u = \sqrt{\frac{r}{a}}\cos\frac{1}{2}f, \tag{4.72}$$

from which the useful relation is readily derived between the eccentric and true anomalies (vii),

$$\sqrt{1-e}\tan\frac{1}{2}f = \sqrt{1+e}\tan\frac{1}{2}u. \tag{4.73}$$

The numerical trouble that might happen to this equation when both tangents are close to the value ∞ is avoided using, for instance (x),

$$\sin\frac{1}{2}(f-u) = \sqrt{\frac{1}{2}(1-\eta)}\sqrt{(r/p)}\sin f = \sqrt{\frac{1}{2}(1-\eta)}\sqrt{(a/r)}\sin u, \tag{4.74}$$

which is also obtained from Eq. (4.72) replacing $\sqrt{2(1-\eta)} = \sqrt{1+e} - \sqrt{1-e}$.

The latter is useful too in expressing the equation of the center in terms of the eccentric anomaly (resp. true anomaly) in closed form. Indeed, using the Kepler equation (4.39),

$$\phi = f - \ell = e\sin u + 2\arcsin\sqrt{\frac{\frac{1}{2}(1-\eta)}{1-e\cos u}}\sin u, \tag{4.75}$$

which, while not listed among Gauss' principal relations, was, in fact, used by Gauss for computing the *greatest* equation of the center avoiding the customary expansions in the eccentricity.

If Eq. (4.75) is rather expressed in terms of the true anomaly, then

$$\phi = \frac{\sigma\eta}{1+\kappa} + 2\arcsin\frac{\sigma}{\sqrt{2(1+\eta)(1+\kappa)}}, \tag{4.76}$$

where σ and κ are the projections of the eccentricity vector in the orbital frame, as given in Eq. (5.21). Then, if σ and κ are written in polar variables, like in Eq. (5.42), the equation of the center is trivially expressed in polar variables.

The alternative relation to Eq. (4.74)

$$\tan\frac{1}{2}(f-u) = \frac{e\sin f}{1+\eta+e\cos f} = \frac{e\sin u}{1+\eta-e\cos u}, \tag{4.77}$$

can be computed either from Eq. (4.74) or from direct trigonometric expansion of $\tan\frac{1}{2}(f-u)$ and the help of Eq. (4.73) after elementary trigonometric manipulations [73]. Then the alternative, equivalent formula to Eq. (4.75) is

$$\phi = e\sin u + 2\tan^{-1}\frac{e\sin u}{\eta+1-e\cos u}, \tag{4.78}$$

and to Eq. (4.76)

$$\phi = \frac{\sigma\eta}{1+\kappa} + 2\tan^{-1}\frac{\sigma}{1+\eta+\kappa}. \tag{4.79}$$

5 The main problem of the artificial satellite

However useful Kepler's laws were in the description of planetary motion, the mass points assumption soon became insufficient, and some observable effects on the motion of the moon were attributed to the flattened figure of the Earth as early as in Laplace's work [260]. The potential derived from Newton laws needs then to be applied to a mass distribution and is obtained from the integration of Laplace's equation, which gives rise to the expansion of the gravitational potential in spherical harmonics. In the case of the Earth, the expansion is dominated by the zonal harmonic of the second degree. For this reason, the truncation of the geopotential to this term alone is customarily called the *main problem* of artificial satellite theory.

The main problem is a suitable model for illustrating the more relevant effects of the dynamics of common low Earth orbits, like the secular precession of the line of nodes and the steady motion of the line apsides. This simplified model was useful for orbit prediction purposes at the beginning of the space era, where the geopotential was not well known. This fact motivated the intensive search for integrable approximations to the main problem dynamics based on the separability of the Hamilton–Jacobi equation, which are known as *intermediary orbits* [620]. Still, presently, the main problem intermediaries are advocated as useful approximations for onboard, short-term, orbit propagation purposes under limited computational resources, a case in which reducing computational time and memory allocation may be of concern [51, 259, 268].

On the other hand, all the integrable approximations of the main problem, even the more sophisticated ones [664], fail in describing the actual dynamics of critically inclined orbits [119]. Alternatively, the intrinsic nature of the critical inclination resonance, which involves the librational dynamics of the argument of the perigee, is clearly disclosed after removing the short-period effects using perturbation theory.

5.1 Geopotential Hamiltonian

Taking a reference frame with the origin at the center of mass of the Earth, the x axis in the direction of the Greenwich meridian, the z axis in the direction of the Earth's rotation axis, and the y axis completing a direct orthogonal frame, the usual expansion of the geopotential in spherical harmonics is given in terms of the spherical coordinates r, λ, and φ, standing for radius, longitude and latitude, respectively. Thus,

$$V = -\frac{\mu}{r} + \mathcal{P}(r, \varphi, \lambda), \tag{5.1}$$

in which μ is the Earth's gravitational parameter and

$$\mathcal{P} = -\frac{\mu}{r} \sum_{n\geq 2} \frac{R_\oplus^n}{r^n} \sum_{m=0}^{n} (C_{n,m} \cos m\lambda + S_{n,m} \sin m\lambda) P_{n,m}(\sin \varphi), \tag{5.2}$$

https://doi.org/10.1515/9783110668513-005

where $R_\oplus$ is the Earth's mean equatorial radius, $P_{n,m}$ are associated Legendre polynomials, and $C_{n,m}$ and $S_{n,m}$ are harmonic coefficients [341].

Recall the rule in Eq. (3.29) for differentiation of a vector in a rotating frame, and assume that the Earth's rotation rate $\boldsymbol{\omega}_\oplus$ takes place in the z axis direction. Then the kinetic energy is

$$T = \frac{1}{2}\frac{\mathrm{d}\boldsymbol{x}}{\mathrm{d}t}\cdot\frac{\mathrm{d}\boldsymbol{x}}{\mathrm{d}t} = \frac{1}{2}(\dot{x}^2+\dot{y}^2+\dot{z}^2) + \omega_\oplus(x\dot{y}-y\dot{x}) + \frac{1}{2}\omega^2(x^2+y^2),$$

and the conjugate momenta to the Cartesian coordinates are computed from the Lagrangian $\mathfrak{L} = T - V$ like

$$X = \frac{\partial\mathfrak{L}}{\partial\dot{x}} = \dot{x} - \omega_\oplus y, \quad Y = \frac{\partial\mathfrak{L}}{\partial\dot{y}} = \dot{y} + \omega_\oplus x, \quad Z = \frac{\partial\mathfrak{L}}{\partial\dot{z}} = \dot{z},$$

which, as follows from Eq. (3.29), are the components of the velocity in the inertial frame. Then the usual construction of the Hamiltonian $\mathcal{H} = \boldsymbol{X}\cdot\dot{\boldsymbol{x}} - \mathfrak{L}$ yields

$$\mathcal{H} = \frac{1}{2}(X^2+Y^2+Z^2) - \omega_\oplus(xY-yX) + V. \tag{5.3}$$

It happens for the Earth that $C_{2,0} = \mathcal{O}(10^{-3})$ whereas the other harmonic coefficients are $\mathcal{O}(C_{2,0}^2)$ or higher. Therefore, the main disturbance of the Keplerian motion affecting artificial satellites in the Earth's close space is due to the first term of the summation in Eq. (5.2). Then the motion in the simplified potential

$$V = -\frac{\mu}{r} - \frac{\mu}{r}C_{2,0}\frac{R_\oplus^2}{r^2}P_{2,0}(\sin\varphi) = -\frac{\mu}{r}\left[1 + J_2\frac{R_\oplus^2}{r^2}\frac{1}{2}(1-3\sin^2\varphi)\right], \tag{5.4}$$

in which $J_2 = -C_{2,0} > 0$, is called the *main problem* of artificial satellite theory, or the J_2 problem. Contrary to the Kepler problem the motion is no longer integrable [102, 312], yet we can obtain a lot of information on the dynamics of this simple model. Note that we can choose units of length and time such that $R_\oplus = 1$ and $\mu = 1$. Therefore, the only essential physical parameter of the main problem is the oblateness coefficient J_2. However, we keep all the parameters explicit in the following derivations because they provide immediate insight, on the one hand, and help in verifying the correctness of analytical derivations by checking dimensions, on the other.

The main problem does not depend on the geocentric longitude, which makes the body-fixed frame formulation unneeded. The Coriolis term $-\omega_\oplus(xY-yX)$ is then removed from Eq. (5.3), in which the kinetic energy is $T = \frac{1}{2}\boldsymbol{X}\cdot\boldsymbol{X}$, and the right ascension of the ascending node $\nu = \Omega = \lambda - \omega_\oplus t$ becomes a cyclic variable. Therefore, the third component of the angular momentum vector N is an integral of the main problem, whose Hamiltonian is thus of just two degrees of freedom.

5.2 Particular solutions

When the main problem Hamiltonian is written in Cartesian variables,

$$\mathcal{M} = \frac{1}{2}(X^2 + Y^2 + Z^2) - \frac{\mu}{r} - J_2 \frac{\mu}{r} \frac{R_\oplus^2}{r^2} \left(\frac{1}{2} - \frac{3}{2} \frac{z^2}{r^2} \right), \tag{5.5}$$

where $r = \sqrt{x^2 + y^2 + z^2}$, the corresponding Hamilton equations provide a simple and efficient formulation for the numerical integration of the Hamiltonian flow. Thus,

$$\begin{aligned} \frac{d^2x}{dt^2} &= -x \frac{\mu}{r^3} \left[1 + J_2 \frac{R_\oplus^2}{r^2} \frac{3}{2} \left(1 - 5 \frac{z^2}{r^2} \right) \right], \\ \frac{d^2y}{dt^2} &= -y \frac{\mu}{r^3} \left[1 + J_2 \frac{R_\oplus^2}{r^2} \frac{3}{2} \left(1 - 5 \frac{z^2}{r^2} \right) \right], \\ \frac{d^2z}{dt^2} &= -z \frac{\mu}{r^3} \left[1 + J_2 \frac{R_\oplus^2}{r^2} \frac{3}{2} \left(3 - 5 \frac{z^2}{r^2} \right) \right], \end{aligned} \tag{5.6}$$

which, in addition, shows the existence of equatorial orbits as well as rectilinear solutions along the x, y, and z axes.

In particular, the manifold of equatorial orbits can be studied from the simplified Hamiltonian obtained making $z = 0$ and $Z = 0$ in Eq. (5.5), which, besides, is written in polar variables

$$\mathcal{M}_{\text{equatorial}} = \frac{1}{2} \left(R^2 + \frac{\Theta^2}{r^2} \right) - \frac{\mu}{r} - \frac{1}{2} J_2 \frac{\mu}{r} \frac{R_\oplus^2}{r^2}, \tag{5.7}$$

to show its radial, integrable character—whose analytical solution involves elliptic integrals and functions [141, 327]. In this particular manifold, $dr/dt = \partial \mathcal{M}_{\text{equatorial}} / \partial R = R$ and $d\theta/dt = \partial \mathcal{M}_{\text{equatorial}} / \partial \Theta = \Theta / r^2$. The latter shows that rectilinear solutions $\Theta = 0$ exist in the equatorial plane, whereas the former shows the existence of circular equatorial orbits.

On the other hand, the main problem Hamiltonian is written in polar variables without constraint to the equatorial case replacing $z/r = \sin I \sin \theta$, as obtained from Eq. (4.11), into Eq. (5.5). That is,

$$\mathcal{M} = \frac{1}{2} \left(R^2 + \frac{\Theta^2}{r^2} \right) - \frac{\mu}{r} - J_2 \frac{\mu}{r} \frac{R_\oplus^2}{r^2} \left(\frac{1}{2} - \frac{3}{2} \sin^2 I \sin^2 \theta \right), \tag{5.8}$$

where $\sin^2 I = 1 - N^2/\Theta^2$. The cyclic character of ν in Eq. (5.8) shows that its conjugate momentum N is an integral of the main problem.

The integral N decouples the Hamiltonian flow into the motion in the orbital plane,

$$\frac{dr}{dt} = R, \tag{5.9}$$

$$\frac{d\theta}{dt} = \frac{\Theta}{r^2} + J_2 \frac{3\mu}{r\Theta} \frac{R_\oplus^2}{r^2} c^2 \sin^2 \theta, \tag{5.10}$$

$$\frac{\mathrm{d}R}{\mathrm{d}t} = \frac{\Theta^2}{r^3} - \frac{\mu}{r^2}\left[1 + J_2\frac{3}{2}\frac{R_\oplus^2}{r^2}(1 - 3s^2\sin^2\theta)\right], \tag{5.11}$$

$$\frac{\mathrm{d}\Theta}{\mathrm{d}t} = -J_2\frac{3}{2}\frac{\mu}{r}\frac{R_\oplus^2}{r^2}s^2\sin 2\theta, \tag{5.12}$$

where we use the abbreviations $c = c(\Theta, N) \equiv \cos I$ and $s = s(\Theta, N) \equiv \sin I$, from the motion of the orbital plane

$$\frac{\mathrm{d}\nu}{\mathrm{d}t} = -\frac{3}{\Theta}\frac{\mu}{r}\frac{R_\oplus^2}{r^2}J_2 c\sin^2\theta.$$

The latter is obtained by indefinite integration once the motion in the orbital plane has been solved. Namely,

$$\nu = \nu_0 - 3\mu R_\oplus^2 J_2 N \int \frac{\sin^2\theta(t)}{\Theta(t)^2 r(t)^3}\,\mathrm{d}t, \tag{5.13}$$

which also shows that the node remains fixed when $N = 0$. That is, meridian planes are invariant manifolds of the main problem. Therefore, polar orbits can be investigated from the simplified Hamiltonian

$$\mathcal{M}_{\text{meridian}} = \frac{1}{2}\left(R^2 + \frac{\Theta^2}{r^2}\right) - \frac{\mu}{r} - J_2\frac{\mu}{r}\frac{R_\oplus^2}{r^2}\left(\frac{1}{2} - \frac{3}{2}\sin^2\theta\right). \tag{5.14}$$

5.3 Secular effects

The integrable cases and invariant manifolds of the main problem of AST discussed in §5.2 exist for any value of J_2. On the other hand, the value of the Earth's J_2 coefficient is very small, what makes the main problem to be naturally viewed as a perturbation of the Kepler problem, in which the elements are no longer constant but vary slowly. This behavior is more clearly seen when using Delaunay variables.

The two first summands on the right side of Eq. (5.8) correspond to the Kepler problem, whereas the last summand is the oblateness disturbing function. In Delaunay variables, the former takes the simple form of Eq. (4.41), while the latter is obtained replacing θ using Eq. (4.40), r from Eq. (4.29), with p given in Eq. (4.21) and e in Eq. (4.38), and recalling that $c = H/G$. However, the true anomaly f will remain an implicit function of ℓ unless we resort to the known expansions of the elliptic motion. In addition, writing the eccentricity in terms of Delaunay variables introduces square roots in the formulation. Moreover, Delaunay variables lack of the insight that is provided by classical orbital elements and usual functions of them.

For the reasons above, the following expressions throughout this monograph will be commonly written in terms of classical orbital elements, as well as in other usual quantities like r or p, with the proviso that they are not to be taken as variables by themselves, but rather as functions of some set of canonical variables. Having available a supplementary table of partial derivatives of these functions with respect to the canonical variables will be of definite help in carrying out the different computations required in the perturbation approach. They are readily obtained from total and logarithmic derivatives like those in the first book of Gauss' *Theoria Motus* [227]. Lists of partial derivatives can be found in [272, 435, 436, 538]. The most common of them are summarized in Table 5.1

Table 5.1: Partial derivatives of some orbital variables.

	$\partial/\partial\ell$	$\partial/\partial L$	$\partial/\partial G$
a	0	$2/(na)$	0
n	0	$-3/a^2$	0
p	0	0	$2\eta/(na)$
e	0	$\eta^2/(na^2e)$	$-\eta/(na^2e)$
η	0	$-\eta/(na^2)$	$1/(na^2)$
c	0	0	$-c/(na^2\eta)$
s	0	0	$(c/s)c/(na^2\eta)$
u	a/r	$\eta\sigma/(na^2e^2)$	$-\sigma/(na^2e^2)$
f	$(a/r)^2\eta$	$(\kappa+2)\sigma/(na^2e^2)$	$-(\kappa+2)\sigma/(na^2e^2\eta)$
r	$a\sigma/\eta$	$\eta^2[2e^2(r/p)-\kappa]/(nae^2)$	$\kappa\eta/(nae^2)$
R	$(a/r)^2\kappa an$	$[(p/r)^2-e^2]\sigma/(ae^2\eta)$	$-\eta^3(a/r)^2\sigma/(ae^2\eta)$
κ	$-(a/r)^2\sigma\eta$	$[\kappa\eta^2-(\kappa+2)\sigma^2]/(na^2e^2)$	$-[\kappa\eta^2-(\kappa+2)\sigma^2]/(na^2e^2\eta)$
σ	$(a/r)^2\kappa\eta$	$\sigma[\eta^2+(\kappa+2)\kappa]/(na^2e^2)$	$-\sigma[\eta^2+(\kappa+2)\kappa]/(na^2e^2\eta)$

Then we rewrite Eq. (5.8) in the compact, meaningful form

$$\mathcal{M} = -\frac{\mu}{2a} - \frac{1}{4}J_2\frac{\mu}{r}\frac{R_\oplus^2}{r^2}(2 - 3s^2 + 3s^2\cos 2\theta), \tag{5.15}$$

where we have the quantities $a \equiv a(L)$, $r \equiv r(\ell, L, G)$, $s \equiv s(G, H)$, and $\theta \equiv \theta(\ell, g, L, G)$, when using Delaunay canonical variables. The corresponding Hamilton equations are

$$\frac{\mathrm{d}\ell}{\mathrm{d}t} = n + nJ_2\frac{3}{4}\frac{R_\oplus^2}{r^2}\frac{p}{r}\frac{1}{\eta^2e^2}[2(2+\kappa)\sigma s^2\sin 2\theta - (\kappa - \kappa^2 - 2\sigma^2)(2 - 3s^2 + 3s^2\cos 2\theta)], \tag{5.16}$$

$$\frac{\mathrm{d}L}{\mathrm{d}t} = -LnJ_2\frac{a^2}{r^2}\frac{R_\oplus^2}{r^2}\frac{3}{4\eta}\left[2\frac{p}{r}s^2\sin 2\theta + (2 - 3s^2 + 3s^2\cos 2\theta)\sigma\right], \tag{5.17}$$

$$\frac{\mathrm{d}g}{\mathrm{d}t} = nJ_2\frac{3}{4}\frac{R_\oplus^2}{r^2}\frac{p}{r}\frac{1}{\eta^3e^2}\left\{2\kappa\frac{p}{r} - 2(2+\kappa)\sigma s^2\sin 2\theta - [3s^2\kappa + (3 - 5c^2)\kappa^2 - 2c^2\sigma^2](1 - \cos 2\theta)\right\}, \tag{5.18}$$

$$\frac{\mathrm{d}G}{\mathrm{d}t} = -GnJ_2\frac{a}{r}\frac{R_\oplus^2}{r^2}\frac{3}{2\eta}s^2\sin 2\theta, \tag{5.19}$$

$$\frac{\mathrm{d}h}{\mathrm{d}t} = -nJ_2\frac{3}{2}\frac{R_\oplus^2}{r^2}\frac{p}{r}\frac{c}{\eta^3}(1-\cos 2\theta), \tag{5.20}$$

where we abbreviated

$$\kappa = e\cos f, \quad \sigma = e\sin f, \tag{5.21}$$

which are the projections of the eccentricity vector in the orbital frame.

As is obvious in Eq. (5.16), the mean motion n is no longer the variation of the mean anomaly ℓ. On the contrary, $n = \mu^2/L(t)^3$ as follows from its definition in Eq. (4.44). On the other hand, we must be aware of the appearance of the eccentricity in denominators of the variations of ℓ and g given by Eqs. (5.16) and (5.18), respectively. In consequence, singularities could appear in the integration of almost circular orbits using these equations. However, one may check that the variation of the mean argument of the latitude $F = \ell + g$, which is given by the sum of Eqs. (5.16) and (5.18), removes the singularity. Therefore, alternative non-singular variables like the semi-equinoctial variables used in §5.7.2 are customarily used in this case, whose variation equations are readily obtained based on the chain rule, as we will do in §6.1.1 or §8.5.

Again, the integral H of the main problem decouples the motion of the node, which is solved by indefinite integration once the reduced differential system given by Eqs. (5.16)–(5.19), is solved.

5.3.1 Picard iterations

Because the differential system (5.16)–(5.19) is of first order, its solution can be approached by Picard iterations [565]. We start the iterations from the Keplerian solution and limit the procedure to a single iteration. Therefore, the validity of the solution will be constrained to a small interval $\Delta t = t_1 - t_0$.

For instance, when the method is applied to the variation equation of h, namely $\mathrm{d}h/\mathrm{d}t = \partial\mathcal{M}/\partial H$, we obtain

$$h(t_1) - h(t_0) = \int_{t_0}^{t_1}\frac{\partial\mathcal{M}}{\partial H}\mathrm{d}t = \frac{1}{n}\frac{\partial}{\partial H}\int_{\ell(t_0)}^{\ell(t_1)}\mathcal{M}\,\mathrm{d}\ell.$$

We limit the time interval Δt to the case in which ℓ advances by 2π, which in the Keplerian approximation is $\Delta t = 2\pi/n$. Then

$$\frac{\Delta h}{\Delta t} = \frac{\partial}{\partial H}\left(\frac{1}{2\pi}\int_{\ell_0}^{\ell_0+2\pi}\mathcal{M}\,\mathrm{d}\ell\right) = \frac{\partial\langle\mathcal{M}\rangle_\ell}{\partial H}. \tag{5.22}$$

In the Keplerian motion assumption used for the first iteration of Picard method, from Eq. (5.15) we obtain

$$\langle \mathcal{M} \rangle_\ell = -\frac{\mu}{2a} - \frac{1}{2\pi} \int_0^{2\pi} J_2 \frac{\mu}{2p} \frac{R_\oplus^2}{r^2} (1 + e\cos f)[1 - 3s^2 \sin^2(f + \omega)]\,\mathrm{d}\ell,$$

which is integrated in closed form with the help of the differential relation (4.64). We obtain

$$\langle \mathcal{M} \rangle_\ell = -\frac{\mu}{2a} + J_2 \frac{\mu}{p} \frac{R_\oplus^2}{p^2} \eta^3 \left(\frac{1}{4} - \frac{3}{4} c^2 \right). \tag{5.23}$$

When this expression is replaced in Eq. (5.22), taking into account that $c = H/G$ is the only function that depends on H, we readily find that in the time in which the mean anomaly advances by 2π, the node evolves under the oblateness perturbation at an average rate

$$\overline{\frac{\mathrm{d}h}{\mathrm{d}t}} = \frac{\Delta h}{\Delta t} = -\frac{3}{2} n J_2 \frac{R_\oplus^2}{p^2} c. \tag{5.24}$$

The trend shown in Eq. (5.24) produces the regression of the nodes of direct inclination orbits ($c > 0$), and the advance of the nodes of retrograde inclination orbits ($c < 0$). Note that, beyond the approximate character of Eq. (5.24), the case $c = 0$ is in agreement with the previously demonstrated fact from Eq. (5.13) that the nodes of polar orbits remain fixed.

Proceeding analogously with the other Hamilton equations, we see that the average variation of L vanishes, which, from the first of Eq. (4.46), shows that, on average, the orbit semimajor axis remains constant. Besides, the average variation of G also vanishes, which, shows that, on average, the orbit remains with constant inclination (because H is an integral of the main problem) and constant eccentricity (because L remains constant on average).

On the other hand, the average variation of the mean anomaly,

$$\overline{\frac{\mathrm{d}\ell}{\mathrm{d}t}} = \frac{\partial \langle \mathcal{M} \rangle_\ell}{\partial L} = n - \frac{3}{4} n J_2 \frac{R_\oplus^2}{p^2} \eta (1 - 3c^2), \tag{5.25}$$

shows that those orbits of the main problem with such inclination that $c = \pm\sqrt{1/3}$ ($I \approx 54.7$ deg for direct orbits) evolve, on average, with Keplerian mean motion.[1] Finally,

$$\overline{\frac{\mathrm{d}g}{\mathrm{d}t}} = \frac{\partial \langle \mathcal{M} \rangle_\ell}{\partial G} = -\frac{3}{4} n J_2 \frac{R_\oplus^2}{p^2} (1 - 5c^2), \tag{5.26}$$

1 Remarkably, the popular GPS orbits are close to this configuration.

which vanishes when $c = \pm\sqrt{1/5}$, thus uncovering a "critical" inclination that "freezes" the periapsis. That is, the periapsis of direct (resp. retrograde) orbits remains fixed, on average, at the inclination $I \approx 63.4$ deg (resp. $I \approx 116.6$ deg). The periapsis of an orbit closer to the equatorial plane than the critical cases advances, whereas the periapsis of an orbit with inclination closer to polar regresses.

5.3.2 Inclination resonances

Orbits that, on average, remain with constant eccentricity and fixed argument of periapsis are known as "frozen orbits" and have different applications for artificial satellite missions [60, 137, 543]. Due to the axial symmetry of the zonal harmonics potential, on average, a satellite on a frozen orbit always reaches the same altitude over the Earth's surface when the subsatellite point gets exactly the same latitude. In consequence, the satellite periodically undergoes the same gravitational pull from the zonal harmonics, which is precisely the description of a physical resonance.

From the dynamical point of view, the phenomenon is identified with a one-to-one commensurability between the draconitic period, the time between two consecutive passages of the satellite through the ascending node, and the anomalistic period, in which the satellite makes two consecutive perigee passages.

Indeed, on account of $g = \theta - f$ and, on average, the anomalies advance at the same rate, Eq. (5.26) is solved for the inclination to obtain

$$\cos^2 I = \frac{1}{5}\left[1 + \frac{4}{3}\frac{p^2}{R_\oplus^2}\frac{1}{J_2}\left(\frac{n_\theta}{n_f} - 1\right)\right], \tag{5.27}$$

where $n_\theta = \Delta\theta/(2\pi)$ and $n_f = \Delta f/(2\pi)$, showing that the one-to-one resonance certainly occurs at the critical inclination. Other resonances $\Delta\theta/\Delta f = j/k$, with k and j integers, would produce different "critical" inclinations. However, the fact that the right side of Eq. (5.27) must be between zero and one imposes the condition

$$1 + 3J_2\frac{R_\oplus^2}{p^2} \geq \frac{\Delta f}{\Delta\theta} \geq 1 - \frac{3}{4}J_2\frac{R_\oplus^2}{p^2}. \tag{5.28}$$

Therefore, for the small value of the Earth's J_2 coefficient, other resonances than the one-to-one resonance that occurs at the critical inclination will be very shallow, as illustrated in Fig. 5.1 for direct inclination orbits when $J_2(R_\oplus/p)^2 = 10^{-3}$; cf. [216, 394].

The accuracy of the results obtained in §5.3.1 is limited to effects of first order of J_2, because we limited the integration of the variation equations to the first iteration of the Picard method, in which the Keplerian approximation is valid. In §5.5 we will show that higher-order effects produce important qualitative variations in the long-term dynamics of the main problem. Nevertheless, the first-order approximation that has been carried out uncovers by itself important relevant effects of the perturbed Keplerian dynamics in the artificial satellite theory. Moreover, the effects of other harmonic terms

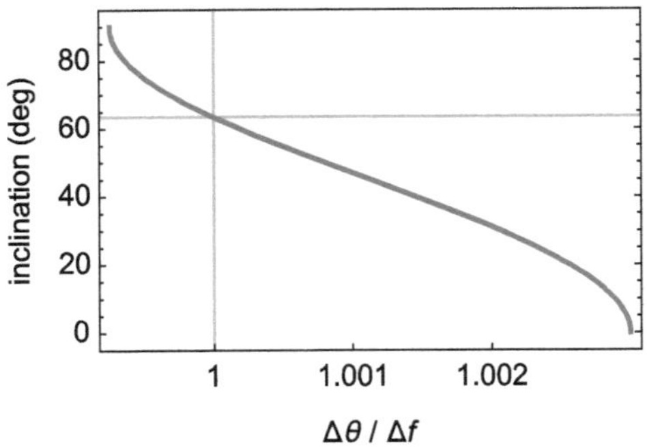

Figure 5.1: Inclination resonances for the Earth, as given by Eqs. (5.27) and (5.28).

of the geopotential which are neglected in the main problem truncation may be comparable to the second-order effects induced by the second zonal harmonic. Their combined effect will be discussed in Chapter 6.

5.4 Intermediaries

Beyond the averaged dynamics discussed in §5.3, there exist a wealth of integrable approximations to the main problem dynamics that preserve its secular behavior. These are the so-called main problem intermediaries [154, 225, 532, 620]. The issue is easily illustrated expanding Eq. (5.8) in the form

$$\mathcal{M} = \mathcal{K} + \mathcal{P}_Q + \mathcal{P}_r + \mathcal{P}_\theta,$$

where $\mathcal{K}$ is the Keplerian (4.10) in polar variables,

$$\mathcal{P}_Q = -J_2 \frac{\mu}{r}\frac{R_\oplus^2}{r^2}\frac{1}{2}, \quad \mathcal{P}_r = J_2 \frac{\mu}{r}\frac{R_\oplus^2}{r^2}\frac{3}{4}s^2, \quad \mathcal{P}_\theta = -J_2 \frac{\mu}{r}\frac{R_\oplus^2}{r^2}\frac{3}{4}s^2 \cos 2\theta.$$

The first approximation to the main problem beyond the Keplerian one is the equatorial main problem $\mathcal{Q} = \mathcal{K} + \mathcal{P}_Q$, whose integrability was discussed in reference to Hamiltonian (5.7). However, simple computations show that $\langle \mathcal{M} \rangle_\ell \neq \langle \mathcal{Q} \rangle_\ell$. That is, the equatorial main problem differs from the full model in long-term effects, and hence is not an acceptable intermediary. Still, equatorial circular orbits have been successfully used as variation orbits in some studies [74].

On the other hand, the truncation $\mathcal{C} = \mathcal{K} + \mathcal{P}_Q + \mathcal{P}_r$ is also free from ϑ, yet it now carries dependence on inclination. Since Θ is constant, the one-degree-of-freedom Hamiltonian $\mathcal{C}$ is integrable. At variance with the equatorial main problem, it can be checked that now $\langle \mathcal{M} - \mathcal{C} \rangle_\ell = \langle \mathcal{P}_\theta \rangle_\ell = 0$. Therefore

$$\mathcal{C} = \frac{1}{2}\left(R^2 + \frac{\Theta^2}{r^2}\right) - \frac{\mu}{r} - J_2 \frac{\mu}{r}\frac{R_\oplus^2}{r^2}\left(\frac{1}{2} - \frac{3}{4}s^2\right) \tag{5.29}$$

is a good intermediary of the main problem Hamiltonian that admits a closed-form solution in terms of elliptic functions [193, 416]. It is known as Cid's radial intermediary [111, 165].

5.4.1 Common intermediaries

Thus, "common" intermediary solutions of the main problem are obtained by ignoring small effects of particular orbits, with the additional requirement that the ignored effects are purely periodic. This simplification of the main problem dynamics must be done skillfully in order to guarantee that the intermediary is integrable and still keeps the secular behavior of the full problem [194].

For instance, if we multiply the last summand of Eq. (5.29) by the unity function

$$\frac{r}{p}(1 + e\cos f) = 1, \tag{5.30}$$

obtained from Eq. (4.29), then Cid's radial intermediary takes the form

$$\mathcal{C} = \frac{1}{2}\left(R^2 + \frac{\Theta^2}{r^2}\right) - \frac{\mu}{r} - J_2\frac{\mu}{p}\frac{R_\oplus^2}{r^2}\left(\frac{1}{2} - \frac{3}{4}s^2\right) - J_2\frac{\mu}{p}\frac{R_\oplus^2}{r^2}e\cos f\left(\frac{1}{2} - \frac{3}{4}s^2\right),$$

where it is easy to check that the last summand averages to zero in the mean anomaly. Therefore, the simpler radial Hamiltonian

$$\mathcal{D} = \frac{1}{2}\left(R^2 + \frac{\Theta^2}{r^2}\right) - \frac{\mu}{r} - J_2\frac{\mu}{p}\frac{R_\oplus^2}{r^2}\left(\frac{1}{2} - \frac{3}{4}s^2\right), \tag{5.31}$$

is also a good intermediary—the "common" version of Deprit's radial intermediary [154]—that, while keeping all the secular terms of the main problem, accepts closed-form integration in elementary functions. Indeed, from the definition of the orbit parameter in Eq. (4.21), we replace $\mu = \Theta^2/p$ in the last summand of Eq. (5.31), which is further reorganized in the form

$$\mathcal{D} = \frac{1}{2}\left(R^2 + \frac{\tilde{\Theta}^2}{r^2}\right) - \frac{\mu}{r}, \tag{5.32}$$

of a *quasi-Keplerian* system with "varied angular momentum"

$$\tilde{\Theta} = \Theta\sqrt{1 + \frac{1}{2}J_2(R_\oplus/p)^2(1 - 3c^2)}, \tag{5.33}$$

which is constant because N and Θ are integrals of $\mathcal{D}$. Deprit's radial intermediary is integrated like the Kepler problem, therefore involving the solution of the Kepler equation.

Radial Hamiltonians are of one degree of freedom due to the cyclic character of θ. Therefore, radial intermediaries are the simplest intermediaries of the main problem. However, two-degrees-of-freedom intermediaries can be derived from the full main problem Hamiltonian using similar simplification procedures to those used in the radial case. For instance, if we multiply the whole disturbing function of the main problem Hamiltonian (5.8) by the unity function (5.30), we readily check that the term $-\frac{1}{2}J_2(\mu/p)(R_\oplus/r)^2(1-3s^2\sin^2\theta)e\cos f$ is purely periodic in the mean anomaly. Neglecting this term we obtain Aksnes' *zonal* intermediary [6],

$$\mathcal{A} = \frac{1}{2}\left(R^2 + \frac{\Theta^2}{r^2}\right) - \frac{\mu}{r} + J_2\frac{\mu}{p}\frac{R_\oplus^2}{r^2}\left(\frac{3}{2}s^2\sin^2\theta - \frac{1}{2}\right), \tag{5.34}$$

which, in view of $\langle\mathcal{M}\rangle_\ell = \langle\mathcal{A}\rangle_\ell$, is also a good intermediary of the main problem that is integrable in elliptic functions [6, 170].

Finally, it is worth noting that there are intermediary solutions of the zonal gravitational potential, without constraint to the main problem, which retain some second-order effects that include the full contribution of J_2. More precisely, solutions of the Laplace equation in oblate spheroidal coordinates derived by Vinti [662–666] admit separability of the Hamilton–Jacobi equation for exact values of J_2 and J_3, and for a value $J_4 = -J_2^2$ that is close to the actual J_4 of the geopotential. As shown in [78], Vinti's problem is equivalent to the generalized problem of two fixed centers [5, 466]—the latter still remaining as the integrable kernel in the core of successful orbit propagation programs [244].

5.4.2 Natural intermediaries

Replacing the main problem Hamiltonian by a different system obtained by neglecting some effects is only justified by the results. Rather, we proceed analytically and look for an infinitesimal contact transformation $(r,\theta,\nu,R,\Theta,N,\epsilon) \mapsto (r',\theta',\nu',R',\Theta',N')$, such that when applied to the main problem it yields a good intermediary in the new variables up to first-order effects of J_2. In this way the intermediary solution is more effective than common intermediaries because it takes account of all the main problem effects, secular and periodic, up to the first order of J_2. This kind of intermediary was called *natural* by Deprit [154], who, in addition, showed that common intermediaries in polar variables can be reset as natural intermediaries.

The naturalization of common intermediaries is achieved by standard application of the Lie transforms method, in which the main problem is the original perturbation Hamiltonian $\mathcal{M} = \mathcal{H}_{0,0} + \epsilon\mathcal{H}_{1,0}$, the intermediary is the new, simplified Hamiltonian $\mathcal{M}' = \mathcal{H}_{0,0} + \epsilon\mathcal{H}_{0,1}$, and $\epsilon = J_2$ is a physical small parameter. Then we only need to solve the first order of the homological equation (2.48).

The solution is most easily obtained in Delaunay variables, in which the Lie derivative is Eq. (4.66), and the homological equation is Eq. (4.67) with $m = 1$, in which both $\widetilde{\mathcal{H}}_{0,1} \equiv \mathcal{H}_{1,0}$ and $\mathcal{H}_{0,1}$ are known.

The naturalization procedure is illustrated for Deprit's radial intermediary, for which we want to compute the infinitesimal contact transformation that, after neglecting effects of order of J_2^2, converts the main problem Hamiltonian (5.8) into the radial Hamiltonian (5.31). In Delaunay variables, the main problem takes the form of Eq. (5.15). Therefore,

$$\mathcal{H}_{0,0} = -\frac{\mu}{2a}, \tag{5.35}$$

$$\mathcal{H}_{1,0} = -\frac{\mu}{r}\frac{R_\oplus^2}{r^2}\frac{1}{2}[1 - 3s^2\sin^2(f + \omega)], \tag{5.36}$$

where a, r, s, f, and ω are functions of the Delaunay variables. The new Hamiltonian is $\mathcal{K} = \mathcal{H}_{0,0} + J_2\mathcal{H}_{0,1} + \mathcal{O}(J_2^2)$ in which, from Eq. (5.31),

$$\mathcal{H}_{0,1} = -\frac{\mu}{p}\frac{R_\oplus^2}{r^2}\left(\frac{1}{2} - \frac{3}{4}s^2\right). \tag{5.37}$$

The form of Eqs. (5.36) and (5.37) shows that the difference $\mathcal{H}_{1,0} - \mathcal{H}_{0,1}$ that enters the right side of Eq. (4.68), is made only of periodic terms of the true anomaly. Therefore, we reformulate the Lie derivative (4.66) in terms of f using the differential relation (4.64), to obtain

$$\mathcal{L}_0 = \frac{G}{r^2}\frac{\partial}{\partial f}. \tag{5.38}$$

In consequence, the homological equation (4.68) turns into

$$\mathcal{W}_m = \int (\widetilde{\mathcal{H}}_{0,m} - \mathcal{H}_{0,m})\frac{r^2}{G}\mathrm{d}f. \tag{5.39}$$

Then, from Eq. (4.21), we replace $\mu = G^2/p$ into Eqs. (5.36) and (5.37), which in turn are replaced into Eq. (5.39) to trivially yield

$$\begin{aligned}\mathcal{W}_1 = -\frac{1}{8}G\frac{R_\oplus^2}{p^2}\{&(4 - 6s^2)e\sin f \\ &+ s^2[3e\sin(f + 2\omega) + 3\sin(2f + 2\omega) + e\sin(3f + 2\omega)]\},\end{aligned} \tag{5.40}$$

or

$$\mathcal{W}_1 = -\frac{\Theta}{8}\frac{R_\oplus^2}{p^2}[2\sigma(2 - 3s^2 - s^2\cos 2\theta) + (3 + 4\kappa)s^2\sin 2\theta], \tag{5.41}$$

in polar variables, in which $s^2 = 1 - N^2/\Theta^2$, $p = \Theta^2/\mu$, and the projections of the eccentricity vector in the orbital frame, previously defined in Eq. (5.21), are now

$$\sigma = \frac{Rp}{\Theta}, \quad \kappa = -1 + \frac{r}{p}. \tag{5.42}$$

The transformation equations are limited to the first relation in Eq. (2.17) for the current case. Then the direct transformation

$$\xi = \xi' - \frac{1}{2}J_2(R_\oplus/p)^2\Delta\xi + \mathcal{O}(J_2^2),$$

with $\xi \in (r,\theta,\nu,R,\Theta,N)$, is given by the first-order corrections,

$$\Delta r = p\left(1 - \frac{3}{2}s^2 - \frac{1}{2}s^2\cos 2\theta\right), \tag{5.43}$$

$$\begin{aligned}\Delta\theta = {} & \left[\frac{3}{2} - \frac{7}{4}s^2 + (2-3s^2)\kappa\right]\sin 2\theta \\ & - \sigma[5 - 6s^2 + (1-2s^2)\cos 2\theta],\end{aligned} \tag{5.44}$$

$$\Delta\nu = c\left[\sigma(3+\cos 2\theta) - \left(\frac{3}{2} + 2\kappa\right)\sin 2\theta\right], \tag{5.45}$$

$$\Delta R = \frac{p\,\Theta}{r^2}\,s^2\sin 2\theta, \tag{5.46}$$

$$\Delta\Theta = -\Theta s^2\left[\left(\frac{3}{2} + 2\kappa\right)\cos 2\theta + \sigma\sin 2\theta\right], \tag{5.47}$$

$$\Delta N = 0, \tag{5.48}$$

in which all the entities are functions of the polar-nodal prime variables. Since the transformation is limited to first-order effects, the same corrections $\Delta\xi$ apply to the inverse transformation

$$\xi' = \xi + \frac{1}{2}J_2(R_\oplus/p)^2\Delta\xi + \mathcal{O}(J_2^2),$$

but now the right sides of Eqs. (5.43)–(5.48) must be evaluated in the original polar variables.

The attentive reader will have noticed that in this particular case

$$\mathcal{H}_{0,1} \neq \langle\mathcal{H}_{1,0}\rangle_\ell \equiv \frac{\mu}{p}\frac{R_\oplus^2}{p^2}\eta^3\frac{1}{4}(3s^2-2),$$

and, therefore, $\mathcal{H}_{0,1}$ is not constrained to belong to the kernel of the Lie derivative. Still, $\mathcal{H}_{0,1}$ holds all the terms of $\mathcal{H}_{1,0}$ pertaining to the kernel of the Lie derivative, and hence $\langle\mathcal{H}_{1,0}\rangle_\ell = \langle\mathcal{H}_{0,1}\rangle_\ell$. It is precisely because of that, that the integrand of Eq. (5.39) is made only of terms pertaining to the image of the Lie derivative, and hence the homological equation was solved without difficulty.

The solution of the quasi-Keplerian Hamiltonian (5.32)–(5.33), which is now written in prime variables, can be approached by a complete Hamiltonian reduction following exactly the same steps as carried out in §4.3 and §4.4 for the solution of the Kepler problem [259]. Alternatively, this quasi-Keplerian Hamiltonian can be reduced to a pure Keplerian system by means of a *torsion* transformation [154, 155].

5.4.3 Torsions for quasi-Keplerian systems

The torsion $(r',\theta',\nu',R',\Theta',N') \mapsto (r^*,\theta^*,\nu^*,R^*,\Theta^*,N^*)$ is a canonical transformation

$$(r',\theta',\nu') = \frac{\partial T}{\partial(R',\Theta',N')}, \quad (R^*,\Theta^*,N^*) = \frac{\partial T}{\partial(r^*,\theta^*,\nu^*)}, \tag{5.49}$$

derived from a generator in mixed variables,

$$T = T(r^*,\theta^*,\nu^*,R',\Theta',N') \equiv r^*R' + W(\theta^*,\nu^*,\Theta',N'),$$

that modifies the angular variables while leaving untouched the radial ones.

Because the node is ignorable in Deprit's radial intermediary we choose $W = \nu^*N' + \tilde{W}(\theta^*,\Theta',N')$. Also, because we want to convert the function $\tilde{\Theta} = \tilde{\Theta}(\Theta',N')$ in Eq. (5.32) into a single variable (the true angular momentum $\Theta^* = \tilde{\Theta}$), we choose $\tilde{W} = \theta^*\tilde{\Theta}(\Theta',N')$. Hence,

$$T = r^*R' + \nu^*N' + \theta^*\tilde{\Theta}(\Theta',N'), \tag{5.50}$$

in which $\tilde{\Theta}(\Theta',N')$ must be replaced by Eq. (5.33) written in the prime variables. Then the torsion transformation Eq. (5.49) is

$$r' = r^*, \tag{5.51}$$

$$\theta' = \theta^*\frac{\partial\tilde{\Theta}}{\partial\Theta'} = \theta^*\left[\frac{\tilde{\Theta}}{\Theta'} - \tilde{J}_2\frac{\Theta'}{\tilde{\Theta}}(2-9c^2)\right], \tag{5.52}$$

$$\nu' = \nu^* + \theta^*\frac{\partial\tilde{\Theta}}{\partial N'} = \nu^* - 3c\tilde{J}_2\frac{\Theta'}{\tilde{\Theta}}\theta^*, \tag{5.53}$$

$$R^* = R', \tag{5.54}$$

$$\Theta^* = \tilde{\Theta} = \Theta'\sqrt{1+\tilde{J}_2(1-3c^2)}, \tag{5.55}$$

$$N^* = N', \tag{5.56}$$

in which we abbreviate $\tilde{J}_2 = \tilde{J}_2(\Theta') \equiv \frac{1}{2}J_2(R_\oplus/p)^2$.

Note that the new, asterisk variables are easily solved from Eqs. (5.52) and (5.53) as explicit functions of the old, prime variables. Conversely, the inverse transformation requires root-finding procedures to solve Θ' from Eq. (5.55), who therefore must remain implicit. Alternatively, since the transformation only needs to be accurate to the first order of J_2, series expansions in powers of J_2 can be made in order to make both transformations explicit. That is, neglecting terms $\mathcal{O}(J_2^2)$ and higher, we obtain

$$\Theta' = \Theta^*\left[1 - \frac{1}{4}J_2\frac{R_\oplus^2}{p^2}(1-3c^2)\right],$$

in which, now, $p = (\Theta^*)^2/\mu$ and $c = N^*/\Theta^*$.

The final Hamiltonian

$$T : \mathcal{D} = \frac{1}{2}\left(R^{*2} + \frac{\Theta^{*2}}{r^{*2}}\right) - \frac{\mu}{r^{*2}}, \tag{5.57}$$

is purely Keplerian in the chart $(r^*, \theta^*, \nu^*, R^*, \Theta^*, N^*)$, and hence is solved like in §4.4.

In summary, the corrections (5.43)–(5.48) remove non-essential periodic effects from the main problem to find an integrable approximation—Deprit's radial intermediary—that keeps all the main problem dynamics up to the first order of J_2, without restriction to the secular terms. In particular, essential periodic effects, which are related to the equation of the center, remain in the intermediary problem.

5.5 Discretization of the flow

Beyond the general behavior predicted by the secular terms discussed in §5.3, or the intermediary solutions in §5.4, other facts of the main problem dynamics must be investigated through numerical explorations, which are local in nature. Still, because the integral provided by the third component of the angular momentum vector decouples the motion, global facts of the reduced dynamics in the instantaneous orbital plane can be obtained with the usual tools of non-linear dynamics, like Poincaré surfaces of section or the continuation of periodic orbits [449].

Indeed, Eqs. (5.9)–(5.12) define a two degrees of freedom, conservative flow in which the integral N plays the role of a simple parameter. Then the dynamics in the parametric plane (N, J_2) can be discussed through propagations of different sets $(r_0, \theta_0, R_0, \Theta_0)$ of initial conditions. In particular, for given values of the parameters, the trajectory in the orbital plane corresponding to a specific energy manifold $\mathcal{M}(r_0, \theta_0, R_0, \Theta_0) = E$ of the Hamiltonian (5.8) is implicitly defined as

$$F_1(r, \theta, R, \Theta; E) \equiv \mathcal{M} - E = 0. \tag{5.58}$$

If we constrain ourselves to the case of bounded motion, this three-dimensional curve intersects the surface $\theta \pmod{2\pi} = 0$ at a given point (r_k, R_k) each time θ advances by 2π—the corresponding value of $\Theta = \Theta(r_k, 0, R_k; E)$ been determined from Eq. (5.58).

In general, consecutive intersections, $k = 1, 2, \ldots$, of the trajectory with the *surface of section* can occur at any place on the (r, R) plane, but for a 2π-periodic orbit they will always happen exactly at the same point—or at a discrete number n of fixed points if the orbit were $2n\pi$-periodic. On the other hand, if the motion in the orbital plane were integrable, then another integral, say $\mathcal{J} = \mathcal{J}(r, \theta, R, \Theta)$, would exist in addition to the Hamiltonian and independent of it. Then, for the same initial conditions as those that determine the energy manifold E we will get $\mathcal{J}(r_0, \theta_0, R_0, \Theta_0) = J$. Therefore, the trajectory in the orbital plane will be also determined as an implicit function of J, say

$$F_2(r, \theta, R, \Theta; J) \equiv \mathcal{J} - J = 0. \tag{5.59}$$

Elimination of Θ between Eqs. (5.59) and (5.58) will give $F_3(r, \theta, R; E, J) = 0$, which shows that points (r_k, R_k) on the surface of section will pertain to the curve implicitly determined by $F_3(r_k, 0, R_k; E, J) = 0$. Conversely, identification of such curves in the surface of section provide hints on integrability [280]. On the other hand, even though the main problem does not accept the third integral $\mathcal{J}$ these invariant curves seem to exist in some regions of phase space, at least for the values of J_2 of interest in astrodynamics [70, 138].

It is worth mentioning that the intersection with the surface of section $\theta = 0$ occurs in the ascending node of the orbit, when it crosses the equatorial plane in the upwards direction. Analogously, intersections with the section $\theta = \pi$ occur in the descending node of the orbit. That is, in both cases $z = 0$ and

$$r_k = \rho_k = \sqrt{x_k^2 + y_k^2}, \quad R_k = P_k = \frac{x_k \dot{x}_k + y_k \dot{y}_k}{\rho_k}.$$

Therefore, the surface of section $\theta = 0$ in polar variables is the same as the surface of section $z = 0$ $(Z > 0)$ in cylindrical variables $(\rho, z, \lambda, P, Z, \Lambda)$, where λ is longitude and its conjugate momentum is $\Lambda = N$, which, since the seminal work of Hénon and Heiles [280] are typical variables used in the representation of surfaces of section of problems with axial symmetry [70, 138, 201, 310, 385]. In consequence, periodic orbits of the main problem in the orbital plane map onto periodic orbits in the rotating meridian plane of the satellite.

On the other hand, while periodic orbits in the orbital plane are not generally periodic in three-dimensional space, a subset of them are certainly periodic. More specifically, when the rate of variation of the node becomes commensurate with the frequency of the periodic orbit in the orbital plane, the orbits become truly periodic in three-dimensional space. Notably, this match occurs more easily when the motion is measured in a frame rotating with the Earth's rotation rate, a case in which three-dimensional periodic orbits repeat their ground trace on the surface of the Earth [137, 386–388, 428, 604].

The discretization of the four-dimensional flow stemming from the main problem Hamiltonian is carried out by numerically propagating Eqs. (5.9)–(5.12), which are sampled each time θ advances by 2π, for different energy manifolds. The reduced flow in the orbital plane is then visualized by means of an atlas of phase portraits, each one corresponding to a different point of the parameter space (N, J_2). An example is presented in Fig. 5.2, where a sequence of phase portraits for different values of the third component of the angular momentum vector is shown. We fixed $J_2 = 0.1$ in order to reduce the integration time needed to visualize long tracks of invariant curves in the computed surfaces. Nevertheless, equivalent figures will be obtained from different J_2 values due to the scale invariance of the main problem [70]. Units of length and time such that $R_\oplus = 1$ and $\mu = 1$ have been chosen, meaning that length is measured in units of the equatorial radius and that one orbital period is traversed in roughly 2π units of time. The energy is fixed to $E = -0.5$ in all cases, which would correspond

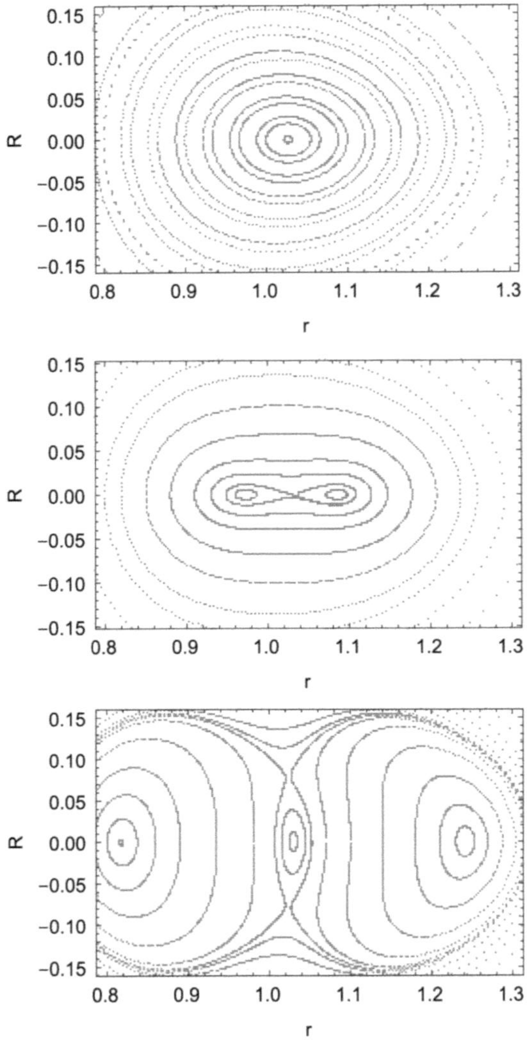

Figure 5.2: Phase portraits (r, R) of the section $\theta \pmod{2\pi} = 0$ for $J_2 = 0.1$, $E = -0.5$, and $N = 0.45$ (upper), 0.44 (center) and 0.43 (lower). Units of length such that $R_{\oplus} = 1$ and time such that $\mu = 1$.

to orbits with semimajor axes slightly longer than the equatorial radius of the oblate body. This is done to enhance the effect of the perturbation on the massless body for illustrative purposes, yet most elliptic orbits corresponding to this case would not be realistic because of their early impact on the surface of the oblate body.

The upper plot of Fig. 5.2 corresponds to the value $N = 0.45$ in the units chosen, and shows a fixed point of the elliptic type $r \approx 1.03$, $R = 0$, corresponding to a stable, circular orbit which is periodic in the (instantaneous) orbital plane. The radii with which other orbits intersect the surface of section $\theta = 0$ vary with time, and shows the advance of the line of apsides, in agreement with the secular motion predicted by

Eq. (5.26). The maximum (resp. minimum) value of r of orbits with rotating periapsis occurs when $R = 0$, that is, when the orbit intersects the section in the apoapsis (resp. periapsis).

The surface of section in the center plot of Fig. 5.2, corresponds to $N = 0.44$, and uncovers important changes in the dynamics. Thus, circular orbits, which now have a slightly higher inclination than in the previous case, are represented by a fixed point of the hyperbolic type with coordinates $r \approx 1.03$, $R = 0$. Besides, two fixed points of the elliptic type exist with $r \approx 1.85$, $R = 0$, and $r \approx 0.97$, $R = 0$. That is, a bifurcation phenomenon happened in which the stability of the circular orbits turned into instability, and two stable elliptic orbits that are periodic in the orbital plane emerged from the bifurcation. These eccentric periodic orbits intersect the surface of section $\theta = 0$ when $R = 0$, and hence their lines of apsides lie on the equatorial plane. Besides, the homoclinic trajectory stemming from the unstable equilibrium separates two regions of librating perigee about the elliptic stable equilibria from the general advance of the perigee outside these libration regions.

Finally, the surface of section corresponding to $N = 0.43$, in which the inclination of the circular orbits is still higher, is presented in the lower plot of Fig. 5.2. Now, five fixed points are clearly noticed in the figure. In particular, circular orbits come back to stability in a new bifurcation phenomenon that produces two new unstable elliptic periodic orbits $r \approx 1.02$, $R \approx \pm 0.08$. That is, each of them crosses the equator ($\theta = 0$) upwards and downwards at the same distance but opposite radial velocity, meaning that the lines of apsides of these unstable periodic orbits lie in the direction $\theta = \pm\frac{\pi}{2}$. Now, the perigees of elliptic orbits evolve with retrograde motion in the region of lower eccentricities, the two regions of perigee libration remain, and the perigees of elliptic orbits with higher eccentricity advance.

The qualitative dynamics remain the same when J_2 is of the order of one thousandth, but then the double bifurcation sequence happens within a very small variation of N in the close vicinity of the critical inclination. The libration region is very thin in that case, and the eccentricity of librating-perigee orbits increases very fast for decreasing values of N.

In summary, the true, numerically propagated dynamics of the main problem exposes important qualitative differences with respect to the predicted behavior of intermediary approximations, yet the differences are bounded to a narrow region in phase space. Indeed, while integrable approximations obtained from separable Hamiltonians are able to detect the existence of orbits with constant eccentricity and fixed argument of the perigee close to the critical inclination, they predict that all of them have the perigee frozen irrespective of their eccentricity. This is not the case, however, as the direct integration of the problem has shown by means of the construction of the Poincaré surfaces of section in Fig. 5.2. Indeed, second-order terms missed by the intermediaries limit radically the number of orbits exhibiting frozen perigee dynamics. In Chapter 6, we will learn how other second-order effects of the geopotential may further change this picture.

5.6 The critical inclination

Beyond the first order of J_2 secular effects computed analytically in §5.3, which miss important features of the real main problem dynamics, we can take advantage of the techniques described in Chapter 2 and apply perturbation theory by Lie transforms to get a higher-order approximation of the main problem dynamics. In view of the simplicity of the Lie derivative and the homological equation in Eqs. (4.66) and (4.67), respectively, the perturbation solution will be approached in Delaunay variables.

To begin with, we write Eq. (5.8) like the usual perturbation Hamiltonian (2.30), in which $\mathcal{H}_{0,0}$ and $\mathcal{H}_{1,0}$ are given by Eqs. (5.35) and (5.36), respectively, $\mathcal{H}_{m,0} = 0$ for $m \geq 2$, and $\varepsilon = J_2$. Once more, we recall that the entities a, r, s, ω, and f are not variables in themselves, but functions of some set of canonical variables, which in the current case are the Delaunay variables.

5.6.1 First order

The procedure starts solving Eq. (4.68) for $m = 1$. The term $\widetilde{\mathcal{H}}_{0,1}$ is the same as $\mathcal{H}_{1,0}$, and the term $\mathcal{H}_{0,1}$ is chosen in such a way that it cancels the terms of $\mathcal{H}_{1,0}$ pertaining to the kernel of the Lie derivative (4.66). These terms are easily identified after expanding Eq. (5.36) like a Fourier series of the mean anomaly [78, 171]. However, to avoid restrictions related to the size of the obtained series, or limitations of the solution to small or moderate values of the eccentricity, $\mathcal{H}_{0,1}$ is more suitably computed in closed form of the eccentricity by direct averaging.

That is, $\mathcal{H}_{0,1} = \langle \mathcal{H}_{1,0} \rangle_{\ell}$, where, on account of the Lie derivative being constrained by definition to the Keplerian flow, we are free to use the differential relation (4.64) to replace the mean anomaly by the true anomaly like the integration variable. In this way, the averaging is readily solved in closed form of the eccentricity. Namely,

$$\mathcal{H}_{0,1} = \frac{1}{2\pi} \int_0^{2\pi} \mathcal{H}_{1,0} \frac{r^2}{a^2 \eta} \, df = \frac{1}{4} \frac{\mu}{a} \frac{R_\oplus^2}{p^2} \eta (3s^2 - 2), \tag{5.60}$$

which, as expected, is the same as the disturbing term of the secular Hamiltonian (5.23), except for the coefficient J_2 which is taken here as the small parameter of the Lie transforms method.

By the same token, the homological equation becomes Eq. (4.69). Replacing $\widetilde{\mathcal{H}}_{0,m} = \mathcal{H}_{1,0}$ by Eq. (5.36) and $\mathcal{H}_{0,m} = \mathcal{H}_{0,1}$ by (5.60), the first-order term of the generating function is readily solved, also in closed form of the eccentricity, to obtain

$$\begin{aligned} \mathcal{W}_1 = -\frac{1}{8} G \frac{R_\oplus^2}{p^2} [&(4 - 6s^2)(\phi + e \sin f) + 3es^2 \sin(f + 2\omega) \\ &+ 3s^2 \sin(2f + 2\omega) + es^2 \sin(3f + 2\omega)] + A_1, \end{aligned} \tag{5.61}$$

in which A_1 is an arbitrary integration "constant" meeting the only condition of being independent of the mean anomaly. That is, $\mathrm{d}A_1/\mathrm{d}\ell = 0$ and A_1 is a function of g, G, and L when expressed in Delaunay variables. While the equation of the center ϕ is a purely periodic function of the mean anomaly, given in Eq. (4.65), the fact that it is not a trigonometric function complicates its closed-form treatment in the computation of higher orders of the perturbation solution, as we will see immediately.

For simplicity in the following computations, we choose here $A_1 = 0$, in this way keeping the number of terms of the generating function at a minimum. However, we highlight that this selection of the arbitrary constant makes $\mathcal{W}_1$ to be made of both long- and short-period terms. Indeed [342, 362, 631],

$$\langle \sin mf \rangle_\ell = 0, \quad \langle \cos mf \rangle_\ell = (-\beta)^m (1 + m\eta), \tag{5.62}$$

where we used the usual definition

$$\beta = \frac{e}{1+\eta} = \sqrt{\frac{1-\eta}{1+\eta}}. \tag{5.63}$$

Then Eq. (5.61) is readily averaged over the mean anomaly, to obtain [363]

$$\langle \mathcal{W}_1 \rangle_\ell = A_1 - \frac{1}{8} G \frac{R_\oplus^2}{p^2} (1 + 2\eta)\beta^2 s^2 \sin 2\omega. \tag{5.64}$$

Therefore, unless A_1 be chosen so that the right side of Eq. (5.64) vanishes, $\mathcal{W}_1$ will depend on hidden long-period terms of the order of e^2 driven by the perigee dynamics. This issue, which is a consequence of the closed-form integration and does not occur when the mean anomaly appears explicitly in the Hamiltonian, will be tackled in more detail in §6.7.

5.6.2 Second order

At second order, the known terms $\widetilde{\mathcal{H}}_{0,2}$ of the homological equation are those of Eq. (2.37), in which $\mathcal{H}_{2,0} \equiv 0$ in the current case. Hence,

$$\widetilde{\mathcal{H}}_{0,2} = \{\mathcal{H}_{1,0}; \mathcal{W}_1\} + \{\mathcal{H}_{0,1}; \mathcal{W}_1\}, \tag{5.65}$$

which bears terms of both the kernel and the image of the Lie derivative (4.66). The former are canceled out of the homological equation (4.68) when choosing $\mathcal{H}_{0,2} = \langle \widetilde{\mathcal{H}}_{0,2} \rangle_\ell$.

After evaluating the Poisson brackets in Eq. (5.65), we identify three types of terms of a different nature. Thus, $\widetilde{\mathcal{H}}_{0,2} = \widetilde{\mathcal{H}}_{0,2,1} + \widetilde{\mathcal{H}}_{0,2,2} + \widetilde{\mathcal{H}}_{0,2,3}$, where

$$\widetilde{\mathcal{H}}_{0,2,1} = -\frac{9}{8} \frac{\mu}{r} \frac{R_\oplus^4}{p^4} \frac{p^2}{r^2} s^2 (5s^2 - 4)\phi \sin(2f + 2\omega), \tag{5.66}$$

$$\widetilde{\mathcal{H}}_{0,2,2} = \frac{3}{16}\frac{\mu}{p}\frac{R_\oplus^4}{p^4}\eta^3\{-\eta(2-3s^2)^2 + s^2(4-5s^2)[3e\cos(f+2\omega) + 3\cos(2f+2\omega) + e\cos(3f+2\omega)]\}, \tag{5.67}$$

$$\widetilde{\mathcal{H}}_{0,2,3} = \frac{3}{128}\frac{\mu}{p}\frac{R_\oplus^4}{p^4}\frac{p^2}{r^2}\sum_{j=0}^{2}\sum_{i=-3}^{3} q_{2j,i}s^{2j}\cos[(2j+i)f + 2jg], \tag{5.68}$$

with the coefficients $q_{2j,i}$ found in Table 5.2.

Table 5.2: Coefficients $q_{2j,i}$ in Eq. (5.68); $q_{4,-3} = q_{4,3} = 0$.

j	i	$q_{2j,i}$
0	0	$-16(4-9s^2+5s^4) - 2(8-8s^2-5s^4)e^2$
	±1	$-\beta[60-124s^2+57s^4+(48-88s^2+30s^4)\eta+(2-3s^2)^2\eta^2]$
	±2	$-e\beta[24-56s^2+31s^4+(8-8s^2-5s^4)\eta]$
	±3	$-e^2(2-3s^2)^2\beta$
1	-3	$e^2(-2+3s^2)\beta$
	-2	$4e^2(14-15s^2)$
	-1	$\beta[3(38-37s^2)+12(8-7s^2)\eta-5(2-3s^2)\eta^2]$
	0	$-8[4-9s^2-(2-s^2)e^2]$
	1	$-\beta[150-221s^2+4(24-35s^2)\eta+(2-3s^2)\eta^2]$
	2	$-4e\beta[22-31s^2+(10-13s^2)\eta]$
	3	$-5e^2(2-3s^2)\beta$
2	-2	$-5e^2$
	-1	$-6e$
	0	$8-2e^2$
	1	$10e$
	2	$3e^2$

After replacing the inverse of the radius in Eq. (5.66) using Eq. (4.29), we check that $\widetilde{\mathcal{H}}_{0,2,1}$ depends on the mean anomaly through terms of the form $\phi\sin mf$, and $\phi\cos mf$, $m = 0,1,\ldots,5$, from which we must identify those pertaining to the kernel of the Lie derivative (4.66). Formulas for the definite integration of these kinds of terms are given in [487]. In particular,

$$\langle\phi\cos mf\rangle_\ell = 0, \quad \langle\phi\sin mf\rangle_\ell = \frac{\eta}{e^m}Q_m(\eta), \quad m \ge 0, \tag{5.69}$$

where the coefficients Q_m related to terms stemming from Eq. (5.66) are detailed in Table 5.3. Note that the apparent singularity for circular orbits is just virtual, as checked when Q_m is expanded in powers of the eccentricity. In the current case the expansions are unnecessary because the troublesome denominators cancel out when the different integrals are grouped together.

Table 5.3: From top to bottom, terms $Q_1, \ldots Q_5$ in Eq. (5.69), cf. [437].

$\frac{3}{2} - \eta - \frac{1}{2}\eta^2$
$-3 + 2\eta + 3\eta^2 - 2\eta^3 - 4\eta^2 \log \frac{2\eta}{1+\eta}$
$\frac{9}{2} - 3\eta - 4\eta^2 + 7\eta^3 - \frac{9}{2}\eta^4 + 16\eta^2 \log \frac{2\eta}{1+\eta}$
$-6 + 4\eta - 2\eta^2 - 16\eta^3 + 28\eta^4 - 8\eta^5 - 8\eta^2(\eta^2 + 5) \log \frac{2\eta}{1+\eta}$
$\frac{15}{2} - 5\eta + \frac{131}{6}\eta^2 + 30\eta^3 - \frac{183}{2}\eta^4 + \frac{149}{3}\eta^5 - \frac{25}{2}\eta^6 + 16\eta^2(3\eta^2 + 5) \log \frac{2\eta}{1+\eta}$

Terms of $\widetilde{\mathcal{H}}_{0,2,2}$ are free from the radius. They only involve trigonometric factors of the form $\cos(mf \pm 2\omega)$, with m integer, whose averaging over the mean anomaly is carried out using Eq. (5.62).

Finally, the coefficient $1/r^2$ in $\widetilde{\mathcal{H}}_{0,2,3}$ makes its averaging elementary with the help of the differential relation between the true and mean anomalies in Eq. (4.64).

After carrying out the required operations, we obtain

$$\begin{aligned} \mathcal{H}_{0,2} = -\frac{\mu}{a}\frac{R_\oplus^4}{p^4}\frac{3}{64}\eta[5(8 - 16s^2 + 7s^4) + 4(2 - 3s^2)^2\eta \\ + (5s^4 + 8s^2 - 8)\eta^2 + 2(15s^2 - 14)s^2e^2\cos 2\omega], \end{aligned} \tag{5.70}$$

which reveals that the long-term dynamics of the main problem is driven by terms of second order of J_2 in the argument of the perigee. This important qualitative effect is missed in the first-order secular Hamiltonian (5.23).

Because we only focus on the contribution of second-order effects of J_2 to the long-term dynamics, we do not deal here with the computation of second-order periodic corrections, which is feasible in closed form, yet nontrivial [363, 407]. On the other hand, the difficulty is easily overcome with the help of Hamiltonian simplification techniques that will be thoroughly discussed in §6.2.1.

5.6.3 The reduced phase space. Frozen orbits

Once the terms $\mathcal{H}_{0,m}$ of the new Hamiltonian have been written in the new, prime variables, after neglecting terms of $\mathcal{O}(J_2^3)$ and higher, we get the new Hamiltonian

$$\widetilde{\mathcal{M}} = \widetilde{\mathcal{M}}(-, g', -, L', G', H') \equiv \mathcal{H}_{0,0} + J_2\mathcal{H}_{0,1} + \frac{1}{2}J_2^2\mathcal{H}_{0,2}, \tag{5.71}$$

with $\mathcal{H}_{0,1}$ and $\mathcal{H}_{0,2}$ given by Eqs. (5.60), and (5.70), respectively, which is only of one degree of freedom in (g', G'). Then orbits of the reduced phase space can be represented either by making g' explicit from $\widetilde{\mathcal{M}} = E$ in Eq. (5.71), or with simple contour plots of this reduced Hamiltonian [559, 600]. Moreover, for a given point (L', H') in the parameter space, the description of the reduced flow is simpler when discussed in terms of the non-dimensional Hamiltonian

$$\mathcal{M}^* = \frac{\widetilde{\mathcal{M}} - \mathcal{H}_{0,0}}{J_2 \mathcal{H}_{0,0}} = \frac{\mathcal{H}_{0,1}}{\mathcal{H}_{0,0}} + \frac{1}{2} J_2 \frac{\mathcal{H}_{0,2}}{\mathcal{H}_{0,0}}, \tag{5.72}$$

in which the energy has been displaced by a constant quantity and the time evolves in the slow scale $\tau = J_2 \mathcal{H}_{0,0} t$, which has units of angular momentum.

A sample of the reduced phase space is presented in Fig. 5.3. Besides, for illustration purposes, L' has been chosen such that $a = R_\oplus$ in which case the effect of the perturbation is magnified, although this choice would make orbits impact the surface of the Earth save the circular case. Moreover, instead of depicting the total angular momentum $G' = G'(g'; H'; L')$ we get more insight by representing the eccentricity $e = e(G'; L')$ as a function of the dynamical parameter

$$\gamma = H'/L', \quad 0 \le \gamma \le \eta \le 1. \tag{5.73}$$

Note that $\gamma = \eta \cos I$. Therefore, γ coincides with the cosine of the inclination of circular orbits, a case in which $G' = L'$.

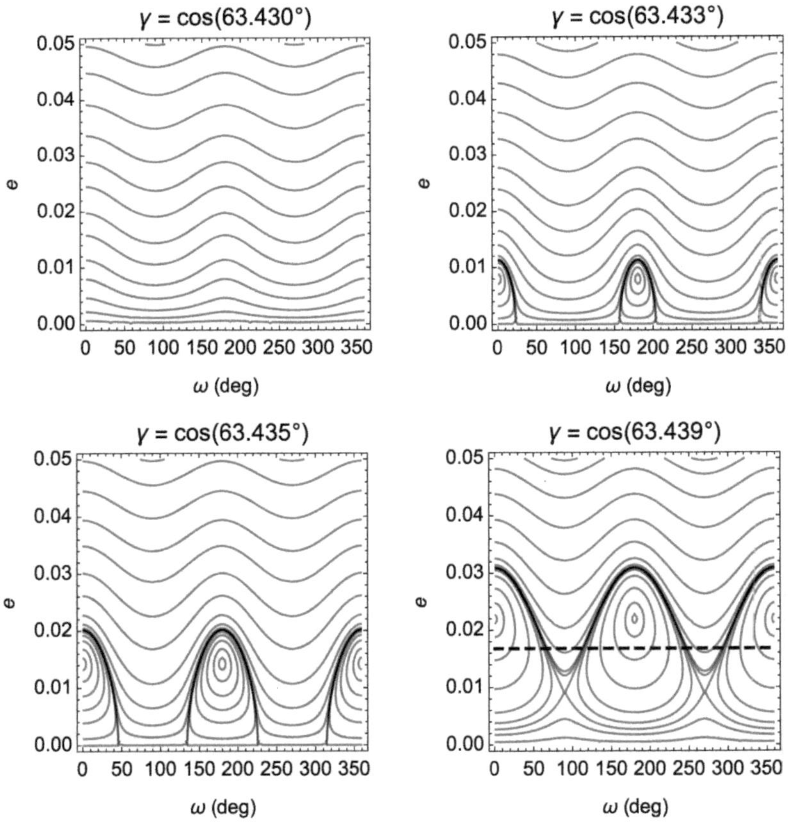

Figure 5.3: Contour plots of the long-term main problem Hamiltonian (5.72) for $L' = \sqrt{\mu R_\oplus}$ and different values $H' = L'\gamma$. Fixed points are frozen orbits.

Thus, the upper-left plot of Fig. 5.3 shows that for large enough values of γ the periapsis of elliptic orbits rotates. When γ decreases to a certain value, two fixed points of the elliptic type appear with abscissas $\omega = 0$ and 180 deg (upper-right and lower-left plots of Fig. 5.3). An additional small decrease of γ results in the appearance of two new fixed points, now of the hyperbolic type, with abscissas $\omega = 90$ and 270 deg (lower-right plot of Fig. 5.3). In the latter plot, the horizontal dashed line displays the value $e = (1 - 5\gamma^2)^{1/2}$, corresponding to the (mean) eccentricity of an eccentric orbit with the critical inclination $I = \arccos(5^{-1/2})$. Orbits below this line have higher inclinations than the critical one, whereas orbits above this line have lower inclinations.

While the behavior depicted in Fig. 5.3 is in general agreement with the real one obtained by the discretization of the flow illustrated in Fig. 5.2, the cylindrical map $(g'G')$ is not suitable for the correct representation of the flow because the argument of the perigee is not defined for circular orbits. Therefore, the bifurcation process that happens when the third component of the angular momentum vector varies is not well perceived in Fig. 5.3. This flaw is normally amended by representing the flow of low eccentricity orbits in the so-called *semi-equinoctial* variables [354], provided by the projections of the eccentricity vector in the nodal frame

$$C = e\cos\omega, \qquad S = e\sin\omega, \tag{5.74}$$

which are free from singularities in the case of circular orbits [130, 171, 355]. The reduced Hamiltonian (5.71) is readily reformulated in these variables by replacing $e^2\cos 2\omega = C^2 - S^2$, $\eta = (1 - C^2 - S^2)^{1/2}$, and $c = \gamma/\eta$ in Eq. (5.72). Corresponding contour plots are presented in Fig. 5.4. Now, the two consecutive bifurcations of the circular orbits, which are represented by the point at the origin, are clearly apparent. Moreover, the contour plots in Fig. 5.4 are similar to the (J_2-scaled) phase portraits in Fig. 5.2, although contrary to the discretization process, here they are rendered almost instantaneously because there is no need of carrying out any numerical integration. The plot in the lower-right corner of Fig. 5.4 shows a dashed circle superimposed, which corresponds to the value $e = (1 - 5\gamma^2)^{1/2}$ highlighted previously in Fig. 5.3. Orbits inside this circle have higher inclinations than the critical one, whereas orbits out of the circle have lower inclinations.

Lower values of γ than those in Fig. 5.4 make the eccentricity of the frozen orbits to grow fast with almost constant inclination. Still, the region of perigee libration is concentrated in a thin strip of eccentricities in which the perigee evolves at a slow rate that only includes second order of J_2 effects at the critical inclination; cf. Eq. (5.83) below. In that case, the eccentricity-vector representation does not provide a clear figure, which, on the contrary is better appreciated in the cylindrical map representation, as illustrated in Fig. 5.5.

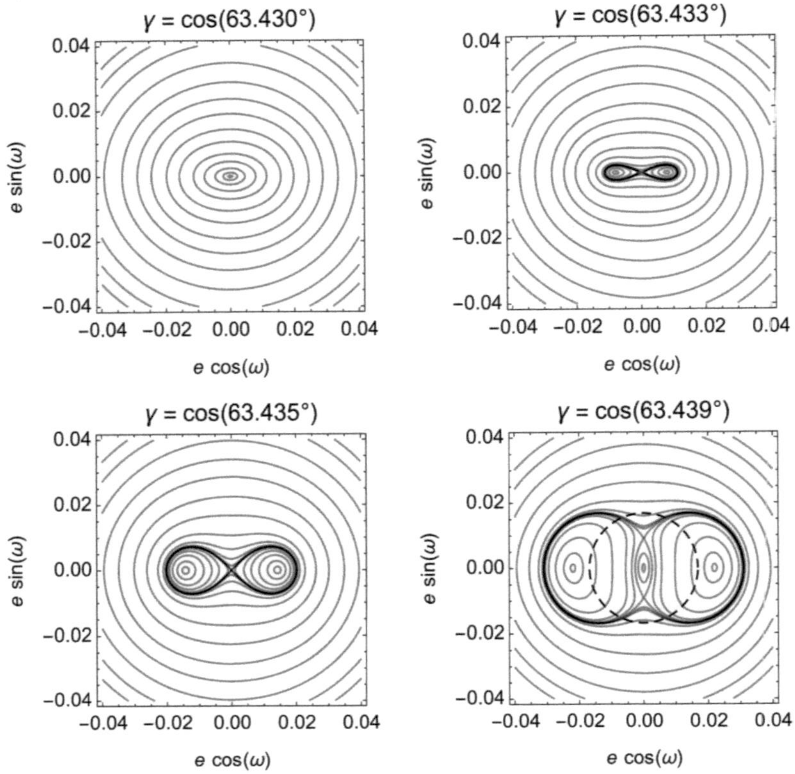

Figure 5.4: Same as Fig. 5.3 in the eccentricity-vector representation.

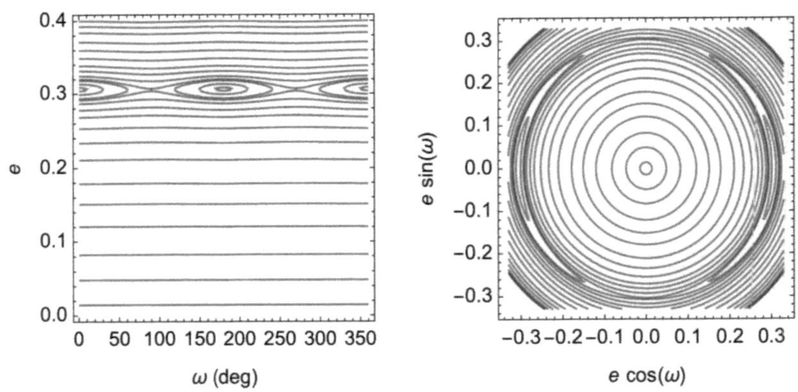

Figure 5.5: Reduced phase space of the main problem for $\gamma = \cos 64.8^\circ$.

5.6.4 Reduced dynamics on the sphere

In fact, the reduced phase space is neither the cylinder (g', G') nor the plane (C, S), but a two-dimensional compact manifold [120, 135]. Disregarding the case of rectilinear orbits, which are not of interest for Earth artificial satellite applications, the flow is properly represented with the non-dimensional coordinates[2] [119, 411, 433]

$$\begin{aligned}\chi_1 &= e\eta s\cos g = \sqrt{1-\eta^2}\sqrt{\eta^2-y^2}\cos g,\\ \chi_2 &= e\eta s\sin g = \sqrt{1-\eta^2}\sqrt{\eta^2-y^2}\sin g,\\ \chi_3 &= \eta^2 - \frac{1}{2}(1+y^2),\end{aligned} \tag{5.75}$$

on the sphere of radius $\rho = \rho(H/L') \equiv \sqrt{\chi_1^2+\chi_2^2+\chi_3^2} = \frac{1}{2}(1-y^2)$. The north pole of the sphere $\chi_1 = \chi_2 = 0, \chi_3 = \rho$, implies $\eta = 1$ and, therefore, represents a circular orbit. Conversely, the south pole $\chi_1 = \chi_2 = 0, \chi_3 = -\rho$ yields $\eta^2 = y^2$, or $G' = |H|$, corresponding to an equatorial elliptic orbit. Other points on the sphere represent families of ellipses with different arguments of the node, whose eccentricity and pericenter evolution is represented by a closed curve.

Noting that $\chi_2^2 = \frac{1}{2}\eta^2 s^2 e^2(1-\cos 2\omega)$, and replacing $s = \sqrt{1-(y/\eta)^2}$, Hamiltonian (5.72) is conveniently written in the form $\mathcal{M}^* = \mathcal{M}^*(\chi_2, \eta; y)$. Namely,

$$\begin{aligned}\mathcal{M}^* = {} & \frac{1}{4\eta^5}(\eta^2-3y^2) + J_2\frac{3R_\oplus^2}{32a^2}\bigg[\chi_2^2(\eta^2-15y^2) + \left(\frac{3}{4}-\eta-\frac{3}{4}\eta^2\right)\eta^4\\ & + y^2\left(\frac{11}{2}+6\eta-\frac{7}{2}\eta^2\right)\eta^2 - y^4\left(\frac{65}{4}+9\eta-\frac{25}{4}\eta^2\right)\bigg]\frac{1}{\eta^{11}}.\end{aligned}$$

Then the differential equations of the flow on the sphere are

$$\frac{d\chi_i}{d\tau} = \{\chi_i; \mathcal{M}^*\} = \{\chi_i; \eta\}\frac{\partial\mathcal{M}^*}{\partial\eta} + \{\chi_i; \chi_2\}\frac{\partial\mathcal{M}^*}{\partial\chi_2}, \quad i = 1,2,3, \tag{5.76}$$

in which $\{\chi_1; \eta\} = -\chi_2/L'$, $\{\chi_2; \eta\} = \chi_1/L'$, $\{\chi_3; \eta\} = 0$, $\{\chi_1; \chi_2\} = 2\eta\chi_3/L'$, $\{\chi_3; \chi_2\} = -2\eta\chi_1/L'$, and

$$\frac{\partial\mathcal{M}^*}{\partial\chi_2} = J_2\frac{3R_\oplus^2}{16a^2}(\eta^2-15y^2)\chi_2\frac{1}{\eta^{11}}, \tag{5.77}$$

$$\begin{aligned}\frac{\partial\mathcal{M}^*}{\partial\eta} = {} & \frac{3}{4\eta^6}(5y^2-\eta^2) - J_2\frac{9R_\oplus^2}{32a^2\eta^{12}}\bigg[(3\eta^2-55y^2)\chi_2^2 + \frac{7}{4}\eta^4 - 2\eta^5 - \frac{5}{4}\eta^6\\ & + y^2\left(\frac{33}{2}\eta^2+16\eta^3-\frac{49}{6}\eta^4\right) - y^4\left(\frac{715}{12}+30\eta-\frac{75}{4}\eta^2\right)\bigg].\end{aligned} \tag{5.78}$$

Particularization of Eq. (5.76) for these values yields

$$L'\frac{d\chi_1}{d\tau} = -\chi_2\left[\frac{\partial\mathcal{M}^*}{\partial\eta} - \chi_3 J_2\frac{R_\oplus^2}{a^2}\frac{3}{8\eta^{10}}(\eta^2-15y^2)\right], \tag{5.79}$$

2 Dimensional versions of these variables have been earlier proposed [120, 592].

$$L'\frac{d\chi_2}{d\tau} = \chi_1\frac{\partial\mathcal{M}^*}{\partial\eta}, \tag{5.80}$$

$$L'\frac{d\chi_3}{d\tau} = -\chi_1\chi_2 J_2\frac{R_\oplus^2}{a^2}\frac{3}{8\eta^{10}}(\eta^2 - 15y^2), \tag{5.81}$$

from which it is readily checked that $\chi_1 d\chi_1/d\tau + \chi_2 d\chi_2/d\tau = -\chi_3 d\chi_3/d\tau$, as it must be from the constraint $\rho\, d\rho = 0$ derived from the constancy of the radius. Therefore, if two of the differential equations are satisfied, then the other one is automatically satisfied.

Replacing $\chi_1 = \chi_2 = 0$ in Eqs. (5.79)–(5.81) shows that circular orbits are always equilibria. Besides, the case $\chi_2 = 0, \chi_1 \neq 0$, corresponds to equilibria with $g' = 0, \pi$, for those values of η satisfying simultaneously the condition $\partial\mathcal{M}^*/\partial\eta = 0$, derived from Eq. (5.80). Thus, from Eq. (5.78),

$$32\eta^6(\eta^2 - 5y^2) = J_2(R_\oplus/a)^2[15\eta^6 + 24\eta^5 + 7(14y^2 - 3)\eta^4 - 192y^2\eta^3 - 9y^2(25y^2 + 22)\eta^2 + 360y^4\eta + 715y^4],$$

which is solved by the Newton–Raphson method starting from the critical inclination $\eta^2 = 5y^2$. The first iteration of the method yields

$$\eta = \sqrt{5}y\left(1 - J_2\frac{R_\oplus^2}{a^2}\frac{1 - 4y^2}{50y^4}\right),$$

from which $G' < \sqrt{5}H$, showing that the bifurcation of the equilibria with $g' = 0, \pi$ happen from a circular orbit with an inclination slightly closer to the equator than the critical inclination.

Analogously, the case $\chi_1 = 0, \chi_2 \neq 0$, corresponds to equilibria with $g' = \pm\frac{\pi}{2}$ for values of η that satisfy the condition

$$\frac{\partial\mathcal{M}^*}{\partial\eta} = \left[\eta^2 - \frac{1}{2}(1 + y^2)\right]J_2\frac{R_\oplus^2}{a^2}\frac{3}{8\eta^{10}}(\eta^2 - 15y^2),$$

derived from Eq. (5.79), where y_3 has been replaced from its definition in Eq. (5.75) and the left side must be replaced by Eq. (5.78) with $\chi_1 = 0$. That is,

$$32\eta^6(\eta^2 - 5y^2) = J_2(R_\oplus/a)^2[35\eta^6 + 24\eta^5 - 7(50y^2 + 7)\eta^4 - 192y^2\eta^3 + 63y^2(5y^2 + 6)\eta^2 + 360y^4\eta + 55y^4],$$

which is solved for η by the Newton–Raphson method starting from the critical inclination $\eta^2 = 5y^2$. The first iteration of the method gives

$$\eta = \sqrt{5}y\left(1 + J_2\frac{R_\oplus^2}{a^2}\frac{9 - 35y^2}{500y^4}\right),$$

from which $G' > \sqrt{5}H$, showing that the bifurcation of the equilibria at $g' = \pm\pi/2$ happens from a circular orbit whose inclination is slightly closer to polar than the critical inclination case.

The stability character of the equilibria, which is guessed from the contour plots in Fig. 5.4, can be computed from the usual linearization of the flow. However, the incorporation of additional effects to the perturbation model introduces drastic changes in the stability, as will be shown in §6.1.3. Therefore, it is not discussed here. The interested reader can find the required details in [120].

The reduced flow on the sphere is readily depicted with standard graphic tools, by assigning colors to the different levels of the (reduced) Hamiltonian integral. Still, refinements in the technique, such as the automatic selection of colors, may be required to ensure enough contrast around isolated but close singularities [116]. This technique is sometimes dubbed "painting" Hamiltonians [270, 546]. An example of this illustrative kind of portrait is presented in Fig. 5.6.

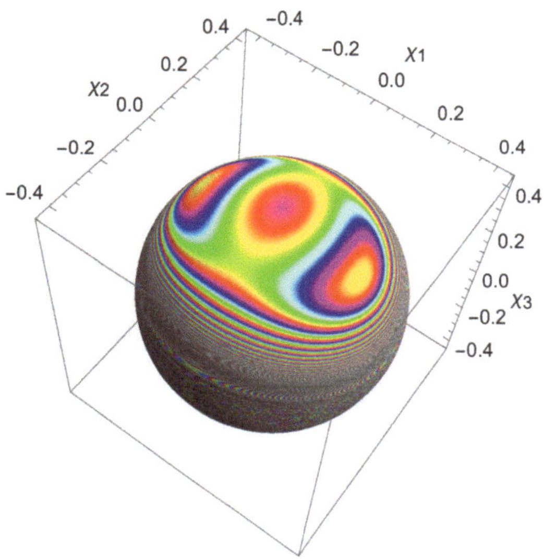

Figure 5.6: Painting the phase space of the main problem.

Finally, it is worth noting that, up to the truncation order, orbits that are frozen in the mean-elements space correspond to orbits of the original space that are periodic in the orbital plane. Indeed, as will be checked in §5.7.1, the short-period corrections needed to recover the original variables take the same value each time the mean anomaly advances by 2π, thus meaning that, save for the argument of the node, the osculating orbital elements of a frozen orbit are repeated periodically.

5.7 Semi-analytical integration

The convenience of removing short-period terms goes beyond the qualitative exploration of the reduced flow and allows for the integration of the main problem in a very efficient semi-analytical way. Indeed, the Hamilton equations of the reduced Hamiltonian (5.71) are

$$\begin{aligned}\frac{d\ell'}{dt} &= n + nJ_2\frac{R_\oplus^2}{p^2}\frac{3}{4}\eta(2-3s^2) + nJ_2^2\frac{R_\oplus^4}{p^4}\frac{3}{128}\eta[15(8-16s^2+7s^4)\\ &\quad + 16(2-3s^2)^2\eta - 5(8-8s^2-5s^4)\eta^2\\ &\quad - (28-30s^2)(3-5\eta^2)s^2\cos 2\omega],\end{aligned} \tag{5.82}$$

$$\begin{aligned}\frac{dg'}{dt} &= nJ_2\frac{R_\oplus^2}{p^2}\frac{3}{4}(4-5s^2) + nJ_2^2\frac{R_\oplus^4}{p^4}\frac{3}{128}\\ &\quad \times [\gamma_{0,0} + \gamma_{0,1}\eta + \gamma_{0,2}\eta^2 + (\gamma_{1,0} + \gamma_{1,2}\eta^2)\cos 2\omega],\end{aligned} \tag{5.83}$$

$$\begin{aligned}\frac{dh'}{dt} &= -nJ_2\frac{R_\oplus^2}{p^2}\frac{3}{2}c - nJ_2^2\frac{R_\oplus^4}{p^4}\frac{3}{32}c[5(8-7s^2)\\ &\quad + 12(2-3s^2)\eta - (4+5s^2)\eta^2 + 2(7-15s^2)e^2\cos 2\omega],\end{aligned} \tag{5.84}$$

$$\frac{dG'}{dt} = G'nJ_2^2\frac{R_\oplus^4}{p^4}\frac{3}{32}s^2(14-15s^2)e^2\sin 2\omega, \tag{5.85}$$

where the inclination polynomials $\gamma_{i,j}$ in Eq. (5.83) are given in Table 5.4, and the quantities n, p, s, η, e, and ω, are functions of the prime variables.

Table 5.4: Inclination polynomials $\gamma_{i,j}$ in Eq. (5.83).

$\gamma_{0,0} = 440 - 860s^2 + 385s^4$	
$\gamma_{0,1} = 192 - 528s^2 + 360s^4$	$\gamma_{1,0} = 56 - 372s^2 + 330s^4$
$\gamma_{0,2} = -56 + 36s^2 + 45s^4$	$\gamma_{1,2} = -56 + 316s^2 - 270s^4$

Because Eqs. (5.82)–(5.85) are free from short-period effects, which are related to the mean anomaly, their numerical integration proceeds with very large step sizes. At each step of the numerical integration, the short-period effects are recovered by analytical evaluation of the first-order corrections computed from Eq. (2.17). To the first order, we get $x = x' + J_2 x_{0,1}$ where $x_{0,1} = \{x; \mathcal{W}_1\}$, and x denotes a particular variable. For direct corrections $x_{0,1}$ must be written in prime variables, whereas for the inverse transformation $x' = x - J_2 x_{0,1}$ the corrections $x_{0,1}$ must be written in original variables.

5.7.1 Short-period corrections in Delaunay variables

In the case of the Delaunay variables, $\mathcal{W}_1$ is replaced by Eq. (5.61). When we choose $A_1 = 0$ the corrections to the Delaunay variables are $H_{0,1} = 0$ and

$$\begin{aligned}\ell_{0,1} = &-\frac{R_\oplus^2}{p^2}\frac{\eta}{32}\Big[6\Big(\frac{4}{e} - e\Big)(2-3s^2)\sin f + 12(2-3s^2)\sin 2f\\ &+ 2e(2-3s^2)\sin 3f - 3es^2\sin(-f+2\omega) - 18s^2\sin 2\omega\end{aligned}$$

$$-3\left(5e+\frac{4}{e}\right)s^2\sin(f+2\omega)-\left(e-\frac{28}{e}\right)s^2\sin(3f+2\omega)$$
$$+18s^2\sin(4f+2\omega)+3es^2\sin(5f+2\omega)\Bigg], \tag{5.86}$$

$$\begin{aligned} g_{0,1} = {} & \frac{R_\oplus^2}{p^2}\frac{1}{32}\Bigg\{24(4-5s^2)\phi+6\left[\frac{4}{e}(2-3s^2)+e(14-17s^2)\right]\sin f \\ & +12(2-3s^2)\sin 2f+2e(2-3s^2)\sin 3f+3es^2\sin(f-2\omega) \\ & -18s^2\sin 2\omega-3\left[\frac{4}{e}s^2+e(8-15s^2)\right]\sin(f+2\omega) \\ & -12(2-5s^2)\sin(2f+2\omega)+\left[\frac{28}{e}s^2-e(8-19s^2)\right]\sin(3f+2\omega) \\ & +18s^2\sin(4f+2\omega)+3es^2\sin(5f+2\omega)\Bigg\}, \end{aligned} \tag{5.87}$$

$$\begin{aligned} h_{0,1} = {} & -\frac{R_\oplus^2}{p^2}\frac{c}{4}[6(\phi+e\sin f)-3e\sin(f+2\omega)-3\sin(2f+2\omega) \\ & -e\sin(3f+2\omega)], \end{aligned} \tag{5.88}$$

$$\begin{aligned} L_{0,1} = {} & L\frac{1}{4}\frac{R_\oplus^2}{p^2}\frac{1}{8\eta^2}\{(4-6s^2)[2(5-3\eta^2-2\eta^3)+3(4+e^2)e\cos f \\ & +6e^2\cos 2f+e^3\cos 3f]+3s^2[e^3\cos(-f+2\omega)+6e^2\cos 2\omega \\ & +3(4+e^2)e\cos(f+2\omega)+4(2+3e^2)\cos(2f+2\omega)+(4+e^2) \\ & \times 3e\cos(3f+2\omega)+6e^2\cos(4f+2\omega)+e^3\cos(5f+2\omega)]\}, \end{aligned} \tag{5.89}$$

$$G_{0,1} = G\frac{R_\oplus^2}{p^2}\frac{s^2}{4}[3e\cos(f+2\omega)+3\cos(2f+2\omega)+e\cos(3f+2\omega)]. \tag{5.90}$$

The inverse corrections are given by Eqs. (5.86)–(5.90) too, but now the quantities in the equations are functions of the original Delaunay variables.

Periodic corrections to the usual orbital elements are sometimes preferred. Because these elements are particular functions of the Delaunay variables, their osculating values are obtained from the mean Delaunay variables using Deprit's recursion (2.15), as described in §2.1. Thus, $\Phi_{0,1} = \Phi_{1,0} + J_2\{\Phi_{0,0}; \mathcal{W}_1\}$ where $\Phi_{0,0}$ is replaced by each orbital element written as a function of the Delaunay variables, and $\Phi_{1,0} = 0$. From Eq. (4.46), it is readily found that

$$\{a;\mathcal{W}_1\} = 2\frac{L_{0,1}}{L}a, \quad \{e;\mathcal{W}_1\} = \frac{\eta^2}{e}\left(\frac{L_{0,1}}{L}-\frac{G_{0,1}}{G}\right), \quad \{I;\mathcal{W}_1\} = \frac{G_{0,1}}{G}\frac{c}{s},$$

whereas $\{\Omega;\mathcal{W}_1\} = h_{0,1}$, $\{\omega;\mathcal{W}_1\} = g_{0,1}$, $\{M;\mathcal{W}_1\} = \ell_{0,1}$.

5.7.2 Short-period corrections in non-singular variables

We note that eccentricity appears in the denominators of Eqs. (5.86) and (5.87), making the periodic corrections $\ell_{0,1}$ and $g_{0,1}$ singular in the case of circular orbits. However, the singularity is of the virtual type [284] and is easily avoided by a simple change of variables.

On the other hand, the fact that $\ell_{0,1}+g_{0,1}$ is free from singularities makes customary the use of the (non-canonical) non-singular variables for zero-eccentricity orbits given by L, h, H, the mean longitude $F = \ell + g$, and the semi-equinoctial variables defined in Eq. (5.74). The needed first-order corrections are

$$\begin{aligned}
F_{0,1} &= \frac{R_\oplus^2}{p^2}\frac{1}{32}\{24(4-5s^2)\phi + 6\beta[22-29s^2+4(4-5s^2)\eta \\
&\quad + (2-3s^2)\eta^2]\sin f + 12e\beta(2-3s^2)\sin 2f + 2e^2\beta(2-3s^2) \\
&\quad \times \sin 3f + 3e^2\beta s^2\sin(f-2\omega) - 18e\beta s^2\sin 2\omega - 3\beta[8-11s^2 \\
&\quad + 4(2-5s^2)\eta - 5s^2\eta^2]\sin(f+2\omega) - 12(2-5s^2)\sin(2f+2\omega) \\
&\quad - \beta[8-47s^2+4(2-5s^2)\eta - s^2\eta^2]\sin(3f+2\omega) \\
&\quad + 18s^2e\beta\sin(4f+2\omega) + 3s^2e^2\beta\sin(5f+2\omega)\}, \\
C_{0,1} &= -\frac{R_\oplus^2}{p^2}\frac{1}{32}\{24(4-5s^2)e\phi\sin\omega + 9e^2(4-5s^2)\cos(f-\omega) \\
&\quad - 2\beta[(20-21s^2)(1+\eta) + (8-12s^2)\eta^2]\cos\omega \\
&\quad - 6[8-10s^2+(10-11s^2)e^2]\cos(f+\omega) - 36c^2e\cos(2f+\omega) \\
&\quad - (8-7s^2)e^2\cos(3f+\omega) + 3(4-13s^2)e^2\cos(f+3\omega) \\
&\quad + 12(1-5s^2)e\cos(2f+3\omega) - 2[14s^2-(2-9s^2)e^2] \\
&\quad \times \cos(3f+3\omega) - 18s^2e\cos(4f+3\omega) - 3s^2e^2\cos(5f+3\omega)\}, \\
S_{0,1} &= \frac{R_\oplus^2}{p^2}\frac{1}{32}\{24(4-5s^2)e\phi\cos\omega + 3e^2(12-13s^2)\sin(f-\omega) \\
&\quad + 2\beta[(20-39s^2)(1+\eta) + 4(2-3s^2)\eta^2]\sin\omega + 6[8-14s^2 \\
&\quad + (6-9s^2)e^2]\sin(f+\omega) + 12(1-3s^2)e\sin(2f+\omega) - 5s^2e^2 \\
&\quad \times \sin(3f+\omega) - 3(4-13s^2)e^2\sin(f+3\omega) - 12(1-5s^2)e \\
&\quad \times \sin(2f+3\omega) + 2[14s^2-(2-9s^2)e^2]\sin(3f+3\omega) \\
&\quad + 18s^2e\sin(4f+3\omega) + 3s^2e^2\sin(5f+3\omega)\}.
\end{aligned}$$

Note that series expansions are mandatory if we want to write explicitly the right sides of these corrections in the non-singular variables F, C, and S.

Alternatively, non-singular first-order corrections to the polar canonical variables can be derived after rewriting the generating function (5.61) in polar variables [317]. Thus,

$$\mathcal{W}_1 = -\Theta\frac{1}{8}\frac{R_\oplus^2}{p^2}[(4-6s^2)\phi + 2\sigma(2-3s^2-s^2\cos 2\theta) + (3+4\kappa)s^2\sin 2\theta],$$

in which the projections of the eccentricity vector in the orbital frame κ and σ are now those in Eq. (5.42). We obtain $N_{0,1} = 0$ and

$$
\begin{aligned}
r_{0,1} &= -\frac{r}{4}\frac{R_\oplus^2}{p^2}(1+\kappa)\left[(2-3s^2)\left(1+\frac{2\eta}{1+\kappa}+\frac{\kappa}{1+\eta}\right)-s^2\cos 2\theta\right],\\
\theta_{0,1} &= \frac{1}{8}\frac{R_\oplus^2}{p^2}\left\{6(4-5s^2)\phi-[6-7s^2+2(4-6s^2)\kappa]\sin 2\theta\right.\\
&\quad\left.+\left[20-24s^2+(4-6s^2)\frac{2+\kappa}{1+\eta}+(4-8s^2)\cos 2\theta\right]\sigma\right\},\\
\nu_{0,1} &= -\frac{1}{4}\frac{R_\oplus^2}{p^2}c[6\phi+2\sigma(3+\cos 2\theta)-(3+4\kappa)\sin 2\theta],\\
R_{0,1} &= \frac{1}{4}\frac{R_\oplus^2}{p^2}\frac{\Theta}{p}\left\{(2-3s^2)\left[\eta+\frac{(1+\kappa)^2}{1+\eta}\right]\sigma-2(1+\kappa)^2s^2\sin 2\theta\right\},\\
\Theta_{0,1} &= \frac{1}{4}\frac{R_\oplus^2}{p^2}\Theta s^2[(3+4\kappa)\cos 2\theta+2\sigma\sin 2\theta].
\end{aligned}
$$

6 Zonal perturbations

The main problem reveals the most relevant features of the artificial satellite problem, and shows the need of taking second-order effects of J_2 into account to get a correct description of qualitative aspects of the long-term dynamics. However, other second-order effects of the geopotential have analogous importance to the terms factored by J_2^2, and, therefore, cannot be neglected in the description of the dynamics of low-Earth orbits. This is the case of the zonal harmonics. More specifically, even zonal harmonics drive secular effects whereas odd zonal harmonics introduce long-period oscillations [130, 341]. This fact was well known, and the triplet papers by Garfinkel, Kozai, and Brouwer in the acclaimed November 1959 issue of the Astronomical Journal provided independent solutions that took the effects of the first few zonal harmonics into account [75, 217, 360].

The effects of zonal harmonics of higher degree than the second are clearly observed in the evolution of low-Earth orbits, and their inclusion in the propagation model extends the time validity of orbit predictions. But beyond these quantitative refinements, they also produce qualitative changes with respect to the main problem dynamics. In particular, they modify the frozen orbits' geometry with respect to the simpler J_2 case, and, therefore, are mandatorily used in the preliminary design of frozen orbits [119, 137, 400]. On the other hand, the effects of the coupling of J_2 with higher-degree zonal harmonics is very small and is customarily neglected in a perturbation approach. In that case, most of the expanded expressions that are typical of perturbation solutions based on the brute-force approach can be avoided. Indeed, up to the order of J_2^2, one can take full benefit in the construction of the perturbation solution from different recursions in the literature [229, 431, 589], and, in particular, from Kaula's seminal recursions [341, 351, 417, 691]. The latter approach is discussed in the reconstruction of Brouwer's completely reduced Hamiltonian from the point of view of the Lie transform approach.

The use of Hamiltonian simplification procedures is also discussed in the computation of second-order periodic corrections of Brouwer's theory, yet these corrections are dispensable in most cases—assumed, of course, that the constants of Brouwer's theory can be properly initialized with a different method [63]. It is shown in this chapter too how the difficulties in the computation of higher orders of the analytical solution in closed form, which are inherent to Brouwer's approach, are avoided when the solution is computed by reverse normalization.

6.1 Zonal problem in mean elements

The disturbing zonal potential is obtained by making $m = 0$ in Eq. (5.2). That is,

$$\mathcal{P} = \frac{\mu}{r} \sum_{n \geq 2} \frac{R_\oplus^n}{r^n} J_n P_{n,0}(\sin\varphi), \tag{6.1}$$

https://doi.org/10.1515/9783110668513-006

where $J_n = -C_{n,0}$. Like in the main problem, the zonal potential does not depend on the geocentric longitude, thus accepting the third component of the angular momentum vector as an integral. Therefore, the zonal problem Hamiltonian is obtained by replacing Eq. (6.1) into Eq. (5.3), from which the Coriolis term is removed.

The disturbing potential (6.1) is written in orbital elements following Kaula's approach [339, 341]. That is,

$$\mathcal{P} = \frac{\mu}{a}\left(\frac{a^2}{r^2}\eta\right)\sum_{i\geq 2} J_i V_i, \tag{6.2}$$

in which

$$V_i = \frac{R_\oplus^i}{p^i}\eta \sum_{j=0}^{i} \mathcal{F}_{i,j}(s) \sum_{k=0}^{i-1} \binom{i-1}{k} e^k \cos^k f \cos[(i-2j)(f+\omega) - i_\pi] \tag{6.3}$$

and

$$\mathcal{F}_{i,j} = \sum_{l=0}^{\min(j,i_0)} \frac{(-1)^{j-l-i_0}}{2^{2i-2l}} \binom{2i-2l}{i}\binom{i}{l}\binom{i-2l}{j-l} s^{i-2l}, \quad i \geq 2l, \quad j \geq l, \tag{6.4}$$

are particularizations for the zonal problem of Kaula inclination functions, whose efficient recursive evaluation has motivated a wealth of research as well as some controversy [232, 251]. The symbol i_0 in Eq. (6.4) and the parity correction i_π in Eq. (6.3) adhere to the index notation convention in [400, 417]. Thus,

$$i^\star = i \bmod 2, \quad i_\pi = \frac{1}{2}\pi i^\star, \quad i_m = \left\lfloor \frac{1}{2}(i-m) \right\rfloor, \tag{6.5}$$

where i, m, are integer numbers and $\lfloor x \rfloor$ is the greatest integer $\leq x$.

Different from the analogous equations in [341, 574], we left the factor $(a/r)^2\eta$ out of the summation in Eq. (6.2) in preparation of a following closed-form integration based on the differential relation between the true and mean anomalies in Eq. (4.64).

Next, the Hamiltonian of the zonal problem is written in the usual form of a perturbation Hamiltonian (2.30). That is,

$$\mathcal{H} = \sum_{m\geq 0} \frac{\epsilon^m}{m!}\mathcal{H}_{m,0}, \tag{6.6}$$

in which the small parameter is $\epsilon = J_2$,

$$\mathcal{H}_{0,0} = -\frac{\mu}{2a}, \tag{6.7}$$

$$\mathcal{H}_{1,0} = \frac{\mu}{a}\left(\frac{a^2}{r^2}\eta\right)V_2, \tag{6.8}$$

$$\mathcal{H}_{2,0} = 2\frac{\mu}{a}\left(\frac{a^2}{r^2}\eta\right)\sum_{i\geq 3}\tilde{J}_i V_i, \tag{6.9}$$

and $\mathcal{H}_{m,0} = 0$ for $m \geq 3$. The terms V_i are given in Eq. (6.3) and $\tilde{J}_i = J_i/J_2^2$ are numeric coefficients of order one. As usual, orbital elements and related quantities are functions of the Delaunay canonical variables.

Like we did in §5.6 with the main problem, the relevant effects of the dynamics of the zonal problem are obtained by reducing the zonal Hamiltonian to a one-degree-of-freedom Hamiltonian in which the short-period effects have been removed up to some truncation order. Because Eq. (6.6) remains as a perturbed Keplerian problem, the Lie derivative and the homological equation are Eq. (4.66) and Eq. (4.68), respectively. In addition, the term $\mathcal{H}_{1,0}$ is the same as Eq. (5.36) for the main problem. Therefore, $\mathcal{H}_{0,1}$ is given by Eq. (5.60) and $\mathcal{W}_1$ is given by Eq. (5.61).

At second order, the known terms of the homological equation are computed from Eq. (2.37), and differ from those in Eq. (5.65) in the term $\mathcal{H}_{2,0}$ given by Eq. (6.9). Therefore, terms of Eq. (6.9) pertaining to the kernel of the Lie derivative (4.66) must be added to those previously computed in Eq. (5.70). These terms are computed in closed form of the eccentricity using the differential relation (4.64). Namely,

$$\langle \mathcal{H}_{2,0} \rangle_\ell = \frac{1}{2\pi} \int_0^{2\pi} \mathcal{H}_{2,0} \frac{r^2}{a^2 \eta} \, \mathrm{d}f = 2\frac{\mu}{a} \sum_{i \geq 3} \tilde{J}_i \langle V_i \rangle_f. \tag{6.10}$$

Averaging V_i over the true anomaly is most easily achieved when Eq. (6.3) is expanded as a Fourier series in f. Following [417], we first expand

$$\cos^k f \cos(\iota f + \alpha) = \cos^k f \cos \iota f \cos \alpha - \cos^k f \sin \iota f \sin \alpha, \tag{6.11}$$

where we abbreviate $\iota \equiv i - 2j$, $\alpha = n\omega - i_\pi$. Then we iterate

$$\cos^k f \cos \iota f = \frac{1}{2}[\cos(\iota + 1)f + \cos(\iota - 1)f] \cos^{k-1} f,$$
$$\cos^k f \sin \iota f = \frac{1}{2}[\sin(\iota + 1)f + \sin(\iota - 1)f] \cos^{k-1} f,$$

to obtain

$$\cos^k f \cos(i - 2j)f = \frac{1}{2^k} \sum_{l=0}^{k} \binom{k}{k-l} \cos(i - 2j - k + 2l)f, \tag{6.12}$$

$$\cos^k f \sin(i - 2j)f = \frac{1}{2^k} \sum_{l=0}^{k} \binom{k}{k-l} \sin(i - 2j - k + 2l)f, \tag{6.13}$$

from which we note that terms in Eq. (6.12) such that $i - 2j - k + 2l = 0$ are the only ones of Eq. (6.3) that are free from the true anomaly. They only occur when $k - i$, and hence $k + i$, are even.

Then, replacing $l = \frac{1}{2}(k - i) + j$ in Eq. (6.12), which in turn is plugged into Eq. (6.3), we obtain

$$\langle V_i \rangle_f = \frac{R_\oplus^i}{p^i} \eta \sum_{j=1}^{i-1} \mathcal{F}_{i,j}(s) \mathcal{G}_{i,j}^*(e) \cos[(i - 2j)\omega - i_\pi], \tag{6.14}$$

in which

$$\mathcal{G}^*_{i,j} = \sum_{k=0}^{i-1} \binom{i-1}{k} \binom{k}{\frac{k+i}{2}-j} \frac{e^k}{2^k} \tag{6.15}$$

are particularizations of Kaula eccentricity functions for the zonal problem multiplied by η^{2i-1}, which in turn are specific cases of Hansen's coefficients [231, 234, 263, 557, 562, 631]. Therefore, Eq. (6.15) is written in Kaula's efficient form,

$$\mathcal{G}^*_{i,j} = \sum_{l=0}^{\tilde{j}-1} \binom{i-1}{q} \binom{q}{l} \frac{e^q}{2^q}, \quad q = 2l + i - 2\tilde{j}, \quad \begin{cases} i \geq 2j \Rightarrow \tilde{j} = j, \\ i < 2j \Rightarrow \tilde{j} = i - j. \end{cases} \tag{6.16}$$

In summary, up to the second order of J_2, the mean-element Hamiltonian, with short-period effects removed, is

$$\mathcal{H} = \mathcal{H}_{0,0} + \epsilon\mathcal{H}_{0,1} + \frac{1}{2}\epsilon^2\mathcal{H}_{0,2}, \tag{6.17}$$

in which $\mathcal{H}_{0,0}$ is the Keplerian (6.7), $\mathcal{H}_{0,1}$ is the same of the main problem in Eq. (5.60), $\mathcal{H}_{0,2}$ is obtained adding Eqs. (5.70) and (6.10), and all the quantities in these expressions are functions of the prime Delaunay variables.

The Hamiltonian (6.17) is partially normalized because, up to the truncation order, it is cyclic in the mean anomaly ℓ' and, in consequence, the Delaunay action L' is a formal integral of the zonal problem in mean elements. Remarkably, the use of Kaula's recursions allows us to replace the expanded Hamiltonian $\mathcal{H} = \mathcal{Q}_{1,0} + \frac{1}{2}\epsilon\mathcal{Q}_{2,0}$ in Appendix A of [119]—which was computed by brute force and fills two quarto-size pages—by the compact expression, cf. [400, 417],

$$\begin{aligned}\mathcal{H} = J_2\frac{\mu}{a}\sum_{i\geq 2}\tilde{J}_i\langle V_i\rangle_f - J_2\frac{\mu}{a}\frac{R_\oplus^4}{p^4}\frac{3\eta}{128}[5(8 - 16s^2 + 7s^4) + 4(2 - 3s^2)^2\eta \\ - (8 - 8s^2 - 5s^4)\eta^2 - 2(14 - 15s^2)s^2e^2\cos 2\omega],\end{aligned} \tag{6.18}$$

where $\mu/a = (G'\eta/p)^2$, $\langle V_i\rangle_f$ is given in Eq. (6.14), and the summation starts now from $i = 2$. Moreover, the zonal Hamiltonian in terms of mean elements, either given by Eq. (6.17) or Eq. (6.18), is easily computed for any number of zonal harmonics using Kaula's recursions, without constraint to the J_2–J_9 model computed in [119].

6.1.1 Averaged flow

Equation (6.18) is a one-degree-of-freedom integrable Hamiltonian. Since $H = H_0$ and $L' = L'_0$ due to the symmetries of the zonal problem and the averaging carried out, respectively, the motions of h' and ℓ' decouple from the reduced (g', G') system. That is, $\mathrm{d}G'/\mathrm{d}t = -\partial\mathcal{H}/\partial g'$, $\mathrm{d}g'/\mathrm{d}t = \partial\mathcal{H}/\partial G'$, from which

$$\frac{\mathrm{d}G'}{\mathrm{d}t} = G' J_2 n \frac{R_\oplus^4}{p^4}\frac{3}{32}(14-15s^2)s^2e^2\sin 2\omega - L'n\sum_{i\geq 3}\frac{J_i}{J_2}\frac{\partial\langle V_i\rangle_f}{\partial\omega}, \tag{6.19}$$

$$\begin{aligned}\frac{\mathrm{d}g'}{\mathrm{d}t} &= n\frac{R_\oplus^2}{p^2}\frac{3}{4}(4-5s^2) + n\sum_{i\geq 3}\frac{J_i}{J_2}\left(\frac{1-2i}{\eta}\langle V_i\rangle_f - \frac{\eta}{e}\frac{\partial\langle V_i\rangle_f}{\partial e} + \frac{c^2}{s\eta}\frac{\partial\langle V_i\rangle_f}{\partial s}\right)\\ &\quad + n\frac{R_\oplus^4}{p^4}\frac{3J_2}{128}[\gamma_{0,0}+\gamma_{0,1}\eta+\gamma_{0,2}\eta^2+(\gamma_{1,0}+\gamma_{1,2}\eta^2)\cos 2\omega],\end{aligned} \tag{6.20}$$

where the inclination polynomials $\gamma_{i,j}$ were previously given in Table 5.4, and

$$\frac{\partial\langle V_i\rangle_f}{\partial\omega} = -\frac{R_\oplus^i}{p^i}\eta\sum_{j=1}^{i-1}(i-2j)\mathcal{F}_{i,j}\mathcal{G}_{i,j}^*\sin[(i-2j)\omega - i_\pi], \tag{6.21}$$

$$\frac{\partial\langle V_i\rangle_f}{\partial e} = \frac{R_\oplus^i}{p^i}\eta\sum_{j=1}^{i-1}\mathcal{F}_{i,j}\frac{\partial\mathcal{G}_{i,j}^*}{\partial e}\cos[(i-2j)\omega - i_\pi], \tag{6.22}$$

$$\frac{\partial\langle V_i\rangle_f}{\partial s} = \frac{R_\oplus^i}{p^i}\eta\sum_{j=1}^{i-1}\frac{\partial\mathcal{F}_{i,j}}{\partial s}\mathcal{G}_{i,j}^*\cos[(i-2j)\omega - i_\pi]. \tag{6.23}$$

Available recursion formulas expedite the computation of the partial derivatives of Kaula's inclination [232, 251] and eccentricity functions [231, 661]. For the lower eccentricities, simplifications in [128, 574, 604] can replace Eqs. (6.19) and (6.20).

Comparison of Eqs. (6.19) and (6.20) with Eqs. (5.83) and (5.85) shows the differences with respect to the main problem dynamics introduced by the Earth's zonal harmonics of higher degree. These additional terms permit one to foresee qualitative changes of the reduced, long-term dynamics. Indeed, different from the main problem alone, the condition $\mathrm{d}G'/\mathrm{d}t = 0$ is no longer achieved, in general, when $\omega = 0, \pi$. On the contrary, the variation of G' vanishes, on average, when $\omega = \pm\frac{1}{2}\pi$, as checked from Eq. (6.21), a case in which $\sin[(i-2j)\omega - i_\pi]$ turns into $\sin 2k\omega$ with $k \in \mathbb{N}$ for i even, and into $\cos(2k+1)\omega$ in the odd case. For the argument of the perigee to remain frozen Eq. (6.20) must vanish too. That is,

$$\begin{aligned}0 &= \frac{3}{4}(4-5s^2) + \frac{R_\oplus^2}{p^2}\frac{3J_2}{128}[\gamma_{0,0}-\gamma_{1,0}+\gamma_{0,1}\eta+(\gamma_{0,2}-\gamma_{1,2})\eta^2]\\ &\quad + \sum_{i\geq 3}\frac{J_i}{J_2}\frac{R_\oplus^{i-2}}{p^{i-2}}\sum_{j=0}^{i}\left[(1-2i)\mathcal{F}_{i,j}\mathcal{G}_{i,j}^* - \mathcal{F}_{i,j}\frac{\eta^2}{e}\frac{\partial\mathcal{G}_{i,j}^*}{\partial e} + \frac{c^2}{s}\frac{\partial\mathcal{F}_{i,j}}{\partial s}\mathcal{G}_{i,j}^*\right]\\ &\quad \times\cos\frac{\pi}{2}(i \mp i^\star - 2j),\end{aligned} \tag{6.24}$$

where the cosine term evaluates to ± 1 depending on the parity of the integer number $i \mp i^\star - 2j$, in which the minus sign corresponds to the case $\omega = +\frac{\pi}{2}$ and the plus sign to the opposite.

Therefore, the eccentricity of a frozen orbit of the zonal problem with $g' = \pm\pi/2$ is computed from the condition equation (6.24) using root-finding procedures to obtain η. Previously, we must replace $s = (1-c^2)^{1/2}$ and $c = \gamma/\eta$, where the dynamical parameter $\gamma = H'/L'$ is the same as previously given in Eq. (5.73). Other equilibria of the reduced dynamics that may exist with different arguments of the frozen perigee are not discussed here.

From Eq. (6.15) we check that $\partial\mathcal{G}^*_{i,j}/\partial e$ in Eq. (6.22) does not introduce divisions by e, and from Eq. (6.4) that $\partial\mathcal{F}_{i,j}/\partial s$ in Eq. (6.23) does not introduce divisions by s. However, these undesired divisors do appear in Eq. (6.20) for the variation of g'. Singularities related to circular orbits can be avoided using the nonsingular, semi-equinoctial variables defined in Eq. (5.74). Thus,

$$\begin{aligned}\frac{\mathrm{d}C}{\mathrm{d}t} &= -\frac{\eta}{L'}\left(\frac{1}{e}\frac{\mathrm{d}G'}{\mathrm{d}t}\right)\cos\omega - \left(e\frac{\mathrm{d}g'}{\mathrm{d}t}\right)\sin\omega,\\ \frac{\mathrm{d}S}{\mathrm{d}t} &= -\frac{\eta}{L'}\left(\frac{1}{e}\frac{\mathrm{d}G'}{\mathrm{d}t}\right)\sin\omega + \left(e\frac{\mathrm{d}g'}{\mathrm{d}t}\right)\cos\omega,\end{aligned}$$

which, in view of $\mathrm{d}G'/\mathrm{d}t = \mathcal{O}(e)$, are free from the eccentricity in denominators. Moreover, when $\omega = \pm\pi/2$

$$\frac{\mathrm{d}C}{\mathrm{d}t} = \mp\left(e\frac{\mathrm{d}g}{\mathrm{d}t}\right)\bigg|_{\omega=\pm\frac{\pi}{2}}, \qquad \frac{\mathrm{d}S}{\mathrm{d}t} = \mp\frac{\eta}{L}\left(\frac{1}{e}\frac{\mathrm{d}G}{\mathrm{d}t}\right)\bigg|_{\omega=\pm\frac{\pi}{2}} \equiv 0.$$

Therefore, the constraint equation for the frozen orbit geometry is obtained multiplying the right side of Eq. (6.24) by the eccentricity, in this way removing the undesired division by the eccentricity.

On the other hand, at difference from the main problem, the odd zonal harmonics introduce singularities for equatorial orbits. If required, both kinds of singularities can be removed using Poincaré nonsingular variables [468],

$$\begin{aligned} x_1 &= L, & y_1 &= \ell + g + h,\\ x_2 &= \sqrt{2}\,\sqrt{L-G}\cos(g+h), & y_2 &= -\sqrt{2}\,\sqrt{L-G}\sin(g+h),\\ x_3 &= \sqrt{2}\,\sqrt{G-H}\cos h, & y_3 &= -\sqrt{2}\,\sqrt{G-H}\sin h. \end{aligned} \tag{6.25}$$

6.1.2 Inclination–eccentricity diagrams of frozen orbits

Rather than looking at the frozen orbits constraint like a polynomial equation in G' for given values L', H' and $g' = \pm\frac{\pi}{2}$, we can view Eq. (6.24) like the curve $e = e(I)$ given by the implicit equation

$$\dot{\omega}\left(e, I; \omega = \pm\frac{1}{2}\pi, a\right) = 0. \tag{6.26}$$

Then, for given mean semimajor axis a and frozen argument of the perigee $\omega = \pm\frac{\pi}{2}$, the eccentricity of the frozen orbit is determined as a function of the inclination. Therefore, inclination–eccentricity diagrams of frozen orbits with $\omega = \pm\frac{\pi}{2}$ are depicted like contour plots of Eq. (6.26) for given values of a. These kinds of diagrams help in locating frozen orbits and disclose the existence of different families of frozen orbits [119]. On the other hand, they provide the analytic analog to the families of periodic orbits that are customarily computed in osculating variables using numerical continuation methods, either in the orbital plane [385, 409, 423] or under the more strict condition of repeating their ground traces [387, 390, 428, 586]. Inclination–eccentricity diagrams provide also an expedite way of exploring the sensitivity of the dynamics with respect to different truncations of the gravitational potential [400].

An example is shown in Fig. 6.1 for the J_2–J_9 truncation of the geopotential. Curves of retrograde-inclination orbits are reflections of the direct-inclination case and are not presented. The semimajor axes have been chosen with the same value as the Earth's equatorial radius to enhance the effect of the perturbations for illustration purposes, although most orbits would impact the surface of the Earth in that case. Numerical values of the zonal harmonic coefficients were taken from the GRACE gravity model [630].

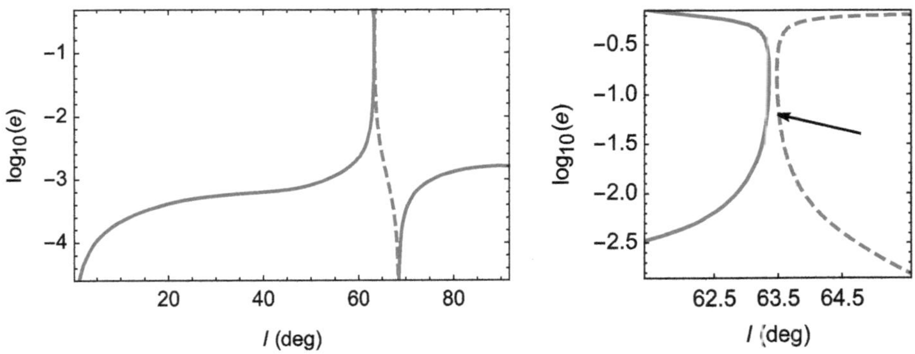

Figure 6.1: Inclination–eccentricity diagram of frozen orbits with $a = R_\oplus$ (J_2–J_9 truncation). Full lines refer to $\omega = 90^\circ$ and dashed lines to $\omega = 270^\circ$. Right: magnification in the vicinity of the critical inclination [400].

The inclination–eccentricity diagram of frozen orbits in Fig. 6.1 illustrates the existence of low-eccentricity Earth frozen orbits in all the range of inclinations save for a narrow region about the critical inclination. Frozen orbits still exist in this small area, but their eccentricities may grow high. For a given semimajor axis (or L') the frozen orbits are grouped in different families parameterized by the third component of the angular momentum vector H. The first family starts with a very low-eccentricity orbit in the equatorial plane ($H = G'$) with the perigee frozen at $\omega = 90^\circ$. Decreasing values of H have the effect of increasing the inclination of the frozen orbits, which undergo small variations of the eccentricity. Eventually, the eccentricity grows high with almost fixed inclination, which happens close to the critical inclination value. We will see later that the orbits of this family are stable. At a certain value $H/L' \approx \cos 63.5^\circ$, which is marked with an arrow in the right plot of Fig. 6.1, two new families of frozen orbits appear with the perigee frozen at $\omega = 270^\circ$. They exist for decreasing values of H, whose effect is to increase the eccentricity of the frozen orbits of one of the bifurcated families with almost fixed inclination, and to increase gradually the inclination of the frozen orbits of the other bifurcated family with decreasing values of the eccentricity. Close to $I \approx 68^\circ$, the orbits of the latter family have very low eccentricities, and, eventually, the frozen perigee turns to $\omega = 90^\circ$. For decreasing values of H the inclination of the frozen orbits continues to increase whereas their eccentricities grow slightly with the perigee frozen at $\omega = 90^\circ$. We will see later that the bifurcated family of low-eccentricity frozen orbits is composed of stable orbits, whereas high-eccentricity frozen orbits with $\omega = 270^\circ$ are unstable. We remark that we intentionally avoided to talk about *circular* orbits because the perigee is not defined in this case and, in consequence, they are excluded from the condition equation used in the construction of the inclination–eccentricity diagrams.

On the other hand, the characteristics of the curves of frozen orbits vary slightly with the degree of the geopotential truncation. In particular, the inclination where the frozen perigee flips by 180° may change by several degrees. This fact is illustrated in Fig. 6.2, where inclination–eccentricity curves corresponding to different truncations of the zonal problem are superimposed in the same diagram. It is also shown in the figure that high-inclination frozen orbits have similar low eccentricities irrespective of the degree of the model, which justifies the use of the simpler J_2–J_3 truncation in the preliminary design of typical missions requiring high-inclination frozen orbits [137, 258]. Therefore, the inexpensive construction of inclination–eccentricity diagrams for different truncations of the zonal potential has been suggested as an alternative criterion to assess the sensitivity of the propagation model for a particular class of orbits [400].

Highly-eccentric non-impact frozen orbits make sense only for large semimajor axis. They concentrate in the close proximity of the critical inclination for any truncation of the gravitational model, because second-order effects are further mitigated by the magnitude of the coefficients $(R_\oplus/a)^m$, which take small values when $m > 2$ and $a \gg R_\oplus$.

6.1.3 Local dynamics: eccentricity-vector diagrams

Relevant differences with respect to the reduced dynamics of the main problem discussed in §5.6.3 are also evident from eccentricity-vector diagrams. They are rendered like contour plots of the mean-element Hamiltonian (6.18), which is readily reformu-

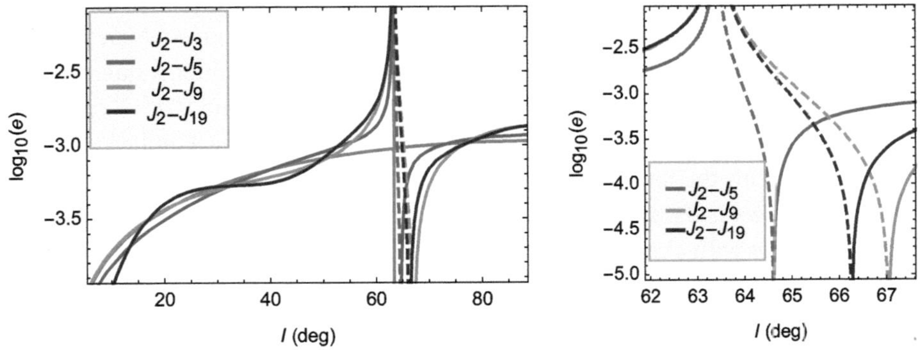

Figure 6.2: Inclination–eccentricity diagram of non-impact frozen orbits ($a = 1.1R_\oplus$) with $\omega = 90^\circ$ (full lines) and $\omega = 270^\circ$ (dashed); cf. [400].

lated in the semi-equinoctial C and S variables using the relations

$$e^j \cos j\omega = \frac{1}{2}[(C - \mathbf{i}S)^j + (C + \mathbf{i}S)^j], \quad e^j \sin j\omega = \frac{1}{2}\mathbf{i}[(C - \mathbf{i}S)^j - (C + \mathbf{i}S)^j],$$

with $\mathbf{i} = \sqrt{-1}$ and j integer.

Thus, as shown in the first row of Fig. 6.3, the J_2–J_3 truncation of the zonal potential breaks the equatorial symmetry of the main problem. Still, the initial bifurcation happens again close to the critical inclination with the change to instability of the circular orbits, and the bifurcated stable eccentric frozen orbits have the perigee at 0 and π. However, the second bifurcation is of a different nature and now happens from a low-eccentricity, unstable orbit with the perigee frozen at 90°. This fact yields subtle differences with respect to the J_2 case in Fig. 5.4, and, contrary to it, we can find long excursions of the eccentricity vector without completing circulatory behavior; while the orbits may reach moderate eccentricities, they eventually approach closely to the circular case. Inclusion of J_4 changes radically the bifurcation geometry, as illustrated in the sequence of plots in the second row of Fig. 6.3. In this case, all the bifurcations happen from orbits with the perigee frozen at 270°. Nevertheless, the perigees of the unstable eccentric orbits that stem from the low-eccentricity unstable orbit at the second bifurcation of the J_2–J_4 case migrate fast from the argument 270° towards 0° and 180°, in the end yielding a picture similar to the main problem case rotated by 90°.

When more zonal harmonics are taken into account, the second bifurcation from a low-eccentricity orbit no longer exists, as shown in the third and fourth rows of Fig. 6.3 for the J_2–J_5 and J_2–J_7 truncations, respectively. Adding more zonal harmonics only produces quantitative variations with respect to the phase portraits in the last row of Fig. 6.3. The number of families of frozen orbits as well as their stability is thus in agreement with the behavior anticipated in §6.1.2.

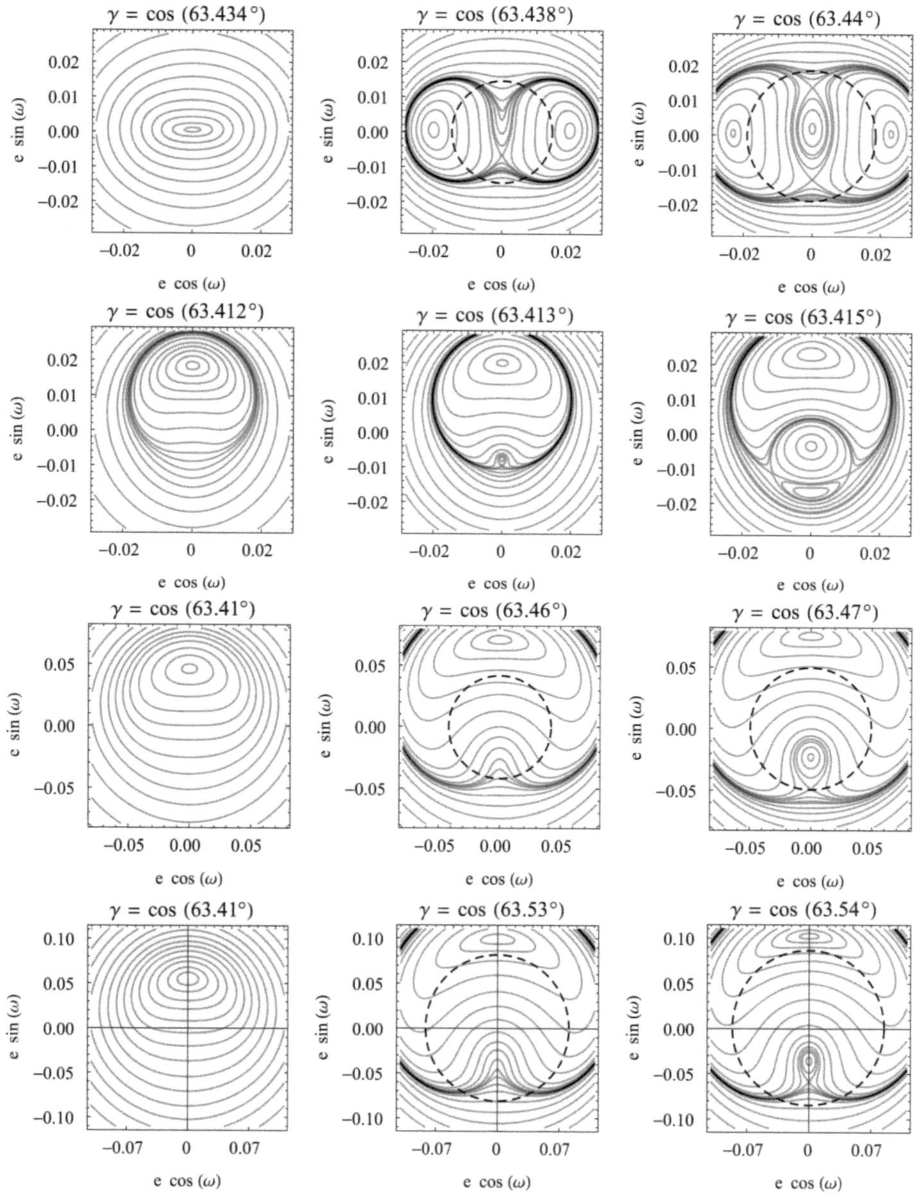

Figure 6.3: Eccentricity-vector plots of low-Earth orbits ($a = 1.1R_\oplus$) for, from top to bottom, the J_2–J_3, J_2–J_4, J_2–J_5, and J_2–J_7 truncations. Note the different scales. The dashed circle $e = (1 - 5\gamma^2)^{1/2}$ marks the critical inclination.

6.2 Hamiltonian simplification

The computation of second-order short-period corrections requires the explicit computation of the second-order term of the generating function from the second-order homological equation. The fact that the equation of the center does not accept a closed-form primitive within the algebra of trigonometric functions [324] generated some controversy around the closed-form solution of $\mathcal{W}_2$. Indeed, the indefinite integration of the equation of the center in closed form relies on the algebraic decomposition of functions of the elliptic motion and the use of the dilogarithmic function [536, 537], a procedure that complicates the perturbation approach. However, the difficulty was only apparent, and it was derived from the particular programming strategies used by researchers involved in the automatization of celestial mechanics computations [171]. In fact, the difficulty had been easily sidestepped by researchers relying on traditional hand computations [8, 363].

Indeed, the known second-order terms involving the equation of the center are those in Eq. (5.66). If it is arranged in the form of a Fourier series,

$$\widetilde{\mathcal{H}}_{0,2,1} = -\frac{\mu}{2a}\frac{R_\oplus^4}{p^4}\frac{9}{8}(5s^2-4)s^2\phi\sum_{j=-1}^{5} q_{|j-2|}e^{|j-2|}\sin(jf+2g),$$

where $q_0 = 3e^2+2$, $q_1 = \frac{3}{4}(e^2+4)$, $q_2 = \frac{3}{2}$, and $q_3 = \frac{1}{4}$, then ϕ shows alone for $j = 0$, thus rendering apparent the difficulty. However, the alternative arrangement

$$\widetilde{\mathcal{H}}_{0,2,1} = -\frac{\mu}{2a}\frac{R_\oplus^4}{p^4}\frac{9}{8}(5s^2-4)s^2\frac{p^2}{r^2}\frac{\phi}{\eta^2}\sum_{j=1}^{3}(2-j^\star)e^{j^\star}\sin(jf+2g),$$

where $j^\star \equiv j \bmod 2$ from the index notation convention in Eq. (6.5), shows that, in fact, we only need to deal with the indefinite integration of functions of the mean anomaly of the form $(p/r)^2\phi\sin(jf+\alpha)$, which are readily integrated by parts [407].

That the terms containing the equation of the center arise in the generating function from the elimination of factors $1/r^2$, becomes clear from the identity $\mathcal{L}_0(\phi/G) + \eta^3/p^2 = 1/r^2$ [155, 156, 271]. That is the reason why the generating function (5.40) of Deprit's radial intermediary in §5.4.2 is free from the equation of the center. In the process of reducing the main problem Hamiltonian to a quasi-Keplerian system the term $1/r^2$ was not removed, thus keeping the essential short-period effects related to the equation of the center in the transformed Hamiltonian. Comparison of the first-order generating functions in Eqs. (5.61) and (5.40) makes this fact evident. Then there naturally emerges the question of whether extending the computations in §5.4.2 to second order may be beneficial in the elimination of short-period terms from the zonal Hamiltonian.

6.2.1 Deprit's elimination of the parallax

In the attempt to transform the zonal Hamiltonian into a quasi-Keplerian system, it is essential that parallactic factors $R_\oplus^m/r^m$ in Eq. (6.1) with $m > 2$ be reduced to the power $m = 2$. After removing other short-period terms, the desirable result would be to join the disturbing effect on the Keplerian motion of the remaining terms in $R_\oplus^2/r^2$ with the total angular momentum in order to obtain a varied angular momentum—as it was already done in Eqs. (5.32) and (5.33) of Deprit's radial intermediary. While this is not feasible, we will see that the procedure furnishes the perturbation approach with notable simplifications.

The formal reduction of parallactic terms has already been achieved with the formulation in Eq. (6.2). Indeed, the short-period terms involved in the parallactic factors of the first-order Hamiltonian (6.8) have already been incorporated to V_2, which, from Eq. (6.3), reads

$$V_2 = (R_\oplus/p)^2\eta[\mathcal{F}_{2,1} + (\mathcal{F}_{2,0} + \mathcal{F}_{2,2})\cos(2f + 2\omega)](1 + e\cos f),$$

where, from Eq. (6.4), $\mathcal{F}_{2,0} = \mathcal{F}_{2,2} = -\frac{3}{8}s^2$, $\mathcal{F}_{2,1} = -\frac{1}{2} + \frac{3}{4}s^2$. After carrying out the pertinent trigonometric reductions, we obtain

$$V_2 = -\frac{1}{8}\frac{R_\oplus^2}{p^2}\eta\{(4 - 6s^2)(1 + e\cos f) + 3s^2 \times [e\cos(f + 2\omega) + 2\cos(2f + 2\omega) + e\cos(3f + 2\omega)]\}. \tag{6.27}$$

Therefore, Eq. (6.8) turns into

$$\mathcal{H}_{1,0} = -\frac{1}{8}\frac{\mu}{p}\frac{R_\oplus^2}{r^2}\{(4 - 6s^2)(1 + e\cos f) + 3s^2[e\cos(f + 2\omega) + 2\cos(2f + 2\omega) + e\cos(3f + 2\omega)]\},$$

from which the new Hamiltonian term is now chosen in such a way that the only short-period effects that remain are those related to the occurrence of $1/r^2$. That is to say, the explicit appearance of the true anomaly is removed by choosing

$$\mathcal{H}_{0,1} = \frac{\mu}{a}\left(\frac{a^2}{r^2}\eta\right)\langle V_2\rangle_f, \tag{6.28}$$

where, from Eq. (6.14), $\langle V_2\rangle_f = (R_\oplus/p)^2\eta[\mathcal{F}_{2,0}\mathcal{G}_{2,0}^*\cos 2\omega + \mathcal{F}_{2,1}\mathcal{G}_{2,1}^*]$, in which, from Eq. (6.16), $\mathcal{G}_{2,0}^* = 0$ and $\mathcal{G}_{2,1}^* = 1$. Hence,

$$\langle V_2\rangle_f = \frac{R_\oplus^2}{p^2}\eta\left(\frac{3}{4}s^2 - \frac{1}{2}\right) \tag{6.29}$$

and

$$\mathcal{H}_{0,1} = \frac{\mu}{p}\frac{R_\oplus^2}{r^2}\left(\frac{3}{4}s^2 - \frac{1}{2}\right), \tag{6.30}$$

which is precisely the first-order term of Deprit's radial intermediary in Eq. (5.37). In consequence, up to an arbitrary integration "constant" that does not depend on ℓ, the first-order term of the generating function is obviously given by Eq. (5.40), which is rewritten in the more compact form

$$\mathcal{W}_1 = \frac{G}{8}\frac{R_\oplus^2}{p^2}\left[(6s^2-4)e\sin f - s^2\sum_{j=1}^{3}3^{\lfloor(4-j)/2\rfloor}e^{|j-2|}\sin(jf+2\omega)\right]. \tag{6.31}$$

At second order, the known terms on the right side of the homological equation are computed from Eq. (2.37). After evaluation of the involved Poisson brackets we obtain

$$\widetilde{\mathcal{H}}_{0,2} = \frac{\mu}{a}\left(\frac{a}{r}\right)^2\eta\left[\frac{R_\oplus^4}{p^4}\eta\sum_{j=0}^{6}\sum_{l=j_1}^{2}q_{j,l}s^{|2l|}\cos(jf+2l\omega)+2\sum_{i\geq 3}\tilde{J}_iV_i\right], \tag{6.32}$$

of which the needed coefficients $q_{j,l} \equiv q_{j,l}(s,e)$ are presented in Table 6.1; cf. [417]. Recall that $j_1 = \lfloor\frac{1}{2}(j-1)\rfloor$ according to the index notation convention in Eq. (6.5).

Table 6.1: Coefficients $q_{j,l} = q_{j,-l}$ in Eq. (6.32).

j	$l = 0$	$l = 1$	$l = 2$
0	$q_{2,0} - \frac{1}{16}(21s^4 - 42s^2 + 20)$	$\frac{3}{64}e^2(14 - 15s^2)$	0
1	$-\frac{1}{32}e(27s^4 - 108s^2 + 64)$	$\frac{7}{16}e(11 - 12s^2)$	0
2	$\frac{3}{64}e^2(5s^4 + 8s^2 - 8)$	$\frac{3}{16}e^2(2 - s^2) + \frac{1}{8}(20 - 21s^2)$	$-\frac{15}{128}e^2$
3		$\frac{3}{16}e(8s^2 - 5)$	$-\frac{9}{64}e$
4		$\frac{3}{32}e^2(13s^2 - 10)$	$\frac{3}{64}(4 - e^2)$
5			$\frac{15}{64}e$
6			$\frac{9}{128}e^2$

In the same way as we did at the first order, the new Hamiltonian term $\mathcal{H}_{0,2}$ is chosen to be composed of those terms of $\widetilde{\mathcal{H}}_{0,2}$ that are free from the explicit appearance of f yet leave untouched the explicit appearance of $1/r^2$ in Eq. (6.32). That is, $\mathcal{H}_{0,2}$ keeps all the terms of $\widetilde{\mathcal{H}}_{0,2}$ pertaining to the kernel of the Lie derivative (4.66), which is now better set in the form of Eq. (5.38), in addition to some additional terms pertaining to its image. This choice is easily made by neglecting all coefficients $q_{j,l}$ in Eq. (6.32) except those with $j = 0$. Namely,

$$\mathcal{H}_{0,2} = \frac{\mu}{a}\left(\frac{a^2}{r^2}\eta\right)\left[\frac{R_\oplus^4}{p^4}\eta(q_{0,0}+2q_{0,1}s^2\cos 2\omega)+2\sum_{i\geq 3}\tilde{J}_i\langle V_i\rangle_f\right], \tag{6.33}$$

where terms $\langle V_i\rangle_f$ were already computed in Eq. (6.14).

The homological equation (5.39) is next solved for $m = 2$. In view of Eqs. (6.32) and (6.33), we obtain

$$\mathcal{W}_2 = G\frac{R_\oplus^4}{p^4}\sum_{j=1}^{6}\frac{1}{j}\sum_{l=j_1}^{2}q_{j,l}s^{|2l|}\sin(jf+2l\omega)+2G\sum_{i\geq 3}\tilde{J}_iW_{2,i}, \tag{6.34}$$

where the integrand of

$$W_{2,i} = \frac{1}{\eta} \int \left(V_i - \langle V_i \rangle_f \right) \mathrm{d}f, \quad i \geq 3, \tag{6.35}$$

is made only of periodic terms in f. Replacing Eqs. (6.3) and (6.14) into Eq. (6.35), we obtain

$$W_{2,i} = \frac{R_\oplus^i}{p^i} \sum_{j=0}^{i} \mathcal{F}_{i,j}(s) \sum_{k=0}^{i-1} \binom{i-1}{k} e^k \mathcal{I}_{i,j,k}(f,\omega), \tag{6.36}$$

where the integrals

$$\mathcal{I}_{i,j,k} = \int \left\{ \cos^k f \cos[\iota(f+\omega) - i_\pi] - \frac{1}{2^k} \binom{k}{\frac{k+\iota}{2}} \cos(\iota\omega - i_\pi) \right\} \mathrm{d}f, \tag{6.37}$$

with $\iota = i - 2j$, are purely periodic functions of f. Thus, Eq. (6.37) is integrated using Eqs. (6.12) and (6.13) to give

$$\mathcal{I}_{i,j,k} = \frac{1}{2^k} \sum_{\substack{l=0 \\ l \neq l^*}}^{k} \binom{k}{l} \frac{1}{2(l - l^*)} \sin[2(l - l^*) f + (i - 2j)\omega - i_\pi], \tag{6.38}$$

where the term $l = l^* \equiv j + \frac{1}{2}(k - i)$ is absent from the summation because it is precisely the one that has canceled out the term $\langle V_i \rangle_f$ in Eq. (6.37).

Regrettably, our initial plan of transforming the zonal problem into a quasi-Keplerian system fails clamorously at the second order of the *elimination of the parallax* because the appearance of $\omega = g'$ in Eq. (6.33) prevents G' from becoming an integral. However, the procedure is still successful because it radically simplifies the zonal problem Hamiltonian by removing non-essential short-period terms. Moreover, the elimination of the parallax can be carried out to any order of the Lie transforms procedure avoiding the appearance of the equation of the center in the generating function, and, in consequence, without leaving the algebra of trigonometric functions [154, 442].

6.2.2 Delaunay normalization

Up to second-order effects, we obtained the simplified Hamiltonian after the elimination of the parallax $\mathcal{K} = \mathcal{K}_{0,0} + \epsilon\mathcal{K}_{1,0} + \frac{1}{2}\epsilon^2\mathcal{K}_{2,0}$, in which $\mathcal{K}_{0,0} = \mathcal{H}_{0,0}$, $\mathcal{K}_{1,0} = \mathcal{H}_{0,1}$, and $\mathcal{K}_{2,0} = \mathcal{H}_{0,2}$, are given in Eqs. (6.7), (6.28), and (6.33), respectively, where all the quantities related to the elliptic motion are now assumed to be functions of the prime Delaunay variables ℓ', g', L', G', and $H' = H$.

Now, the elimination of the remaining short-period terms from the Hamiltonian is straightforward by means of a *Delaunay normalization* [156]. At first order, we choose $\mathcal{K}_{0,1} = \langle \mathcal{K}_{1,0} \rangle_{\ell'}$, namely

$$\mathcal{K}_{0,1} = \frac{1}{2\pi} \int_0^{2\pi} \mathcal{K}_{1,0}\, \mathrm{d}\ell' = \frac{1}{2\pi} \int_0^{2\pi} \mathcal{K}_{1,0} \frac{r^2}{a^2 \eta}\, \mathrm{d}f = \frac{\mu}{a} \langle V_2 \rangle_f, \tag{6.39}$$

where $\langle V_2 \rangle_f$ is given in Eq. (6.29). Because $\widetilde{\mathcal{K}}_{0,1} = \mathcal{K}_{1,0} = \mathcal{K}_{0,1}(a/r)^2\eta$, the integrand in the homological equation (4.69) cancels out and the term $\mathcal{V}_1$ is trivially solved. Hence,

$$\mathcal{V}_1 = \frac{\phi}{n} \mathcal{K}_{0,1} = L\phi \langle V_2 \rangle_f. \tag{6.40}$$

The known terms at second order are computed from Eq. (2.37) with $\mathcal{H}$ replaced by $\mathcal{K}$. We obtain

$$\begin{aligned} \widetilde{\mathcal{K}}_{0,2} = &-\frac{\mu}{a} \frac{R_\oplus^4}{p^4} \frac{1}{16} \eta^2 (2 - 3s^2)^2 \left[3 + \frac{a^2}{r^2} (1 + 2\beta \cos f + e\beta \cos 2f) \right] + \frac{\mu}{a} \frac{R_\oplus^4}{p^4} \\ &\times \left(\frac{a^2}{r^2} \eta \right) \eta (q_{0,0} + 2q_{0,1} s^2 \cos 2\omega) + 2\frac{\mu}{a} \left(\frac{a^2}{r^2} \eta \right) \sum_{i \geq 3} \tilde{J}_i \langle V_i \rangle_f. \end{aligned} \tag{6.41}$$

Then we choose $\mathcal{K}_{0,2} = \langle \widetilde{\mathcal{K}}_{0,2} \rangle_{\ell'}$, which is averaged with the help of the differential relation (4.64) to obtain

$$\mathcal{K}_{0,2} = \frac{\mu}{a} \frac{R_\oplus^4}{p^4} \eta \left[q_{0,0} - \frac{1 + 3\eta}{16} (3s^2 - 2)^2 + 2q_{0,1} s^2 \cos 2\omega \right] + 2\frac{\mu}{a} \sum_{i \geq 3} \tilde{J}_i \langle V_i \rangle_f. \tag{6.42}$$

That is, up to the second order of J_2, the new Hamiltonian in mean elements, with short-period effects removed, is

$$\mathcal{Q} = \mathcal{Q}_{0,0} + J_2 \mathcal{Q}_{1,0} + \frac{1}{2} J_2^2 \mathcal{Q}_{2,0}, \tag{6.43}$$

where $\mathcal{Q}_{0,0}$ is the Keplerian (6.7) in double-prime variables, the term $\mathcal{Q}_{1,0} = \mathcal{K}_{0,1}(G'', L''; H)$ is given in Eq. (6.39), and $\mathcal{Q}_{2,0} = \mathcal{K}_{0,2}(g'', G'', L''; H)$ is obtained from Eq. (6.42).

As expected, the partially normalized Hamiltonian (6.43) is formally the same as the mean-element Hamiltonian in prime variables previously obtained in Eq. (6.17). The benefit of the detour taken with the preliminary elimination of the parallax is that the computation of $\mathcal{V}_2$ from Eq. (4.69) is now straightforward, yielding

$$\begin{aligned} \mathcal{V}_2 = \phi &\left[G' \frac{R_\oplus^4}{p^4} (q_{0,0} + 2q_{0,1} s^2 \cos 2\omega) - \frac{L'}{\eta} \langle V_2 \rangle_f^2 + 2L' \sum_{i \geq 3} \tilde{J}_i \langle V_i \rangle_f \right] \\ &- G' \frac{R_\oplus^4}{p^4} \frac{\beta}{32} (2 - 3s^2)^2 (4 \sin f + e \sin 2f). \end{aligned} \tag{6.44}$$

6.2.3 Short-period corrections

The first-order corrections of the elimination of the parallax are exactly the same as those of Deprit's radial intermediary given in Eqs. (5.43)–(5.48). As we will discuss later in §6.6, the most relevant second-order terms are those related to the initialization of the constants of a perturbation theory, and, in particular, the computation of the mean semimajor axis [469]. Then we limit ourselves here to the computation of the periodic corrections of the inverse transformation up to second order in the case of the Delaunay action $L = \sqrt{\mu a}$.

In the case of the elimination of the parallax, $L' = L + J_2 L_{0,1} + \frac{1}{2} J_2^2 L_{0,2}$, where, from Eqs. (2.17) and (2.27), $L_{0,1} = \partial \mathcal{W}_1/\partial \ell$, $L_{0,2} = \partial \mathcal{W}_2/\partial \ell - \{L_{0,1}; \mathcal{W}_1\}$, with $\mathcal{W}_1$ and $\mathcal{W}_2$ given by Eqs. (6.31) and (6.34), respectively. The partial derivatives with respect to the mean anomaly are readily computed using the differential relation (4.64). Hence,

$$L_{0,1} = \frac{1}{8} L \frac{R_\oplus^2}{r^2} \frac{1}{\eta^2} \left[(4 - 6s^2) e \cos f + s^2 \sum_{j=1}^{3} 3^{\lfloor 2-j/2 \rfloor} j e^{|j-2|} \cos(jf + 2\omega) \right],$$

from which

$$\{L_{0,1}; \mathcal{W}_1\} = L \frac{R_\oplus^4}{p^4} \frac{p^2}{r^2} \frac{1}{\eta^4} \sum_{i=0}^{2} \sum_{j=0}^{8} \sum_{k=0}^{2} q_{2i,j,k} s^{2i} e^{2k+j^\star} \cos[(2i + j - 4)f + 2i\omega], \tag{6.45}$$

where $j^\star = j \bmod 2$, from Eq. (6.5), and $q_{2i,j,k}$ are the inclination polynomials in Table 6.2. Finally,

$$\begin{aligned} \frac{\partial \mathcal{W}_2}{\partial \ell} = G \frac{a^2}{r^2} \eta \Bigg\{ & \frac{R_\oplus^4}{p^4} \sum_{j=1}^{6} \sum_{l=j_1}^{2} q_{j,l} s^{|2l|} \cos(jf + 2l\omega) + 2 \sum_{i \ge 3} \sum_{j=0}^{i} \sum_{k=0}^{i-1} \tilde{J}_i \frac{R_\oplus^i}{p^i} \mathcal{F}_{i,j} \\ & \times \binom{i-1}{k} \frac{e^k}{2^k} \sum_{\substack{l=0 \\ l \ne l^*}}^{k} \binom{k}{l} \cos[2(l - l^*)f + (i - 2j)\omega - i_\pi] \Bigg\}, \end{aligned}$$

where the coefficients $q_{j,l}$ are given in Table 6.1, and we recall from Eq. (6.38) that $l^* \equiv j + (k - i)/2$.

As regards the Delaunay normalization, the generating function of the first order is $\mathcal{V}_1 = \frac{1}{4} G' (R_\oplus/p)^2 (3s^2 - 2)\phi$, as follows from Eqs. (6.40) and (6.29), from which Eq. (2.17) yields the first-order corrections to the polar-nodal variables. Namely, $\Theta_{0,1} = 0$, $N_{0,1} = 0$, and

$$\begin{aligned} r_{0,1} &= -\frac{1}{2} r (3s^2 - 2)[2\eta + \kappa(1 + \kappa)/(1 + \eta)], \\ \theta_{0,1} &= \frac{1}{2} [3(5s^2 - 4)\phi + (3s^2 - 2)(2 + \kappa)\sigma/(1 + \eta)], \\ \nu_{0,1} &= 3c\phi, \\ R_{0,1} &= \frac{1}{2} (\Theta/p)\sigma(3s^2 - 2)[\eta + (1 + \kappa)^2/(1 + \eta)], \end{aligned}$$

Table 6.2: Non-null coefficients $q_{2i,j,k}$ in Eq. (6.45).

(j,k)	$i=0$	$i=1$	$i=2$
0, 2	$q_{0,8,2}$	$q_{2,8,2}$	$q_{4,8,2}$
1, 1	$q_{0,7,1}$	$q_{2,7,1}$	$q_{4,7,1}$
2, 1	$q_{0,6,1}$	$\frac{3}{8}(1-3s^2)$	$\frac{177}{128}$
2, 2	$q_{0,6,2}$	$\frac{3}{8}(5-6s^2)$	$\frac{3}{32}$
3, 0	$q_{0,5,0}$	$\frac{1}{16}(5-21s^2)$	$\frac{117}{64}$
3, 1	$q_{0,5,1}$	$\frac{1}{64}(214-267s^2)$	$\frac{9}{8}$
4, 0	$\frac{1}{32}(-3s^4+12s^2+8)$	$\frac{7}{8}(2-3s^2)$	$\frac{21}{32}$
4, 1	$\frac{1}{64}(261s^4-120s^2+32)$	$\frac{1}{4}(5-9s^2)$	$\frac{177}{64}$
4, 2	$\frac{3}{256}(101s^4-40s^2-8)$	$-\frac{3}{64}(31s^2-26)$	$\frac{69}{256}$
5, 0	$\frac{1}{64}(63s^4-12s^2+32)$	$-\frac{3}{16}(27s^2-19)$	$\frac{93}{64}$
5, 1	$\frac{1}{64}(207s^4-96s^2+4)$	$-\frac{3}{64}(9s^2-2)$	$\frac{3}{2}$
6, 1	$\frac{3}{128}(67s^4-32s^2+16)$	$-\frac{3}{8}(10s^2-7)$	$\frac{153}{128}$
6, 2	$\frac{3}{8}s^2(2s^2-1)$	$\frac{3}{8}(s^2-1)$	$\frac{9}{32}$
7, 1	$\frac{3}{32}(9s^4-6s^2+2)$	$-\frac{27}{64}(3s^2-2)$	$\frac{27}{64}$
8, 2	$\frac{3}{512}(27s^4-24s^2+8)$	$-\frac{9}{128}(3s^2-2)$	$\frac{27}{512}$

where the right sides are in double-prime variables for direct corrections, and in prime variables for inverse corrections. Recall that the latter must be subtracted according to Eq. (2.27).

Finally, we compute $L'' = L' + J_2 L'_{0,1} + \frac{1}{2}J_2^2 L'_{0,2}$, in which $L'_{0,1} = \partial\mathcal{V}_1/\partial\ell'$, and $L'_{0,2} = \partial\mathcal{V}_2/\partial\ell' - \{L'_{0,1};\mathcal{V}_1\}$, using the generating function terms given in Eqs. (6.40) and (6.44). Once more, the partial derivatives with respect to the mean anomaly are computed using the differential relation (4.64), to obtain

$$L'_{0,1} = \frac{1}{4}L'\frac{R_\oplus^2}{p^2}\eta(3s^2-2)\left(\frac{a^2}{r^2}\eta-1\right),$$

$$\{L'_{0,1};\mathcal{V}_1\} = -L'\frac{R_\oplus^4}{p^4}(2-3s^2)^2\frac{1}{128\eta^4}\sum_{j=0}^{4}Q'_j(e)\cos jf,$$

$$\begin{aligned}\frac{\partial\mathcal{V}_2}{\partial\ell'} &= -L'\frac{R_\oplus^4}{p^4}\frac{a^2}{r^2}\frac{\beta\eta^2}{16}(2-3s^2)^2(2\cos f + e\cos 2f) + \left(\frac{a^2}{r^2}\eta-1\right)\\ &\quad\times L'\left[\eta\frac{R_\oplus^4}{p^4}(q_{0,0}+2q_{0,1}s^2\cos 2\omega) - \frac{1}{\eta}\langle V_2\rangle_f^2 + 2\sum_{i\ge3}\tilde{J}_i\langle V_i\rangle_f\right],\end{aligned}$$

in which the terms Q'_j are given in Table 6.3 and the $q_{0,j}$ in Table 6.1.

Table 6.3: Coefficients Q' of the Delaunay transformation of L'.

$Q'_0 = 105 - 60\eta^2 - 18\eta^3 + 3\eta^4 + 2\eta^5$	$Q'_3 = 12e(2 - \eta^2 - \eta^3)$
$Q'_1 = 4\beta(42 + 42\eta - 5\eta^2 - 14\eta^3 - 5\eta^4)$	$Q'_4 = (1 - \eta)^2(3 + 6\eta + 5\eta^2 + 2\eta^3)$
$Q'_2 = 4(21 - 16\eta^2 - 7\eta^3 + \eta^4 + \eta^5)$	

6.3 Brouwer's solution by complete reduction

As we already discussed in §6.1.1, the long-term Hamiltonian (6.43), or the one in Eq. (6.17), is integrable. Still, it is not separable and, in the style of Brouwer [75], we compute an additional Lie transformation in order to remove the long-period terms, in this way obtaining a completely reduced Hamiltonian.

6.3.1 Secular terms

The removal of long-period terms from Eq. (6.43), which are related to the perigee dynamics, will be achieved by a new Lie transformation to triple-prime variables. Because the problem is now independent of ℓ'', the generating function of the new Lie transformation is also chosen independent of it, say $\mathcal{U} = \sum_{m\geq 0}(J_2^m/m!)\,\mathcal{U}_{m+1}$ with $\mathcal{U}_{m+1} \equiv \mathcal{U}_{m+1}(-, g'', -, L'', G'', H'')$. As a consequence, it is immediately checked from Eq. (4.66) that the Lie derivative $\mathcal{L}_0(\mathcal{U}_m)$ always vanishes along the Keplerian flow stemming from $\mathcal{Q}_{0,0}$. Therefore, the homological equation (4.67) becomes

$$0 = \widetilde{\mathcal{Q}}_{0,m} - \mathcal{Q}_{0,m}, \tag{6.46}$$

and $\mathcal{U}_m$ cannot be computed at the step m.

However, since $\mathcal{Q}_{1,0}$ is already free from $\omega = g''$, there is no difficulty in choosing $\mathcal{Q}_{0,1} = \mathcal{Q}_{1,0}$ at first order, and proceed to the second order leaving $\mathcal{U}_1$ undetermined. At second order, after computing $\widetilde{\mathcal{Q}}_{0,2}$ from Eq. (2.37), we rewrite Eq. (6.46) in the form $\{\mathcal{U}_1; \mathcal{Q}_{1,0}\} = \frac{1}{2}(\mathcal{Q}^*_{0,2} - \mathcal{Q}_{0,2})$ with $\mathcal{Q}^*_{2,0} = \mathcal{Q}_{2,0}$. Analogously, at third order, from repeated applications of Deprit's fundamental recursion (2.15) we obtain $\{\mathcal{U}_2; \mathcal{Q}_{1,0}\} = \frac{1}{2}(\mathcal{Q}^*_{0,3} - \mathcal{Q}_{0,3})$, in which $\mathcal{Q}^*_{0,3} = \{\mathcal{Q}_{0,2}; \mathcal{U}_1\} + \{\mathcal{Q}_{1,1}; \mathcal{U}_1\} + \{\mathcal{Q}_{2,0}; \mathcal{U}_1\} + \mathcal{Q}_{3,0}$.

In this way, we obtain the new homological equation

$$\mathcal{L}_1(\mathcal{U}_{m-1}) \equiv \{\mathcal{U}_{m-1}; \mathcal{Q}_{1,0}\} = \frac{1}{m}(\mathcal{Q}^*_{0,m} - \mathcal{Q}_{0,m}), \tag{6.47}$$

where $\mathcal{Q}^*_{0,m}$ stands for the computable terms up to the step m of the procedure, and

$$\mathcal{L}_1(\mathcal{U}_{m-1}) \equiv \frac{3}{4} n \frac{R_\oplus^2}{p^2}(4 - 5s^2)\frac{\partial \mathcal{U}_{m-1}}{\partial g''}, \tag{6.48}$$

thus shifting by one the actual order of the perturbation approach with respect to the order at which the generating function of the elimination of the periapsis can be determined. Because the right hand of Eq. (6.47) is a trigonometric polynomial in g'', the kernel of the Lie operator $\mathcal{L}_1$ is composed of terms that are free from g''.[1]

Once more, the homological equation is conveniently solved by indefinite integration, yet, regrettably, the divisor $\Delta = 4-5s^2$ harms the transformation equations of the long-period elimination and prevents one from application of the solution to orbits in the close vicinity of the critical inclination. One way of avoiding computer overflow during runtime when propagating critically-inclined orbits with analytical solutions of this kind, is to replace the critical divisor Δ by the alternative form: $1/\Delta = 0$ when $\Delta = 0$, and $1/\Delta = [1-\exp(-100\Delta^2)]/\Delta$ otherwise, which provides the correct value out of the critical inclination and never becomes singular [121, 299].[2]

Thus, at second order we obtain

$$\mathcal{U}_1 = \frac{2}{3}\frac{p^2}{R_\oplus^2}\frac{1}{n(4-5s^2)}\int(\mathcal{Q}_{2,0}-\mathcal{Q}_{0,2})\mathrm{d}g'', \tag{6.49}$$

where the new Hamiltonian term $\mathcal{Q}_{0,2}$ is composed of those terms of $\mathcal{Q}_{2,0}$ pertaining to the kernel of $\mathcal{L}_1$. That is, $\mathcal{Q}_{0,2} = \langle\mathcal{Q}_{2,0}\rangle_\omega$. In view of Eq. (6.42), we obtain

$$\mathcal{Q}_{0,2} = \frac{\mu}{a}\frac{R_\oplus^4}{p^4}\eta\left[q_{0,0} - \frac{1+3\eta}{16}(3s^2-2)^2\right] + 2\frac{\mu}{a}\sum_{k\geq 2}\tilde{J}_{2k}\langle\langle V_{2k}\rangle_f\rangle_\omega, \tag{6.50}$$

in which the coefficient $q_{0,0}$ is given in Table 6.1, and, from Eq. (6.14), we obtain

$$\langle\langle V_{2k}\rangle_f\rangle_\omega = (R_\oplus/p)^{2k}\eta\mathcal{F}_{2k,k}(s)\mathcal{G}^*_{2k,k}(e), \tag{6.51}$$

where $i^\star = i \bmod 2$, and $i_0 = \lfloor\frac{1}{2}i\rfloor$ from the index convention in Eq. (6.5), which shows that only the even zonal harmonics—$i = 2k$, $i_\pi = 0$ in Eq. (6.14)—can have terms belonging to the kernel of the Lie derivative (6.48). In consequence, they are the only terms of the potential contributing secular effects.

After expressing $\mathcal{Q}_{0,i}$, $i = 0,1,2$ in triple-prime variables, up to $\mathcal{O}(J_2^2)$, we obtain the secular, completely reduced Hamiltonian

$$\mathcal{S} = \mathcal{S}(-,-,-,L''',G''',H''') \equiv \mathcal{Q}_{0,0} + J_2\mathcal{Q}_{0,1} + \frac{1}{2}J_2^2\mathcal{Q}_{0,2},$$

whose Hamilton equations are trivially integrable. Indeed, the momenta are constant, $L''' = L_0''$, $G''' = G_0'''$, $H''' = H_0$, whereas their conjugate angles evolve linearly with

1 An alternative, equivalent procedure is to reorganize the one-degree-of-freedom Hamiltonian (6.43) in the form $\mathcal{Q} = \mathcal{Q}'_{0,0} + J_2\mathcal{Q}'_{1,0}$, in which $\mathcal{Q}'_{0,0} = \mathcal{Q}_{0,0} + J_2\mathcal{Q}_{1,0}$ and $\mathcal{Q}'_{1,0} = \frac{1}{2}J_2\mathcal{Q}_{2,0}$, with the same assumption: that the generating function does not depend on ℓ.

2 This approach is attributed to R. H. Smith in [297].

time,

$$\ell''' = \ell_0''' + \frac{\partial \mathcal{S}}{\partial L'''}t, \quad g''' = g_0''' + \frac{\partial \mathcal{S}}{\partial G'''}t, \quad h''' = h_0''' + \frac{\partial \mathcal{S}}{\partial H'''}t,$$

with the constant, secular frequencies

$$\begin{aligned}
\frac{\partial \mathcal{S}}{\partial L'''} &= n + nJ_2\frac{R_\oplus^2}{p^2}\frac{3}{4}\eta(2-3s^2) + nJ_2^2\frac{R_\oplus^4}{p^4}\frac{3}{8}\eta \\
&\quad \times \left[\frac{5}{16}\eta^2(5s^4+8s^2-8) + \eta(2-3s^2)^2 + \frac{15}{16}(7s^4-16s^2+8)\right] \\
&\quad - nJ_2^2\sum_{i\geq 3}\tilde{J}_i(1-i^\star)\frac{R_\oplus^i}{p^i}\eta\mathcal{F}_{i,i_0}\left[3\mathcal{G}_{i,i_0}^* - \frac{\eta^2}{e}\frac{\partial \mathcal{G}_{i,i_0}^*}{\partial e}\right], \\
\frac{\partial \mathcal{S}}{\partial G'''} &= nJ_2\frac{R_\oplus^2}{p^2}\frac{3}{4}(4-5s^2) + \frac{3}{128}nJ_2^2\frac{R_\oplus^4}{p^4}[\eta^2(45s^4+36s^2-56) \\
&\quad + 24\eta(3s^2-2)(5s^2-4) + 5(77s^4-172s^2+88)] + nJ_2^2\sum_{i\geq 3}\tilde{J}_i \\
&\quad \times (1-i^\star)\frac{R_\oplus^i}{p^i}\left[(1-2i)\mathcal{F}_{i,i_0}\mathcal{G}_{i,i_0}^* + \frac{c^2}{s}\frac{\partial \mathcal{F}_{i,i_0}}{\partial s}\mathcal{G}_{i,i_0}^* - \mathcal{F}_{i,i_0}\frac{\eta^2}{e}\frac{\partial \mathcal{G}_{i,i_0}^*}{\partial e}\right], \\
\frac{\partial \mathcal{S}}{\partial H'''} &= -nJ_2\frac{R_\oplus^2}{p^2}\frac{3}{2}c + nJ_2^2\frac{R_\oplus^4}{p^4}\frac{3c}{32}[\eta^2(5s^2+4) + 12\eta(3s^2-2) \\
&\quad + 5(7s^2-8)] - nJ_2^2\sum_{i\geq 3}\tilde{J}_i(1-i^\star)\frac{R_\oplus^i}{p^i}\frac{c}{s}\frac{\partial \mathcal{F}_{i,i_0}}{\partial s}\mathcal{G}_{i,i_0}^*.
\end{aligned}$$

The osculating solution is obtained after applying the long-period corrections to the secular terms, thus obtaining mean elements, which in turn must be modulated with the short-period corrections computed in §6.2.3.

6.3.2 Long-period corrections

The integrand of Eq. (6.49) is composed of the periodic terms

$$\mathcal{Q}_{2,0} - \langle \mathcal{Q}_{2,0}\rangle_\omega = 2\frac{\mu}{a}\frac{R_\oplus^4}{p^4}\eta q_{0,1}s^2\cos 2\omega + 2\frac{\mu}{a}\sum_{i\geq 3}\tilde{J}_i(\langle V_i\rangle_f - \langle\langle V_i\rangle_f\rangle_\omega),$$

in which the coefficient $q_{0,1}$ is given in Table 6.1 and, from Eqs. (6.14) and (6.51),

$$\langle V_i\rangle_f - \langle\langle V_i\rangle_f\rangle_\omega = \frac{R_\oplus^i}{p^i}\eta\sum_{\substack{j=0\\ j\neq i/2}}^{i-1}\mathcal{F}_{i,j}\mathcal{G}_{i,j}^*\cos[(i-2j)\omega - i_\pi].$$

Hence,

$$\mathcal{U}_1 = \frac{(4/3)G}{4-5s^2}\left[\frac{R_\oplus^2}{p^2}\frac{q_{0,1}}{2}s^2\sin 2\omega + \sum_{i\geq 3}\tilde{J}_i\frac{R_\oplus^{i-2}}{p^{i-2}}\sum_{\substack{j=0\\ j\neq i/2}}^{i-1}\mathcal{F}_{i,j}\mathcal{G}_{i,j}^*\frac{\sin[(i-2j)\omega - i_\pi]}{i-2j}\right]. \tag{6.52}$$

The transformation from secular to mean elements $x'' = x''' + J_2 x''_{0,1} + \mathcal{O}(J_2^2)$, $x \in (\ell, g, h, L, G, H)$ is obtained as usual, computing the long-period corrections $x''_{0,1}$ from the first row of Eq. (2.17). In particular, $L''_{0,1} = 0$, $H''_{0,1} = 0$, and

$$\ell''_{0,1} = \frac{\partial \mathcal{U}_1}{\partial L''} = \frac{1}{16}\frac{R_\oplus^2}{p^2}\eta^3 \frac{14-15s^2}{4-5s^2} s^2 \sin 2\omega + \frac{4}{3(4-5s^2)} \sum_{i\geq 3} \tilde{J}_i \frac{R_\oplus^{i-2}}{p^{i-2}} \sum_{\substack{j=0\\ j\neq i/2}}^{i} \mathcal{F}_{i,j} \frac{\eta^3}{e} \frac{\partial \mathcal{G}^*_{i,j}}{\partial e} \frac{\sin[(i-2j)\omega - i_\pi]}{i-2j},$$

$$g''_{0,1} = \frac{\partial \mathcal{U}_1}{\partial G''} = \frac{1}{16}\frac{R_\oplus^2}{p^2}\left[\frac{112-408s^2+520s^4-225s^6}{2(4-5s^2)^2} e^2 - \frac{14-15s^2}{4-5s^2} s^2\right] \times \sin 2\omega + \frac{4}{3(4-5s^2)} \sum_{i\geq 3} \tilde{J}_i \frac{R_\oplus^{i-2}}{p^{i-2}} \sum_{\substack{j=0\\ j\neq i/2}}^{i} \left[\frac{c^2}{s}\frac{\partial \mathcal{F}_{i,j}}{\partial s}\mathcal{G}^*_{i,j} - \mathcal{F}_{i,j}\frac{\eta^2}{e}\frac{\partial \mathcal{G}^*_{i,j}}{\partial e} + \left(\frac{14-15s^2}{4-5s^2} + 4 - 2i\right)\mathcal{F}_{i,j}\mathcal{G}^*_{i,j}\right] \frac{\sin[(i-2j)\omega - i_\pi]}{i-2j},$$

$$h''_{0,1} = \frac{\partial \mathcal{U}_1}{\partial H} = -\frac{1}{16}\frac{R_\oplus^2}{p^2}\frac{56-120s^2+75s^4}{(4-5s^2)^2} c e^2 \sin 2\omega - \frac{4c}{3(4-5s^2)} \times \sum_{i\geq 3} \tilde{J}_i \frac{R_\oplus^{i-2}}{p^{i-2}} \sum_{\substack{j=0\\ j\neq i/2}}^{i} \left(\frac{10\mathcal{F}_{i,j}}{4-5s^2} + \frac{1}{s}\frac{\partial \mathcal{F}_{i,j}}{\partial s}\right)\mathcal{G}^*_{i,j} \frac{\sin[(i-2j)\omega - i_\pi]}{i-2j},$$

$$G''_{0,1} = -\frac{\partial \mathcal{U}_1}{\partial g''} = -\frac{4G''}{3(4-5s^2)}\left[\frac{R_\oplus^2}{p^2}\frac{3}{64} e^2 (14-15s^2) s^2 \cos 2\omega + \sum_{i\geq 3} \tilde{J}_i \frac{R_\oplus^{i-2}}{p^{i-2}} \sum_{\substack{j=0\\ j\neq i/2}}^{i} \mathcal{F}_{i,j}\mathcal{G}^*_{i,j} \cos[(i-2j)\omega - i_\pi]\right],$$

where the entities on the right sides are functions of the secular (triple-prime) Delaunay variables. The inverse corrections, $x''' = x'' - J_2 x''_{0,1} + \mathcal{O}(J_2^2)$, are computed analogously expressing the quantities on the right sides of the long-period corrections $x''_{0,1}$ as functions of the double-prime Delaunay variables.

6.4 Reverse normalization. Long periods removed first

Alternatively to Brouwer's approach to compute a completely reduced Hamiltonian by sequentially eliminating the short- and long-period terms, one can reverse the sequence and remove first the long-period terms. This approach may seem unnatural, for the Lie derivative (4.66) does not depend on the argument of the perigee. However, because the disturbing potential of the zonal problem (6.2) only depends on g through $\theta = f + g$, as clearly noticed from Eq. (6.3), a smart choice of the new Hamiltonian terms, as well as proper selection of the arbitrary integration constants in which the generating function depends upon, allows the reverse approach not only to be feasible, but advantageous when one pursues a higher-order normalization of the zonal Hamiltonian in closed form of the eccentricity [408].

6.4.1 Normalization of the total angular momentum

We start from the original zonal Hamiltonian (6.6). New terms $\mathcal{H}_{0,m}$ are chosen in such a way that not only they cancel out the terms of the homological equation pertaining to the kernel of the Lie derivative, as is mandatory, but they also cancel the additional terms, if any, that are not purely periodic on g. That is, $\partial\mathcal{H}_{0,m}/\partial g = 0$. In this way g is made cyclic, up to the truncation order, and, in consequence, G is turned into a formal integral of the new, partially normalized Hamiltonian.

Therefore, at first order, we choose $\mathcal{H}_{0,1}$ to be composed of the terms of $\mathcal{H}_{1,0}$ that are free from g. Thus, from Eq. (6.8),

$$\mathcal{H}_{0,1} = \frac{\mu}{a}\left(\frac{a^2}{r^2}\eta\right)\langle V_2\rangle_g, \tag{6.53}$$

where, on account of Eq. (6.27),

$$\langle V_2\rangle_g = \frac{R_\oplus^2}{p^2}\eta\frac{1}{4}(3s^2-2) + \frac{R_\oplus^2}{p^2}\eta\frac{1}{4}(3s^2-2)e\cos f. \tag{6.54}$$

Comparison with Eq. (6.30) shows that the new Hamiltonian term (6.53) certainly carries all the terms pertaining to the kernel of the Lie derivative. In particular, it is composed of the same terms as the first-order Hamiltonian of the elimination of the parallax in addition to one term of the image—the one factored by $\cos f$.

The homological equation is then solved using Eq. (5.39), from which we readily obtain

$$\mathcal{W}_1 = -\frac{1}{8}G\frac{R_\oplus^2}{p^2}s^2\sum_{i=1}^{3}3^{\lfloor 2-i/2\rfloor}e^{|i-2|}\sin(if+2\omega) + C_1. \tag{6.55}$$

The first-order term of the generating function is now simpler than Eq. (6.31) although it carries an additional, "constant" term $C_1 \equiv C_1(g,L,G,H)$. Incidentally, except for the term C_1, Eq. (6.55) is precisely the generating function of Cid's radial intermediary (5.29); cf. [111, 154].

At second order $\mathcal{L}_0(\mathcal{W}_2) = \widetilde{\mathcal{H}}_{0,2} - \mathcal{H}_{0,2}$, in which the computable terms $\widetilde{\mathcal{H}}_{0,2}$ are given in Eq. (2.37). Namely,

$$\widetilde{\mathcal{H}}_{0,2} = \mathcal{H}_{2,0} + \{\mathcal{H}_{1,0};\mathcal{W}_1\} + \{\mathcal{H}_{0,1};\mathcal{W}_1\}, \tag{6.56}$$

where $\mathcal{H}_{2,0}$ is given in Eq. (6.9) and the computation of the Poisson brackets now involves the partial derivatives of the still unknown function C_1.

We want $\mathcal{H}_{0,2}$ to cancel the terms of $\widetilde{\mathcal{H}}_{0,2}$ that are free from g. As regards $\mathcal{H}_{2,0}$, they are simply

$$\langle\mathcal{H}_{2,0}\rangle_g = 2\frac{\mu}{a}\left(\frac{a^2}{r^2}\eta\right)\sum_{i\geq 3}\tilde{J}_i\langle V_i\rangle_g.$$

We readily check from Eq. (6.3) that odd terms $\langle V_{2i-1}\rangle_g$ always vanish, and even terms $\langle V_{2i}\rangle_g$ stem from the values of the summation index $j = \frac{1}{2}i$. Hence,

$$\langle\mathcal{H}_{2,0}\rangle_g = 2\frac{\mu}{a}\left(\frac{a^2}{r^2}\eta\right)\sum_{i\geq 2}\tilde{J}_{2i}\frac{R_\oplus^{2i}}{p^{2i}}\eta\mathcal{F}_{2i,i}\sum_{k=0}^{2i-1}\binom{2i-1}{k}e^k\cos^k f. \tag{6.57}$$

Still, $\langle\mathcal{H}_{2,0}\rangle_g$ does not carry all the terms of the kernel contributed by $\mathcal{H}_{2,0}$. To check that, we compute

$$\mathcal{H}_{2,0} - \langle\mathcal{H}_{2,0}\rangle_g = 2nG\frac{a^2}{r^2}\sum_{i\geq 3}\tilde{J}_i(V_i - \langle V_i\rangle_g), \tag{6.58}$$

where $V_i - \langle V_i\rangle_g$ is obtained by simply avoiding the value $j = \frac{1}{2}i$. Then, making use of the differential relation (4.64), we readily obtain

$$\langle\mathcal{H}_{2,0} - \langle\mathcal{H}_{2,0}\rangle_g\rangle_\ell = \sum_{i\geq 3}K_i, \quad K_i = 2\frac{\mu}{a}\tilde{J}_i\langle V_i - \langle V_i\rangle_g\rangle_f,$$

where, analogously to Eq. (6.14),

$$K_i = 2\frac{\mu}{a}\tilde{J}_i\frac{R_\oplus^i}{p^i}\eta\sum_{\substack{j=1\\ j\neq i/2}}^{i}\mathcal{F}_{i,j}\mathcal{G}_{i,j}^*\cos[(i-2j)\omega - i_\pi], \quad i\geq 3. \tag{6.59}$$

Terms K_i, of this type must be nullified in the homological equation with a proper choice of the arbitrary function C_1.

These kinds of offending terms also stem from the evaluation of the Poisson brackets in Eq. (6.56). Indeed, after identifying the terms of $\{\mathcal{H}_{1,0};\mathcal{W}_1\} + \{\mathcal{H}_{0,1};\mathcal{W}_1\}$ that are free from g, which will be left in the new Hamiltonian, the remaining terms, which do depend on g, take the form of a trigonometric polynomial in f factored by $1/r^2$. The term of zeroth degree of this trigonometric polynomial takes the particular form

$$K_2 = n\frac{a^2}{r^2}\eta\left[\frac{3}{2}\frac{R_\oplus^2}{p^2}(5s^2-4)\frac{\partial C_1}{\partial g} - G\frac{R_\oplus^4}{p^4}\frac{3}{32}(15s^2-14)s^2e^2\cos 2g\right],$$

and must be canceled out to avoid the appearance of non-periodic terms in f in the generating function.

All the offending terms K_i, $i \geq 2$, are easily nullified by properly choosing

$$C_1 = \frac{4G}{3(5s^2-4)} \sum_{i\geq 3} \tilde{J}_i \frac{R_\oplus^{i-2}}{p^{i-2}} \sum_{\substack{j=0 \\ j\neq i/2}}^{i} \mathcal{F}_{i,j}\mathcal{G}^*_{i,j} \frac{\sin[(i-2j)\omega - i_\pi]}{i-2j}$$
$$+ G\frac{R_\oplus^2}{p^2}\frac{1}{32}\frac{15s^2-14}{5s^2-4}s^2e^2 \sin 2g. \tag{6.60}$$

When we do that, the first-order term of the generating function in Eq. (6.55) becomes completely defined, and the known terms $\widetilde{\mathcal{H}}_{0,2}$ in Eq. (6.56) can be recomputed. We find $\{\mathcal{H}_{1,0}; \mathcal{W}_1\} + \{\mathcal{H}_{0,1}; \mathcal{W}_1\} = \widetilde{\mathcal{H}}_{0,2,\mathrm{K}} + \widetilde{\mathcal{H}}_{0,2,\mathrm{I}}$, where

$$\widetilde{\mathcal{H}}_{0,2,\mathrm{I}} = \frac{\mu}{p}\frac{R_\oplus^4}{p^4}\frac{p^2}{r^2}\frac{1}{512}\sum_{i=1}^{2}\sum_{j=-1}^{4}(j+i_0)Q_{i,j}s^{2i}\cos[(j+i_0)f + 2ig], \tag{6.61}$$

$i_0 = \lfloor \frac{1}{2} i \rfloor$ from Eq. (6.5), and the coefficients $Q_{i,j}$ are given in Table 6.4. Then $\widetilde{\mathcal{H}}_{0,2,\mathrm{I}}$ is composed only of terms pertaining to the image of the Lie derivative. On the other hand,

$$\widetilde{\mathcal{H}}_{0,2,\mathrm{K}} = \frac{\mu}{p}\frac{R_\oplus^4}{p^4}\frac{p^2}{r^2}\frac{3}{64}s^2[8(4s^2-3) + (23s^2-16)e^2] - \frac{\mu}{p}\frac{R_\oplus^4}{p^4}\frac{p^2}{r^2}\frac{s^2}{(5s^2-4)^2}$$
$$\times \frac{3}{512}\{[(975s^6 - 2250s^4 + 1728s^2 - 448)e^2 - 4(5s^2-4)(795s^4$$
$$- 1198s^2 + 448)]e\cos f - 4e^2(775s^6 - 1710s^4 + 1236s^2 - 288)$$
$$\times \cos 2f + e^3(825s^6 - 1990s^4 + 1616s^2 - 448)\cos 3f\}, \tag{6.62}$$

which is free from the argument of the periapsis, is composed of terms of both the kernel and the image of the Lie derivative.

Table 6.4: Coefficients $Q_{i,j}$ in Eqs. (6.61) and (6.64).

j	$i = 1$	$i = 2$
−1	$12\frac{84-188s^2+105s^4}{4-5s^2}e^3$	$-3\frac{208-430s^2+225s^4}{(4-5s^2)^2}e^3$
0	0	$-60\frac{38-87s^2+50s^4}{(4-5s^2)^2}e^2$
1	$48\frac{148-340s^2+195s^4}{4-5s^2}e - 24(14-15s^2)e^3$	$-4\frac{122-135s^2}{4-5s^2}e - 8\frac{61-135s^2+75s^4}{(4-5s^2)^2}e^3$
2	$96(8-9s^2) + 24\frac{52-116s^2+65s^4}{4-5s^2}e^2$	$24 - 12\frac{23-25s^2}{4-5s^2}e^2$
3	$64(7-8s^2)e - 4(2-3s^2)\frac{14-15s^2}{4-5s^2}e^3$	$24e - 3\frac{14-15s^2}{4-5s^2}e^3$
4	$12(6-7s^2)e^2$	$6e^2$

Then we can safely choose

$$\mathcal{H}_{0,2} = \langle\widetilde{\mathcal{H}}_{0,2}\rangle_g = \widetilde{\mathcal{H}}_{0,2,\mathrm{K}} + \langle\mathcal{H}_{2,0}\rangle_g, \tag{6.63}$$

from Eqs. (6.62) and (6.57), and solve $\mathcal{L}_0(\mathcal{W}_2) = \mathcal{H}_{2,0} + \widetilde{\mathcal{H}}_{0,2,\mathrm{I}} - \langle\mathcal{H}_{2,0}\rangle_g$, in closed form of the eccentricity, using Eq. (5.39). We obtain

$$\mathcal{W}_2 = \frac{1}{n}\int \widetilde{\mathcal{H}}_{0,2,\mathrm{I}}\frac{r^2}{a^2\eta}\mathrm{d}f + \frac{1}{n}\int\left(\mathcal{H}_{2,0} - \langle\mathcal{H}_{2,0}\rangle_g - \sum_{i\geq 3}K_i\right)\frac{r^2}{a^2\eta}\mathrm{d}f,$$

into which we plug Eqs. (6.61), (6.58), and (6.59), to obtain

$$\begin{aligned}\mathcal{W}_2 = {} & G\frac{R_\oplus^4}{p^4}\frac{1}{512}\sum_{i=1}^{2}\sum_{j=-1}^{4} Q_{i,j}s^{2i}\sin[(j+i_0)f + 2ig] \\ & + 2G\sum_{i\geq 3}\tilde{J}_i\frac{R_\oplus^i}{p^i}\sum_{\substack{j=0\\ j\neq i/2}}^{i}\mathcal{F}_{i,j}\sum_{k=0}^{i-1}\binom{i-1}{k}e^k\mathcal{J}_k + C_2.\end{aligned} \tag{6.64}$$

The integrals $\mathcal{J}_k = \int\cos^k f\cos[(i-2j)(f+g) - i_\pi]\,\mathrm{d}f$ are solved with the help of Eqs. (6.11)–(6.13), to obtain

$$\mathcal{J}_k = \frac{1}{2^k}\sum_{l=0}^{k}\binom{k}{k-l}\frac{\sin[(i-2j)\theta - (k-2l)f - i_\pi]}{i-2j-k+2l}, \tag{6.65}$$

but the integration "constant" $C_2 \equiv C_2(g, L, G, H)$ will remain undetermined until the next order of the perturbation approach. That is, like we did with C_1, C_2 will be determined by imposing the requirement that the third-order term of the generating function be free from secular or mixed terms in the true anomaly.

It is worth checking that the choices $\mathcal{H}_{0,1}$ and $\mathcal{H}_{0,2}$ we made in Eqs. (6.53) and (6.63), respectively, are not the unique ones that satisfy the purpose of eliminating g from the original Hamiltonian. Indeed, as checked in Eqs. (6.53) and (6.54), the term $\langle\mathcal{H}_{1,0}\rangle_g$ still carries terms of the image of the Lie derivative that can be pruned away. Thus, the alternative selection of $\mathcal{H}_{0,1}$ in the form of Eq. (6.30), like in the elimination of the parallax, also fulfills our requirements. In that case, the generating function is obviously the one of the elimination of the parallax in Eq. (6.31), which has one more trigonometric term than Eq. (6.55).

The second-order term of the new Hamiltonian can be analogously simplified. In particular, as checked in Eq. (6.57), the term $\langle\mathcal{H}_{2,0}\rangle_g$ still carries different terms of the image of the Lie derivative that can be pruned away. Thus, we use Eq. (6.12) to write

$$\cos^k f = \frac{1-k^\star}{2^k}\binom{k}{k_0} + \frac{1}{2^{k-1}}\sum_{j=0}^{k_1}\binom{k}{k-j}\cos(k-2j)f, \tag{6.66}$$

where $k_0 = \lfloor\frac{1}{2}k\rfloor$, $k_1 = \lfloor\frac{1}{2}(k-1)\rfloor$ and $k^\star = k \bmod 2$, from Eq. (6.5). Then odd k values always kill the first term on the right side of Eq. (6.66), thus yielding trigonometric terms in f, which pertain to the image of the Lie derivative. On the contrary, after rearranging the summation, even values $k = 2n$ yield

$$\cos^{2n} f = \frac{1}{2^{2n}}\binom{2n}{n} + \frac{1}{2^{2n-1}}\sum_{j=1}^{n}\binom{2n}{n-j}\cos 2jf, \tag{6.67}$$

which always carries one term pertaining to the kernel. Therefore, $\langle\mathcal{H}_{2,0}\rangle_g$ is replaced by

$$\langle\mathcal{H}_{2,0}\rangle_g' = 2\frac{\mu}{p}\frac{p^2}{r^2}\sum_{i\geq 2}\tilde{J}_{2i}\frac{R_\oplus^{2i}}{p^{2i}}\mathcal{F}_{2i,i}\mathcal{G}_{2i,i}^*, \tag{6.68}$$

where, from Eq. (6.16),

$$\mathcal{G}_{2i,i}^* = \sum_{j=0}^{i-1}\binom{2i-1}{2j}\binom{2j}{j}\frac{e^{2j}}{2^{2j}}. \tag{6.69}$$

On the other hand, terms of $\widetilde{\mathcal{H}}_{0,2,\mathrm{K}}$ pertaining to the image of the Lie derivative can be removed analogously, replacing Eq. (6.62) by

$$\widetilde{\mathcal{H}}_{0,2,\mathrm{K}}' = -\frac{\mu}{p}\frac{R_\oplus^4}{p^4}\frac{p^2}{r^2}\frac{3}{8}\left[3 - 4s^2 + \left(2 - \frac{23}{8}s^2\right)e^2\right]s^2.$$

The new selection $\mathcal{H}_{0,2} = \widetilde{\mathcal{H}}_{0,2,\mathrm{K}}' + \langle\mathcal{H}_{2,0}\rangle_g'$, namely

$$\mathcal{H}_{0,2} = \frac{\mu}{p}\frac{p^2}{r^2}\left\{2\sum_{i\geq 2}\tilde{J}_{2i}\frac{R_\oplus^{2i}}{p^{2i}}\mathcal{F}_{2i,i}\mathcal{G}_{2i,i}^* - \frac{R_\oplus^4}{p^4}\left[\frac{9}{8} - \frac{3}{2}s^2 + \left(\frac{3}{4} - \frac{69}{64}s^2\right)e^2\right]s^2\right\}, \tag{6.70}$$

clearly fulfills the requisites of having eliminated the perigee while keeping all the terms of the kernel of the Lie derivative, and is much simpler than Eq. (6.63).

The simplicity of Eq. (6.70) is not, of course, for free and it is handicapped with a concomitant generating function much more involved than Eq. (6.64). Intermediate Hamiltonians between Eqs. (6.63) and (6.70) give rise to a whole family of radial intermediaries. The radial character of these partially normalized Hamiltonians is immediately seen when they are reformulated in polar variables replacing $e\cos f = p/r$ and $e = \sqrt{\kappa^2 + \sigma^2}$, with κ and σ given in Eq. (5.42), and recalling that $p = \Theta^2/\mu$.

6.4.2 Normalization of the semimajor axis

The short-period elimination after the periapsis has been removed becomes now a quite simple operation. The short-period Hamiltonian $\mathcal{K} = \mathcal{K}_{0,0} + J_2\mathcal{K}_{1,0} + \frac{1}{2}J_2^2\mathcal{K}_{2,0}$ is composed of the Keplerian $\mathcal{K}_{0,0} = \mathcal{H}_{0,0}$, the first-order term $\mathcal{K}_{1,0} = \mathcal{H}_{0,1}$ given by Eq. (6.53), and the second-order term $\mathcal{K}_{2,0} = \mathcal{H}_{0,2}$ given by Eq. (6.63). All the quantities are now functions of the prime Delaunay variables.

Then, at first order, the homological equation (4.68) is solved by choosing

$$\mathcal{K}_{0,1} = \frac{\mu}{p}\eta^3\frac{R_\oplus^2}{p^2}\left(-\frac{1}{2}+\frac{3}{4}s^2\right) \tag{6.71}$$

and computing $\mathcal{U}_1 = (1/n)\mathcal{K}_{0,1}(\phi + e\sin f)$.

At second order, the known terms $\widetilde{\mathcal{K}}_{0,2}$ are given by Eq. (6.56), which in addition to the term $\mathcal{K}_{2,0}$ needs the computation of the Poisson brackets,

$$\begin{aligned}\{\mathcal{K}_{1,0};\mathcal{U}_1\} + \{\mathcal{K}_{0,1};\mathcal{U}_1\} = {}& -\frac{3}{16}\frac{R_\oplus^4}{p^4}\frac{G^2}{p^2}(2-3s^2)^2\eta^4 - \frac{3}{64}\frac{R_\oplus^4}{p^4}\frac{G^2}{r^2}(2-3s^2)^2 \\ & \times\left[2(4+e^2) + \beta(15+12\eta+\eta^2)\cos f\right. \\ & \left.+ 2(3-2\eta-\eta^2)\cos 2f + \beta e^2\cos 3f\right].\end{aligned} \tag{6.72}$$

The term $\mathcal{K}_{0,2}$ is chosen by removing the short-period effects from $\widetilde{\mathcal{K}}_{0,2}$. That is, $\mathcal{K}_{0,2} = \langle\{\mathcal{K}_{1,0};\mathcal{U}_1\}\rangle_\ell + \langle\{\mathcal{K}_{0,1};\mathcal{U}_1\}\rangle_\ell + \langle\widetilde{\mathcal{H}}_{0,2,\mathrm{K}}\rangle_\ell + \langle\langle\mathcal{H}_{2,0}\rangle_g\rangle_\ell$. The averaging is once more carried out in closed form of the eccentricity using the differential relation (4.64). Thus, from Eq. (6.57),

$$\langle\langle\mathcal{H}_{2,0}\rangle_g\rangle_\ell = 2\frac{\mu}{a}\sum_{i\geq 2}\tilde{J}_{2i}\frac{R_\oplus^{2i}}{p^{2i}}\eta\mathcal{F}_{2i,i}\sum_{k=0}^{2i-1}\binom{2i-1}{k}e^k\frac{1}{2\pi}\int_0^{2\pi}\cos^k f\,\mathrm{d}f,$$

which, on account of Eqs. (6.66)–(6.67) and (6.16), yields

$$\langle\langle\mathcal{K}_{2,0}\rangle_g\rangle_\ell = 2\frac{\mu}{a}\eta\sum_{i\geq 2}\tilde{J}_{2i}\frac{R_\oplus^{2i}}{p^{2i}}\mathcal{F}_{2i,i}\mathcal{G}_{2i,i}^{*}. \tag{6.73}$$

Analogously, from Eqs. (6.62) and (6.72), we obtain

$$\langle\widetilde{\mathcal{H}}_{0,2,\mathrm{K}}\rangle_\ell = \frac{\mu}{a}\frac{R_\oplus^4}{p^4}\frac{3\eta}{64}s^2[8(4s^2-3) + (23s^2-16)e^2], \tag{6.74}$$

$$\langle\{\mathcal{K}_{1,0}+\mathcal{K}_{0,1};\mathcal{U}_1\}\rangle_\ell = -\frac{\mu}{a}\frac{R_\oplus^4}{p^4}\frac{3\eta}{32}(2-3s^2)^2(5+2\eta-\eta^2). \tag{6.75}$$

Hence,

$$\begin{aligned}\mathcal{K}_{0,2} = {}& -\frac{\mu}{a}\frac{R_\oplus^4}{p^4}\eta\frac{3}{64}[5(7s^4-16s^2+8) + 4(3s^2-2)^2\eta \\ & + (5s^4+8s^2-8)\eta^2] + 2\frac{\mu}{a}\eta\sum_{i\geq 2}\tilde{J}_{2i}\frac{R_\oplus^{2i}}{p^{2i}}\mathcal{F}_{2i,i}\mathcal{G}_{2i,i}^{*},\end{aligned} \tag{6.76}$$

which, as expected, is the same completely reduced Hamiltonian term previously obtained in Eq. (6.50).

The homological equation (4.68) is then solved to give the second-order term of the generating function,

$$\begin{aligned}\mathcal{U}_2 = {} & \frac{1}{n}(\langle\widetilde{\mathcal{H}}_{0,2,\mathrm{K}}\rangle_\ell + \langle\langle\mathcal{H}_{2,0}\rangle_g\rangle_\ell + \langle\{\mathcal{K}_{1,0} + \mathcal{K}_{0,1};\mathcal{U}_1\}\rangle_\ell)\phi \\ & + \frac{1}{n}\int\left(\widetilde{\mathcal{H}}_{0,2,\mathrm{K}}\frac{r^2}{a^2\eta} - \langle\widetilde{\mathcal{H}}_{0,2,\mathrm{K}}\rangle_\ell\right)\mathrm{d}f \\ & + \frac{1}{n}\int\left(\langle\mathcal{H}_{2,0}\rangle_g\frac{r^2}{a^2\eta} - \langle\langle\mathcal{H}_{2,0}\rangle_g\rangle_\ell\right)\mathrm{d}f \\ & + \frac{1}{n}\int\left(\{\mathcal{K}_{1,0} + \mathcal{K}_{0,1};\mathcal{U}_1\}\frac{r^2}{a^2\eta} - \langle\{\mathcal{K}_{1,0} + \mathcal{K}_{0,1};\mathcal{U}_1\}\rangle_\ell\right)\mathrm{d}f, \end{aligned} \tag{6.77}$$

which in view of Eqs. (6.62) and (6.74), (6.57) and (6.73), and (6.72) and (6.75), is integrated without difficulty in closed form of the eccentricity.

6.4.3 Alfriend and Coffey's elimination of the perigee

Rather than starting from a standard normalization of the total angular momentum, Alfriend and Coffey took the approach of eliminating the perigee from the simplified Hamiltonian obtained after the elimination of the parallax [14]. That is,

$$\mathcal{H} = \mathcal{H}_{0,0} + J_2\mathcal{H}_{1,0} + \frac{1}{2}J_2^2\mathcal{H}_{2,0} + \mathcal{O}(J_2^3),$$

where $\mathcal{H}_{0,0}$ is the Keplerian, $\mathcal{H}_{1,0}$ is given in Eq. (6.30), and $\mathcal{H}_{2,0}$ is given in Eq. (6.33). Namely,

$$\begin{aligned}\mathcal{H}_{2,0} = {} & \frac{\mu}{p}\frac{p^2}{r^2}\left[\frac{R_\oplus^4}{p^4}(q_{0,0} + 2q_{0,1}s^2\cos 2\omega)\right. \\ & \left. + 2\sum_{i\geq 3}\tilde{J}_i\frac{R_\oplus^i}{p^i}\sum_{j=0}^{i}\mathcal{F}_{i,j}(s)\mathcal{G}_{i,j}^*(e)\cos[(i-2j)\omega - i_\pi]\right],\end{aligned}$$

where $q_{0,0}$ and $q_{0,1}$ are given in Table 6.1.

Now, $\mathcal{H}_{1,0}$ is already free from the argument of the periapsis. Therefore, $\mathcal{H}_{0,1} = \mathcal{H}_{1,0}$ and $\mathcal{W}_1 = C_1$. At second order, the computable terms $\widetilde{\mathcal{H}}_{0,2}$ in Eq. (2.37) that enter the homological equation are $\widetilde{\mathcal{H}}_{0,2} = \mathcal{H}_{2,0} + 2\{\mathcal{H}_{1,0}; C_1\}$, in which the evaluation of the involved Poisson bracket yields

$$\begin{aligned}\{\mathcal{H}_{1,0}; C_1\} = {} & \frac{1}{4}\frac{R_\oplus^2}{p^2}\frac{G^2}{r^2}(2-3s^2)(2e\sin f + e^2\sin 2f)\frac{1}{\eta^3}\frac{\partial C_1}{\partial L} \\ & + \frac{1}{4}\frac{R_\oplus^2}{p^2}\frac{G}{r^2}\left[3(4-5s^2) + (2-3s^2)\left(\frac{2}{e}\cos f + \cos 2f\right)\right]\frac{\partial C_1}{\partial g}.\end{aligned}$$

Alfriend and Coffee impose the condition that the new Hamiltonian be free from long-period effects. In view of C_1 going to be a periodic function of g, the partial derivatives $\partial C_1/\partial L$ and $\partial C_1/\partial g$ will be also periodic in g, and hence $\langle\{\mathcal{H}_{1,0}; C_1\}\rangle_g = 0$. Then

$$\mathcal{H}_{0,2} = \langle \mathcal{H}_{2,0}\rangle_g = \frac{\mu}{p}\frac{p^2}{r^2}\left[\frac{R_\oplus^4}{p^4}q_{0,0} + 2\sum_{i\geq 2}\tilde{J}_{2i}\frac{R_\oplus^{2i}}{p^{2i}}\mathcal{F}_{2i,i}(s)\mathcal{G}^*_{2i,i}(e)\right], \tag{6.78}$$

which is of similar complexity to that of Eq. (6.70). Then

$$\begin{aligned}\widetilde{\mathcal{H}}_{0,2} - \mathcal{H}_{0,2} = {} & \frac{1}{2}\frac{R_\oplus^2}{p^2}\frac{G^2}{r^2}(2-3s^2)(2e\sin f + e^2\sin 2f)\frac{1}{\eta^3}\frac{\partial C_1}{\partial L} \\ & + \frac{1}{2}\frac{R_\oplus^2}{p^2}\frac{G}{r^2}(2-3s^2)\left(\frac{2}{e}\cos f + \cos 2f\right)\frac{\partial C_1}{\partial g} \\ & + \frac{G}{r^2}\frac{R_\oplus^2}{p^2}\frac{3}{2}(4-5s^2)\frac{\partial C_1}{\partial g} + 2\frac{G^2}{r^2}\frac{R_\oplus^4}{p^4}q_{0,1}s^2\cos 2\omega \\ & + 2\frac{G^2}{r^2}\sum_{i\geq 3}\tilde{J}_i\frac{R_\oplus^i}{p^i}\sum_{\substack{j=0\\ j\neq i/2}}^{i}\mathcal{F}_{i,j}\mathcal{G}^*_{i,j}\cos[(i-2j)\omega - i_\pi],\end{aligned} \tag{6.79}$$

where the partial derivative of the integration constant with respect to g has been intentionally split into two parts to show that the last three terms must mutually cancel to prevent the appearance of offending secular terms in the generating function. The canceling is effectively achieved with the choice

$$\begin{aligned}C_1 = -\frac{4G}{3(4-5s^2)}\Bigg\{ & \frac{R_\oplus^2}{p^2}\frac{3}{64}(14-15s^2)s^2e^2\frac{\sin 2\omega}{2} \\ & + \sum_{i\geq 3}\tilde{J}_i\frac{R_\oplus^{i-2}}{p^{i-2}}\sum_{\substack{j=0\\ j\neq i/2}}^{i}\mathcal{F}_{i,j}\mathcal{G}^*_{i,j}\frac{\sin[(i-2j)\omega - i_\pi]}{i-2j}\Bigg\}.\end{aligned} \tag{6.80}$$

After replacing Eq. (6.80) into Eq. (6.79), the homological equation (5.39) is integrated to give

$$\mathcal{W}_2 = \frac{R_\oplus^2}{p^2}\frac{2-3s^2}{4}\left[\left(\frac{4}{e}\sin f + \sin 2f\right)\frac{\partial C_1}{\partial g} - (4e\cos f + e^2\cos 2f)\frac{G}{\eta^3}\frac{\partial C_1}{\partial L}\right], \tag{6.81}$$

in which

$$\begin{aligned}\frac{\partial C_1}{\partial g} = -\frac{4G}{3(4-5s^2)}\Bigg\{ & \frac{R_\oplus^2}{p^2}\frac{3}{64}(14-15s^2)s^2e^2\cos 2\omega \\ & + \sum_{i\geq 3}\tilde{J}_i\frac{R_\oplus^{i-2}}{p^{i-2}}\sum_{\substack{j=0\\ j\neq i/2}}^{i}\mathcal{F}_{i,j}\mathcal{G}^*_{i,j}\cos[(i-2j)\omega - i_\pi]\Bigg\},\end{aligned}$$

$$\frac{\partial C_1}{\partial L} = -\frac{4\eta}{3(4-5s^2)}\left\{\frac{R_\oplus^2}{p^2}\frac{3}{32}(14-15s^2)s^2\eta^2\frac{\sin 2\omega}{2}\right.$$
$$\left. + \sum_{i\geq 3}\tilde{J}_i\frac{R_\oplus^{i-2}}{p^{i-2}}\sum_{\substack{j=0\\ j\neq i/2}}^{i}\mathcal{F}_{i,j}\frac{\eta^2}{e}\frac{\partial\mathcal{G}_{i,j}^*}{\partial e}\frac{\sin[(i-2j)\omega - i_\pi]}{i-2j}\right\}.$$

The consequent elimination of short-period terms in a Delaunay normalization follows the same steps as in §6.4.2, producing the same first-order Hamiltonian as the one in Eq. (6.71); yet we have the simpler first-order generating function

$$\mathcal{U}_1 = \frac{1}{n}\mathcal{K}_{0,1}\phi. \tag{6.82}$$

At second order $\widetilde{\mathcal{K}}_{0,2}$ is given in Eq. (2.37), and we choose the new Hamiltonian term $\mathcal{K}_{0,2} = \langle\widetilde{\mathcal{K}}_{0,2}\rangle_\ell = \langle\mathcal{K}_{2,0}\rangle_\ell + \langle\{\mathcal{K}_{1,0};\mathcal{U}_1\}\rangle_\ell + \langle\{\mathcal{K}_{0,1};\mathcal{U}_1\}\rangle_\ell$, where

$$\langle\{\mathcal{K}_{1,0};\mathcal{U}_1\}\rangle_\ell + \langle\{\mathcal{K}_{0,1};\mathcal{U}_1\}\rangle_\ell = -\frac{\mu}{a}\frac{R_\oplus^4}{p^4}\eta\frac{1}{16}(2-3s^2)^2(1+3\eta)$$

and

$$\langle\mathcal{K}_{2,0}\rangle_\ell = -\frac{\mu}{a}\frac{R_\oplus^4}{p^4}\frac{\eta}{16}\left[20-42s^2+21s^4+\frac{3}{4}(8-8s^2-5s^4)e^2\right]$$
$$+2\frac{\mu}{a}\eta\sum_{i\geq 2}\tilde{J}_{2i}\frac{R_\oplus^{2i}}{p^{2i}}\mathcal{F}_{2i,i}(s)\mathcal{G}_{2i,i}^*(e),$$

the addition of both terms providing the same $\mathcal{K}_{0,2}$ first obtained in Eq. (6.50), with the standard Brouwer approach, and then in Eq. (6.76) without the need of resorting to the elimination of the parallax simplification.

6.5 Higher orders of the perturbation solution. A test case

To illustrate how the use of canonical simplifications help to disentangle the computations required in higher-order solutions, and how these higher orders refine the accuracy of an analytical solution computed in Brouwer's style, we focus now on the J_2 problem alone and extend Brouwer's perturbation solution to higher orders. Due to the particular value of the Earth's J_2 coefficient, working in standard floating-point arithmetic [332] the analytical solution reaches the numerical precision at order five, which means neglecting terms of $\mathcal{O}(10^{-18})$ and higher in the perturbation approach [408]. To avoid long listings, we only present explicit expressions up to order three for the generating function, from which the computation of the periodic corrections only involves the evaluation of Poisson brackets, as described in §2.1.3. Still, since the elimination of the perigee is affected by an offset between the orders with which the Hamiltonian and the generating function are fully determined, we provide also explicit expressions of the fourth-order Hamiltonian terms of each transformation.

6.5.1 Preliminary simplification. Elimination of the parallax

The first two orders already have been computed in §6.2.1. Instead of the physical small parameter J_2, we use now a formal small parameter ε and abbreviate, for convenience,

$$\tilde{\epsilon} \equiv -\frac{1}{4}C_{2,0}\frac{R_\oplus^2}{p^2}, \tag{6.83}$$

which is not constant as far as $p = p(G)$ and g is not yet a cyclic variable. Thus, Eq. (6.30) turns into

$$\mathcal{H}_{0,1} = \tilde{\epsilon}\frac{\mu}{r}\frac{p}{r}(3s^2 - 2), \tag{6.84}$$

and Eq. (6.31) into

$$\mathcal{W}_1 = G\tilde{\epsilon}\sum_{l=0}^{1}\sum_{k=1}^{2l+1} P_{1,l,k}(s)e^{k^*}s^{2l}\sin(kf + 2l\omega), \tag{6.85}$$

where $P_{1,0,1} = 3s^2 - 2$, $P_{1,1,1} = P_{1,1,2} = -\frac{3}{2}$, $P_{1,1,3} = -\frac{1}{2}$, and $k^\star = k \bmod 2$ from the index convention in Eq. (6.5).

At second order, after neglecting higher-order zonal harmonics from Eq. (6.33) we arrange the Hamiltonian in the general form

$$\mathcal{H}_{0,i} = \tilde{\epsilon}^i\frac{\mu}{r}\frac{p}{r}\sum_{j=0}^{i_0}\sum_{k=0}^{i_0-j} e^{2k+2j}\rho_{i,j,k}(s)s^{2j}\cos 2j\omega, \tag{6.86}$$

with $i_0 = \lfloor\frac{1}{2}i\rfloor$ from the index convention in Eq. (6.5). The coefficients $\rho_{2j,k}$ are provided in Table 6.5, jointly with the inclination polynomials of previous and following orders.

Proceeding analogously, we rearrange Eq. (6.34) in the general form

$$\mathcal{W}_i = G\tilde{\epsilon}^i\sum_{l=0}^{i}\sum_{k=2l-i}^{2l+i}\sum_{j=0}^{i_0} P_{i,l,k,j}(s)e^{2j+k^*}s^{2l}\sin(kf + 2l\omega), \tag{6.87}$$

where the inclination polynomials $P_{2,l,k,j}$ are given in Table 6.6.

At third order, Deprit's recursion (2.15) yields the computable terms

$$\widetilde{\mathcal{H}}_{0,3} = \{\mathcal{H}_{0,1} + 2\mathcal{H}_{1,0}, \mathcal{W}_2\} + \{\mathcal{H}_{0,2} + \mathcal{H}_{1,1} + \mathcal{H}_{2,0}, \mathcal{W}_1\} + \mathcal{H}_{3,0}, \tag{6.88}$$

where $\mathcal{H}_{3,0} = 0$ for the main problem. Terms $1/r^m$ with $m > 2$ are rewritten using the identity (5.30). Then, after expressing $\widetilde{\mathcal{H}}_{0,3}$ in the form of a Fourier series in f, we choose the simplified Hamiltonian by removing the trigonometric terms that depend explicitly on the true anomaly. The new Hamiltonian term is obtained from Eq. (6.86), in which, now, $i = 3$, and the inclination polynomials $\rho_{3,j,k}$ are provided in Table 6.5;

Table 6.5: Inclination polynomials $p_{i,j,k}$ in Eq. (6.86) up to $i = 4$. Adapted by permission from Springer [440].

1,0,0 :	$3s^2 - 2$
2,0,0 :	$-21s^4 + 42s^2 - 20$
2,0,1 :	$\frac{3}{4}(5s^4 + 8s^2 - 8)$
2,1,0 :	$-\frac{3}{2}(15s^2 - 14)$
3,0,0 :	$3(420s^6 - 987s^4 + 756s^2 - 208)$
3,0,1 :	$\frac{3}{4}(2715s^6 - 4542s^4 + 2232s^2 - 464)$
3,1,0 :	$-\frac{9}{8}(285s^4 - 208s^2 - 32)$
4,0,0 :	$-6(9210s^8 - 33366s^6 + 43719s^4 - 25212s^2 + 5344)$
4,0,1 :	$\frac{3}{16}(865425s^8 - 1315296s^6 + 120048s^4 + 510336s^2 - 155008)$
4,0,2 :	$-\frac{27}{64}(65985s^8 - 164400s^6 + 125824s^4 - 32960s^2 + 3712)$
4,1,0 :	$-\frac{3}{8}(786705s^6 - 1970794s^4 + 1602800s^2 - 432480)$
4,1,1 :	$-\frac{27}{16}(5045s^6 - 3734s^4 - 4464s^2 + 2688)$
4,2,0 :	$\frac{27}{32}(14235s^4 - 25280s^2 + 11144)$

Table 6.6: Non-zero coefficients $P_{2,l,k,j}$ in Eq. (6.87); $P_{2,2,5,0} = P_{2,2,4,0}$.

0,1,0 : $-\frac{1}{4}(27s^4 - 108s^2 + 64)$	1,2,0 : $20 - 21s^2$	1,4,1 : $\frac{3}{8}(13s^2 - 10)$	2,4,0 : $\frac{3}{4}$
0,2,1 : $\frac{3}{16}(5s^4 + 8s^2 - 8)$	1,2,1 : $-\frac{3}{2}(s^2 - 2)$	2,2,1 : $-\frac{15}{16}$	2,4,1 : $-\frac{3}{16}$
1,1,0 : $-7(12s^2 - 11)$	1,3,0 : $8s^2 - 5$	2,3,0 : $-\frac{3}{4}$	2,6,1 : $\frac{3}{16}$

cf. [118, 440]. The homological equation (5.39) gives $\mathcal{W}_3$ in the form of Eq. (6.87) with $i = 3$, with the inclination polynomials $P_{3,l,k,j}$ provided in Table 6.7.

From Deprit's recursion (2.15), the known terms at fourth order are

$$\widetilde{\mathcal{H}}_{0,4} = \{\mathcal{H}_{0,1} + 3\mathcal{H}_{1,0}, \mathcal{W}_3\} + \{\mathcal{H}_{0,2} + 2\mathcal{H}_{1,1} + 3\mathcal{H}_{2,0}, \mathcal{W}_2\} + \{\mathcal{H}_{0,3} + \mathcal{H}_{1,2} + \mathcal{H}_{2,1} + \mathcal{H}_{3,0}, \mathcal{W}_1\} + \mathcal{H}_{4,0}, \tag{6.89}$$

in which $\mathcal{H}_{4,0} = 0$ for the J_2-problem. After evaluation, the inverse powers of r with exponents higher than 2 are reduced to $1/r^2$ using Eq. (5.30). Then the usual trigonometric reduction is carried out to cast $\widetilde{\mathcal{H}}_{0,4}$ in the form of a multivariate Fourier series in the true anomaly and the argument of the periapsis, from which the new Hamiltonian term $\mathcal{H}_{0,4}$ is chosen by removing from $\widetilde{\mathcal{H}}_{0,4}$ the terms that depend explicitly on the true anomaly f. The result is also arranged in the general form of Eq. (6.86), with corresponding coefficients on Table 6.5.

The procedure ends changing osculating by prime variables in the terms $\mathcal{H}_{0,m}$. However, we avoid the prime notation in what follows for brevity except when there may be risk of confusion.

Table 6.7: Non-zero coefficients $P_{3,l,k,j}$ in Eq. (6.87); $P_{3,3,7,0} = P_{3,3,5,0}$.

$0,1,0: \frac{1}{8}(23523s^6 - 39174s^4 + 20128s^2 - 4416)$	$2,3,1: -\frac{3}{32}(73s^2 - 126)$
$0,1,1: \frac{3}{32}(3393s^6 - 6222s^4 + 3328s^2 - 736)$	$2,4,0: \frac{1}{4}(276s^2 - 233)$
$0,2,1: \frac{1}{8}(5790s^6 - 8549s^4 + 3448s^2 - 600)$	$2,4,1: \frac{1}{16}(6704 - 6951s^2)$
$0,3,1: \frac{1}{32}(543s^6 - 1018s^4 + 608s^2 - 160)$	$2,5,0: \frac{1}{20}(4455s^2 - 3934)$
$1,-1,1: \frac{3}{64}(6921s^4 - 12384s^2 + 5408)$	$2,5,1: \frac{9}{160}(175s^2 - 82)$
$1,1,0: \frac{1}{16}(-74025s^4 + 111232s^2 - 40992)$	$2,6,1: \frac{1}{32}(3813s^2 - 3398)$
$1,1,1: \frac{3}{64}(13043s^4 - 30208s^2 + 16064)$	$2,7,1: \frac{9}{224}(417s^2 - 374)$
$1,2,0: \frac{1}{2}(-3069s^4 + 4976s^2 - 1952)$	$3,3,1: -\frac{153}{64}$
$1,2,1: -\frac{3}{4}(1833s^4 - 3326s^2 + 1462)$	$3,4,1: -\frac{9}{8}$
$1,3,0: \frac{1}{16}(-21289s^4 + 44928s^2 - 21856)$	$3,5,0: \frac{135}{16}$
$1,3,1: \frac{3}{64}(263s^4 + 224s^2 - 288)$	$3,5,1: -\frac{27}{64}$
$1,4,1: \frac{1}{8}(-2076s^4 + 5763s^2 - 3208)$	$3,6,0: \frac{15}{2}$
$1,5,1: \frac{9}{320}(435s^4 + 256s^2 - 448)$	$3,6,1: \frac{51}{8}$
$2,1,1: \frac{3}{32}(2427s^2 - 2354)$	$3,7,1: \frac{81}{64}$
$2,2,1: \frac{1}{32}(31226 - 33783s^2)$	$3,8,1: \frac{207}{64}$
$2,3,0: -2(699s^2 - 652)$	$3,9,1: \frac{27}{64}$

6.5.2 Partial normalization. Long-period elimination

Up to fourth order, the Hamiltonian after the elimination of the parallax is

$$\mathcal{K} = \sum_{i=0}^{4} \frac{\varepsilon^i}{i!} \mathcal{K}_{i,0} \tag{6.90}$$

where $\mathcal{K}_{0,0}$ is the Keplerian term, and, for $i > 0$ the Hamiltonian terms adopt the general, compact form of Eq. (6.86). That is, $\mathcal{K}_{i,0} = \mathcal{H}_{0,i}$ in which, now, the quantities p, r, e, s, and ω are functions of the prime Delaunay variables.

Next, the long-period terms are eliminated like in [14]. The two first orders of Alfriend and Coffey's elimination the perigee have already been computed in §6.4.3. Thus, Eq. (6.84) remains unaltered while $\mathcal{U}_1 \equiv C_1$ is obtained by dropping the contribution of the zonal harmonics of higher degree from Eq. (6.80). In the new scale due to the current use of a formal small parameter ε, opposite to the physical parameter J_2, we obtain

$$\mathcal{U}_1 = G\tilde{\epsilon}\frac{15s^2 - 14}{8(5s^2 - 4)} s^2 e^2 \sin 2\omega. \tag{6.91}$$

Analogously, from Eq. (6.78),

$$\mathcal{K}_{0,2} = \tilde{\epsilon}^2 \frac{\mu}{r}\frac{p}{r}\left[\frac{3}{4}e^2(5s^4 + 8s^2 - 8) - 21s^4 + 42s^2 - 20\right], \tag{6.92}$$

and from Eq. (6.81), up to an integration constant,

$$\mathcal{U}_2 = \tilde{\epsilon}^2 G \frac{15s^2 - 14}{4(5s^2 - 4)}(3s^2 - 2)s^2\left[4e\sin(f + 2\omega) + e^2\sin(2f + 2\omega)\right] + C_2, \tag{6.93}$$

where $C_2 \equiv C_2(g, L, G, H)$ will be determined at the next order by imposing the necessary conditions to get a generating function purely periodic in f.

The computable terms $\widetilde{\mathcal{K}}_{0,3}$ of the third order of the perturbation approach are the same as those of Eq. (6.88), rewritten in $\mathcal{K}$ instead of $\mathcal{H}$, in which the intermediate term of Deprit's triangle (2.16) $\mathcal{K}_{1,1} = \mathcal{K}_{0,2} - \{\mathcal{K}_{0,1}; \mathcal{U}_1\}$ needs to be computed first. Then, after evaluation of the involved Poisson brackets and the consequent use of Eq. (5.30) to reduce the exponents of the radius to only the case $1/r^2$, $\widetilde{\mathcal{K}}_{0,3}$ is reformulated as a Fourier series in f, from which the new Hamiltonian term $\mathcal{K}_{0,3}$ is obtained by selecting those terms that are free from both f and g.

In this process we need to evaluate the Poisson bracket $\{3\mathcal{K}_{1,0}, \mathcal{U}_2\}$, as follows from Eq. (6.88). On account of $\mathcal{K}_{1,0} \equiv \mathcal{K}_{1,0}(\ell, -, -, L, G, H)$ and $C_2 \equiv C_2(-, g, -, L, G, H)$, the Poisson bracket involves partial differentiation of the yet undetermined integration "constant" C_2, on which $\mathcal{U}_2$ depends upon, with respect to the Delaunay variables g and L.

Once $\widetilde{\mathcal{K}}_{0,3}$ is computed, we proceed first by identifying the terms $T_{3,l}$ pertaining to the kernel of the Lie derivative. That is, $\partial T_{3,l}/\partial\ell = 0$. We find

$$\sum_{l=0}^{3} T_{3,l} = \tilde{\epsilon}^3 \frac{G^2}{r^2} \sum_{l=0}^{2} \sum_{j=0}^{2-l} \frac{\gamma_{3,j,l}(s) e^{2j+2l}}{(5s^2 - 4)^{j+2l_0}} s^{2l} \cos 2l\omega + 9\tilde{\epsilon}\frac{G}{r^2}\frac{\partial C_2}{\partial g}(5s^2 - 4), \tag{6.94}$$

where the inclination polynomials $\gamma_{3,j,l}$ are given in Table 6.8; cf. [440]. The term $T_{3,0}$, the first of the summation in Eq. (6.94), will be canceled of the homological equation out by making the new Hamiltonian to have an analogous term. On the contrary, as far as the new Hamiltonian must not depend on g, the other terms of the kernel of the Lie derivative cannot be canceled with this artifact. Nevertheless, in view of C_2 being yet formal, these terms of the kernel are removed from the homological equation by choosing C_2 in such a way that $T_{3,1} + T_{3,2} + T_{3,3}(C_2) = 0$. Straightforward integration yields

$$C_2 = -\tilde{\epsilon}^2 G \sum_{l=1}^{2} \sum_{j=0}^{2-l} \frac{\gamma_{3,j,l}(s)}{18l(5s^2 - 4)^{j+2l-1}} e^{2j+2l} s^{2l} \sin 2l\omega, \tag{6.95}$$

which resolves the arbitrariness in the definition of the second-order term of the generating function of the long-period elimination in Eq. (6.93).

Now, we can compute the partial derivatives of C_2 involved in $\widetilde{\mathcal{K}}_{0,3}$, which is in this way fully materialized, and solve the homological equation

$$\mathcal{U}_i = \frac{1}{n}\int(\widetilde{\mathcal{K}}_{0,i} - \mathcal{K}_{0,i})\mathrm{d}\ell = \frac{1}{n}\int(\widetilde{\mathcal{K}}_{0,i} - \mathcal{K}_{0,i})\frac{r^2}{a^2\eta}\,\mathrm{d}f, \tag{6.96}$$

Table 6.8: Inclination polynomials $\gamma_{i,j,l}$. Terms $\gamma_{i,0,0} = \rho_{i,0,0}$ are in Table 6.5.

2,1,0:	$\frac{3}{4}(5s^4+8s^2-8)$
3,1,0:	$\frac{3}{8}(26475s^8-65880s^6+58068s^4-22496s^2+3712)$
3,2,0:	$-\frac{9}{16}s^2(15s^2-14)(450s^6-925s^4+590s^2-112)$
3,0,1:	$-\frac{27}{4}(125s^4-207s^2+88)$
3,1,1:	$\frac{9}{16}(15s^2-14)(45s^4+36s^2-56)$
3,0,2:	$-\frac{9}{16}(15s^2-14)^2(15s^2-13)$
4,1,0:	$\frac{3}{16}(4298775s^{10}-9984900s^8+5832072s^6+2075520s^4-2816384s^2+620032)$
4,2,0:	$-\frac{9}{64}(7601625s^{12}-29437800s^{10}+44628520s^8-33274368s^6+12549504s^4-2249728s^2+178176)$
4,3,0:	$\frac{9}{8}s^2(15s^2-14)(6750s^{10}-10125s^8-5670s^6+17078s^4-9576s^2+1568)$
4,0,1:	$-\frac{3}{8}(16407375s^{10}-68238100s^8+112604400s^6-92289344s^4+37587200s^2-6084096)$
4,1,1:	$\frac{3}{8}(28676250s^{12}-126428625s^{10}+227723100s^8-213964120s^6+110351824s^4-29582080s^2+3225600)$
4,2,1:	$-\frac{3}{32}(15s^2-14)(2328750s^{12}-8703375s^{10}+13317150s^8-10848180s^6+5157560s^4-1450624s^2+200704)$
4,0,2:	$\frac{3}{16}(1235250s^8-4334325s^6+5686350s^4-3304956s^2+717728)$
4,1,2:	$\frac{3}{32}(15s^2-14)^2(1800s^6+2655s^4-8208s^2+3928)$
4,0,3:	$-\frac{3}{32}(15s^2-14)^3(825s^4-1445s^2+634)$

in which $\mathcal{K}_{0,3} = T_{3,0}$ takes the general form

$$\mathcal{K}_{0,i} = \tilde{\epsilon}^i \frac{\mu}{r}\frac{p}{r} \sum_{j=0}^{i-1} \frac{\gamma_{i,j,0}(s)}{(5s^2-4)^j} e^{2j}, \tag{6.97}$$

with $i = 3$ in the current case. We obtain

$$\mathcal{U}_3 = \frac{G\tilde{\epsilon}^3}{(5s^2-4)^3} \sum_{j=0}^{2} \sum_{l=0}^{2} \sum_{\substack{k=2l_0-2 \\ k\neq 0}}^{2l_{-1}+2} \Gamma_{3,l,k,j} e^{2j-k^*} s^{2l} \sin(kf + 2l\omega) + C_3, \tag{6.98}$$

with $l_{-1} \equiv \lfloor \frac{1}{2}(l+1) \rfloor$ and $l_0 \equiv \lfloor \frac{1}{2} l \rfloor$ from Eq. (6.5), and the 18 non-zero inclination polynomials $\Gamma_{3,l,k,j}$ are listed in Table 6.9. Once again, the integration constant $C_3 \equiv C_3(g, L, G)$ will remain undetermined until the next order.

At fourth order, the procedure starts again filling Deprit's triangle (2.16) with the computable terms $\mathcal{K}_{1,2} = \mathcal{K}_{0,3} - \{\mathcal{K}_{0,2}; \mathcal{U}_1\}$, and $\mathcal{K}_{2,1} = \mathcal{K}_{1,2} - \{\mathcal{K}_{1,0}; \mathcal{U}_2\} - \{\mathcal{K}_{1,1}; \mathcal{U}_1\}$, which are needed in the computation of $\widetilde{\mathcal{K}}_{0,4}$ from Eq. (6.89), where $\mathcal{H}$ is replaced by $\mathcal{K}$. Again, this process involves partial differentiation of C_3 with respect to the Delaunay variables g and L. After reducing the inverse powers of the radius to coefficients $1/r^2$ with the

Table 6.9: Non-zero inclination polynomials $\Gamma_{3,l,k,j}$ in Eq. (6.98). $\Gamma_{3,0,2,1} = \frac{1}{4}\Gamma_{3,0,1,1}$, $\Gamma_{3,1,-1,2} = 4\Gamma_{3,1,-2,2}$, $4\Gamma_{3,1,2,2} = \Gamma_{3,1,1,2}$, $16\Gamma_{3,1,4,1} = 2\Gamma_{3,1,3,1} = \Gamma_{3,1,2,0}$, and $30\Gamma_{3,2,4,2} = 5\Gamma_{3,2,3,2} = 3\Gamma_{3,2,2,1}$.

0,1,1	$-\frac{1}{16}s^4(3s^2-2)(5s^2-4)(15s^2-14)^2$
0,1,2	$-\frac{1}{32}s^2(15s^2-14)(6075s^8-15960s^6+14556s^4-5104s^2+448)$
0,2,2	$-\frac{1}{64}s^2(15s^2-14)(2925s^8-7710s^6+7064s^4-2496s^2+224)$
1,−2,2	$-\frac{3}{64}(3s^2-2)(5s^2-4)(15s^2-14)(45s^4+36s^2-56)$
1,1,1	$\frac{3}{2}(5s^2-4)^2(3105s^6-7251s^4+5598s^2-1424)$
1,1,2	$-\frac{9}{16}(5s^2-4)(15s^2-14)(85s^6-22s^4-96s^2+48)$
1,2,0	$3(3s^2-2)^2(5s^2-4)^2(15s^2-14)$
1,2,1	$\frac{3}{8}(5s^2-4)^2(1890s^6-4317s^4+3258s^2-808)$
2,1,2	$-\frac{1}{32}(15s^2-14)^2(135s^4-242s^2+112)$
2,2,1	$-\frac{5}{32}(3s^2-2)(5s^2-4)(15s^2-14)^2$
2,2,2	$-\frac{1}{32}(15s^2-14)^2(30s^4-55s^2+26)$

help of Eq. (5.30), and carrying out the usual trigonometric reduction, we identify five terms of $\widetilde{\mathcal{K}}_{0,4}$ pertaining to the kernel of the Lie derivative. Namely,

$$\sum_{l=0}^{4} T_{4,l} = \tilde{\epsilon}^4 \frac{G^2}{r^2} \sum_{l=0}^{3} \sum_{j=0}^{3-l} \frac{e^{2j+2l}\gamma_{4,j,l}s^{2l}}{(5s^2-4)^{j+2l_{-1}}} \cos 2l\omega + \tilde{\epsilon}\frac{G}{r^2}12(5s^2-4)\frac{\partial C_3}{\partial g},$$

where the inclination polynomials $\gamma_{4,j,l}$ are also displayed in Table 6.8.

The term $T_{4,0}$ of the summation does not depend on g and is easily nullified with an equivalent term of the new Hamiltonian, which takes again the form of Eq. (6.97). The remaining terms of the kernel are removed from the homological equation by choosing C_3 in such a way that $T_{4,1} + T_{4,2} + T_{4,3} + T_{4,4}(C_3) = 0$. We readily obtain

$$C_3 = -\tilde{\epsilon}^3 G \sum_{l=1}^{3} \sum_{j=0}^{3-l} \frac{\gamma_{4,j,l}}{24(5s^2-4)^{j+1+2l_{-1}}} e^{2j+2l} s^{2l} \sin 2l\omega. \tag{6.99}$$

The procedure ends replacing prime by double-prime Delaunay variables in the new Hamiltonian terms $\mathcal{K}_{0,i}$, $i = 0, \dots, 4$, which are free from long-period effects and only depend on short-period effects through the coefficient $1/r^2$.

6.5.3 Complete Hamiltonian reduction. Short-period elimination

Because the argument of the perigee is cyclic in the Hamiltonian in double-prime variables, up to the truncation order, G'' is constant and so it is p in Eq. (6.83). Therefore,

$\tilde{\epsilon}$ is constant in the double-prime variables and is taken hereafter as the small parameter of the perturbation approach. Then the Hamiltonian after the long-period elimination is written like $\mathcal{Q} = \sum_{i=0}^{4}(\tilde{\epsilon}^i/i!)\mathcal{Q}_{i,0}$, where $\mathcal{Q}_{0,0}$ is the Keplerian term, and, from Eq. (6.97),

$$\mathcal{Q}_{i,0} = \frac{\mu}{p}\frac{p^2}{r^2}\sum_{j=0}^{i-1}\left(\frac{e^2}{5s^2-4}\right)^j \gamma_{i,j,0}(s), \quad i > 0,$$

where p, r, e, and s are now functions of the double-prime Delaunay variables, and the inclination polynomials $\gamma_{i,j,0}$ were previously given in Tables 6.8 and 6.5 up to the fourth order of J_2.

The first order of the Delaunay normalization has already been computed, and is obtained by rescaling Eq. (6.71), for the Hamiltonian, and Eq. (6.82) for the generating function, taking into account that now we are using $\tilde{\epsilon}$ as the small parameter of the perturbation arrangement. Namely,

$$\mathcal{Q}_{0,1} = \frac{\mu}{p}\eta^3(3s^2-2), \quad \mathcal{V}_1 = G(3s^2-2)\phi. \tag{6.100}$$

At second order, after reducing as usual the exponents of the inverse powers of r, the computable terms yield

$$\widetilde{\mathcal{Q}}_{0,2} = -3\frac{\mu}{p}(3s^2-2)^2\eta^4 - \frac{\mu}{p}\frac{p^2}{r^2}\left[6(s^2-1)(5s^2-4) \right.$$
$$\left. - \frac{3}{4}(5s^4+8s^2-8)e^2 + \beta(3s^2-2)^2(2e\cos f + e^2\cos 2f)\right],$$

from which the new Hamiltonian is chosen by averaging short-period terms. Thus, $\mathcal{Q}_{0,2} = \langle\widetilde{\mathcal{Q}}_{0,2}\rangle_\ell$. The averaging is carried out in closed form with the help of the differential relation (4.64), and is rearranged in the general form

$$\mathcal{Q}_{0,i} = \frac{\mu}{p}\frac{\eta^3}{(5s^2-4)^{i-1}}\sum_{j=0}^{2i-2}\eta^j\Lambda_{i,j}, \tag{6.101}$$

where division by $5s^2-4$ does not happen in this particular case ($i = 2$) due to analogous factors in the inclination polynomials $\lambda_{2,j}$, as checked in Table 6.10; cf. [440]. As expected, the term $\mathcal{Q}_{0,2}$ is the same as $\mathcal{K}_{0,2}$ in Eq. (6.76) after removing zonal harmonics of higher degree than 2, and making a rescaling due to the different small parameter.

The homological equation takes the form of (4.69), where $\mathcal{H}$ is replaced by $\mathcal{Q}$ and $\mathcal{W}$ by $\mathcal{V}$, in which the integrand is purely periodic in f. Therefore, it is readily solved in closed form of the eccentricity. At second order we obtain

$$\mathcal{V}_2 = G\phi\sum_{j=0}^{1}\Phi_{2,j}e^{2j} - G\beta(3s^2-2)^2\sum_{j=1}^{2}\Lambda_{2,j}e^j\sin jf,$$

with $\Phi_{2,0} = 6(1-s^2)(5s^2-4)$, $\Phi_{2,1} = \frac{3}{4}(5s^4+8s^2-8)$, $\Lambda_{2,1} = 2$, and $\Lambda_{2,2} = \frac{1}{2}$.

Table 6.10: Inclination polynomials $\lambda_{i,j}$ in Eq. (6.101).

2,0:	$-\frac{15}{4}(7s^4-16s^2+8)(5s^2-4)$
2,1:	$-3(3s^2-2)^2(5s^2-4)$
2,2:	$-\frac{3}{4}(5s^4+8s^2-8)(5s^2-4)$
3,0:	$\frac{45}{16}(28700s^{10}-107205s^8+158960s^6-118492s^4+45152s^2-7168)$
3,1:	$\frac{135}{4}(3s^2-2)(5s^2-4)^2(7s^4-16s^2+8)$
3,2:	$-\frac{9}{8}(28675s^{10}-98005s^8+130852s^6-87164s^4+30176s^2-4608)$
3,3:	$\frac{45}{4}(3s^2-2)(5s^2-4)^2(5s^4+8s^2-8)$
3,4:	$-\frac{9}{16}s^2(15s^2-14)(450s^6-925s^4+590s^2-112)$
4,0:	$\frac{9}{64}(27768125s^{14}-347238500s^{12}+1247118600s^{10}-2156830160s^8+2074755680s^6$ $-1140109440s^4+335476224s^2-41000960)$
4,1:	$\frac{135}{16}(5s^2-4)(362775s^{12}-1629460s^{10}+3049320s^8-3064144s^6+1755680s^4$ $-547072s^2+72704)$
4,2:	$\frac{45}{32}(6075125s^{14}-18069200s^{12}+9483000s^{10}+26897576s^8-48702032s^6+34262976s^4$ $-11442432s^2+1499136)$
4,3:	$-\frac{45}{8}(5s^2-4)(395775s^{12}-1651940s^{10}+2876216s^8-2709296s^6+1479456s^4$ $-450816s^2+60416)$
4,4:	$\frac{45}{64}(7246125s^{14}-32291700s^{12}+57674440s^{10}-51359984s^8+22168224s^6-2743168s^4$ $-914944s^2+221184)$
4,5:	$-\frac{63}{16}(5s^2-4)(79125s^{12}-299100s^{10}+432600s^8-289840s^6+78944s^4+1280s^2-3072)$
4,6:	$\frac{9}{8}s^2(15s^2-14)(6750s^{10}-10125s^8-5670s^6+17078s^4-9576s^2+1568)$

At third order, the computable terms $\widetilde{\mathcal{Q}}_{0,3}$ are those in Eq. (6.88) with $\mathcal{H}$ replaced by $\mathcal{Q}$. After the usual reduction to powers of $1/r^2$, we find terms that depend on the equation of the center and terms that do not. The first are

$$T_\phi = -3\frac{\mu}{p}(3s^2-2)(5s^4+8s^2-8)\frac{p^2}{r^2}(2e\sin f + e^2\sin 2f)\phi, \tag{6.102}$$

where $p^2/r^2 = (1+e\cos f)^2$ from Eq. (4.29). Then, after carrying out the usual trigonometric reduction, we find that these terms give rise to terms exactly of the same type as those that were first found in §5.6.2. They are analogously averaged, to obtain

$$\langle T_\phi\rangle_\ell = \frac{\mu}{p}\frac{3}{2}(3s^2-2)(5s^4+8s^2-8)\eta^3(2\eta^3+\eta^2-3).$$

Among the remaining terms we find terms that are free from short-period effects and terms with coefficients $1/r^2$. The former are directly incorporated into the new Hamiltonian, and the latter are trivially averaged with the help of the differential relation (4.64). The new Hamiltonian term $\mathcal{Q}_{0,3}$ is then obtained by adding the partial results and takes the general form of Eq. (6.101) with the coefficients $\lambda_{3,j}$ in Table 6.10.

The integration of the homological equation (4.69) is standard except for the term T_ϕ in Eq. (6.102), which needs integration by parts. The final combination of the integrals of the different types of terms yields the solution

$$\mathcal{V}_3 = \frac{G\phi}{(5s^2-4)^2}\sum_{j=0}^{2}\sum_{k=0}^{-4j_1}\Phi_{3,j,k}\eta^k e^j\cos jf + G(3s^2-2)\beta^2\sum_{j=1}^{4}\sum_{k=0}^{4-4j_1}\Lambda_{3,j,k}\eta^{k-1}e^j\sin jf, \tag{6.103}$$

with the inclination polynomials $\Phi_{3,j,k}$ and $\Lambda_{3,j,k}$ of Table 6.11.

Table 6.11: Non-vanishing inclination polynomials $\Phi_{3,j,k}$ (upper) and $\Lambda_{3,j,k}$ (lower) in Eq. (6.103). $\Phi_{3,0,1}=0$, $\Phi_{3,2,0}=\frac{1}{4}\Phi_{3,1,0}=\Phi_{3,0,3}$, $\Lambda_{3,1,4}=4\Lambda_{3,2,4}=-\Lambda_{3,1,3}$, and $\frac{4}{11}\Lambda_{3,2,0}=8\Lambda_{3,4,0}=\Lambda_{3,3,0}=\frac{1}{3}\Lambda_{3,1,0}$.

$0,0:\ \frac{9}{16}(146500s^{10}-538025s^8+784240s^6-574668s^4+215520s^2-33792)$	
$0,2:\ -\frac{3}{8}(109125s^{10}-372775s^8+499020s^6-332724s^4+114080s^2-16896)$	
$0,3:\ \frac{3}{2}(3s^2-2)(5s^2-4)^2(5s^4+8s^2-8)$	
$0,4:\ -\frac{9}{16}s^2(15s^2-14)(450s^6-925s^4+590s^2-112)$	
$1,0:\ -3(3s^2-2)^2$	$1,3:\ -\frac{9}{2}(5s^4+8s^2-8)$
$1,1:\ \frac{1}{2}(189s^4-816s^2+488)$	$2,1:\ \frac{1}{4}(99s^4-348s^2+200)$
$1,2:\ \frac{1}{2}(153s^4-1032s^2+656)$	$2,2:\ \frac{1}{8}(207s^4-840s^2+496)$

At fourth order, the computable terms are those of Eq. (6.89), and need the previous computation of the intermediate terms $\mathcal{Q}_{1,2}$ and $\mathcal{Q}_{2,1}$ from Deprit's triangle (2.16). Terms of $\widetilde{\mathcal{Q}}_{0,4}$ depending on the equation of the center are of the same types as those found at third order and, therefore, are analogously integrated.

The new Hamiltonian term $\mathcal{Q}_{0,4}=\langle\widetilde{\mathcal{Q}}_{0,4}\rangle_\ell$ is also arranged in the general form of Eq. (6.101) with corresponding inclination polynomials $\lambda_{4,j}$ in Table 6.10.

6.5.4 Secular frequencies

Thus, up to the truncation order, we have the completely reduced Hamiltonian

$$\mathcal{S}=\mathcal{S}(-,-,-,L''',G''',H''')\equiv\sum_{m\geq 0}\frac{\epsilon^m}{m!}\mathcal{S}_m, \tag{6.104}$$

in which the terms $\mathcal{S}_m$, $m=0,1,\ldots,$ are obtained from corresponding terms $\mathcal{Q}_{0,m}$ after changing double-prime by triple-prime variables, and we removed the tilde from the small parameter for brevity. The trivial integration of Hamilton equations yields the analytical solution for the secular terms

$$\ell''' = \ell_0''' + n_M t, \quad g''' = g_0''' + n_\omega t \quad h''' = h_0''' + n_\Omega t,$$

where $n_M = \partial\mathcal{S}/\partial L'''$, $n_\omega = \partial\mathcal{S}/\partial G'''$, $n_\Omega = \mathcal{S}/\partial H'''$, $L''' = L_0'''$, $G''' = G_0'''$, $H''' = H_0'''$, and the initialization constants $(\ell_0''', g_0''', h_0''', L_0''', G_0''', H_0''')$ of the analytical theory are computed from corresponding initial conditions after carrying out the different inverse transformations. Details on the particular form taken by the frequencies n_M, n_ω, and n_Ω, can be found in [405].

On the other hand, as mentioned in §5.3, the singularities of Delaunay variables may make the analytical perturbation solution unsuitable in different cases, and a perturbation solution in a set of nonsingular variables may then be preferred. If that is the case, we only need to reformulate the perturbation solution in the desired set of variables without need of computing it from the original Hamiltonian written in the non-singular variables [171, 468].

For instance, the non-canonical set of variables (F, C, S, h, L, H) discussed in §5.7.2, avoids singularity issues related to low-eccentricity orbits, and is customarily used in different applications. In these variables, the secular terms of the main problem are given by $H = H_0$, $L = L_0$, and

$$\begin{aligned}
F &= F_0 + n_F t,\\
C &= e\cos(g_0 + n_\omega t) = C_0 \cos n_\omega t - S_0 \sin n_\omega t,\\
S &= e\sin(g_0 + n_\omega t) = S_0 \cos n_\omega t + C_0 \sin n_\omega t,\\
h &= h_0 + n_\Omega t,
\end{aligned} \tag{6.105}$$

where $F_0 = \ell_0 + g_0$, $C_0 = e\cos g_0$, $S_0 = e\sin g_0$, $e = (1 - G_0^2/L_0^2)^{1/2}$, $n_F = n_M + n_\omega$, and the triple-prime notation has been omitted for simplicity.

After standard partial derivation, we arrange the secular frequencies in the compact form

$$n_F = n + n\sum_{m=1}^{\tilde{m}} \frac{\epsilon^m}{(5s^2-4)^m} \sum_{i=0}^{2m-1} \Psi_{m,i}(s)\eta^i, \tag{6.106}$$

$$n_\omega = n\sum_{m=1}^{\tilde{m}} \frac{\epsilon^m}{(5s^2-4)^m} \sum_{i=0}^{2m-2} \omega_{m,i}(s)\eta^i, \tag{6.107}$$

$$n_\Omega = nc\sum_{m=1}^{\tilde{m}} \frac{\epsilon^m}{(5s^2-4)^m} \sum_{i=0}^{2m-2} \Omega_{m,i}(s)\eta^i, \tag{6.108}$$

where $n = \mu^2/L_0^3$, from Eq. (4.44), and the inclination polynomials $\Psi_{m,i}$, $\omega_{m,i}$, and $\Omega_{m,i}$ are provided in Table 6.12 up to the third order of the perturbation approach. We remark that divisors $5s^2 - 4$ cancel out for the two first orders, as checked in Table 6.12, and only appear starting from the third order [408].

Table 6.12: From top to bottom, inclination polynomials $\Psi_{i,j}$, $\omega_{i,j}$, and $\Omega_{i,j}$ in Eqs. (6.106)–(6.108). The coefficients $\omega_{i,0}$ are the same as $\Psi_{i,0}$ and are not shown. After [408].

1,0:	$-3(5s^2-4)^2$
1,1:	$-3(3s^2-2)(5s^2-4)$
2,0:	$\frac{15}{8}(5s^2-4)^2(77s^4-172s^2+88)$
2,1:	$\frac{9}{8}(5s^2-4)^2(155s^4-256s^2+104)$
2,2:	$\frac{3}{8}(5s^2-4)^2(189s^4-156s^2+8)$
2,3:	$\frac{15}{8}(5s^2-4)^2(5s^4+8s^2-8)$
3,0:	$-\frac{15}{32}(2439500s^{12}-11312175s^{10}+21772080s^8-22346500s^6+12956400s^4-4043136s^2+533248)$
3,1:	$-\frac{45}{32}(5s^2-4)(62300s^{10}-260365s^8+431504s^6-356508s^4+147552s^2-24576)$
3,2:	$\frac{3}{16}(1835625s^{12}-7723875s^{10}+13291500s^8-12015300s^6+6064176s^4-1644928s^2+192256)$
3,3:	$\frac{15}{16}(5s^2-4)(18175s^{10}-85105s^8+153172s^6-136540s^4+61408s^2-11264)$
3,4:	$\frac{3}{32}(213750s^{12}-1441125s^{10}+3537000s^8-4313100s^6+2835280s^4-967808s^2+135424)$
3,5:	$\frac{21}{32}s^2(5s^2-4)(15s^2-14)(450s^6-925s^4+590s^2-112)$
2,1:	$9(3s^2-2)(5s^2-4)^3$
2,2:	$\frac{3}{8}(5s^2-4)^2(45s^4+36s^2-56)$
3,1:	$-\frac{45}{4}(5s^2-4)^3(168s^6-497s^4+460s^2-136)$
3,2:	$\frac{3}{16}(2150625s^{12}-9409875s^{10}+16968300s^8-16218180s^6+8729136s^4-2535808s^2+315136)$
3,3:	$-\frac{15}{4}(5s^2-4)^3(105s^6+39s^4-228s^2+104)$
3,4:	$\frac{3}{32}(438750s^{12}-1771125s^{10}+2865000s^8-2345100s^6+999760s^4-199808s^2+12544)$
1,0:	$-6(5s^2-4)$
2,0:	$\frac{15}{2}(5s^2-4)^2(7s^2-8)$
2,1:	$18(3s^2-2)(5s^2-4)^2$
2,2:	$\frac{3}{2}(5s^2-4)^2(5s^2+4)$
3,0:	$-\frac{15}{8}(215250s^{10}-823025s^8+1255040s^6-953760s^4+361088s^2-54464)$
3,1:	$-\frac{45}{4}(5s^2-4)^3(63s^4-124s^2+56)$
3,2:	$\frac{3}{8}(430125s^{10}-1553550s^8+2222340s^6-1570224s^4+546432s^2-74624)$
3,3:	$-\frac{15}{4}(5s^2-4)^3(45s^4+28s^2-40)$
3,4:	$\frac{3}{8}(50625s^{10}-168375s^8+215900s^6-130800s^4+35840s^2-3136)$

6.5.5 Sample application. The PRISMA orbit

We present a test of the analytical solution for a low-eccentricity orbit at 500 km of altitude and in sun-synchronous configuration, in agreement with the mission requisites of the PRISMA mission [553]. In particular, the following initial conditions in Cartesian coordinates are chosen (km for position and km/s for velocity):

$$\begin{aligned} x_0 &= -4178.63775517221, & \dot{x}_0 &= 5.84458519389825, \\ y_0 &= 1571.13919300305, & \dot{y}_0 &= -0.579214366053911, \\ z_0 &= 5224.69084171088, & \dot{z}_0 &= 4.85361424021968. \end{aligned} \tag{6.109}$$

For the gravitational model we take the values $\mu = 398600.4415\,\mathrm{km}^3/\mathrm{s}^2$, $R_\oplus = 6378.1363\,\mathrm{km}$, and $J_2 = 0.001082634$. The reference orbit is then computed by numerical integration of the equations of motion in Cartesian coordinates in Eq. (5.6), which is carried out in extended precision to guarantee that all the figures remaining after truncation to double precision are exact.

The time history of the orbital elements is shown in the following figures for an interval of 3 days, in which the satellite orbits the Earth about 45.6 times. As shown in Fig. 6.4, the semimajor axis a experiences small oscillations of $\sim$ 10 km amplitude with half the orbital period about its average value of $\sim$ 6880 km (left plot), while the inclination of the orbital plane $I \approx 97.42$ deg remains almost constant, with small oscillations of less than 20 arc seconds of amplitude and the same period as the semimajor axis (right plot).

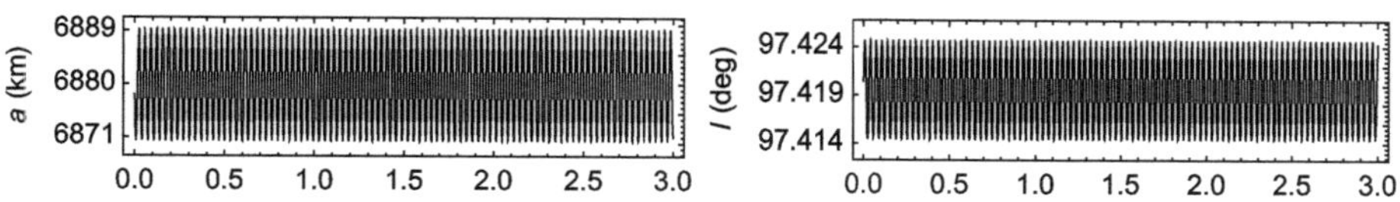

Figure 6.4: Time histories of the semimajor axis (left plot) and inclination (right plot) of the true PRISMA orbit. Abscissas are in days.

The behavior of the eccentricity e is quite similar, as shown in Fig. 6.5, with small periodic oscillations of variable amplitude with respect to its average value $\sim 2 \times 10^3$. On the contrary, the instantaneous argument of the perigee ω undergoes notable oscillations of variable amplitude that may reach about 50 degree. These periodic oscillations modulate a linear decreasing trend of about 1 deg/day, which is colored in white superimposed to the time history of the argument of the perigee. The right plot of Fig. 6.5 shows the evolution of the eccentricity vector, where the arrow marks the point corresponding to the initial conditions. The dashed circle in this last plot illustrates the long-period circulatory motion of the average eccentricity vector, which is in full agreement with the long-term behavior of high-inclination orbits previously discussed in §5.6.3.

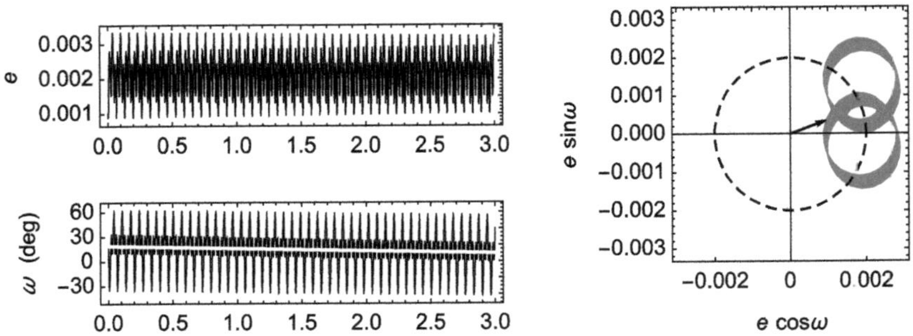

Figure 6.5: Eccentricity vector of the true PRISMA orbit along 3 days.

The right ascension of the ascending node of the PRISMA orbit grows almost linearly at the Sun-synchronous rate ~ 1 deg/day that is modulated with small periodic oscillations of similar amplitude to those observed in the inclination. The same behavior is observed in the mean distance to the node $F = M + \omega$, yet its rate is notably faster, by more than 15 orbits per day, and the amplitude of the periodic oscillations is about 3.6 arc minutes in this case.

The reference orbit is now compared with those provided by different truncations of the analytical solution. To begin with, we illustrate the importance of making an accurate initialization of the constants of the perturbation theory.

6.5.5.1 Initialization of the constants

First of all, we completely disregard the periodic corrections and take the secular terms of the analytical solution like if they where osculating elements. Then, transforming the initial conditions in Eq. (6.109) into semi-equinoctial variables, we obtain

$$\begin{aligned} F_0 &= 0.8726646200250181, & L_0 &= 52360.56175616003, \\ C_0 &= 0.9396928336552479 \times 10^{-3}, & S_0 &= 0.3420158197412482 \times 10^{-3}, \\ h_0 &= 2.9349734000392003, & H_0 &= -6762.329846647862, \end{aligned} \tag{6.110}$$

were F_0 and h_0 are given in radians, C_0 and S_0 are non-dimensional, and L_0 and H_0 have dimensions km^2/s. Next, the secular frequencies of the motion are computed up to the second order of J_2 using Eqs. (6.106)–(6.108), to get

$$\begin{aligned} n_F &= 1.105341787346819 \times 10^{-3}, \\ n_\omega &= -7.080920112885583 \times 10^{-7}, \\ n_\Omega &= 1.994353947362547 \times 10^{-7}, \end{aligned} \tag{6.111}$$

in radians per second, producing an anomalistic period (from perigee to perigee) $T = 2\pi/(n_F - n_\omega) = 94.68$ min. With these values, the evolution of the secular terms is computed by simple evaluation of Eq. (6.105) at the desired output time t.

The errors of the solution in triple-prime variables are depicted in Fig. 6.6, where, in addition to the expected periodic errors (of the same amplitude as the periodic oscillations of the orbital elements of the true orbit), a general shift of the errors of any orbital element from the zero average is apparent. On the other hand, as it is checked by visual inspection of Fig. 6.6, the initial error is zero in all cases due to the use of the same initial conditions both in the true, numerically integrated solution and in the analytical propagation of the secular terms.

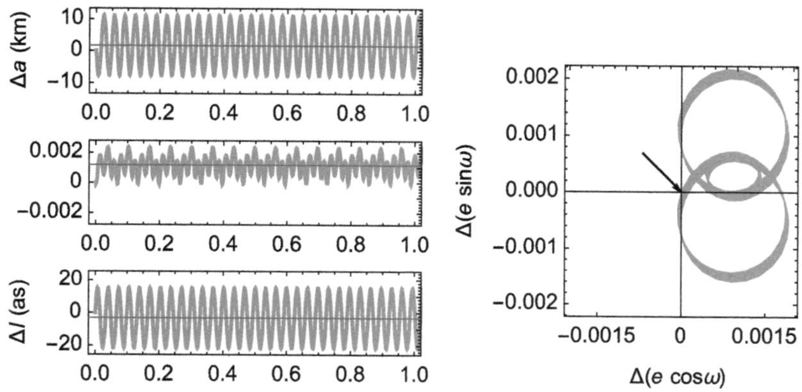

Figure 6.6: Errors of the secular solution when the initial conditions are not converted to secular values. Abscissas are in days.

Moreover, in addition to the general shift form the zero average, the errors of F and Ω undergo a secular drift which is evident in their time histories in Fig. 6.7. This is mainly due to the inaccurate computation $L''' = L$, which corrupts the accuracy of all the secular frequencies, most notably of n_M and, in consequence, n_F. The latter produces an important linear growth of the error of the mean distance to the node of about 2 deg/day that is predominant over the periodic perturbations since the beginning of the propagation, as clearly observed in the upper plot of Fig. 6.7. The secular increase of the errors of Ω is also evident in the lower plot of Fig. 6.7. However, on account of n_Ω being only $\mathcal{O}(J_2)$, its effect is less harmful than in the case of n_F, which is $\mathcal{O}(1)$, and the errors of Ω grow only at a the small rate of about 3.5 as/day. Along the short propagation interval shown in Fig. 6.7 the errors of Ω are dominated by the periodic terms, but the secular component of the errors will prevail in longer intervals.

Both types of effects in the errors, the shift from the zero average and the secular increase, are obvious consequences of the fact that, because the initial conditions were not transformed into secular elements, we are actually comparing two different orbits, and hence the secular terms do not reflect the average behavior of the true solution. Because of that, the position predicted by the secular solution can differ from the actual one in hundreds of kilometers after just one-day propagation. This is illustrated in Fig. 6.8, where the norm of the position error—root-sum-square (RSS) of the

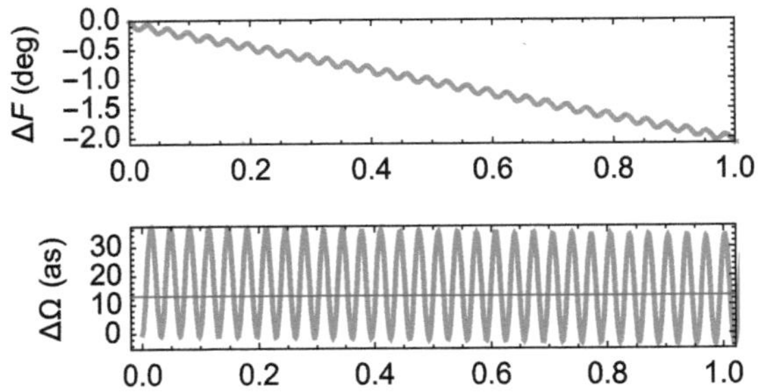

Figure 6.7: Errors of the secular propagation of F and Ω when the initial conditions are not converted to secular values. Abscissas are in days.

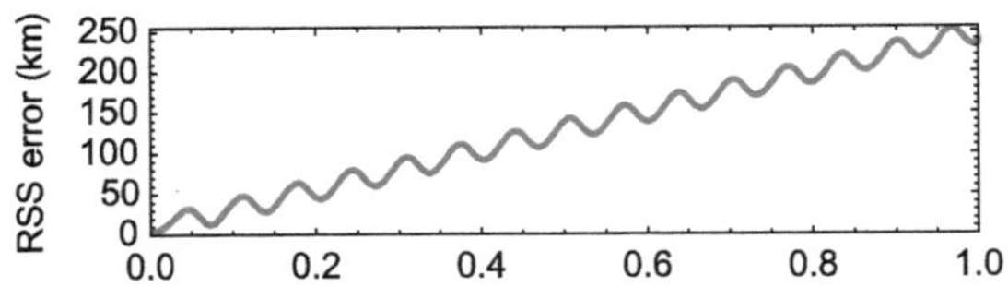

Figure 6.8: Root-sum-square error of the Cartesian coordinates of the secular approximation with no conversion of the initial conditions. Abscissas are in days.

difference between the position of the integrated exact equations (5.6) and the position obtained from the analytical propagation of the secular terms with Eq. (6.105)—is displayed.

The next step is to transform the osculating initial conditions in Eq. (6.110) consecutively to prime, double-prime, and triple-prime variables to obtain the correct values of the constants (initial elements and frequencies) that initialize the secular solution. That is, the inverse transformations of the elimination of the parallax, the elimination of the perigee, and the Delaunay normalization, are applied sequentially, up to the first order of J_2, to obtain

$$\begin{aligned} F_1 &= 0.8716628560891988, & L_1 &= 52366.94663215522, \\ C_1 &= 0.1841678296708005 \times 10^{-2}, & S_1 &= 0.7152507807642872 \times 10^{-3}, \\ h_1 &= 2.935061847045128, & H_1 &= H_0, \end{aligned} \tag{6.112}$$

from which we compute the secular frequencies

$$\begin{aligned} n_F &= 1.104938198224251 \times 10^{-3}, \\ n_\omega &= -7.075076094488982 \times 10^{-7}, \\ n_\Omega &= 1.992424728390034 \times 10^{-7}. \end{aligned} \tag{6.113}$$

Relative differences with the values in Eqs. (6.110) and (6.111) are $\mathcal{O}(J_2)$. In this case also we do not recover the periodic effects after the analytical propagation of the secular terms, and just compare the secular propagation with the true orbit. Therefore, the periodic errors are expected to oscillate with the same amplitude as before. However, the secular solution gets now much closer to the average values of the actual orbit.

The beneficial effect of transforming osculating variables into secular elements is illustrated in Fig. 6.9 for the eccentricity vector. Now, the initial errors no longer vanish, as it correctly must happen because we are comparing osculating elements with secular elements. The amplitude of the periodic errors remain similar in the two cases, as expected. This behavior is also checked, of course, with the other orbital elements. In addition, the secular components of the errors are now mostly eliminated. This is illustrated for the right ascension of the ascending node and the mean longitude in Fig. 6.10, where the errors of the secular propagation with and without conversion from osculating to secular elements are superimposed. Again, we note that the amplitudes of the periodic components of the errors remain unaltered.

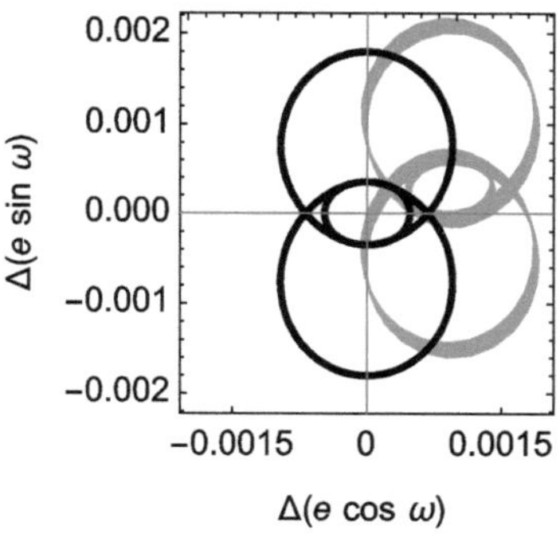

Figure 6.9: Errors of the eccentricity vector computed from the secular terms with (black curve), and without conversion of the initial conditions (gray curve).

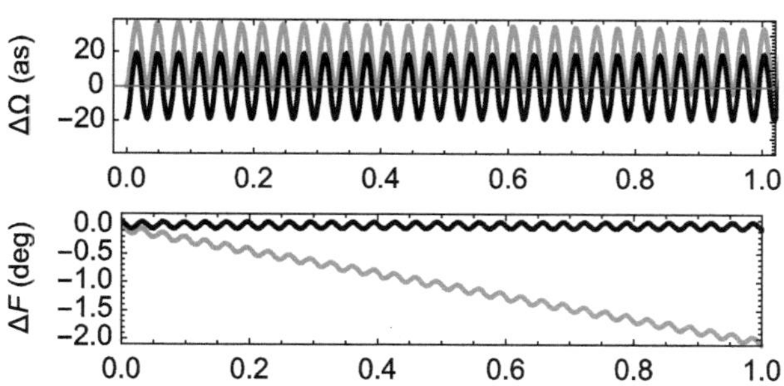

Figure 6.10: Errors of Ω and *F* with proper initialization (black curves), superimposed to analogous graphics in Fig. 6.7 (gray curves). Abscissas are in days.

The radical improvements obtained in the secular propagation by the proper computation of the secular frequencies is illustrated in Fig. 6.11, where the time history

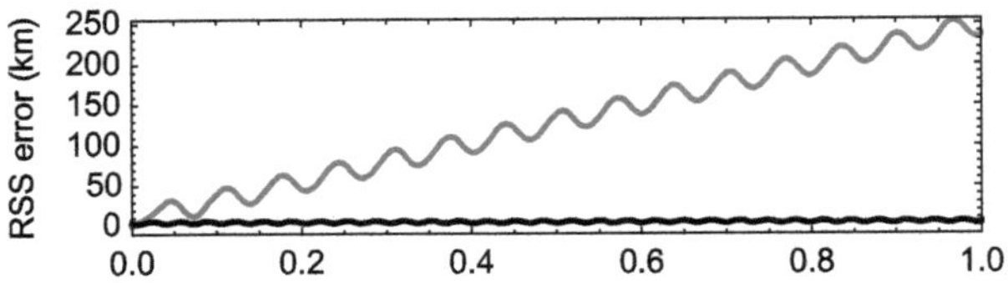

Figure 6.11: RSS errors when the initial conditions are properly initialized (black curve) superimposed to the results in Fig. 6.8 (gray curve). Abscissas are in days.

of the RSS errors of the Cartesian coordinates is superimposed to the previous case. Now, the errors of the secular terms enjoy periodic nature, with amplitudes that remain bounded along the whole propagation interval to just a few km.

6.5.5.2 Accuracy of the periodic corrections

Once it has been illustrated the importance of properly transforming the osculating initial conditions into triple-prime variables, we check that the periodic corrections provide the expected accuracy from the truncation order of the perturbation solution. The test consists in computing the initialization constants of the analytical solution for different states of the true solution. If the theory were exact, every point along the whole propagation interval would produce exactly the same constant values for each of the three integrals of the secular problem L''' and G''', and H''', as well as for the secular frequencies n_F, n_ω and n_Ω. However, the perturbation solution only retains the periodic corrections up to the truncation order of the small parameter—J_2 in the current case. Then it would be expected that the values of the formal integrals may oscillate within the accuracy of the neglected terms.

For instance, if we use the printed expressions in §6.5, which are accurate up to the third order of J_2, we would expect oscillations of $\mathcal{O}(J_2^4) \approx 10^{-12}$ relative to the values of the initialization constants. And this is exactly what we find, as illustrated with Fig. 6.12, where the relative errors of $L_0'''(t_i)$ as computed with respect to their arithmetic mean along the three-days interval are depicted in the upper plot. The lower plot shows the relative errors of the secular values $h_0'''(t_i)$ with respect to their linear fit $h''' = 2.93506 + 0.000717275t$. Only the first three orbits of the three-days interval are presented in the figure for clarity.

When the conversion from osculating to secular elements is applied to other elements, like the classical Keplerian set, the accuracy obtained may be different from that expected from the truncation order of the perturbation solution. This is due to the error propagation through the operations needed to compute the desired variable as a function of the Delaunay canonical variables. Thus, for instance, because $\Delta a = \mathcal{O}(\Delta L)$, as readily obtained from the first of Eq. (4.46), the accuracy in the computation of the secular semi-major axis should be the same as expected from the truncation order of the perturbation solution. On the contrary the propagation of the truncation errors of G and L in the computation of the secular eccentricity from the second equation of

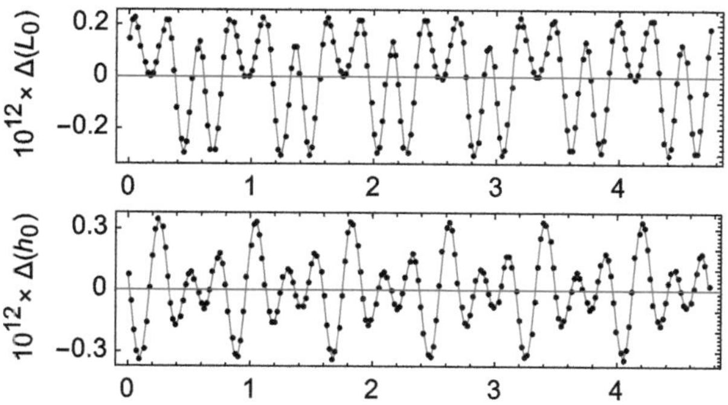

Figure 6.12: Relative errors of the secular elements L_0''' (upper) and h_0''' (lower) when the inverse corrections are accurate to $\mathcal{O}(J_2^3)$. Abscissas are in hours.

Eq. (4.46) would divide the error by the eccentricity value. In fact, this would be equivalent to having computed directly the periodic corrections to the orbital elements using analogous equations to Eq. (2.17). For instance, the Poisson bracket $\{e, \mathcal{W}\}$, which would be needed in the computation of the secular value of the eccentricity, involves the computation $\partial e/\partial G = -\eta/(eL)$, showing the magnification of the eccentricity errors for low-eccentricity orbits.

The improvements obtained in the computation of the secular terms for increasing truncation orders of the perturbation solution is further illustrated in Fig. 6.13 for the semimajor axis, where the errors obtained with the fourth- and fifth-order truncations of the periodic corrections are displayed together—the latter amounting to thousandths of a micrometer for a PRISMA orbit and reaching the numerical accuracy of the computer [408].

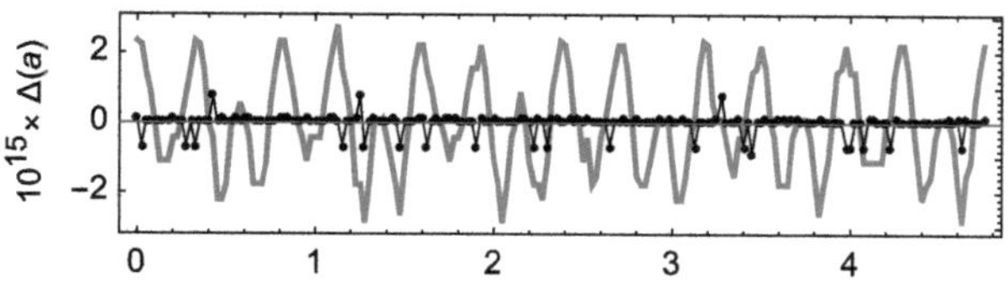

Figure 6.13: Relative errors of $a(L_0''')$ when using the fourth- (gray line) and fifth-order truncation (black dots) of the inverse transformation. Abscissas are in hours.

6.5.5.3 Long-term performance

Lastly we assess the performance of the analytical solution in a long-term propagation. That is, after the proper initialization of the constants of the perturbation solution, each time $t = t_i$ the secular terms are evaluated, they are sequentially corrected with the direct periodic corrections of the Delaunay normalization, to get double-prime variables, of the elimination of the perigee, to get prime variables, and of the elimination of the parallax, to finally get osculating variables. The latter are compared with the corresponding ephemeris of the reference orbit to get the RSS errors. The propagation interval encompasses one year, which amounts to more than 5500 PRISMA orbits. For comparison, we label the different solutions by (*i*:*s*:*d*), with *i* standing for the order of the inverse transformation, *s* that of the secular terms, and *d* the order of the direct transformation.

When the (1:2:1) solution is used—the first-order periodic correction and second-order secular terms, like in Brouwer's original case—the RSS errors caused by the truncation of the secular terms dominate the propagation in the long term. They grow at a linear rate of about half a km per day from the initial error of one meter, which is due to the truncation to the first order of the periodic corrections, up to a RSS error of about 160 km at the end of the propagation interval. Things improve notably with the second order of the inverse transformation—the (2:2:1) solution—in which case the upgraded secular frequencies reduce the rate of the RSS errors to $\sim$ 1.1 m/day. At the end of the propagation, the error is now about half a km, showing once again the importance of the accurate initialization of the constants of the analytical theory. The second order of the direct transformation—the (2:2:2) solution—does not show improvements over the (2:2:1) case in the long term due to the fast growth of the secular errors. On the contrary, the solution is penalized with increased computational burden at each step of the output.

Additional benefits are observed when the secular frequencies are computed up to the third order—the (2:3:1) solution—which reduces the secular rate of the RSS errors to $\sim$ 14 cm/day with a negligible increase of the computational burden because the improved, third-order frequencies are evaluated just once. For the same reason, the solution is not penalized when initializing the constants with third-order corrections—the (3:3:1) solution. In that case, the RSS errors grow at the slow rate of just a few mm/day, whose accumulation only reaches a couple of meters after one year. Since the direct periodic corrections are still computed from the first-order truncation, the amplitude of the periodic oscillations of the RSS errors of the (3:3:1) solution is of about one meter, and hence the errors remain of similar magnitude along the one-year propagation. The (4:4:1) solution completely buries the secular trend of the RSS errors under the periodic components, and they only start to show up when the second-order direct corrections are taken into account—the (4:4:2) solution. The secular trend becomes dominant again, yet at a tiny rate of just a few μm/day, when the direct corrections are truncated at the third order—the (4:4:3) solution. The RSS errors remain bounded at the micrometer level along the one-year interval with the (5:5:3) solution, in which the constants of the analytical solution are initialized to the numerical precision of the computer. Finally, we found that while some improvements are achieved beyond that truncation of the direct periodic corrections, they are only apparent at the beginning of the propagation and do not have observable effects in the long term. Figure 6.14 illustrates the improvements in accuracy achieved with different truncations of the analytical perturbation solution. The plots are rendered in a logarithmic scale to enhance comparison.

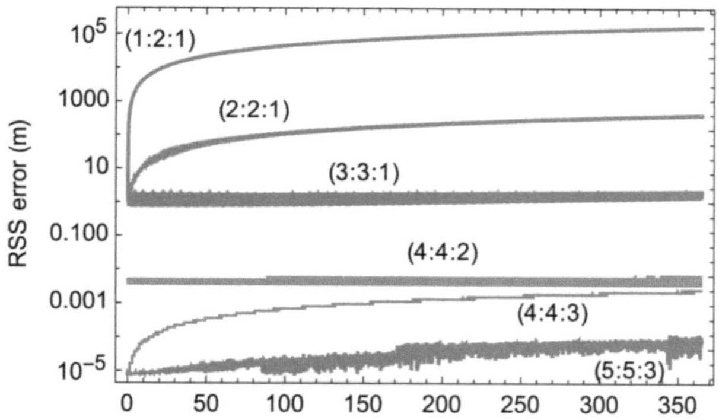

Figure 6.14: Root-sum-square error of different truncations of the main problem solution. Abscissas are in days.

On the other hand, the size of the trigonometric series representing the periodic corrections grows notably with the order of the perturbation approach. This means that the evaluation time of the analytical solution grows correspondingly. In the current test case, we found that the time spent in the evaluation of the third-order corrections is one order of magnitude larger than in the evaluation of the much simpler first-order corrections. We must note, however, that the application of optimization strategies for the evaluation of the periodic corrections [115, 272, 407], which has not been carried out here, might bias the comparisons towards more benign figures.

6.6 Initialization issues. Breakwell and Vagners' approach

It is worth to remark that a characteristic feature of perturbation solutions is that they always leave in error, however small, the frequencies of the angle variables. For instance, the mean anomalistic motion $n_M \approx 1.1 \times 10^{-3}$ rad/s of the PRISMA orbit discussed in §6.5.5, will unavoidably bear an error of $\sim 10^{-19}\,\mathrm{s}^{-1}$ due to the number representation in double-precision floating-point arithmetic [332]. In consequence, the linear growth of the (secular) mean anomaly along the $\approx 3\times10^7$ s that comprise a year would yield an error of the order of 10^{-12} radians, which for the PRISMA semimajor axis of 6878.137 km would produce a concomitant in-track error of several micrometers. These figures are in perfect agreement with the results obtained with the (5:5:3) truncation of the analytical solution, and they grow accordingly for lower-order truncations of the perturbation solution.

On the other hand, initializing the constants of the analytical solution from a given truncation of the periodic corrections is inconsistent with the propagation of the secular terms obtained from a higher-order truncation. This was soon noticed in the case of Brouwer's (1:2:1) solution [75], and errors in the in-track direction that were unexpected from the second-order truncation of the secular terms were repeatedly reported in the literature [57, 252, 469]. The difficulty, of course, disappears when the inverse periodic corrections are computed up to the same order as the secular terms. Alternatively, since the most sensible constant to be initialized is the secular mean motion, as already discussed in §1.2.1, different strategies were proposed to avoid the tiresome computation of the generating function term up to the same order as the secular terms [63, 90, 469, 675]. We only describe the smart, inexpensive procedure pointed out by Breakwell and Vagners [63], which is specially useful in the computation of a second-order solution to the artificial satellite problem when other perturbations in addition to the oblateness are taken into account.

Thus, for given initial conditions $(\boldsymbol{x}_0, \boldsymbol{X}_0)$ the energy of the corresponding orbit is computed *exactly* from the original Hamiltonian $E_0 = \mathcal{H}(\boldsymbol{x}_0, \boldsymbol{X}_0)$. This value of the energy differs from the value of the Hamiltonian in the triple-prime variables by a remainder $\mathcal{R}$ which is due to the truncation order of the secular Hamiltonian (6.104). Namely,

$$E_0 = -\frac{\mu^2}{2L_0'^2} + \sum_{m=1}^{k} \frac{\epsilon^m}{m!} \mathcal{S}_m(L_0''', G_0''', H) + \mathcal{R}(\ell_0''', g_0''', L_0''', G_0''', H), \tag{6.114}$$

where $\mathcal{R} = \mathcal{O}(\epsilon^{k+1})$ when the transformation from original to primed variables is carried out up to the order $m = k$. If this transformation is computed only to the order $m = k - 1$, then, while the terms of the summation in Eq. (6.114) will not degrade the order of the remainder, because they are multiplied by ϵ^m with $m \geq 1$, the Keplerian term certainly will. This fact can be used to improve the initial value of the Delaunay action in the triple-prime variables by computing $\hat{L}_0$ from

$$\frac{\mu^2}{2\hat{L}_0^2} = -E_0 + \sum_{m=1}^{k} \frac{\epsilon^m}{m!} \mathcal{S}_{0,m}(L_0''', G_0'', H). \tag{6.115}$$

After replacing $L_0''' = \hat{L}_0$ in Eq. (6.114) the remainder will certainly be $\mathcal{O}(\epsilon^{k+1})$. When the constants of the analytical solution are initialized using the improved value $\hat{L}_0$, it can be checked that solutions $(k - 1{:}k{:}k - 1)$ enjoy comparable accuracy to the more rigorous solutions $(k{:}k{:}k - 1)$.

6.7 Centered elements

As mentioned in §5.6.1, when the perturbation solution is approached in closed form, the generating function of the short-period elimination can bear long-period effects related to the arbitrary integration constant that appears in its computation. If this happens, the next order of the Hamiltonian in mean elements will be deprived of some long-period terms. While this is irrelevant in a semi-analytical approach in which the osculating elements are recovered at each step of the numerical integration by applying the short-period corrections, there are different applications in which the evolution of the mean elements is significative by itself. In those cases, it is important that the computed values of the mean elements be as close as possible to the average values of the osculating elements. To achieve that, the mean-element Hamiltonian must bear all the long-period terms of the solution. That is, the first-order term of the generating function, which enters in the computations of the second-order term of the mean-element Hamiltonian through Deprit's recursion (2.15), must be completely free from hidden long-period terms. Note that this condition is not sufficient when the perturbation theory is extended to higher orders, a case in which additional considerations must be made to obtain the desired "centered" elements [197, 488].

Hidden long-period terms are carried, in particular, by the first-order term of the generating function of the elimination of the parallax in Eq. (6.31). Indeed, on account of the equation of the center averaging to zero, we immediately find that the average of $\mathcal{W}_1$ in Eq. (6.31) or, equivalently, in Eq. (5.40) over the mean anomaly is exactly the same as the average over the mean anomaly of the first term of the generating function of the removal of short-period effects in Eq. (5.61). Then we directly choose from Eq. (5.64)

$$A_1 = \frac{1}{8} G \frac{R_\oplus^2}{p^2} (1 + 2\eta)\beta^2 s^2 \sin 2\omega, \tag{6.116}$$

like the arbitrary "constant" that must be added to Eq. (6.31) in order to release $\mathcal{W}_1$ from hidden long-period terms. This additional term contributes 18 new trigonometric terms to Eq. (6.32), all of which pertain to the image of the Lie derivative save for the term

$$\Delta\mathcal{H}_{0,2} = \frac{\mu}{a} \frac{R_\oplus^4}{p^4} \frac{a^2}{r^2} \frac{3}{16} (4 - 5s^2)(1 + 2\eta)\eta^2\beta^2 s^2 \cos 2\omega, \tag{6.117}$$

which pertains to the kernel of the Lie derivative. It must be added to Eq. (6.33) in order to keep all the long-period effects in the simplified Hamiltonian after the elimination of the parallax [442].

On the other hand, the first-order term of the generating function of the Delaunay normalization is free from long-period terms, as readily checked in Eq. (6.40). Therefore, the only new second-order Hamiltonian term is (6.117), which, after having averaged over the mean anomaly,

$$\Delta\mathcal{K}_{0,2} = \langle \Delta\mathcal{H}_{0,2} \rangle_\ell = \frac{\mu}{a} \frac{R_\oplus^4}{p^4} \frac{3}{16} (4 - 5s^2)(1 + 2\eta)\eta\beta^2 s^2 \cos 2\omega, \tag{6.118}$$

where all the entities are now functions of the double-prime Delaunay variables, is directly added to Eq. (6.42).

7 Tesseral perturbations

While secular and long-period perturbations are associated with the effects of zonal harmonics, the effects of sectorial and tesseral harmonics are generally limited to short-period perturbations of small amplitude. The time dependency introduced by the tesseral harmonics into the Hamiltonian is customarily removed by the simple expedient of moving to a rotating frame. However, the problem no longer enjoys symmetry with respect to the Earth's rotation axis, and the number of degrees of freedom is increased by one with respect to the zonal case. Still, for low-Earth orbits the Keplerian remains as the zeroth order of the Hamiltonian, and hence tesseral effects are easily removed in closed form of the eccentricity in this particular case [220]. On the contrary, the homological equation becomes a partial differential equation for a general orbit. Closed-form normalization is also feasible, yet it may require the use of numerical averaging techniques [471, 540]. Alternatively, closed-form normalization of the tesseral harmonics can be achieved by analytical relegation techniques [169, 538]. However, at least for sub-synchronous orbits, the relegation of tesseral harmonics exhibits a dependency on the eccentricity [602] that may make the closed-form relegation of tesseral terms vacuous. In that case, the standard normalization of tesseral terms expanded in powers of the eccentricity is definitely simpler than computing a higher order of the relegation algorithm and may well be also more effective [438].

On the other hand, tesseral terms can induce long-period effects in resonant configurations in which the mean motion of the satellite is commensurate with the Earth's rotation rate. Then the tesseral potential needs to be unavoidably expanded in powers of the eccentricity to identify the resonant terms that must remain in the simplified long-term Hamiltonian. In that case, the mean semimajor axis is no longer an integral, and, on the contrary, undergoes long-period oscillations of moderate amplitude.

7.1 Tesseral potential in orbital elements

The disturbing part of the geopotential (5.2) is written in orbital elements like

$$\mathcal{P} = -\frac{\mu}{a}\frac{a^2\eta}{r^2}\sum_{i\geq 2}\sum_{j=0}^{i} V_{i,j}(p,e,f,\omega,\nu,I), \tag{7.1}$$

where

$$\begin{aligned} V_{i,j} = \eta\frac{R_\oplus^i}{p^i}(1+e\cos f)^{i-1}\sum_{k=0}^{i}\mathcal{F}_{i,j,k}(I) \\ \times\left\{S_{i,j}\sin\left[(i-2k)f+\psi_{i,j,k}\right]+C_{i,j}\cos\left[(i-2k)f+\psi_{i,j,k}\right]\right\}, \end{aligned} \tag{7.2}$$

with

$$\psi_{i,j,k} = (i-2k)\omega + j\nu - (i-j)_\pi, \tag{7.3}$$

https://doi.org/10.1515/9783110668513-007

in which $\nu = \Omega - \omega_\oplus t$ is the longitude of the node in the rotating frame, $(i-j)_\pi = \frac{\pi}{2}[(i-j) \bmod 2]$, following the index convention in Eq. (6.5), and, as follows from the original definition of Kaula inclination functions [341],

$$\mathcal{F}_{i,j,k} = \sum_{l=0}^{\min(k,i_j)} \sum_{m=0}^{j} \sum_{q=\min(i-j-2l+m,k-l)}^{i} \binom{j}{m}\binom{i-j-2l+m}{q} \\ \times \binom{j-m}{k-l-q} \frac{(-1)^{q-i_j}}{2^{2i-2l}} \frac{(2i-2l)!}{l!(i-l)!(i-j-2l)!} c^m s^{i-j-2l}, \tag{7.4}$$

with $i_j = \lfloor \frac{1}{2}(i-j) \rfloor$. We recall that $c \equiv \cos I$, $s \equiv \sin I$. Taking $j = 0$ in Eq. (7.4) we recover the particularization in Eq. (6.4) for the zonal problem. The efficient evaluation of the inclination functions as well as their derivatives can be achieved with different procedures [251].

After replacing the binomial expansion of $(1 + e\cos f)^{i-1}$ in Eq. (7.2), and making additional rearrangements using Eqs. (6.12) and (6.13), we obtain

$$V_{i,j} = \frac{R_\oplus^i}{p^i} \eta \sum_{k=0}^{i} \mathcal{F}_{i,j,k}(I) \sum_{m=0}^{i-1} \binom{i-1}{m} \frac{e^m}{2^m} \sum_{l=0}^{m} \binom{m}{l} \\ \times \big[(S_{i,j} \cos \psi_{i,j,k} - C_{i,j} \sin \psi_{i,j,k}) \sin(i-2k-m+2l)f \\ + (C_{i,j} \cos \psi_{i,j,k} + S_{i,j} \sin \psi_{i,j,k}) \cos(i-2k-m+2l)f \big]. \tag{7.5}$$

7.2 Low-Earth orbits. Garfinkel's perturbation approach

The time dependency of the tesseral terms of the geopotential is masked in Eq. (7.3) by the use of the longitude of the node in the rotating frame ν. This time dependency is also avoided in Hamiltonian (5.3) when using a modified set of Delaunay variables in which $h = \nu \equiv \Omega - \omega_\oplus t$, whereas $H = xY - yX = (\mu/a)\eta c/n$ still remains as te conjugate momentum to h. Using these canonical variables and with the previous definition of $V_{i,j}$ in Eq. (7.5), the Hamiltonian (5.3) is written in the form

$$\mathcal{H} = -\frac{\mu}{2a} - \frac{\mu}{a}\frac{\omega_\oplus}{n}\eta c - \frac{\mu}{a}\frac{a^2\eta}{r^2}\left(V_{2,0} + \sum_{j=1}^{2} V_{2,j} + \sum_{i\geq 3}\sum_{j=0}^{i} V_{i,j} \right). \tag{7.6}$$

The period of common low-Earth orbits is about an hour and a half, while the Earth needs $\approx$ 24 hours to complete a rotation about its polar axis. In consequence, $\omega_\oplus/n = \mathcal{O}(J_2^{1/2})$, whereas, for $i > 2$, $C_{i,j}$ and $S_{i,j}$ are $\mathcal{O}(J_2^2)$ or smaller. In this particular case Eq. (7.6) can be rearranged as the usual perturbation Hamiltonian (2.30) in which, now, $\varepsilon \sim \omega_\oplus/n \sim J_2^{1/2}$ is a formal, rather than a physical small parameter, and, cf. [220, 221, 539],

$$\mathcal{H}_{0,0} = -\frac{\mu}{2a}, \tag{7.7}$$

$$\mathcal{H}_{1,0} = -H\omega_\oplus, \tag{7.8}$$

$$\mathcal{H}_{2,0} = -(2!)\frac{\mu}{a}\frac{a^2\eta}{r^2}V_{2,0}, \tag{7.9}$$

$$\mathcal{H}_{3,0} = 0, \tag{7.10}$$

$$\mathcal{H}_{4,0} = -(4!)\frac{\mu}{a}\frac{a^2\eta}{r^2}\left(\sum_{j=1}^{2} V_{2,j} + \sum_{i\geq 3}\sum_{j=0}^{i} V_{i,j}\right), \tag{7.11}$$

and $\mathcal{H}_{m,0} = 0$, $m \geq 5$.

7.2.1 Elimination of the parallax of tesseral terms

As has been done with the zonal problem, non-essential short-period tesseral effects are first removed using the elimination of the parallax simplification. Again, the Lie derivative is given by Eq. (4.66), and the alternative expression in Eq. (5.38) will be used when found convenient.

On account of $\mathcal{H}_{1,0}$ being free from the mean anomaly, the first-order homological equation $\mathcal{L}_0(\mathcal{W}_1) = \mathcal{H}_{1,0} - \mathcal{H}_{0,1}$ is solved by choosing $\mathcal{H}_{0,1} = \mathcal{H}_{1,0}$ and $\mathcal{W}_1 = 0$. At second order $\mathcal{L}_0(\mathcal{W}_2) = \widetilde{\mathcal{H}}_{0,2} - \mathcal{H}_{0,2}$, where, from Eq. (2.37), $\widetilde{\mathcal{H}}_{0,2} = \mathcal{H}_{2,0}$. Then we choose $\mathcal{H}_{0,2}$ such that it cancels the terms of Eq. (7.9) that do not depend explicitly on the true anomaly. Namely,

$$\mathcal{H}_{0,2} = -(2!)\frac{\mu}{a}\frac{a^2\eta}{r^2}\langle V_{2,0}\rangle_f. \tag{7.12}$$

Because $\langle V_{i,j}\rangle_f$ terms are those with $l = k + \frac{1}{2}(m - i)$ in Eq. (7.5), we obtain

$$\langle V_{i,j}\rangle_f = \frac{R_\oplus^i}{p^i}\eta\sum_{k=0}^{i-1}\mathcal{F}_{i,j,k}(I)\mathcal{G}_{i,k}^*(e)(C_{i,j}\cos\psi_{i,j,k} + S_{i,j}\sin\psi_{i,j,k}), \tag{7.13}$$

where $\mathcal{G}_{i,k}^*$ was defined in Eq. (6.15), from which

$$\langle V_{2,0}\rangle_f = C_{2,0}\frac{R_\oplus^2}{p^2}\eta\left(\frac{3}{4}s^2 - \frac{1}{2}\right). \tag{7.14}$$

As expected from the formal nature of the small parameter in the current approach, this result differs from Eq. (6.29)—in the same way as Eq. (7.12) differs from Eq. (6.30)—in the coefficient $C_{2,0}$. Like in the zonal case, the homological equation is solved from (5.39), to give $\mathcal{W}_2 = -(2!)GW_{2,0}$, where integration of the terms

$$W_{i,j} = \frac{1}{\eta}\int(V_{i,j} - \langle V_{i,j}\rangle_f)\mathrm{d}f,$$

is approached following analogous steps as we did when integrating Eq. (6.35). We obtain

$$W_{i,j} = \frac{R_\oplus^i}{p^i} \sum_{k=0}^{i} \mathcal{F}_{i,j,k}(I) \sum_{m=0}^{i-1} \binom{i-1}{m} e^m \mathcal{I}_{i,j,k,m}(f,\nu,\omega), \tag{7.15}$$

where

$$\mathcal{I}_{i,j,k,m} = \frac{1}{2^m} \sum_{\substack{l=0\\ l\neq l^*}}^{m} \binom{m}{l} \bigg[(C_{i,j} \sin\psi_{i,j,k} - S_{i,j}\cos\psi_{i,j,k}) \frac{\cos 2(l-l^*)f}{2(l-l^*)}$$
$$+ (C_{i,j}\cos\psi_{i,j,k} + S_{i,j}\sin\psi_{i,j,k}) \frac{\sin 2(l-l^*)f}{2(l-l^*)} \bigg]$$

and $l^* = k + \frac{1}{2}(m-i)$. It can be checked that $\mathcal{I}_{i,0,k,m}$ matches $\mathcal{I}_{i,k,m}$ in Eq. (6.38), as expected, save for the appearance of the harmonic coefficients $C_{i,0}$ due to the use of a formal small parameter in the current approach.

In particular, we find that $\mathcal{W}_2$ is the same as $\mathcal{W}_1$ in Eq. (6.31), except for the factorial 2 derived from the different perturbation arrangement, and the coefficient $-C_{2,0}$ stemming from the use of a formal small parameter. That is,

$$\mathcal{W}_2 = C_{2,0} \frac{G}{4} \frac{R_\oplus^2}{p^2} \left[(4-6s^2) e \sin f + s^2 \sum_{i=1}^{3} 3^{\lfloor 2-i/2 \rfloor} e^{|i-2|} \sin(if + 2\omega) \right], \tag{7.16}$$

which is free from the longitude of the node in the rotating frame h. Next, we take $\mathcal{H}_{1,1} = \mathcal{H}_{0,2}$, from Deprit's recursion (2.15), and proceed to the third order of the perturbation approach.

The known terms $\widetilde{\mathcal{H}}_{0,3}$ are those of Eq. (6.88), which, since $\mathcal{W}_1 = 0$, $\mathcal{H}_{0,1} = \mathcal{H}_{1,0}$, and $\mathcal{H}_{3,0}$ vanishes from its definition in Eq. (7.10), yields

$$\widetilde{\mathcal{H}}_{0,3} = 3\{\mathcal{H}_{1,0}, \mathcal{W}_2\} = 3\omega_\oplus \frac{\partial \mathcal{W}_2}{\partial h} = 0.$$

Hence, $\mathcal{H}_{0,3} = \mathcal{H}_{1,2} = \mathcal{H}_{2,1} = 0$, $\mathcal{W}_3 = 0$, and we proceed to the fourth order.

From Eq. (6.89), we obtain $\widetilde{\mathcal{H}}_{0,4} = 3\{\mathcal{H}_{0,2} + \mathcal{H}_{2,0}, \mathcal{W}_2\} + \mathcal{H}_{4,0}$, where the terms obtained from the evaluation of the Poisson brackets are the same as those obtained in the second order of §6.2.1 except for a scaling factor $4!/2! = 12$ derived from the different perturbation arrangement. Thus,

$$\widetilde{\mathcal{H}}_{0,4} = \frac{4!}{2!} \frac{\mu}{a} \frac{a^2 \eta}{r^2} \left[C_{2,0}^2 \frac{R_\oplus^4}{p^4} \eta \sum_{j=0}^{6} \sum_{l=j_1}^{2} q_{j,l} s^{|2l|} \cos(jf + 2l\omega) \right.$$
$$\left. + 2\sum_{i\geq 3} V_{i,0} + 2\sum_{i\geq 2}\sum_{j=1}^{i} V_{i,j} \right], \tag{7.17}$$

where $j_1 = \lfloor\frac{1}{2}(j-1)\rfloor$ from Eq. (6.5), and $\mathcal{H}_{0,4}$ is chosen to be made of those terms of $\widetilde{\mathcal{H}}_{0,4}$ that are free from the explicit appearance of f. That is,

$$\mathcal{H}_{0,4} = \frac{4!}{2!}\frac{\mu}{a}\frac{a^2\eta}{r^2}\left[C_{2,0}^2\frac{R_\oplus^4}{p^4}\eta(q_{0,0}+2q_{0,1}s^2\cos 2\omega) + 2\sum_{i\geq 3}\langle V_{i,0}\rangle_f + 2\sum_{i\geq 2}\sum_{j=1}^{i}\langle V_{i,j}\rangle_f\right], \tag{7.18}$$

thus adding the tesseral contribution to the equivalent term of the zonal problem in Eq. (6.33). Next, from (5.39) we obtain

$$\mathcal{W}_4 = 4!G\left(\frac{C_{2,0}^2}{2!}\frac{R_\oplus^4}{p^4}\sum_{j=1}^{6}\sum_{l=j_1}^{2}\frac{q_{j,l}}{j}s^{|2l|}\sin(jf+2l\omega) + \sum_{i\geq 3}W_{i,0} + \sum_{i\geq 2}\sum_{j=1}^{i}W_{i,j}\right),$$

where $W_{i,j}$ is given by Eq. (7.15).

At fifth order, the only non-vanishing term obtained from Deprit's recursion (2.15) is

$$\widetilde{\mathcal{H}}_{0,5} = 5\{\mathcal{H}_{1,0},\mathcal{W}_4\} = 5\omega_\oplus\frac{\partial\mathcal{W}_4}{\partial h} = 5!\omega_\oplus G\sum_{i\geq 2}\sum_{j=1}^{i}\frac{\partial W_{i,j}}{\partial h},$$

whose terms are all periodic in f, as follows from Eq. (7.15). Then we choose $\mathcal{H}_{0,5} = 0$.

As usual, the periodic corrections to obtain osculating elements from prime elements and vice versa are derived from the generating function using Eq. (2.17).

7.2.2 Delaunay normalization. *m*-daily terms

After the Hamiltonian simplification process we have $\mathcal{K} = \sum_{m\geq 0}(\varepsilon^m/m!)\mathcal{K}_{m,0}$, in which $\mathcal{K}_{0,0}$ and $\mathcal{K}_{1,0}$ continue to be the Keplerian and Coriolis terms in Eqs. (7.7) and (7.8), respectively, $\mathcal{K}_{2,0} = \mathcal{H}_{0,2}$ is given by Eq. (7.12), $\mathcal{K}_{3,0} = 0$, $\mathcal{K}_{4,0} = \mathcal{H}_{0,4}$ is given by Eq. (7.18), and $\mathcal{K}_{5,0} = 0$. The quantities $a, r, \eta\ldots$, in the new Hamiltonian terms are functions of the prime Delaunay variables.

The removal of short-period terms still remaining in the simplified Hamiltonian $\mathcal{K}$ is readily achieved by means of the Delaunay normalization. At first order,

$$\mathcal{K}_{0,1} = \mathcal{K}_{1,0} = -H'\omega_\oplus, \tag{7.19}$$

and $\mathcal{U}_1$ is chosen null like in the previous simplification. At second order $\widetilde{\mathcal{K}}_{0,2} = \mathcal{K}_{2,0}$, from which we choose $\mathcal{K}_{0,2} = \langle\mathcal{K}_{2,0}\rangle_\ell = -(2!)(\mu/a)\langle V_{2,0}\rangle_f$. Namely,

$$\mathcal{K}_{0,2} = 2!\frac{\mu}{a}(-C_{2,0})\frac{R_\oplus^2}{p^2}\eta\left(\frac{3}{4}s^2 - \frac{1}{2}\right), \tag{7.20}$$

which is the same as Eq. (6.39) except for the 2 factorial, stemming from the different arrangement of the perturbation Hamiltonian, and the explicit appearance of the harmonic coefficient $-C_{2,0} = J_2$ due to the current use of a formal small parameter. Then we compute $\mathcal{U}_2$ from Eq. (4.69), which yields $\mathcal{U}_2 = (\phi/n)\mathcal{K}_{0,2}$, which is equivalent to Eq. (6.40) by the same considerations as above.

At third order all the computable terms vanish, $\widetilde{\mathcal{K}}_{0,3} = 0$. Hence, $\mathcal{K}_{0,3} = 0$ and $\mathcal{U}_3 = 0$.

At fourth order we obtain $\widetilde{\mathcal{K}}_{0,4} = \{\mathcal{K}_{2,0} + \mathcal{K}_{0,2}; \mathcal{U}_2\} + \mathcal{K}_{4,0}$, where

$$\begin{aligned}\{\mathcal{K}_{2,0} + \mathcal{K}_{0,2}; \mathcal{U}_2\} = C_{2,0}^2 \frac{4!}{2!} \frac{\mu}{a} \frac{R_\oplus^4}{p^4} \frac{\eta^2}{16} (2 - 3s^2)^2 \\ \times \left[3 + \frac{a^2}{r^2}(1 + 2\beta\cos f + e\beta\cos 2f)\right]\end{aligned} \tag{7.21}$$

is the same as the corresponding term in Eq. (6.41) except for a scale factor. Then we choose $\mathcal{K}_{0,4} = \langle\widetilde{\mathcal{K}}_{0,4}\rangle_\ell$, to obtain

$$\begin{aligned}\mathcal{K}_{0,4} &= \frac{4!}{2!} \frac{\mu}{a} C_{2,0}^2 \frac{R_\oplus^4}{p^4} \eta \left[q_{0,0} - \frac{1 + 3\eta}{16}(2 - 3s^2)^2 + 2q_{0,1} s^2 \cos 2\omega\right] \\ &\quad + 4!\frac{\mu}{a} \sum_{i\geq 3} \langle V_{i,0}\rangle_f + 4!\frac{\mu}{a} \sum_{i\geq 2} \sum_{j=1}^{i} \langle V_{i,j}\rangle_f,\end{aligned} \tag{7.22}$$

which adds the tesseral perturbations to its equivalent term of the zonal problem in Eq. (6.42). The homological equation is trivially solved to give $\mathcal{U}_4 = (\phi/n)\mathcal{K}_{0,4}$.

At fifth order, the computable terms are

$$\widetilde{\mathcal{K}}_{0,5} = 5\{\mathcal{K}_{1,0}; \mathcal{U}_4\} = -5\omega_\oplus\{H'; \mathcal{U}_4\} = 5!\frac{\omega_\oplus}{n} \frac{\mu}{a} \phi \sum_{i\geq 2} \sum_{j=1}^{i} \frac{\partial\langle V_{i,j}\rangle_f}{\partial h'},$$

which only depend on the mean anomaly through ϕ, and, therefore, are purely periodic in ℓ. Hence, we choose $\mathcal{K}_{0,5} = 0$ like we did in the previous simplification.

Up to the truncation order of the theory, the transformation equations from mean to prime elements and vice versa are derived as usual from the generating function using Eq. (2.17). Alternative short-period corrections can be found, for instance, in [256] for the classical set of orbital elements.

The procedure ends by replacing prime by double-prime variables in Eqs. (7.19), (7.20), and (7.22), to get

$$\mathcal{Q} = \sum_{m\geq 0}^{5} \frac{\varepsilon^m}{m!} \mathcal{K}_{0,m} + \mathcal{O}(J_2^3). \tag{7.23}$$

Therefore, up to the second order of J_2, the only difference between the Hamiltonian (7.23) and the zonal problem Hamiltonian (6.43) in mean elements, arises from the terms contributed by the double summation in Eq. (7.22). On account of Eq. (7.13), these additional periodic terms take the form

$$\mathcal{P}_{g'',h''} = \frac{\mu}{a}\eta \sum_{i\geq 2} \frac{R_{\oplus}^{i}}{p^{i}} \sum_{j=1}^{i} \sum_{k=0}^{i-1} \mathcal{F}_{i,j,k}\mathcal{G}_{i,k}^{*}(C_{i,j}\cos\psi_{i,j,k} + S_{i,j}\sin\psi_{i,j,k}), \tag{7.24}$$

where, because $j \neq 0$, all the arguments $\psi_{i,j,k}$ are related to the rotation of the Earth, as checked in Eq. (7.3).

In particular, taking $i = 2k$ in Eq. (7.3) we obtain $\psi_{2k,j,k} = j\nu - (2k-j)_{\pi}$, which shows that even-degree tesseral harmonics yield periodic terms that depend exclusively on some integer multiple of the longitude of the node in the rotating frame. Moreover, writing the eccentricity polynomials $\mathcal{G}_{i,k}^{*}(e)$ defined in Eq. (6.15) in the more detailed form

$$\mathcal{G}_{i\,\mathrm{even},k}^{*} = \binom{0}{\frac{i}{2}-k} + \sum_{l=1}^{i/2} \binom{i-1}{2l}\binom{2l}{l+\frac{i}{2}-k}\frac{e^{2l}}{2^{2l}},$$

$$\mathcal{G}_{i\,\mathrm{odd},k}^{*} = \sum_{l=1}^{(i-1)/2} \binom{i-1}{2l-1}\binom{2l-1}{l+\frac{i-1}{2}-k}\frac{e^{2l-1}}{2^{2l-1}},$$

we immediately see that the only eccentricity polynomials having terms independent of the eccentricity are those $\mathcal{G}_{2k,k}^{*}$. These terms cause the most relevant tesseral effects on low-Earth orbits [339]. Namely,

$$\widetilde{\mathcal{P}}_{g'',h''} = \frac{\mu}{a}\eta \sum_{k\geq 1} \frac{R_{\oplus}^{2k}}{p^{2k}} \sum_{j=1}^{2k} \mathcal{F}_{2k,j,k}(C_{2k,j}\cos\psi_{2k,j,k} + S_{2k,j}\sin\psi_{2k,j,k}). \tag{7.25}$$

For $k = 1, j = 2$, Kaula [341] talked about a *semi-daily* perturbation due to the equatorial ellipticity, and, by extension, other authors name these kinds of terms m-daily terms [99, 335].

7.2.3 Elimination of the node. Long-term Hamiltonian

The tesseral Hamiltonian with short-period effects removed is still of two degrees of freedom. However, since it is referred to the rotating frame, the longitude of the node ν evolves much faster than the argument of the perigee ω, whose evolution will determine the long-term dynamics. Therefore, a new Lie transformation is applied to further reduce the Hamiltonian in double-prime variables by removing $\nu = h''$, up to the truncation order. Thus, renaming $\mathcal{Q}_{m,0} = \mathcal{K}_{0,m}$ in Eq. (7.23), we start from the perturbation Hamiltonian $\mathcal{Q} = \sum_{m\geq 0}(\varepsilon^m/m!)\mathcal{Q}_{m,0}$.

Like when dealing with the removal of long-period terms in §6.3.1, the generating function $\mathcal{V}$ of the transformation that removes the node does not depend on the mean anomaly. Therefore, $\mathcal{L}_0(\mathcal{V}_m) = 0$ and the computation of $\mathcal{V}$ is delayed by one order from the computation of the new Hamiltonian. Since, besides, none of the terms $\mathcal{Q}_{m,0}$ depends on ν except $\mathcal{Q}_{4,0}$, we take $Q_{0,i} = Q_{i,0}$ for $i = 1, 2, 3$, and $\mathcal{V}_1 = \mathcal{V}_2 = 0$. At fourth order, we find that the computable terms are $\widetilde{\mathcal{Q}}_{0,4} = 4\{\mathcal{Q}_{1,0}, \mathcal{V}_3\} + \mathcal{Q}_{4,0}$, and we choose $\mathcal{Q}_{0,4} = \langle \mathcal{Q}_{4,0} \rangle_{h''}$. Thus,

$$\mathcal{Q}_{0,4} = -\frac{4!}{2!}\frac{\mu}{a}C_{2,0}^2\frac{R_\oplus^4}{p^4}\eta\left[q_{0,0} - \frac{1+3\eta}{16}(2-3s^2)^2 + 2q_{0,1}s^2\cos 2\omega\right] + 4!\frac{\mu}{a}\sum_{i\geq 3}\langle V_{i,0}\rangle_f, \tag{7.26}$$

which, save for scaling factors, is the same as the second-order term (6.42) of the mean-element Hamiltonian of the zonal problem—yet in different variables.

Then the homological equation $\mathcal{L}_0(\mathcal{V}_4) \equiv 0 = 4\{\mathcal{Q}_{1,0}, \mathcal{V}_3\} + \mathcal{Q}_{4,0} - \mathcal{Q}_{0,4}$ is solved for $\mathcal{V}_3$, to obtain

$$\mathcal{V}_3 = \frac{1}{4\omega_\oplus}\int(\mathcal{Q}_{0,4} - \mathcal{Q}_{4,0})\,\mathrm{d}h'' = -\frac{3!}{\omega_\oplus/n}\frac{G}{\eta}\sum_{i\geq 2}\sum_{j=1}^{i}\int\langle V_{i,j}\rangle_f\,\mathrm{d}\nu,$$

which, from Eq. (7.13), yields

$$\mathcal{V}_3 = \frac{3!G}{\omega_\oplus/n}\sum_{i\geq 2}\sum_{j=1}^{i}\frac{R_\oplus^i}{p^i}\sum_{k=0}^{i-1}\mathcal{F}_{i,j,k}\mathcal{G}_{i,k}^*\left(S_{i,j}\frac{\cos\psi_{i,j,k}}{j} - C_{i,j}\frac{\sin\psi_{i,j,k}}{j}\right). \tag{7.27}$$

The periodic corrections stemming from Eq. (7.27) are third-order corrections as far as the divisor $\omega_\oplus/n \sim \varepsilon$ and tesseral harmonic coefficients are $\mathcal{O}(\varepsilon^4)$.

At fifth order, the computable terms are $\widetilde{\mathcal{Q}}_{0,5} = 5\{\mathcal{Q}_{1,0}, \mathcal{V}_4\} + 10\{\mathcal{Q}_{2,0}, \mathcal{V}_3\}$, and, analogously to the previous case, the fifth-order homological equation is solved for $\mathcal{V}_4$, to yield

$$\mathcal{V}_4 = \frac{1}{5\omega_\oplus}\int(\mathcal{Q}_{0,5} - 10\{\mathcal{Q}_{2,0}, \mathcal{V}_3\})\,\mathrm{d}h'',$$

where, because $\{\mathcal{Q}_{2,0}, \mathcal{V}_3\}$ only contributes periodic terms in h'', we choose $\mathcal{Q}_{0,5} = \langle\{\mathcal{Q}_{2,0}, \mathcal{V}_3\}\rangle_{h''} = 0$.

The transformation equations of the elimination of the node are then computed as usual from the formulas in §2.1.3.

7.3 Exact integration to the second order of J_2

The assumption $\omega_\oplus/n = \mathcal{O}(J_2^{1/2})$ does not apply in the general case. Garfinkel's approach might still be used for sub-synchronous nonresonant motion, but it would need to be extended to much higher orders, thus making the perturbation procedure impractical. Indeed, save for low-Earth orbits, the ratio $\omega_\oplus/n$ will be commonly $\mathcal{O}(1)$, being exactly one at the geosynchronous regime. Then the zeroth-order Hamiltonian cannot be taken in general as the Keplerian term alone, but the Keplerian in the rotating frame. That is, in Delaunay variables,

$$\mathcal{H}_{0,0} = -\frac{\mu^2}{2L^2} - \omega_\oplus H.$$

In consequence, the Lie derivative (2.49) is $\mathcal{L}_0 = n\partial/\partial\ell - \omega_\oplus\partial/\partial h$, from which the homological equation (2.48) becomes

$$n\frac{\partial\mathcal{W}_m}{\partial\ell} - \omega_\oplus\frac{\partial\mathcal{W}_m}{\partial h} = \widetilde{\mathcal{H}}_{0,m} - \mathcal{H}_{0,m}, \tag{7.28}$$

now involving the solution of partial differential equations, contrary to plain integrals.

In spite of the apparent simplicity of Eq. (7.28), the fact that the dependency of the geopotential on the mean anomaly is not explicit, but implicit through the true anomaly, complicates finding a closed-form solution to the homological equation. Therefore, expansions of the elliptic motion are customarily carried out to make explicit the mean anomaly in the tesseral Hamiltonian. Once the mean anomaly shows in the geopotential, the solution of Eq. (7.28) is straightforward, allowing for the elimination of periodic terms related to tesseral harmonics [256, 489] without limitation to the case of nonresonant motion [434].

On the other hand, the homological equation can be reduced to integrals [540]. Indeed, Eq. (7.28) is a first-order linear partial differential equation that, at least for a perturbation solution that neglects third-order effects of J_2 and higher, can be solved by the method of characteristics [147, 565]. Moreover, up to $\mathcal{O}(J_2^2)$ there is no coupling between zonal and tesseral harmonics, and the elimination of tesseral effects can be approached separately in a preliminary step [121].

Thus, we write the geopotential Hamiltonian $\mathcal{H} = \mathcal{H}_{0,0} + \varepsilon\mathcal{H}_{1,0} + \frac{1}{2}\varepsilon^2\mathcal{H}_{2,0}$, with the small parameter formal, and

$$\mathcal{H}_{0,0} = -\frac{\mu}{2a} - \omega_\oplus H, \tag{7.29}$$

$$\mathcal{H}_{1,0} = -\frac{\mu}{a}\frac{a^2\eta}{r^2}V_{2,0}, \tag{7.30}$$

$$\mathcal{H}_{2,0} = -2\frac{\mu}{a}\frac{a^2\eta}{r^2}\left(\sum_{i\ge3} V_{i,0} + \sum_{i\ge2}\sum_{j=1}^{i} V_{i,j}\right), \tag{7.31}$$

where the functions $V_{i,j}$ are those previously defined in Eq. (7.5).

At the first order of the perturbation approach we left the Hamiltonian unaltered by choosing $\mathcal{H}_{0,1} = \mathcal{H}_{1,0}$ and $\mathcal{W}_1 = 0$. At second order, $\widetilde{\mathcal{H}}_{0,2} = \mathcal{H}_{2,0}$, and we remove the tesseral effects by choosing $\mathcal{H}_{0,2} = \langle\mathcal{H}_{2,0}\rangle_h$. That is, the new Hamiltonian term limits to the zonal harmonics contribution.

In order to solve the homological equation (7.28) for $m = 2$, we transform the partial differential equation into the Lagrange–Charpit equations,

$$\frac{\mathrm{d}h}{-\omega_\oplus} = \frac{\mathrm{d}\ell}{n} = \frac{\mathrm{d}\mathcal{W}_2}{-2nG(a/r)^2 \sum_{i\geq 2}\sum_{j=1}^{i} V_{i,j}}. \tag{7.32}$$

We first solve the ordinary differential equation formed by the first two terms of Eq. (7.32), $n\,\mathrm{d}h = -\omega_\oplus \mathrm{d}\ell$, whose solution is given by the characteristic curves

$$h = h_0 - (\omega_\oplus/n)\ell. \tag{7.33}$$

Next, $\nu = h$ is replaced in Eq. (7.3) by the characteristic curves to obtain $\psi_{i,j,k} = (i - 2k)\omega + jh_0 - j(\omega_\oplus/n)\ell - (i-j)_\pi$, and Eq. (7.5) is rearranged in the form

$$V_{i,j} = \frac{R_\oplus^i}{p^i}\eta \sum_{k=0}^{i} \mathcal{F}_{i,j,k}(I) \sum_{m=0}^{i-1} \binom{i-1}{m} \frac{e^m}{2^m} \sum_{l=0}^{m} \binom{m}{l}$$
$$\times \left[S_{i,j}\sin(\alpha_{i,j,k,l,m} + \beta_{i,j,k}) + C_{i,j}\cos(\alpha_{i,j,k,l,m} + \beta_{i,j,k})\right],$$

in which $\alpha_{i,j,k,l,m} = (i - 2k - m + 2l)f - j(\omega_\oplus/n)\ell$ is the only argument that depends on the mean anomaly, whereas $\beta_{i,j,k} = (i - 2k)\omega + jh_0 - (i-j)_\pi$.

Finally, $\mathcal{W}_2$ is solved from the last two terms of Eq. (7.32) by indefinite integration along the characteristic curves. Replacing $\mathrm{d}\ell$ from Eq. (4.64), we obtain

$$\mathcal{W}_2 = -2L \sum_{i\geq 2}\sum_{j=1}^{i} \frac{R_\oplus^i}{p^i}\eta \sum_{k=0}^{i} \mathcal{F}_{i,j,k}(I) \sum_{m=0}^{i-1} \binom{i-1}{m} \frac{e^m}{2^m} \sum_{l=0}^{m} \binom{m}{l}$$
$$\times \Bigg\{ [S_{i,j}\cos\beta_{i,j,k} - C_{i,j}\sin\beta_{i,j,k}] \int \sin\alpha_{i,j,k,l,m}\mathrm{d}f$$
$$+ [S_{i,j}\sin\beta_{i,j,k} + C_{i,j}\cos\beta_{i,j,k}] \int \cos\alpha_{i,j,k,l,m}\mathrm{d}f \Bigg\}, \tag{7.34}$$

where h_0 is replaced in $\beta_{i,j,k}$ after having been solved from Eq. (7.33).

The periodic corrections $\Delta x = \frac{1}{2}\{x; \mathcal{W}_2\}$, in which x denotes each of the Delaunay variables, are then obtained from Eq. (2.17). They also involve the integrals $I_{\sin} = \int \sin\alpha_{i,j,k,l,m}\mathrm{d}f$, $I_{\cos} = \int \cos\alpha_{i,j,k,l,m}\mathrm{d}f$, which depend simultaneously on the mean and true anomalies, whose analytical solutions are not known to exist. Analytical approximations to these integrals can be computed either making expansions of the integrands in the eccentricity or integrating them by parts in the mean anomaly—the latter providing equivalent series to those given by the classical relegation of the node, which requires a distinction between super-synchronous $n/\omega_\oplus$ and sub-synchronous cases (see [471] for details).

Alternatively, these integrals can be evaluated numerically to arbitrary precision at the desired output point $f = f_p$. Namely,

$$I_{\cos}(f_p) = I_{\cos}(f_0) + \int_{f_0}^{f_p} \cos\alpha_{i,j,k,l,m}\,\mathrm{d}f,$$

and an analogous integration for the sine case. This last procedure avoids issues related to series convergence, as well as the need for separating sub- and super-synchronous cases [471]. Note, however, that, in spite of the lower limit, f_0 being arbitrary and not seeming to be of concern, in fact it is, because the computation of the constants $I_{\sin}(f_0)$, $I_{\cos}(f_0)$ requires one in turn to know approximate analytical solutions of $I_{\sin}(f)$ and $I_{\cos}(f)$.

7.4 Relegation of tesseral effects

Rather than solving exactly the homological equation at each step m of the perturbation approach, one might be satisfied with finding an approximation to $\mathcal{W}_m$ of $\mathcal{O}(\varepsilon^m)$, of the form

$$\mathcal{W}_m = \sum_{l\geq 0} \delta^l \mathcal{W}_{m,l}, \tag{7.35}$$

where $\delta = n/\omega_\oplus < 1$ for super-synchronous orbits, and $\delta = \omega_\oplus/n < 1$ for sub-synchronous orbits—the synchronous case requiring specific treatment.

A rule for the selection of the maximum value of the index l is to choose

$$l_{\max} = \left\lfloor \frac{\log \varepsilon}{\log \delta} \right\rfloor - 1, \tag{7.36}$$

so that $\delta^{l_{\max}+1} = \mathcal{O}(\varepsilon)$ in agreement with the required approximation. Thus, for instance, in the usual case $\varepsilon = \mathcal{O}(J_2) = \mathcal{O}(10^{-3})$, for an orbit with $a = 10500$ km we get $\delta \approx 1/8$ and hence $l_{\max} = 2$, whereas for a typical GPS orbit $\delta \approx 1/2$, thus requiring eight terms in the summation (7.35).

In what follows, we focus on the case $\delta = \omega_\oplus/n$ of sub-synchronous orbits, although the same philosophy can be applied to the super-synchronous case.

7.4.1 Basic algorithm

Replacing Eq. (7.35) in the homological equation (7.28), we get

$$\sum_{l\geq 0} \delta^l \left(n \frac{\partial \mathcal{W}_{m,l}}{\partial \ell} - \omega_\oplus \frac{\partial \mathcal{W}_{m,l}}{\partial h} \right) = \widetilde{\mathcal{H}}_{0,m} - \mathcal{H}_{0,m},$$

which is rearranged in the form

$$\frac{\partial \mathcal{W}_{m,0}}{\partial \ell} - \frac{1}{n}(\widetilde{\mathcal{H}}_{0,m} - \mathcal{H}_{0,m}) + \sum_{l\geq 1} \delta^l \left(\frac{\partial \mathcal{W}_{m,l}}{\partial \ell} - \frac{\partial \mathcal{W}_{m,l-1}}{\partial h} \right) = 0, \tag{7.37}$$

and solved iteratively by equating equal powers of δ. Namely,

$$\mathcal{W}_{m,0} = \frac{1}{n}\int (\widetilde{\mathcal{H}}_{0,m} - \mathcal{H}_{0,m})\,\mathrm{d}\ell, \quad \mathcal{W}_{m,l} = \int \frac{\partial \mathcal{W}_{m,l-1}}{\partial h}\,\mathrm{d}\ell \quad (l > 0).$$

Note that this procedure cannot be used to properly deal with the terms $\mathcal{P}_{g,h}$ in Eq. (7.24), which, unless being moved to the new Hamiltonian, would introduce mixed secular terms in the integration of the generating function. These terms are, nonetheless, easily removed in a following elimination of the node, as it was effectively done in §7.2.3. Alternatively, terms $\mathcal{W}_m$ of the generating function are often split into two parts

$$\mathcal{W}_m = \mathcal{W}_m^{\dagger} + \sum_{l\geq 0} \delta^l \mathcal{W}_{m,l}^{*}, \tag{7.38}$$

where the dagger part is used to integrate the terms of $\mathcal{P}_{g,h}$ without further iterations. Now, replacing Eq. (7.38) into Eq. (7.28) yields

$$-\omega_{\oplus}\frac{\partial \mathcal{W}_m^{\dagger}}{\partial h} + \sum_{l\geq 0}\delta^l\left(n\frac{\partial \mathcal{W}_{m,l}^{*}}{\partial \ell} - \omega_{\oplus}\frac{\partial \mathcal{W}_{m,l}^{*}}{\partial h}\right) = \widetilde{\mathcal{H}}_{0,m} - \mathcal{H}_{0,m},$$

and hence $\mathcal{W}_m^{\dagger} = (1/\omega_{\oplus})\int \mathcal{P}_{g,h}\,\mathrm{d}h$, whereas

$$\mathcal{W}_{m,0}^{*} = \frac{1}{n}\int (\widetilde{\mathcal{H}}_{0,m} - \mathcal{H}_{0,m} - \mathcal{P}_{g,h})\,\mathrm{d}\ell, \quad \mathcal{W}_{m,l}^{*} = \int \frac{\partial \mathcal{W}_{m,l-1}}{\partial h}\,\mathrm{d}\ell \quad (l > 0).$$

The closed-form integration of terms $\mathcal{W}_{m,l}^{*}$, $l \geq 0$, is feasible following the procedures described in [324], where it is shown that the general case reduces to the integration of three types of integrals, to be solved in the radius, in the true anomaly, and in the eccentric anomaly, respectively.

7.4.2 Tesseral relegation with low eccentricity

Typically, terms $\mathcal{W}_{m,l}^{*}$ grow in size with each iteration l making the relegation process unpractical for a relative low number of iterations. This fact limits practical application of the sub-synchronous relegation algorithm to orbits relatively close to the Earth's surface, in which case $\delta \ll 1$. On the other hand, slight modifications of the relegation algorithm allow one to extend its application to orbits with any semimajor axis, yet constrained to the case of low eccentricities [438, 441].

First of all, the homological equation (7.28) is reformulated in terms of the true anomaly, contrary to the mean one, making use of the differential relation (4.64), in which the radius is further replaced using the conic equation (4.29). This turns Eq. (7.28) into

$$(1 + e\cos f)^2\frac{n}{\eta^3}\frac{\partial \mathcal{W}_m}{\partial f} - \omega_{\oplus}\frac{\partial \mathcal{W}_m}{\partial h} = \widetilde{\mathcal{H}}_{0,m} - \mathcal{H}_{0,m}. \tag{7.39}$$

Now, we replace Eq. (7.35) by the analogous one,

$$\mathcal{W}_m = \sum_{l\geq 0} e^l \mathcal{W}_{m,l}, \tag{7.40}$$

and the maximum number of iterations is estimated again from Eq. (7.36) with δ replaced by e. Besides, without carrying out any expansion of the elliptic motion, we rearrange the right side of Eq. (7.39) in the form

$$\widetilde{\mathcal{H}}_{0,m} - \mathcal{H}_{0,m} = \sum_{l\geq 0} e^l P_{m,l}(f, g, h, L, G, H). \tag{7.41}$$

Finally, both Eqs. (7.40) and (7.41) are plugged into Eq. (7.39), to obtain

$$\sum_{l\geq 0} e^l \mathcal{L}_0^*(\mathcal{W}_{m,l}) = \sum_{l\geq 0} e^l F_l, \tag{7.42}$$

in which

$$\mathcal{L}_0^* = n\frac{\partial}{\partial f} - \eta^3 \omega_\oplus \frac{\partial}{\partial h}, \tag{7.43}$$

$$F_0 = \eta^3 P_{m,0}, \tag{7.44}$$

$$F_1 = \eta^3 P_{m,1} - 2n\cos f\,\frac{\partial \mathcal{W}_{m,0}}{\partial f}, \tag{7.45}$$

$$F_l = \eta^3 P_{m,l} - 2n\cos f\,\frac{\partial \mathcal{W}_{m,l-1}}{\partial f} - n\cos^2 f\,\frac{\partial \mathcal{W}_{m,l-2}}{\partial f}, \qquad l > 1. \tag{7.46}$$

Terms of the generating function are no longer obtained by indefinite integration, but from the solution of a partial differential equation. Like in the case of exact integration in §7.3, we can solve Eq. (7.42) by the Lagrange–Charpit method, now along the characteristic curves $h = h_0 - (\omega_\oplus/n)\eta^3 f$. On the other hand, as far as the terms on the right side of the successive partial differential equations are of trigonometric nature, the solution is readily obtained in view of

$$\mathcal{L}_0^*[\sin(if + jh + \gamma)] = (in - j\eta^3\omega_\oplus)\cos(if + jh + \gamma),$$

with i, j integers. Now, there is no need of special treatment of the terms $\mathcal{P}_{g'',h''}$, given in Eq. (7.24), because the procedure does not rely any longer on indefinite integration in the mean anomaly, but on the solution of partial differential equations involving both the true anomaly and the longitude of the node in the rotating frame. On the other hand, the occurrence of divisors $in - j\eta^3\omega_\oplus$ in the generating function makes it singular for resonances, $i/j = \eta^3\omega_\oplus/n$.

7.4.3 Sample applications

To illustrate the importance of taking the effects of tesseral terms into account in the perturbation solution, we provide two examples of satellites evolving under de perturbation of a 2×2 truncation of the geopotential.

The first example deals with a Galileo-type orbit, with orbit parameters $a = 29600$ km, $e = 0.001$, $I = 56^\circ$, $\Omega = 0$, $\omega = 270^\circ$, and $f = 90^\circ$. The reference orbit is provided by the one-month propagation of these initial conditions in the full 2×2 potential, which amounts to about 51 orbits. When it is compared with the propagation of the same initial conditions in the model with the tesseral effects removed, that is, the main problem, we found that the error of the semimajor axis reaches more than 40 meter, yet, as expected, the errors are of periodic nature, as shown in the top plot of Fig. 7.1.

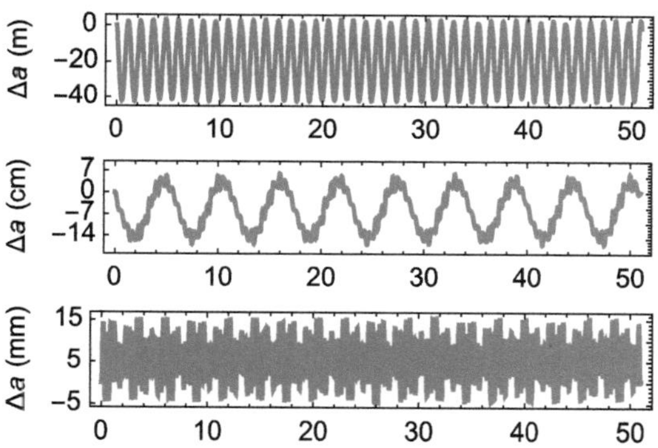

Figure 7.1: Galileo-type orbit (2×2 geopotential). Semimajor axis errors when ignoring tesseral effects (top plot), and when they are relegated without (center plot) and with one iteration (bottom plot). Abscissas label the number of orbits.

Periodic corrections derived from the application of the relegation algorithm without iterations—which are applied both to the initial conditions (corrections obtained from the inverse transformation) and to the ephemeris computed from the numerical propagation of the simplified 2×0 model (corrections obtained from the direct transformation)—reduce the amplitude of the errors to the centimeter level, as observed in the center plot of Fig. 7.1, meaning a 99.5 % improvement with respect to the maximum amplitude of the errors in the previous case. One single iteration of the relegation algorithm, which is applied again to both the inverse and the direct transformations, further reduces the amplitude of the semimajor axis errors to just a few millimeters, as displayed in the lower plot of Fig. 7.1. This means an additional 91 % improvement relative to the maximum amplitude of the previous case. No relevant improvements are observed with additional iterations of the relegation, which was, in fact, expected. Indeed, replacing the 2×2 truncation of the geopotential by the main problem means neglecting the third order of J_2 in the perturbation approach, which is consistent with the fact that the maximum amplitude of the semimajor axis' error is $\sim 10^{-9}$ times smaller than the semimajor axis' value itself.

In the case of the eccentricity, even though the improvements are not so impressive when the periodic corrections of the relegation are applied without iterations, they still achieve 59.5 % of the maximum amplitude with respect to the case with no corrections (first and second from the upper plots of Fig. 7.2). Eccentricity errors notably are reduced with a first iteration of the relegation (second from the lower plot of Fig. 7.2), reaching a 99.4 % improvement with respect to the case with no corrections. Improvements in the eccentricity errors are still observed when using a second iteration of the relegation algorithm, as shown in Fig. 7.2, but this is not the case of other orbital elements for which no relevant improvements are appreciated after the first iteration of the relegation. As mentioned in §6.5.5, this particular behavior of the eccentricity is not in contradiction with the order of the perturbation approach, and it is derived from the fact that the periodic corrections to the eccentricity involve the computation of the Poisson bracket $\{e, \mathcal{W}\}$, which has the eccentricity as divisor. Then the series (7.40) needs to be extended to one order higher than in the other cases. Also because of this, the final improvement relative to the eccentricity value is of the order of J_2^3/e.

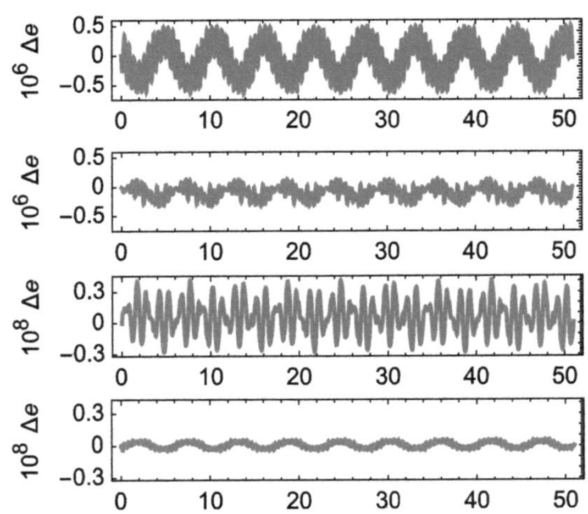

Figure 7.2: Galileo-type orbit. From top to bottom, eccentricity errors when ignoring tesseral effects, and when they are relegated without, with one, and with two iterations. Abscissas label the number of orbits.

It is worth noting that a similar case happens to the inclination, whose periodic corrections involve the computation of the Poisson bracket $\{I, \mathcal{W}\}$ that introduces the sine of the inclination in denominators. However, in view of the high inclination of Galileo orbits, the sine of the inclination is of the order of the unity in the current example, and, in consequence, the corresponding periodic corrections do not undergo any negative effect. Indeed, a single iteration of the relegation algorithm in the computation of both the direct and the inverse corrections is enough to reduce the amplitude of the inclination errors to just a few tens of microarcsecond. That is, the relative error is of the order of 10^{-10}, which is in agreement with the truncation order of the perturbation theory.

On the other hand, since the relegation of tesseral effects has been computed only to the first order of the perturbation model, which we recall is equivalent to the second order of J_2, there is an effective truncation of the perturbation solution to $\mathcal{O}(J_2^2)$ that limits its accuracy. This truncation introduces a secular trend in the perturbation solution and prevents the purely periodic nature of the errors. Indeed, as shown in Fig. 7.3, a small secular drift is clearly observed in the evolution of the mean argument of the latitude $F = M+\omega$. This drift is reduced from about 1.3 arcseconds/orbital period (~ 185 meter per orbit in the in-track direction) when tesseral effects are ignored, to ~ 3.7 milliarcsecond/orbital period (~ half a meter per orbit in the in-track direction) when the relegation is applied without iteration, to the very low rate of less than half a milliarcsecond/orbital period (~ 5 cm per orbit in the in-track direction) when one iteration of the relegation is applied. This secular trend also corrupts the propagation of the other orbital elements, but, due to its smallness, it is clearly exceeded by the periodic errors and hence is not appreciated in the corresponding plots.

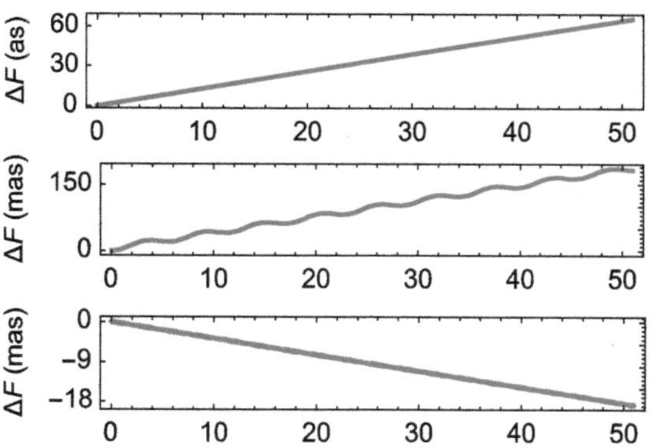

Figure 7.3: Galileo-type orbit. Mean argument of the latitude errors when ignoring tesseral effects (upper plot), and when they are relegated without (center plot), and with one iteration (lower plot). Abscissas label the number of orbits.

For the second example we choose the case that showed less error improvement in [602], where the authors used a different version of the tesseral relegation algorithm. Namely, a highly inclined and elliptic orbit with $a = 18520$ km, $e = 0.35$, $I = 100°$, which we complement, as before, with $\Omega = 0$, $\omega = 270°$, and $f = 90°$. Again, the initial conditions are propagated for one month, which amounts to about 103 orbits in the current case due to the smaller semimajor axis. Because the eccentricity is now much larger than in the previous example, and in view of the relegation algorithm in Eq. (7.42) being scaled by the eccentricity, improvements obtained by successive iterations of the relegation are now expected to continue beyond the two iterations of the previous example. More precisely, on account of $e^7 = \mathcal{O}(J_2)$ we would expect successive improvements in the periodic corrections up to the relegation order $l_{\max} = 6$.

Inclusion of higher-order effects of the relegation is particularly important in the conversion of the initial osculating elements into the variables resulting from the relegation, which is carried out with the *inverse* transformation of the Lie transforms procedure. More precisely, as discussed in §6.5.5 and §6.6, an inaccuracy in the conversion of the initial semimajor axis results in a slightly incorrect mean motion that causes an associated secular trend in the errors of the mean anomaly. This is illustrated in Fig. 7.4, in which we clearly observe how the relegation improves the perturbation solution when reducing the inaccuracies in the inverse transformation using an increasing number of iterations. Thus, the secular trend is about 2 arcseconds per orbit when no corrections are made, which amounts to ~ 160 meter per orbit in the along-track direction (error curve with black dots in Fig. 7.4). The slope of the secular errors reduces gradually with the different number of iterations of the relegation, until it becomes very small with the sixth iteration, in which the error is less than 1 arcsecond times orbital period, or about 80 cm per orbit in the along-track direction (black dashed curve in Fig. 7.4). Of course, increasing the number of iterations of the relegation algorithm in the computation of the periodic corrections of the *direct* transformation also improves the propagation of the perturbation solution. However, this increased accuracy only modifies the amplitude of the periodic components of the errors and is less relevant in the long term.

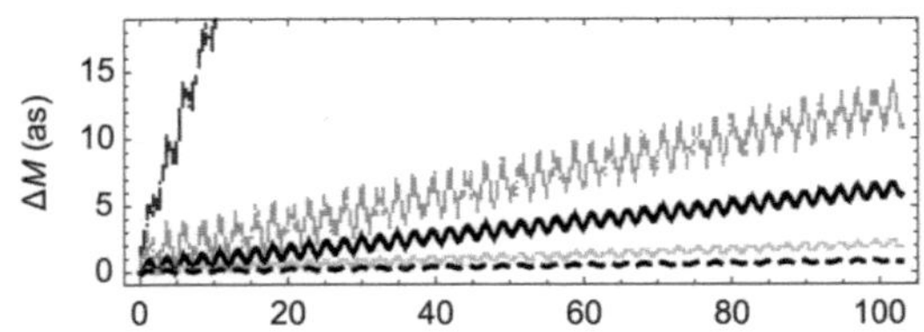

Figure 7.4: Mean anomaly error of the perturbation solution. Test orbit with a = 18520 km, e = 0.35, I = 100°. Tesseral effects ignored (black dots), and when they are relegated without (gray dots), and with two (black line), four (gray line), and six iterations (black dashed line). Abscissas are orbits.

On the other hand, as mentioned before, the secular errors derived from the early truncation of the perturbation solution to the second order of J_2 cannot be reduced unless the whole perturbation approach is extended to higher orders of J_2, irrespective of the number of iterations that are carried out with the relegation algorithm in the computation of the generating function. However, the computation of higher orders would lead to involved computations related to the coupling of zonal and tesseral harmonics that would happen starting at the third order.

An example of the improvements achieved with different iterations of the relegation is presented in Fig. 7.5 for the semimajor axis. Despite errors improving slower than in the Galileo case, due to the higher eccentricity of the orbit, two iterations of the relegation are enough to get a 90 % improvement in the errors of the semimajor axis. Analogous rates of improvement are found in this particular example for the other orbital elements, which yield a 98 % improvement in a, e, and i when the sixth iteration of the relegation is compared to the case in which no tesseral direct corrections are used.

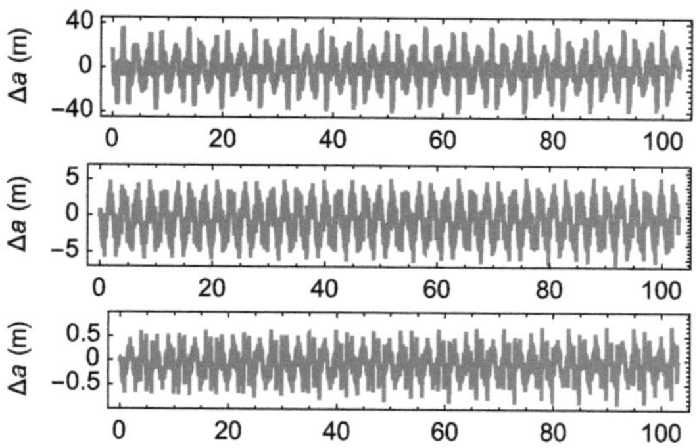

Figure 7.5: Eccentric orbit. Semimajor axis errors of the relegation without (top) and with two (center) and six iterations (bottom plot). Abscissas are orbital periods.

7.5 Tesseral resonances

Tesseral effects must be treated differently when the orbital period and the Earth's rotational period are commensurable. That is,

$$Q\left(\omega_{\oplus} - \frac{\mathrm{d}\Omega}{\mathrm{d}t}\right) = P\left(\frac{\mathrm{d}M}{\mathrm{d}t} + \frac{\mathrm{d}\omega}{\mathrm{d}t}\right), \tag{7.47}$$

where P and Q are mutually prime integers. In that case, the subsatellite point will repeat its ground trace,[1] in this way undergoing the same forces periodically in the typical resonance phenomenon. For exact resonances, the dynamics is essentially that of the pendulum, producing long-term oscillations in the pair of variables related to the semimajor axis and the longitude of the ascending node. When the commensurability is only approximate, resonant effects superimpose to a secular rate in the longitude of the ascending node [228].

1 In fact, since these rates are not constant, second derivatives should be taken into account. In particular, that of the mean anomaly [203].

In particular, since the rate of variation of Ω and ω is small, resonance effects will be apparent when commensurabilities between the mean motion of the orbiter and the rotation rate of the Earth take place—that is, when $\omega_\oplus Q \approx nP$. In that case, divisors $nP - \omega_\oplus Q$ that may appear in the removal of short-period tesseral effects will prevent convergence of the perturbation solution for orbits close to the resonant regime [77]. The fact that similar offending divisors do not occur in the exact tesseral normalization discussed in §7.3 just means that a computer program based on it will not break in that case, but the real behavior of a resonant orbit is in no way seized by that perturbation solution. In fact, a trigonometric function whose argument involves a resonant combination of the mean anomaly and the longitude of the node in the rotating frame will evolve slowly for orbits in the corresponding resonant regime. In that case, tesseral resonant terms must remain in the mean-element Hamiltonian after the short-period elimination—an operation that now gets a clear different meaning from the usual normalization procedure based on the removal of the mean anomaly.

Still, the perturbation approach is useful in dealing with resonant orbits, as is illustrated below for typical orbital regimes of global navigation satellite systems.

7.5.1 Resonant terms of the geopotential

To remove all the short-period effects from a tesseral perturbation Hamiltonian, it is unavoidable to trace tesseral resonances in the mean anomaly. Then, instead of using the closed-form expression of the disturbing function in Eq. (7.1) with the $V_{i,j}$ functions given by Eq. (7.5), we rather resort to the classical expressions of Kaula [341] in which trigonometric functions of the true anomaly are replaced by traditional expansions in the mean anomaly [78, 557]. Thus,

$$\frac{r^n}{a^n}\exp(\mathbf{i}mf) = \sum_{k=-\infty}^{\infty} X_k^{m,n}(e)\exp(\mathbf{i}kM),$$

where the Hansen coefficients $X_k^{m,n} = \frac{1}{2\pi}\int_0^{2\pi}(r/a)^n \exp[\mathbf{i}(mf - kM)]\,\mathrm{d}M$, are commonly expressed in terms of Bessel functions, and are computed using efficient recursion in the literature [231, 234, 562].

Then Eq. (7.1) is written explicitly in the mean anomaly like

$$\mathcal{P} = -\frac{\mu}{a}\sum_{i\geq 2}\sum_{j=0}^{i}\mathfrak{V}_{i,j}, \tag{7.48}$$

where now

$$\mathfrak{V}_{i,j} = \frac{R_\oplus^i}{p^i}\sum_{k=0}^{i}\mathcal{F}_{i,j,k}(I)\sum_{l=-\infty}^{\infty}G_{i,k,l}(e)(C_{i,j}\cos\Psi_{i,j,k,l} + S_{i,j}\sin\Psi_{i,j,k,l}), \tag{7.49}$$

in which, using the Delaunay angles $\ell = M$, $g = \omega$, $h = \Omega - \omega_{\oplus}t$,

$$\Psi_{i,j,k,l} = (i - 2k + l)\ell + jh + (i - 2k)g - (i - j)_{\pi}. \tag{7.50}$$

Kaula eccentricity functions $G_{i,k,l} = X_{i-2k+l}^{-i-1,i-2k}$ are computed as follows [231, 341, 631]. When $l + i = 2k$,

$$G_{i,k,2k-i} = \frac{1}{\eta^{2i-1}} \sum_{j=0}^{k^*-1} \binom{i-1}{2j+i-2k^*} \binom{2j+i-2k^*}{j} \left(\frac{e}{2}\right)^{2j+i-2k^*}, \tag{7.51}$$

in which $k^* = k$ if $k \le i/2$, and $k^* = i - k$ otherwise. When $l + i \neq 2k$,

$$G_{i,k,l} = \frac{2^i(-\beta)^{|l|}}{(1+\eta)^i} \sum_{j=0}^{\infty} P_{i,k^*,l^*,j} Q_{i,k^*,l^*,j} \beta^{2j},$$

with $k^* = k$ and $l^* = l$ if $k \le i/2$, and $k^* = i - k$ and $l^* = -l$ otherwise, and

$$P_{i,k,l,j} = \sum_{m=0}^{r} \frac{1}{m!} \binom{2k-2i}{r-m} \left[\frac{2k-i-l}{2}(1+\eta)\right]^m,$$

$$Q_{i,k,l,j} = \sum_{m=0}^{h} \frac{1}{m!} \binom{-2k}{h-m} \left[\frac{i+l-2k}{2}(1+\eta)\right]^m,$$

in which $h = j$ and $r = j + l$ if $l > 0$, and $h = j - l$ and $r = j$ otherwise.

The time derivative of $\Psi_{i,j,k,l}$ is arranged in the form

$$\frac{d\Psi_{i,j,k,l}}{dt} = (i - 2k + l)\left(\frac{d\ell}{dt} + \frac{dg}{dt}\right) + j\frac{dh}{dt} - l\frac{dg}{dt},$$

which, in view of Eq. (7.47), shows that arguments $\Psi_{i,j,k,l}$ with

$$l = \tilde{l} \equiv (P/Q)j - i + 2k \tag{7.52}$$

will evolve slowly when close to a Q:P-tesseral resonance. That is, terms $\Psi_{i,j,k,\tilde{l}} = j[(P/Q)(\ell + g) + h] - \tilde{l}g - (i - j)_{\pi}$ no longer contribute short-period effects in spite of the term in square brackets depending on the fast angles ℓ and h.

The resonant term is combined to a single variable by means of the canonical transformation $(\lambda, \gamma, \delta, \Lambda, \Gamma, \Delta) \mapsto (\ell, g, h, L, G, H)$ given by

$$\lambda = \ell + g + \frac{Q}{P}h, \quad \gamma = g, \quad \delta = h \quad \Lambda = L, \quad \Gamma = G - L, \quad \Delta = H - \frac{Q}{P}L, \tag{7.53}$$

where λ differs from the mean longitude in a fraction $1 - Q/P$ of the argument of the node, and $-L \le \Gamma \le 0$. That the transformation (7.53) is canonical is checked from the differential form $\Lambda\, d\lambda + \Delta\, d\delta + \Gamma\, d\gamma = L\, d\ell + G\, dg + H\, dh$ [243]. Different alternatives to Eq. (7.53) may, of course, be explored [325, 434].

Plugging the inverse transformation of Eq. (7.53) into Eq. (7.50) we obtain $\Psi_{i,j,k,l} = (i - 2k + l)\lambda + [j - (i - 2k + l)Q/P]\delta - l\gamma - (i - j)_{\pi}$, which for Q:P-resonant coefficients in

Eq. (7.52) yields

$$\Psi_{i,j,k,\tilde{l}} = j(P/Q)\lambda - \tilde{l}\gamma - (i-j)_\pi. \tag{7.54}$$

Therefore, tesseral resonant terms depend only on the longitude of the stroboscopic node λ and the argument of the perigee γ.

The form of the critical arguments (7.54) suggests one to redefine the harmonic coefficients like

$$S_{i,j} = J_{i,j}\sin(P/Q)j\lambda_{i,j}, \quad C_{i,j} = J_{i,j}\cos(P/Q)j\lambda_{i,j},$$

which in the particular case $P = Q$ agrees with previous definitions in the literature [106, 341]. Hence, Eq. (7.49) is conveniently replaced by the alternative form

$$\mathfrak{V}_{i,j} = \frac{R_\oplus^i}{p^i}\sum_{k=0}^{i}\mathcal{F}_{i,j,k}(I)\sum_{l=-\infty}^{\infty}G_{i,k,l}(e)J_{i,j}\cos[\Psi_{i,j,k,l} - (P/Q)j\lambda_{i,j}]. \tag{7.55}$$

7.5.2 Short-period elimination

In the new P:Q-resonant variables, the perturbation Hamiltonian in Eqs. (7.29)–(7.31) reads

$$\mathcal{H}_{0,0} = -\frac{\mu}{2a} - \omega_\oplus\left(\Delta + \frac{Q}{P}\Lambda\right), \tag{7.56}$$

$$\mathcal{H}_{1,0} = -\frac{\mu}{a}\mathfrak{V}_{2,0}, \tag{7.57}$$

$$\mathcal{H}_{2,0} = -2\frac{\mu}{a}\left(\sum_{i\geq 3}\mathfrak{V}_{i,0} + \sum_{i\geq 2}\sum_{j=1}^{i}\mathfrak{V}_{i,j}\right), \tag{7.58}$$

where now $a = \Lambda^2/\mu$, $\eta = 1 - |\Gamma|/\Lambda$, and $c = [(Q/P)\Lambda + \Delta]/(\Lambda - |\Gamma|)$.

The zeroth-order Hamiltonian (7.56) yields the Lie derivative

$$\mathcal{L}_0 = \{\ \ ,\mathcal{H}_{0,0}\} \equiv \left(n - \omega_\oplus\frac{Q}{P}\right)\frac{\partial}{\partial\lambda} - \omega_\oplus\frac{\partial}{\partial\delta}. \tag{7.59}$$

Then the homological equation in the new variables is the partial differential equation

$$\left(n - \omega_\oplus\frac{Q}{P}\right)\frac{\partial\mathcal{W}_m}{\partial\lambda} - \omega_\oplus\frac{\partial\mathcal{W}_m}{\partial\delta} = \widetilde{\mathcal{H}}_{0,m} - \mathcal{H}_{0,m}, \tag{7.60}$$

in which the short-period elimination is achieved by selecting the new Hamiltonian terms free from δ, that is $\mathcal{H}_{0,m} = \langle\widetilde{\mathcal{H}}_{0,m}\rangle_\delta$.

It is clear now that the terms

$$\langle\mathfrak{V}_{i,j}\rangle_\delta = \frac{R_\oplus^i}{p^i}\sum_{k=0}^{i}\mathcal{F}_{i,j,k}\sum_{l=-\infty}^{\infty}G_{i,k,l}J_{i,j}\cos\Psi^*_{i,j,k,\tilde{l}}, \tag{7.61}$$

with $\Psi^*_{i,j,k,\tilde{l}} = j(P/Q)(\lambda - \lambda_{i,j}) - \tilde{l}\gamma - (i-j)_\pi$, which are kept in the long-term Hamiltonian, entail only long-period effects related to γ and λ. Note that Eq. (7.61) also applies to the case of zonal terms, in which $\langle \mathfrak{V}_{i,0}\rangle_\delta = \langle \mathfrak{V}_{i,0}\rangle_\ell$ as expected.

Therefore, at first order,

$$\widetilde{\mathcal{H}}_{0,1} = \mathcal{H}_{1,0} = -C_{2,0}\frac{\mu}{a}\frac{R_\oplus^2}{p^2}\sum_{k=0}^{2}\mathcal{F}_{2,0,k}(I)\sum_{l=-\infty}^{\infty}G_{2,k,l}(e)\cos\Psi_{2,0,k,l},$$

in which

$$\Psi_{2,0,k,l} = (2-2k+l)\left(\lambda - \frac{Q}{P}\delta\right) - l\gamma. \tag{7.62}$$

The new Hamiltonian term $\mathcal{H}_{0,1} = \langle\widetilde{\mathcal{H}}_{0,1}\rangle_\delta$ is chosen using Eq. (7.61) with $i = 2$ and $j = 0$. Since $\tilde{l} = 2k-2$, we get

$$\mathcal{H}_{0,1} = -\frac{\mu}{a}\langle\mathfrak{V}_{2,0}\rangle_\delta = -C_{2,0}\frac{\mu}{a}\frac{R_\oplus^2}{p^2}\sum_{k=0}^{2}\mathcal{F}_{2,0,k}(I)G_{2,k,2k-2}(e)\cos\Psi_{2,0,k,2k-2},$$

where $\Psi_{2,0,0,-2} = 2\gamma$, $\Psi_{2,0,1,0} = 0$, $\Psi_{2,0,2,2} = -2\gamma$, from Eq. (7.62); $\mathcal{F}_{2,0,0} = \mathcal{F}_{2,0,2} = -\frac{3}{8}s^2$, $\mathcal{F}_{2,0,1} = -\frac{1}{2} + \frac{3}{4}s^2$, from Eq. (7.4); and $G_{2,0,-2} = G_{2,2,2} = 0$, $G_{2,1,0} = 1/\eta^3$, from Eq. (7.51). As expected, $\mathcal{H}_{0,1}$ is exactly the same as given in Eq. (5.60), which is composed only of secular terms.

Next, the first-order homological equation (7.60),

$$\mathcal{L}_0(\mathcal{W}_1) = -C_{2,0}\frac{\mu}{a}\frac{R_\oplus^2}{p^2}\sum_{k=0}^{2}\mathcal{F}_{2,0,k}\sum_{\substack{l=-\infty\\ l\neq\tilde{l}}}^{\infty}G_{2,k,l}\cos\Psi_{2,0,k,l},$$

is readily solved in view of $\mathcal{L}_0(\sin\Psi_{2,0,k,l}) = (2-2k+l)n\cos\Psi_{2,0,k,l}$, as obtained from Eqs. (7.59) and (7.62). Therefore

$$\mathcal{W}_1 = -C_{2,0}L\frac{R_\oplus^2}{p^2}\sum_{k=0}^{2}\mathcal{F}_{2,0,k}(I)\sum_{\substack{l=-\infty\\ l\neq\tilde{l}}}^{\infty}G_{2,k,l}(e)\frac{\sin\Psi_{2,0,k,l}}{2-2k+l}. \tag{7.63}$$

It is worth noting that there are no tesseral effects hitherto and, therefore, $\mathcal{W}_1$ is just an expanded version of Eq. (5.61) in power series of the eccentricity, save for an integration constant. Indeed it is easy to check from Eqs. (7.63) and (7.62) that $\langle\mathcal{W}_1\rangle_\ell = 0$, whereas it was not at all the case of Eq. (5.61), as shown by Eq. (5.64).

At second order $\mathcal{L}_0(\mathcal{W}_2) = \widetilde{\mathcal{H}}_{0,2} - \mathcal{H}_{0,2}$, where the computable terms $\widetilde{\mathcal{H}}_{0,2}$ are the usual ones given by Eq. (2.37). Because Poisson brackets are invariant with respect to canonical transformations, they can be evaluated in the most convenient set of canonical variables. In consequence, we resort to previously computed values in §5.6. In particular, $\mathcal{W}_1$ is obtained by adding to Eq. (5.61) the integration constant A_1 in Eq. (6.116) that guarantees that $\langle \mathcal{W}_1 \rangle_\ell = 0$, as happens with Eq. (7.63). As far as we are using here a formal small parameter, both Eqs. (5.61) and (6.116) must be previously multiplied by the oblateness coefficient $C_{2,0}$.

Then, to the already known terms previously computed in Eqs. (5.66), (5.67), and (5.68), which must now to be multiplied by the physical small parameter $C_{2,0}^2$, we need to add the term $\mathcal{H}_{2,0}$, given in Eq. (7.58), as well as the additional terms derived from the non-null value of the integration constant A_1. That is,

$$\{\mathcal{H}_{1,0} + \mathcal{H}_{0,1}; A_1\} = C_{2,0}^2 \frac{3}{16} \frac{\mu}{p} \frac{R_\oplus^4}{p^4} \left\{ \eta^3 s^2 (4 - 5s^2) \beta^2 (1 + 2\eta) \cos 2g + \frac{1}{16} \frac{p^2}{r^2} \right.$$
$$\left. \times \sum_{j=0}^{2} \sum_{i=-3}^{3} q'_{2j,i} s^{2j} \cos[(2j + i + j_5) f + 2jg] \right\}, \tag{7.64}$$

where the coefficients $q'_{2j,i}$ are given in Table 7.1, and $j_5 = \lfloor \frac{1}{2}(j - 5) \rfloor$ as follows from the index convention in Eq. (6.5).

Table 7.1: Coefficients $q'_{2j,i}$ in Eq. (7.64). $q'_{0,-3} = q'_{0,3} = q'_{4,-2} = 0$.

j	i	$q'_{2j,i}$
0	−2	$-5e^2 s^4 \beta^3$
	−1	$-24 s^4 \beta^2 (1 - \eta)$
	0	$-s^2 \beta^3 [16 + 23s^2 + (48 - 66s^2)\eta + (32 - 61s^2)\eta^2]$
	1	$-8s^2 \beta^2 [4 + (8 - 15s^2)\eta]$
	2	$-2s^2 \beta [2(4 - s^2) + (8 - 23s^2)\eta - (16 - 31s^2)\eta^2]$
1	−3	$-2\beta^3 e^2 (2 - 3s^2)$
	−2	$-8(2 - 3s^2)\beta^2 (1 - \eta)$
	−1	$2\beta^3 [2 + s^2 + 6(6 - 7s^2)\eta + (26 - 31s^2)\eta^2]$
	0	$16(4 - 5s^2)\beta^2 (1 + 2\eta)$
	1	$2\beta[42 - 59s^2 + 2(6 - 7s^2)\eta - (30 - 37s^2)\eta^2]$
	2	$24(2 - 3s^2)(1 - \eta)$
	3	$6e(2 - 3s^2)(1 - \eta)$
2	−3	$-e^2 \beta^3$
	−1	$\beta^3 (21 + 18\eta + \eta^2)$
	0	$40\beta^2 (2 + \eta)$
	1	$3\beta(39 - 6\eta - 5\eta^2)$
	2	$72(1 - \eta)$
	3	$15e(1 - \eta)$

Next, we choose $\mathcal{H}_{0,2} = \langle \widetilde{\mathcal{H}}_{0,2} \rangle_\delta = \langle \mathcal{H}_{2,0} \rangle_\delta + \langle \{\mathcal{H}_{1,0} + \mathcal{H}_{0,1} \mathcal{W}_1\} \rangle_\delta$, where the last term is obtained by adding to Eq. (5.70), which must first be multiplied by $C_{2,0}^2$, the part of

Eq. (7.64) that is free from short-period effects. We finally find

$$\mathcal{H}_{0,2} = -2\frac{\mu}{a}\left(\sum_{i\geq 3}\langle\mathfrak{V}_{i,0}\rangle_\delta + \sum_{i\geq 2}\sum_{j=1}^{i}\langle\mathfrak{V}_{i,j}\rangle_\delta\right) - C_{2,0}^2\frac{\mu}{p}\frac{R_\oplus^4}{p^4}\frac{3}{64}\eta^3 \tag{7.65}$$
$$\times\{5(8-16s^2+7s^4)+4(2-3s^2)^2\eta-(8-8s^2-5s^4)\eta^2$$
$$+2[4(4-5s^2)\eta^2-5(6-7s^2)(1+\eta)^2]\beta^2s^2\cos 2\omega\},$$

with the terms $\langle\mathfrak{V}_{i,j}\rangle_\delta$ computed from Eq. (7.61).

7.5.3 The 2:1 resonance. GPS orbits

One notable instance of orbits in deep 2:1 tesseral resonance is the case of GPS orbits. Since these orbits have high altitudes, we only deal with a gravitational model that is truncated to the fifth degree and order. Besides, because of the low eccentricity of this type of orbit, the expansions in the mean anomaly are truncated to the fourth power of the eccentricity for perturbations of the first order of J_2, and just to the second order of the eccentricity in the case of effects of the second order of J_2 [434].

The transformation Eq. (7.53) is now particularized for $\lambda = \ell+g+2h$ and $\Delta = H-2L$. The long-term Hamiltonian $\mathcal{K} = \mathcal{H}_{0,0}+\mathcal{H}_{0,1}+\frac{1}{2}\mathcal{H}_{0,2}$, which is obtained from Eq. (5.60) and (7.65) after writing everything in terms of the new variables, is rather written like the perturbation of an intermediary [223],

$$\mathcal{K} = \mathcal{I}_{2:1}(-,-,\lambda,\Delta,\Gamma,\Lambda) + e\mathcal{P}_{2:1}(-,\gamma,\lambda,\Delta,\Gamma,\Lambda). \tag{7.66}$$

In particular, the GPS intermediary takes the form [434]

$$\mathcal{I}_{2:1} = -\frac{\mu}{2a} - \omega_\oplus(\Delta+2\Lambda) + \frac{\mu}{2a}\frac{1}{16}$$
$$\times\left\{\frac{R_\oplus^2}{a^2}8J_{2,0}P_{2,0}X_{2,0} + \frac{R_\oplus^3}{a^3}60J_{3,2}P_{3,2}X_{3,2}\sin(\lambda-\lambda_{3,2}) - \frac{3}{2}\frac{R_\oplus^4}{a^4}\right.$$
$$\times[2J_{2,0}^2(Q_{2,0}+Q_{2,2}e^2+Q_{2,4}e^4) - J_{4,0}P_{4,0}X_{4,0} - 560J_{4,4}P_{4,4}X_{4,4}$$
$$\left.\times\cos 2(\lambda-\lambda_{4,4})] + 105\frac{R_\oplus^5}{a^5}J_{5,2}P_{5,2}X_{5,2}\sin(\lambda-\lambda_{5,2})\right\}, \tag{7.67}$$

where $P_{i,j}$, $Q_{i,j}$ are the inclination polynomials on the left side of Table 7.2, and $X_{i,j}$ are the eccentricity polynomials on the right side of the same table, which also includes the sectoral harmonic of degree and order five, which will be used in the next example in §7.5.4.

Visual inspection of Eq. (7.67) clearly shows the relevance of the different resonant terms of the intermediary $\mathcal{I}_{2:1}$, in which the resonant harmonic coefficients $J_{3,2}$, $J_{4,4}$, and $J_{5,2}$, are multiplied by numeric coefficients that are exceedingly greater than those multiplying the zonal harmonic coefficients.

Table 7.2: Inclination (left) and eccentricity polynomials (right) in Eq. (7.67). Adapted by permission from Springer: [434].

$P_{2,0} = 1 - 3c^2$	$X_{2,0} = 1 + \frac{3}{2}e^2 + \frac{15}{8}e^4$
$P_{3,2} = s(1+c)(1-3c)$	$X_{3,2} = 1 + 2e^2$
$P_{4,0} = 3 - 30c^2 + 35c^4$	$X_{4,0} = 1 + 5e^2$
$P_{4,4} = s^2(1+c)^2$	$X_{4,4} = 1 + e^2$
$P_{5,2} = s(1+c)(1-3c-9c^2+15c^3)$	$X_{5,2} = 1 + \frac{13}{2}e^2$
$P_{5,5} = s^2(1+c)^3$	$X_{5,5} = 1 - \frac{3}{2}e^2$
$Q_{2,0} = 1 - 8c^2 + 19c^4$	
$Q_{2,2} = \frac{7}{4} - \frac{41}{2}c^2 + \frac{243}{4}c^4$	
$Q_{2,4} = \frac{13}{8} - 36c^2 + \frac{1027}{8}c^4$	

As a first approach, we may neglect the eccentricity contribution in second-order terms of the resonant Hamiltonian (7.66) for orbits with very low eccentricity. Thus, on the one hand, save for $X_{2,0}$, the eccentricity polynomials in Table 7.2 are simplified to unity. On the other hand, the perturbation of the intermediary is neglected because it is multiplied by e. Hence, $\mathcal{K} \approx \mathcal{I}_{2:1}(\lambda, \Lambda; \Gamma, \Delta)$, and the problem is reduced to a one-degree-of-freedom Hamiltonian in the longitude of the stroboscopic node λ and its conjugate momentum Λ, which only depends on the semimajor axis. Therefore, the reduced integrable flow can be visualized with simple contour plots of Eq. (7.67).

A sample visualization of the reduced flow in the (λ, a) plane is presented in Fig. 7.6 for the values of the (dynamical) parameters Δ and Γ corresponding to a low-eccentricity GPS-type orbit with mean elements $a = 26560$ km, $e = 0.001$, $i = 55$ deg. The position that one of the GPS satellites would occupy is superimposed in Fig. 7.6 to the contour plots of the intermediary. In particular, we used the initial conditions of GPS-53 (E2) given in [26].

On the other hand, the eccentricity of actual GPS orbits can be close to one hundredth and its effect cannot be neglected in general. Then the perturbation of the intermediary $\mathcal{P}_{2:1}$ must be taken into account, at least for those terms that do not depend on e. Namely

$$\begin{aligned}
\mathcal{P}_{2:1} = \frac{\mu}{2a}\frac{3}{4}\bigg\{ & \frac{R_\oplus^3}{a^3}J_{3,0}T_{3,0}\sin y + \frac{5}{4}\frac{R_\oplus^5}{a^5}J_{5,0}T_{5,0}\sin y \\
& + \frac{R_\oplus^2}{a^2}J_{2,2}[6T_{2,2,-1}\cos(y - \lambda + \lambda_{2,2}) - T_{2,2,1}\cos(y + \lambda + \lambda_{2,2})] \\
& + \frac{R_\oplus^4}{a^4}J_{4,2}\left[\frac{75}{4}T_{4,2,-1}\cos(y - \lambda + \lambda_{4,2}) - \frac{5}{2}T_{4,2,1}\cos(y + \lambda - \lambda_{4,2})\right] \\
& - 630\frac{R_\oplus^5}{a^5}J_{5,4}T_{5,4}\sin(y - 2\lambda + 2\lambda_{5,4})\bigg\} + \mathcal{O}(e),
\end{aligned} \tag{7.68}$$

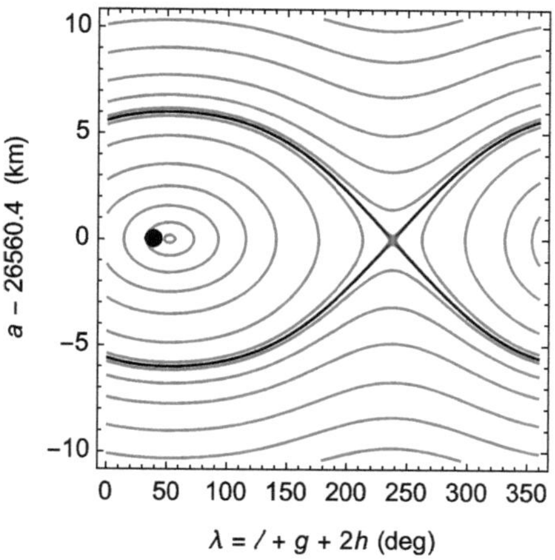

Figure 7.6: Sample phase space of the intermediary (7.67) with the position of a GPS satellite superimposed (black spot).

where the inclination polynomials $T_{j,k,l}$ are given in Table 7.3. For the actual values of the Earth's harmonic coefficients, the most relevant effect on the perturbation is due to the $J_{2,2}$ term.

Table 7.3: Inclination polynomials in Eq. (7.68). Adapted by permission from Springer: [434].

$T_{2,2,-1} = s^2$	$T_{2,2,1} = (1+c)^2$	$T_{3,0} = (1-5c^2)s$
$T_{4,2,-1} = s^2(1-7c^2)$	$T_{4,2,1} = (1+c)^2(1-7c+7c^2)$	
$T_{5,0} = (1-14c^2+21c^4)s$	$T_{5,4} = s^3(1+c)(1-5c)$	

Now the GPS long-term Hamiltonian (7.66) remains of two degrees of freedom, yet the integration of its Hamilton equations evolves with very long step sizes and is very effective. The short-period tesseral effects can be recovered analytically at each integration step from the transformation equations of the short-period elimination. In fact, the numerical integration fits quite well when superimposed to the contour plot of Fig. 7.6; cf. [434]. Alternatively, the long-term dynamics can be explored with the usual tools of non-linear dynamics such as Poincaré surfaces of section [187].

7.5.4 The 5:3 resonance. Galileo disposal orbits

Galileo operational satellites move in almost circular, 17 to 10 repeat groundtrack orbits at an altitude of about 23,222 km over the surface of the Earth and 56° of nominal inclination.[2] Due to the high altitude, tesseral resonance effects stemming from the

2 http://www.esa.int/Our_Activities/Navigation/Galileo

repetition of the ground trace are insignificant, but this orbital regime is not so far from the 5 to 3 commensurability with the Earth's rotation period that may affect defunct satellites disposed a few hundred kilometers above the operational orbits.

Focusing on the 5:3-tesseral resonance, the stroboscopic canonical transformation (7.53) yields $\lambda = \ell + g + (5/3)h$ and $\Delta = H - (5/3)L$. Then the long-term Hamiltonian is again rearranged in the form of a perturbed intermediary $\mathcal{K} = \mathcal{I}_{5:3}(-,-,\lambda,\Delta,\Gamma,\Lambda) + e\mathcal{P}_{5:3}(-,\gamma,\lambda,\Delta,\Gamma,\Lambda)$, where

$$\begin{aligned}\mathcal{I}_{5:3} = -\frac{\mu}{2a} - \omega_\oplus\left(\Delta + \frac{5}{3}\Lambda\right) + \frac{\mu}{2a}\bigg\{&\frac{R_\oplus^2}{a^2}\frac{1}{2}J_{2,0}P_{2,0}X_{2,0} \\ &- \frac{3}{32}\frac{R_\oplus^4}{a^4}\left[2J_{2,0}^2(Q_{2,0} + Q_{2,2}e^2 + Q_{2,4}e^4) - J_{4,0}P_{4,0}X_{4,0}\right] \\ &+ \frac{4725}{16}\frac{R_\oplus^5}{a^5}J_{5,5}P_{5,5}X_{5,5}\cos(\lambda + \lambda_{5,5})\bigg\},\end{aligned} \tag{7.69}$$

in which $P_{j,k}$, $Q_{j,k}$, and $X_{j,k}$ were previously given in Table 7.2.

The eccentricity of Galileo orbits is now much smaller than in the GPS case, and, therefore, the disturbing effects of the perturbation $\mathcal{P}_{5:3}$ are no longer relevant, thus yielding an integrable system.

The pendulum-type dynamics of 5:3-resonant orbits is illustrated in Fig. 7.7 for the dynamical parameters Δ and Γ corresponding to a Galileo-type orbit ($I = 56^\circ$, $e = 0.001$) 400 km above the nominal constellation. Now, the oscillations of the semimajor axis in the libration region reduce their maximum amplitude to about ±1.2 km, that is, about one fifth of the case of the 1:2 resonance. For the latter, the amplitude of the oscillations of the semimajor axis in the libration region can reach ∼ ±6 km, as observed in Fig. 7.6.

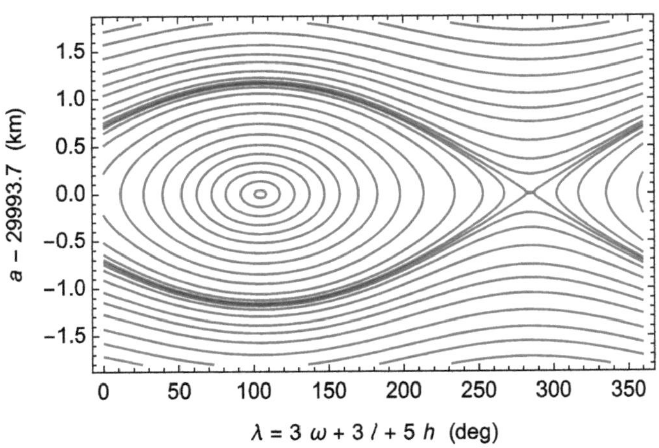

Figure 7.7: Sample phase space of the intermediary (7.69) showing the 5 to 3 tesseral resonance ($I = 56^\circ$, $e = 0.001$) [443].

8 Lunisolar perturbations

Effects of the Moon and the Sun on close-Earth orbits are clearly observable and must be taken into account when modeling their dynamics. The artificial satellite problem is then set in the framework of the restricted n-body problem. However, in those cases in which lunisolar effects are small compared to the gravitational attraction of the Earth, the problem can still be cast in the form of perturbed Keplerian motion about the Earth, and, therefore, be approached by Hamiltonian perturbations. To this aim, the third-body disturbing function is customarily expanded in the ratio of the distances from the central body to the satellite and to the perturber as an infinite series in Legendre polynomials. In most cases, the disturbing function of the Sun can be truncated to the first term of the series, yet the correct modeling of the lunar disturbing function may need higher-order truncations in some orbital regimes.

On the other hand, because the coupling of third-body and geopotential perturbations only yields higher-order effects, both perturbations are customarily treated separately. For third-body perturbations, the solution can be computed in closed form of the eccentricity of the orbits of the satellite and the disturbing bodies. However, because the eccentricities of the orbits of both the Moon and the Sun relative to the Earth are small, it is customary to simplify the perturbation series by making expansions in the eccentricity of each third-body orbit.

Lunisolar perturbations make the Hamiltonian time-dependent because they depend on the positions of these bodies, which need to be taken either from an ephemeris database or from analytical ephemeris.[1] However, dealing explicitly with time is avoided in the construction of the perturbation solution by moving to the extended phase space.

Lunisolar resonances are common and profuse, and hence motion under third-body perturbations is preferably integrated semi-analytically. Out of resonances, in addition to removing the mean anomaly of the satellite, removing the mean anomalies of the Moon and the Sun by perturbations notably speeds the semi-analytical integration. A general picture of the perturbed dynamics can then be obtained from the fast propagation of different trajectories.

8.1 The third-body potential

Under the assumption of point masses, m, $m_\oplus$, and $m_\star$, for the satellite, the Earth, and a disturbing body, respectively, the acceleration of the satellite in an inertial frame is

1 Low-precision ephemerides, like those in [482] extracted from [109, 110], suffice for this purpose.

https://doi.org/10.1515/9783110668513-008

given by Newton's gravitational law,

$$\frac{\mathrm{d}^2\boldsymbol{\rho}}{\mathrm{d}t^2} = -\mathcal{G}m_\oplus \frac{\boldsymbol{\rho}-\boldsymbol{\rho}_\oplus}{\|\boldsymbol{\rho}-\boldsymbol{\rho}_\oplus\|^3} - \mathcal{G}m_\star \frac{\boldsymbol{\rho}-\boldsymbol{\rho}_\star}{\|\boldsymbol{\rho}-\boldsymbol{\rho}_\star\|^3}, \tag{8.1}$$

where $\boldsymbol{\rho}, \boldsymbol{\rho}_\oplus$, and $\boldsymbol{\rho}_\star$, are the position vectors of the satellite, the Earth, and the disturbing body, respectively, and $\mathcal{G}$ is the gravitational constant. Analogously, for the Earth

$$\frac{\mathrm{d}^2\boldsymbol{\rho}_\oplus}{\mathrm{d}t^2} = -\mathcal{G}m_\star \frac{\boldsymbol{\rho}_\oplus-\boldsymbol{\rho}_\star}{\|\boldsymbol{\rho}_\oplus-\boldsymbol{\rho}_\star\|^3} - \mathcal{G}m \frac{\boldsymbol{\rho}_\oplus-\boldsymbol{\rho}}{\|\boldsymbol{\rho}_\oplus-\boldsymbol{\rho}\|^3}, \tag{8.2}$$

and a similar equation applies for the acceleration of the disturbing body.

Denoting $\boldsymbol{r} = \boldsymbol{\rho} - \boldsymbol{\rho}_\oplus$, $r = \|\boldsymbol{r}\|$, $\boldsymbol{r}_\star = \boldsymbol{\rho}_\star - \boldsymbol{\rho}_\oplus$, $r_\star = \|\boldsymbol{r}_\star\|$, and subtracting Eqs. (8.1) and (8.2), we obtain

$$\frac{\mathrm{d}^2\boldsymbol{r}}{\mathrm{d}t^2} = -\mathcal{G}(m+m_\oplus)\frac{\boldsymbol{r}}{r^3} - \mathcal{G}m_\star\left(\frac{\boldsymbol{r}-\boldsymbol{r}_\star}{\|\boldsymbol{r}-\boldsymbol{r}_\star\|^3} + \frac{\boldsymbol{r}_\star}{r_\star^3}\right), \tag{8.3}$$

which shows that the acceleration of the satellite relative to the Earth can be derived from the potential $\mathcal{V} = -\mu/r + \mathcal{V}_\star$ in which

$$\mathcal{V}_\star = -\frac{\mu_\star}{r_\star}\chi_\star\left(\frac{r_\star}{\|\boldsymbol{r}-\boldsymbol{r}_\star\|} - \frac{\boldsymbol{r}\cdot\boldsymbol{r}_\star}{r_\star^2}\right), \tag{8.4}$$

where $\mu_\star = \mathcal{G}(m_\star + m_\oplus)$, and $\chi_\star = m_\star/(m_\star + m_\oplus)$.

8.1.1 Expansion of the potential for a close-Earth satellite

For a close-Earth satellite $r/r_\star \ll 1$, and we write

$$\|\boldsymbol{r}-\boldsymbol{r}_\star\| = r_\star\sqrt{1 - 2(r/r_\star)\cos\psi_\star + (r/r_\star)^2},$$

where $\psi_\star$ is the angle encompassed by the directions from the Earth of the satellite $\hat{\boldsymbol{r}} = \boldsymbol{r}/r$ and the disturbing body $\hat{\boldsymbol{r}}_\star = \boldsymbol{r}_\star/r_\star$. That is,

$$\cos\psi_\star = \hat{\boldsymbol{r}}\cdot\hat{\boldsymbol{r}}_\star = \frac{xx_\star + yy_\star + zz_\star}{rr_\star}. \tag{8.5}$$

In that case, the right side of Eq. (8.4) is conveniently replaced by the Legendre polynomial expansion,

$$\mathcal{V}_\star = -\chi_\star\frac{n_\star^2 a_\star^3}{r_\star}\sum_{i\geq 2}\frac{r^i}{r_\star^i}P_i(\cos\psi_\star), \tag{8.6}$$

where the gravitational parameter of the third body is written in terms of the semimajor axis $a_\star$ and the mean motion $n_\star$ of its orbit relative to the Earth, and the Legendre polynomials P_i are given by the usual binomial expansion of Rodrigues' formula [314, 320, 569],

$$P_i(\cos\psi_\star) = \frac{1}{2^i}\sum_{l=0}^{i_0}(-1)^l\binom{i}{l}\binom{2i-2l}{i}\cos^{i-2l}\psi_\star, \tag{8.7}$$

where $i_0 = \lfloor\frac{1}{2}i\rfloor$ from Eq. (6.5). The term $-\mu_\star/r_\star$ has been neglected in Eq. (8.6) because it brings about a null contribution to the satellite acceleration in Eq. (8.3).

The reasoning used in deriving the disturbing potential Eq. (8.4) does not change when adding more masses to the system. Therefore, the lunisolar disturbing potential of a close-Earth satellite will be made of both $\mathcal{V}_\odot$ and $\mathcal{V}_☾$, each of which is given by Eq. (8.4). For most applications, the Legendre polynomial expansion of the former can be truncated to the second degree, but the latter may require one to take up to the Legendre polynomial P_6 into account in the case of orbits of common space telescopes [436]. Besides, $\chi_☾ = 1/82.28$ whereas $\chi_\odot$ is taken the unity.

8.1.2 The disturbing potential in the apsidal frame

Short-period terms of the expanded third-body disturbing potential (8.6) are effectively isolated when the direction of the massless body is given by its components in the apsidal frame $(O, \hat{\boldsymbol{e}}, \hat{\boldsymbol{b}}, \boldsymbol{n})$ defined in §4.5.1. Because third-body perturbations are conveniently handled in terms of the eccentric anomaly [158, 302, 342], we follow the steps in [425] and rearrange Eq. (8.6) like

$$\mathcal{V}_\star = -\frac{\mu}{2a}\chi_\star\frac{n_\star^2}{n^2}\frac{a_\star^3}{r_\star^3}\frac{a}{r}\sum_{i\geq 2}\frac{1}{2^{i-1}}\frac{a^{i-2}}{r_\star^{i-2}}V_i, \tag{8.8}$$

with

$$V_i = \sum_{l=0}^{i_0}(-1)^l\binom{i}{l}\binom{2i-2l}{i}\frac{r^{2l+1}}{a^{2l+1}}\left(\frac{r}{a}\cos\psi_\star\right)^{i-2l}. \tag{8.9}$$

Analogously to the geopotential case, we left the coefficient a/r out of the summation in Eq. (8.8) to ease closed-form integration, as will become evident later.

Now, Eq. (8.9) is written in the apsidal frame by first replacing Eq. (4.50) into Eq. (8.5) and use the geometric relations (4.71) to obtain

$$\frac{r}{a}\cos\psi_\star = (\hat{\boldsymbol{e}}\cdot\hat{\boldsymbol{r}}_\star)(\cos u - e) + (\hat{\boldsymbol{b}}\cdot\hat{\boldsymbol{r}}_\star)\eta\sin u, \tag{8.10}$$

from which, using the binomial expansion,

$$\begin{aligned}\left(\frac{r}{a}\cos\psi_\star\right)^{i-2l} &= \sum_{k=0}^{i-2l}\binom{i-2l}{k}(\hat{\boldsymbol{e}}\cdot\hat{\boldsymbol{r}}_\star)^k(\cos u - e)^k \\ &\quad\times(\hat{\boldsymbol{b}}\cdot\hat{\boldsymbol{r}}_\star)^{i-2l-k}\eta^{i-2l-k}\sin^{i-2l-k}u,\end{aligned} \tag{8.11}$$

where

$$(\cos u - e)^k = \sum_{m=0}^{k} \binom{k}{m} (-1)^m e^m \cos^{k-m} u. \tag{8.12}$$

Next, from Eq. (4.26),

$$\frac{r^{2l+1}}{a^{2l+1}} = (1 - e\cos u)^{2l+1} = \sum_{j=0}^{2l+1} \binom{2l+1}{j} (-1)^j e^j \cos^j u. \tag{8.13}$$

Hence,

$$\begin{aligned} V_i = {} & \sum_{l=0}^{i_0} \binom{i}{l} \binom{2i-2l}{i} \sum_{j=0}^{2l+1} \binom{2l+1}{j} \sum_{k=0}^{i-2l} \binom{i-2l}{k} \sum_{m=0}^{k} \binom{k}{m} e^{j+m} (\hat{\boldsymbol{e}} \cdot \hat{\boldsymbol{r}}_\star)^k \\ & \times (-1)^{j+l+m} \eta^{i-2l-k} (\hat{\boldsymbol{b}} \cdot \hat{\boldsymbol{r}}_\star)^{i-2l-k} \cos^{j+k-m} u \sin^{i-2l-k} u. \end{aligned} \tag{8.14}$$

8.2 Long-term motion. The extended phase space

The lunisolar perturbations Hamiltonian is $\mathcal{H} = -\frac{1}{2}(\mu/a) + \mathcal{V}_{☾} + \mathcal{V}_{\odot}$, in which solar and lunar ephemerides are known functions of time. While Deprit's perturbation algorithm by Lie transforms also applies to time-dependent Hamiltonians [151], descriptions in Chapter 2 were constrained to the case of conservative Hamiltonians. Therefore, the time dependency is avoided using the homogeneous formalism [558, 621] by introducing a new coordinate $\tau = t$ and conjugate momentum $T = -\mathcal{H} + \text{const}$, such that, in the *extended* phase space $(\ell, g, h, \tau, L, G, H, T)$, the Hamiltonian

$$\mathcal{H} = -\frac{\mu}{2a} + T + \mathcal{V}_{☾}(\ell, g, h, \tau, L, G, H, -) + \mathcal{V}_{\odot}(\ell, g, h, \tau, L, G, H, -) \tag{8.15}$$

is conservative.

The Hamiltonian (8.15) accepts the usual perturbation arrangement in Eq. (2.30), in which the small parameter ε is formal and, because the disturbing effects of the Sun and the Moon on close-Earth orbits are of the same order,

$$\mathcal{H}_{0,0} = -\frac{\mu}{2a}, \tag{8.16}$$

$$\mathcal{H}_{1,0} = T, \tag{8.17}$$

$$\mathcal{H}_{2,0} = 2!(\mathcal{V}_{☾} + \mathcal{V}_{\odot}), \tag{8.18}$$

and $\mathcal{H}_{m,0} = 0$ for $m \geq 3$. The third-body disturbing potentials are replaced by their Legendre polynomials expansion in Eq. (8.6) with $\star \equiv \odot$ and $\star \equiv ☾$, respectively. The factorial 2 that scales $\mathcal{H}_{2,0}$ has been prepended to compensate the corresponding divisor in the Hamiltonian arrangement.

Since $\mathcal{H}_{0,0}$ continues to be the Keplerian, the Lie derivative remains Eq. (4.66). However, Poisson brackets that eventually appear in the Lie transforms procedure must be evaluated in the extended phase space to properly take the time dependency into account. On the other hand, with the perturbation arrangement of Eqs. (8.16)–(8.18), time-dependency issues are relegated to the third order of the perturbation approach. By now, we limit ourselves to this particular case and constrain the removal of short-period effects from Hamiltonian (8.15) up to the second order of ε.

Thus, at the first order we choose $\mathcal{H}_{0,1} = T$ and $\mathcal{W}_1 = 0$. At the second order the known terms are $\widetilde{\mathcal{H}}_{0,2} = \mathcal{H}_{2,0}$. Then, from Eq. (8.18), we choose

$$\mathcal{H}_{0,2} = 2\langle \mathcal{V}_{☾} \rangle_\ell + 2\langle \mathcal{V}_{\odot} \rangle_\ell, \tag{8.19}$$

and compute $\mathcal{W}_2$ from the homological equation (4.68). Namely,

$$\mathcal{W}_2 = \frac{2}{n}\int (\mathcal{V}_{☾} - \langle \mathcal{V}_{☾} \rangle_\ell)\mathrm{d}\ell + \frac{2}{n}\int (\mathcal{V}_{\odot} - \langle \mathcal{V}_{\odot} \rangle_\ell)\mathrm{d}\ell. \tag{8.20}$$

8.2.1 Short-period elimination by Lie transforms

The term $\mathcal{H}_{0,2}$ in Eq. (8.19) is computed in closed form with the help of the differential relation (4.62). Namely, $\langle \mathcal{V}_\star \rangle_\ell = \langle \mathcal{V}_\star (r/a) \rangle_u$, which now makes the reasons for having left the term a/r out of the summation in Eq. (8.8) evident. We obtain

$$\langle \mathcal{V}_\star \rangle_\ell = -\chi_\star \frac{\mu}{2a}\frac{n_\star^2}{n^2}\frac{a_\star^3}{r_\star^3}\sum_{i\geq 2}\frac{1}{2^{i-1}}\frac{a^{i-2}}{r_\star^{i-2}}\langle V_i \rangle_u. \tag{8.21}$$

To compute $\langle V_i \rangle_u$, we convert the trigonometric functions of u into trigonometric polynomials in u. From standard relations between exponentials and circular functions [341], we obtain

$$\begin{aligned}\cos^{j+k-m} u \sin^{i-2l-k} u = {} & \frac{(-\mathbf{i})^{i-k-2l}}{2^{i+j-m-2l}}\sum_{q=0}^{i-k-2l}\sum_{t=0}^{j+k-m}\binom{i-k-2l}{q}(-1)^q \\ & \times \binom{j+k-m}{t}[\cos 2(\tilde{q}-q)u + \mathbf{i}\sin 2(\tilde{q}-q)u],\end{aligned} \tag{8.22}$$

where $\tilde{q} = \frac{1}{2}(i+j-m) - l - t$. That all the terms of Eq. (8.22) are periodic save for those in which $q = \tilde{q}$, which in turn requires $i+j-m$ to be even, is now evident. Hence,

$$\begin{aligned}\langle V_i \rangle_u = {} & \sum_{l=0}^{i_0}\binom{i}{l}\binom{2i-2l}{i}\sum_{j=0}^{2l+1}\binom{2l+1}{j}\sum_{k=0}^{i-2l}\binom{i-2l}{k}\sum_{m=0}^{k}\binom{k}{m} \\ & \times \Xi_{j+k-m}^{i-2l-k}(-1)^{j+l+m}e^{j+m}(\hat{\boldsymbol{e}}\cdot\hat{\boldsymbol{r}}_\star)^k(\eta\hat{\boldsymbol{b}}\cdot\hat{\boldsymbol{r}}_\star)^{i-k-2l},\end{aligned} \tag{8.23}$$

where the coefficient $\Xi_{j+k-m}^{i-2l-k} = \langle \cos^{j+k-m} u \sin^{i-2l-k} u \rangle_u$ is obtained making $q = \tilde{q}$ in the right side of Eq. (8.22). Note that $i - k$ must be even to remove the imaginary unit. Therefore, the calculation of Eq. (8.21) is completed, and the long-term Hamiltonian is obtained after replacing original by prime variables in Eq. (8.19).

The generating function (8.20) is now computed from Eq. (4.70) either for the lunar or for the solar perturbations. That is, $\mathcal{W}_2 = 2\mathcal{W}_{\odot} + 2\mathcal{W}_{☾}$, where

$$\mathcal{W}_\star = -\chi_\star L \frac{n_\star^2}{n^2} \frac{a_\star^3}{r_\star^3} \sum_{i\geq 2} \frac{1}{2^i} \frac{a^{i-2}}{r_\star^{i-2}} \left[\langle V_i \rangle_u e \sin u + \int (V_i - \langle V_i \rangle_u) \mathrm{d}u \right]. \tag{8.24}$$

The integrand $V_i - \langle V_i \rangle_u$ consists of purely periodic terms in u, which are obtained from Eq. (8.14) by simply avoiding the value $q = \tilde{q}$ in the corresponding summation of Eq. (8.22). Then the required integration is readily solved; cf. [425].

8.2.2 Averaged flow in vectorial elements

Due to the abundance of resonances between the mean motions of Moon and Sun, and the rate of variation of the orbital elements of the satellite [128, 240], further simplifications of the third-body disturbing potential are not generally pursued. Rather, the flow stemming from the mean-element Hamiltonian

$$\mathcal{K} = \mathcal{K}_{0,0} + \mathcal{K}_{1,0} + \frac{1}{2}\mathcal{K}_{2,0}, \tag{8.25}$$

where $\mathcal{K}_{0,0} = -\mu/(2a)$ is now constant, $\mathcal{K}_{1,0} = T$, and $\mathcal{K}_{2,0} = \mathcal{H}_{0,2}$ is given by Eqs. (8.19) and (8.21) after rewritten in prime (mean) variables, is numerically integrated. At each time of the numerical integration, which proceeds with long steps and is very efficient, osculating elements may be recovered by evaluation of the analytical short-period corrections derived in the usual way from the generating function $\mathcal{W} = \mathcal{W}_{☾} + \mathcal{W}_{\odot}$ obtained from Eq. (8.24).

When high orders of the Legendre polynomials expansion of the disturbing potential are needed, the integration of the variations of the vectorial elements defining the apsidal frame is specially efficient [425]. In particular, the variations of the mean vectorial elements $\boldsymbol{\eta} = \boldsymbol{G}/L' = \eta\boldsymbol{n}$, $\boldsymbol{e} = e\hat{\boldsymbol{e}}$, have been previously given in Eqs. (4.60)–(4.61). In order to use them here, we only need to make $\mathcal{Q} \equiv -(1/L')\langle \mathcal{V}_\star \rangle_\ell$ for each of the disturbing bodies, using Eq. (8.21), and write $\hat{\boldsymbol{e}}$ and $\boldsymbol{n}$ in Eq. (8.23) in terms of the nondimensional magnitudes $\boldsymbol{e}$ and $\boldsymbol{\eta}$, respectively.

To do that, we replace

$$e^{j+m} (\hat{\boldsymbol{e}} \cdot \hat{\boldsymbol{r}}_\star)^k (\eta \hat{\boldsymbol{b}} \cdot \hat{\boldsymbol{r}}_\star)^{i-k-2l} = (\boldsymbol{e} \cdot \boldsymbol{e})^{\frac{j+m-k}{2}} (\boldsymbol{e} \cdot \hat{\boldsymbol{r}}_\star)^k [\eta^2 (\hat{\boldsymbol{b}} \cdot \hat{\boldsymbol{r}}_\star)^2]^{\frac{i-k}{2}-l}$$

in Eq. (8.23) and remove the dependence on the unit vector $\hat{\boldsymbol{b}}$ using the identity

$$\|\hat{\boldsymbol{r}}_\star\|^2 = (\hat{\boldsymbol{e}} \cdot \hat{\boldsymbol{r}}_\star)^2 + (\hat{\boldsymbol{b}} \cdot \hat{\boldsymbol{r}}_\star)^2 + (\boldsymbol{n} \cdot \hat{\boldsymbol{r}}_\star)^2 = 1. \tag{8.26}$$

Namely, $\eta^2(\hat{\boldsymbol{b}}\cdot\hat{\boldsymbol{r}}_\star)^2 = X - (\hat{\boldsymbol{e}}\cdot\hat{\boldsymbol{r}}_\star)^2$, where $X = X(\boldsymbol{e},\boldsymbol{\eta},\hat{\boldsymbol{r}}_\star) \equiv 1 - e^2 + \xi^2 - \zeta^2$, with $\xi = \boldsymbol{e}\cdot\hat{\boldsymbol{r}}_\star$, $\zeta = \boldsymbol{\eta}\cdot\hat{\boldsymbol{r}}_\star$. After a new binomial expansion, we readily obtain

$$\begin{aligned}\langle V_i\rangle_u = {} & \sum_{l=0}^{i_0}\binom{i}{l}\binom{2i-2l}{i}\sum_{j=0}^{2l+1}\binom{2l+1}{j}\sum_{k=0}^{i-2l}\binom{i-2l}{k}\sum_{m=0}^{k}\binom{k}{m}\Xi_{j+k-m}^{i-2l-k}\\ & \times\sum_{s=0}^{\frac{i-k}{2}-l}\binom{\frac{i-k}{2}-l}{s}(-1)^{j+l+m+s}e^{j+m-k-2s}\xi^{2s+k}X^{\frac{i-k}{2}-l-s},\end{aligned} \tag{8.27}$$

which only depends on $\boldsymbol{e}$, and $\boldsymbol{\eta}$, as desired, as well as on the disturbing body direction $\hat{\boldsymbol{r}}_\star$. The first few terms $\langle V_i\rangle_u$ are displayed in Table 8.1.

Table 8.1: Some terms $\langle V_i\rangle_u$, given by Eq. (8.27)—after [425].

$\langle V_2\rangle_u = 1 - 3(2e^2 - 5\xi^2 + \zeta^2)$
$\langle V_3\rangle_u = -\frac{5}{2}\xi[3(1 - 8^2) - 15\zeta^2 + 35\xi^2]$
$\langle V_4\rangle_u = \frac{3}{4}\{3 - 20e^2 + 80e^4 + 10[7(1 - 10e^2)\xi^2 - (3 - 10e^2)\zeta^2] + 35(21\xi^4 - 14\zeta^2\xi^2 + \zeta^4)\}$
$\langle V_5\rangle_u = -\frac{3}{4}\xi\{35(1 - 8e^2 + 40e^4) + 490[(1 - 12e^2)\xi^2 - (1 - 4e^2)\zeta^2] + 147(33\xi^4 - 30\zeta^2\xi^2 + 5\zeta^4)\}$
$\langle V_6\rangle_u = \frac{1}{4}\{5(5 - 42e^2 + 168e^4 - 560e^6) + 105[3(3 - 28e^2 + 168e^4)\xi^2 - (5 - 28e^2 + 56e^4)\zeta^2] + 315[(5 - 14e^2)\zeta^4 - 18(3 - 14e^2)\zeta^2\xi^2 + 33(1 - 14e^2)\xi^4] + 231(429\xi^6 - 495\zeta^2\xi^4 + 135\zeta^4\xi^2 - 5\zeta^6)\}$
$\langle V_7\rangle_u = -\frac{9}{16}\xi\{35(5 - 48e^2 + 224e^4 - 896e^6) + 105[11(3 - 32e^2 + 224e^4)\xi^2 - 3(15 - 96e^2 + 224e^4)\zeta^2] + 231[143(1 - 16e^2)\xi^4 - 110(3 - 16e^2)\zeta^2\xi^2 + 15(5 - 16e^2)\zeta^4] + 429(715\xi^6 - 1001\zeta^2\xi^4 + 385\zeta^4\xi^2 - 35\zeta^6)\}$
$\langle V_8\rangle_u = \frac{5}{64}\{7(35 - 360e^2 + 1728e^4 - 5376e^6 + 16128e^8) + 252[11(5 - 54e^2 + 288e^4 - 1344e^6)\xi^2 - (35 - 270e^2 + 864e^4 - 1344e^6)\zeta^2] + 1386[143(1 - 12e^2 + 96e^4)\xi^4 - 66(5 - 36e^2 + 96e^4)\zeta^2\xi^2 + (35 - 180e^2 + 288e^4)\zeta^4] + 12012[143(1 - 18e^2)\xi^6 + 33(5 - 18e^2)\xi^2\zeta^4 - 429(1 - 6e^2)\xi^4\zeta^2 - (7 - 18e^2)\zeta^6] + 6435(2431\xi^8 - 4004\zeta^2\xi^6 + 2002\zeta^4\xi^4 - 308\zeta^6\xi^2 + 7\zeta^8)\}$

The variation equations of the averaged flow, Eqs. (4.60)–(4.61), are then derived from Eqs. (8.21) and (8.27) taking into account that

$$\begin{aligned}&\nabla_{\boldsymbol{\eta}}(\boldsymbol{e}\cdot\boldsymbol{e}) = \boldsymbol{0}, &&\nabla_{\boldsymbol{\eta}}(\boldsymbol{e}\cdot\hat{\boldsymbol{r}}_\star) = \boldsymbol{0}, &&\nabla_{\boldsymbol{\eta}}X = -2(\boldsymbol{\eta}\cdot\hat{\boldsymbol{r}}_\star)\hat{\boldsymbol{r}}_\star,\\ &\nabla_{\boldsymbol{e}}(\boldsymbol{e}\cdot\boldsymbol{e}) = 2\boldsymbol{e}, &&\nabla_{\boldsymbol{e}}(\boldsymbol{e}\cdot\hat{\boldsymbol{r}}_\star) = \hat{\boldsymbol{r}}_\star, &&\nabla_{\boldsymbol{e}}X = -2\boldsymbol{e} + 2(\boldsymbol{e}\cdot\hat{\boldsymbol{r}}_\star)\hat{\boldsymbol{r}}_\star.\end{aligned}$$

We borrow from [425]

$$\frac{\mathrm{d}\boldsymbol{\eta}}{\mathrm{d}t} = n\epsilon_\star\sum_{i\geq 2}\frac{1}{2^i}\frac{a^{i-2}}{r_\star^{i-2}}[\gamma_i(\boldsymbol{\eta}\times\hat{\boldsymbol{r}}_\star) + \rho_i(\boldsymbol{e}\times\hat{\boldsymbol{r}}_\star)], \tag{8.28}$$

$$\frac{\mathrm{d}\boldsymbol{e}}{\mathrm{d}t} = n\epsilon_\star\sum_{i\geq 2}\frac{1}{2^i}\frac{a^{i-2}}{r_\star^{i-2}}[\gamma_i(\boldsymbol{e}\times\hat{\boldsymbol{r}}_\star) + \rho_i(\boldsymbol{\eta}\times\hat{\boldsymbol{r}}_\star) + 4\rho_{i-1}(\boldsymbol{\eta}\times\boldsymbol{e})], \tag{8.29}$$

in which $\epsilon_\star = \chi_\star (n_\star/n)^2 (a_\star/r_\star)^3$, and

$$\beta_i = \sum_{l=0}^{i_0} \binom{i}{l}\binom{2i-2l}{i} \sum_{j=0}^{2l+1} \binom{2l+1}{j} \sum_{k=0}^{i-2l} \binom{i-2l}{k} \sum_{m=0}^{k} \binom{k}{m} \Xi_{j+k-m}^{i-2l-k}$$
$$\sum_{s=0}^{\frac{i-k}{2}-l} \binom{\frac{i-k}{2}-l}{s} \frac{e^{j+m-k-2s}\zeta^{2s+k-1}}{(-1)^{j+l+m+s}} X^{\frac{i-k}{2}-l-s-1} B_{i,k,l,s} \tag{8.30}$$

represents γ_i when $B_{i,k,l,s} \equiv \xi\zeta(k+l+s-i)$, and ρ_i when $B_{i,k,l,s} \equiv (k+2s)X + (i-k-2l-2s)\xi^2$. The complexity of Eq. (8.30) is only apparent, as shown by the first polynomials γ_i and ρ_i listed in Table 8.2.

Table 8.2: Some polynomials ρ_i, γ_i, given by Eq. (8.30). Credit: [425], reproduced with permission © ESO.

$$\begin{aligned}
\rho_1 &= -3,\\
\rho_2 &= 30\xi,\\
\gamma_2 &= -6\zeta,\\
\rho_3 &= -\tfrac{15}{2}[1-8e^2+5(7\xi^2-\zeta^2)],\\
\gamma_3 &= 75\xi\zeta,\\
\rho_4 &= 105[1-10e^2+7(3\xi^2-\zeta^2)]\xi,\\
\gamma_4 &= -15[3-10e^2+7(7\xi^2-\zeta^2)]\zeta,\\
\rho_5 &= -\tfrac{105}{4}\{1-8e^2+40e^4+14[3(1-12e^2)\xi^2-(1-4e^2)\zeta^2]+21(33\xi^4-18\xi^2\zeta^2+\zeta^4)\},\\
\gamma_5 &= 735(1-4e^2+9\xi^2-3\zeta^2)\xi\zeta,\\
\rho_6 &= \tfrac{63}{2}\{5(3-28e^2+168e^4)+30[11(1-14e^2)\xi^2-3(3-14e^2)\zeta^2]+33(143\xi^4-110\xi^2\zeta^2+15\zeta^4)\}\xi,\\
\gamma_6 &= -\tfrac{105}{2}\{5-28e^2+56e^4+6[9(3-14e^2)\xi^2-(5-14e^2)\zeta^2]+33(33\xi^4-18\xi^2\zeta^2+\zeta^4)\}\zeta,\\
\rho_7 &= -\tfrac{315}{16}\{5-48e^2+224e^4-896e^6+9[11(3-32e^2+224e^4)\xi^2-(15-96e^2+224e^4)\zeta^2]+33[143(1\\
&\quad -16e^2)\xi^4-66(3-16e^2)\xi^2\zeta^2+3(5-16e^2)\zeta^4]+429(143\xi^6-143\xi^4\zeta^2+33\xi^2\zeta^4-\zeta^6)\},\\
\gamma_7 &= \tfrac{189}{8}\{15(15-96e^2+224e^4)+110[11(3-16e^2)\xi^2-3(5-16e^2)\zeta^2]+143(143\xi^4-110\xi^2\zeta^2\\
&\quad +15\zeta^4)\}\xi\zeta,\\
\rho_8 &= \tfrac{495}{8}\{7(5-54e^2+288e^4-1344e^6)+77[13(1-12e^2+96e^4)\xi^2-3(5-36e^2+96e^4)\zeta^2]\\
&\quad +1001[13(1-18e^2)\xi^4-26(1-6e^2)\xi^2\zeta^2+(5-18e^2)\zeta^4]+715(221\xi^6-273\xi^4\zeta^2+91\xi^2\zeta^4-7\zeta^6)\}\xi,\\
\gamma_8 &= -\tfrac{315}{8}\{35-270e^2+864e^4-1344e^6+11[33(5-36e^2+96e^4)\xi^2-(35-180e^2+288e^4)\zeta^2]\\
&\quad +143[143(1-6e^2)\xi^4-22(5-18e^2)\xi^2\zeta^2+(7-18e^2)\zeta^4]+715(143\xi^6-143\xi^4\zeta^2+33\xi^2\zeta^4-\zeta^6)\}\zeta.
\end{aligned}$$

The use of vectorial elements increases the dimension of the differential system to be numerically integrated from four to six. However, the improvements in the propagation with respect to traditional formulations based on angular elements are awesome due to the fact that ρ_i and γ_i only involve arithmetic operations, contrary to trigonometric functions, which are evaluated by hardware [425]. The constraints $\boldsymbol{e}\cdot\boldsymbol{\eta} = 0$, and

$\boldsymbol{e}\cdot\boldsymbol{e}+\boldsymbol{\eta}\cdot\boldsymbol{\eta} = 1$ can be used to integrate less variations, thus avoiding redundancy. Alternatively, these scalar relations are used to test the quality of the numerical integration of the whole vectorial system [290].

Finally, it is worth mentioning that when the third-body disturbing potential is truncated to the first term P_2 of the Legendre polynomials expansion the variation of $\boldsymbol{e}$ is scaled by the eccentricity, as shown by the fact that $\rho_2 = 30\boldsymbol{e}\cdot\hat{\boldsymbol{r}}_\star$ in Table 8.2. In consequence the variation of the eccentricity vector vanishes for (mean) circular orbits, a fact that was already noted in [128] (see also [575]). However, this particular solution is illusory and ceases to exist as soon as P_3 is taken into account [20], a fact that is made clear by simple inspection of ρ_3 in Table 8.2.

The mean-element propagation is completed with the integration of either a slow time element or the mean anomaly. For the latter,

$$\frac{\mathrm{d}\ell'}{\mathrm{d}t} = \frac{\partial\mathcal{K}}{\partial L'} = n + \frac{\partial\langle\mathcal{V}_\star\rangle_\ell}{\partial L'}, \tag{8.31}$$

using Eqs. (8.21) and Eq. (8.27), we get

$$\begin{aligned}\frac{\mathrm{d}\ell'}{\mathrm{d}t} &= n - n\chi_\star\frac{n_\star^2}{n^2}\frac{a_\star^3}{r_\star^3}\sum_{i\geq 2}\frac{1}{2^i}\frac{a^{i-2}}{r_\star^{i-2}}\sum_{l=0}^{i_0}\binom{i}{l}\binom{2i-2l}{i}\sum_{j=0}^{2l+1}\binom{2l+1}{j}\\ &\quad\times\sum_{k=0}^{i-2l}\binom{i-2l}{k}\sum_{m=0}^{k}\binom{k}{m}\Xi_{j+k-m}^{i-2l-k}\sum_{s=0}^{\frac{i-k}{2}-l}\binom{\frac{i-k}{2}-l}{s}\frac{\Upsilon_{i,j+m}^{l,2s+k}}{(-1)^{j+l+m+s}},\end{aligned}$$

in which, denoting $d = \frac{1}{2}(i-b) - c$ and $t = a - b$,

$$\Upsilon_{i,a}^{b,c} = [ae^{t-2} + (2i-2d-a)e^t]X^d(\boldsymbol{e}\cdot\hat{\boldsymbol{r}}_\star)^b + 2de^{t-2}X^{d-1}(\boldsymbol{e}\cdot\hat{\boldsymbol{r}}_\star)^{b+2}.$$

To avoid troubles in the integration of this element in the case of circular orbits, the integration of ℓ' can be replaced by the integration of the variation of a nonsingular timing element like $\Psi = \ell + g + h$ [291].

8.2.3 Sample application. The case of high Earth orbits

In many cases, the integration of the mean-elements equations can be useful in itself without need of computing the ephemeris, as is the case of the design of end-of-life disposal strategies for compliance with space law. We illustrate the performance of the mean-element solution in the presence of strong lunisolar perturbations with a challenging example taken from [436]: The 100-years propagation of the orbit of Simbol-X satellite, an abandoned project on an X-ray telescope [267]. The initial conditions used both in the numerical integration of the mean-element equations and the propagation of the non-averaged model (osculating elements) are $a = 106247.136454$ km,

$e = 0.75173$, $I = 5.2789$ deg, $\Omega = 49.351$ deg, and $\omega = -179.992$ deg. For the aims of such a kind of propagation the initial value of the mean anomaly is irrelevant and was set to zero. The numerical propagation of an orbit with these characteristics—highly eccentric and with an apogee radius that can reach half the Earth–Moon distance, thus undergoing important perturbations from the Moon's gravitational pull—is also challenging, and the numerical reference to which to compare the mean-element results was integrated with a Störmer–Cowell method of order eight and a step size of 60 seconds, referred to July 1, 2014, at 20.7208333 h [267].

Results in Fig. 8.1 show the time history of the numerically integrated reference superimposed to the mean-element solution. The latter was derived from a mean-element model that takes up to P_6 Legendre polynomial in the expansion of the lunar disturbing potential while solar perturbations are truncated to the second degree. Oblateness effects are also taken into account, yet limited to first-order effects of J_2 [436].

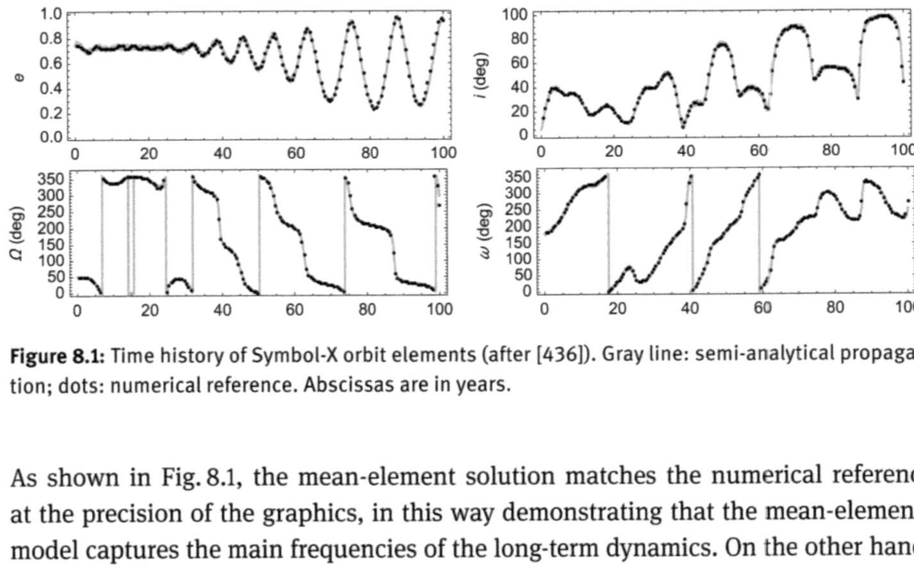

Figure 8.1: Time history of Symbol-X orbit elements (after [436]). Gray line: semi-analytical propagation; dots: numerical reference. Abscissas are in years.

As shown in Fig. 8.1, the mean-element solution matches the numerical reference at the precision of the graphics, in this way demonstrating that the mean-elements model captures the main frequencies of the long-term dynamics. On the other hand, the osculating semimajor axis experiences important variations, of hundreds of km, as opposite to the constant value of the averaged one [436]. These large, irregular variations are a consequence of the different lunisolar resonances undergone by Simbol-X, which, due to its orbital period of 4 days, is also affected by a 7:1 mean-motion resonance. However, this mean-motion resonance, which would prevent the complete removal of the mean anomaly that has been carried out in the mean-element Hamiltonian, only accumulates in much longer time intervals.

The evolution of Moser elements $\boldsymbol{f} = \boldsymbol{\eta} + \boldsymbol{e}$ and $\boldsymbol{g} = \boldsymbol{\eta} - \boldsymbol{e}$ on the unit sphere are depicted in Fig. 8.2 for the same test orbit. Red curves correspond to the mean-element solution and black ones to the reference orbit. Dashed arrows point to the

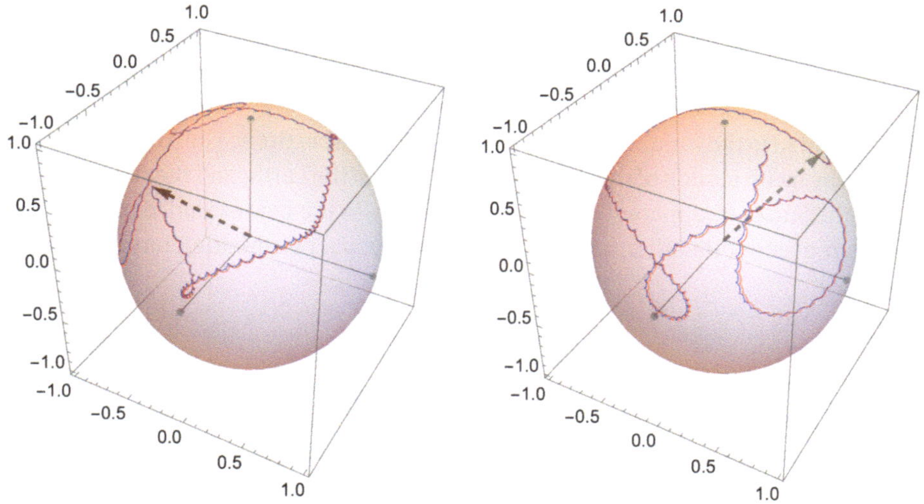

Figure 8.2: Simbol-X: Moser elements $\boldsymbol{f}$ (left) and $\boldsymbol{g}$ (right) on the unit sphere.

starting point of the propagation. Semi-annual oscillations of small amplitude that modulate the main path, which is governed by the Moon orbit dynamics, are due to solar perturbations and are clearly noted in Fig. 8.2.

Tests in [435] for a variety of orbits showed that the numerical integration of the mean-element equations performs between one and two orders of magnitude faster than the direct numerical integration of the non-averaged model. Still, these figures are not definitive and may vary if the tolerance of the numerical integration is relaxed or based on a different solver [23]. On the other hand, the version that propagates vectorial elements instead of the trigonometric functions used in [436] commonly reduces the propagation time to less than one half the time needed by the latter, and it is particularly efficient in the case of highly eccentric orbits [425].

8.3 Third-body's mean anomaly averaging

Lunisolar resonances may happen between the mean motion of the third body and the rate of variation of the slow evolving angles of the satellite's orbit [128]. Far away from them, the mean-element Hamiltonian can be dramatically simplified by carrying out a new Lie transformation that eliminates the periodic effects related to the mean anomaly of the third body. In this way, the integration of the double-averaged, long-term Hamiltonian by numerical methods is even much faster and efficient. This procedure is described below, basing our assumptions on the orbital characteristics of Galileo orbits, whose period is about 14 hours and, therefore are free from third-body mean-motion resonances, and whose long-term dynamics was briefly discussed in §7.5.4 for geopotential perturbations.

We start from the Hamiltonian in mean elements (8.25), which is written in Delaunay variables using Eqs. (4.52) and (4.53), and move to a further extended phase space $(\ell', g', h', \ell_☾, \theta_\odot, t, L', G', H', L_☾, \Theta_\odot, T)$, by introducing two arbitrary momenta $L_☾$ and $\Theta_\odot$ conjugated to the variables $\ell_☾ = n_☾ t$ and $\theta_\odot = n_\odot t$, respectively. The constant $n_☾$ represents the mean motion of the Moon in its orbit about the Earth, and the constant $n_\odot$ is that of the Sun in its apparent orbit about the Earth. In the new extended phase space the mean-element Hamiltonian is arranged in the usual perturbation form, in which

$$\mathcal{K}_{0,0} = -\frac{\mu}{2a} + n_☾ L_☾ + n_\odot \Theta_\odot, \quad \mathcal{K}_{1,0} = T, \quad \mathcal{K}_{2,0} = 2\langle \mathcal{V}_☾ \rangle_\ell + 2\langle \mathcal{V}_\odot \rangle_\ell.$$

Therefore, the Lie derivative (2.49) is now $\mathcal{L}_0 = n\partial/\partial\ell' + n_☾ \partial/\partial\ell_☾ + n_\odot \partial/\partial\theta_\odot$. However, since the mean anomaly of the satellite has already been removed from the Hamiltonian, we can disregard partial differentiation with respect to ℓ'. Moreover, as far as there is no coupling between lunar and solar effects, the solution of the other two partial differentials can be approached separately. In that case, the closed-form integration is achieved solving the homological equation (4.69) with the help of the true anomaly of either the Sun or the Moon.

As a necessary preliminary step in the solution of the homological equation, the third-body disturbing potential in mean elements given in Eq. (8.21) must be reformulated by replacing the lunisolar ephemeris $\boldsymbol{r}_\star$ in Eqs. (8.23) and (8.21), by corresponding perturbations in the orbital elements of the third body.

8.3.1 Moon and Sun disturbing effects

At difference from the inclination with respect to the equator, the orbit of the Moon maintains an almost constant inclination with respect to the ecliptic, a plane over which the longitude of the Moon's ascending node varies close to linearly. Therefore, the ecliptic is conveniently used like an intermediate plane in the calculation of the orbital elements of the Moon [366]. Thus,

$$(x_☾, y_☾, z_☾)^\tau = R_1(-\varepsilon)\, R_3(-N)\, R_1(-J)\, R_3(-\theta_☾)\, (r_☾, 0, 0)^\tau, \tag{8.32}$$

where $\theta_☾ = \gamma + f_☾$, with $f_☾$ the true anomaly of the Moon and γ the argument of the perigee of the Moon orbit referred to the ecliptic, N and J are the longitude of the node and the inclination of the Moon's orbit over the ecliptic, respectively, and $\varepsilon \approx 23.5$ deg is the obliquity of the ecliptic. The Moon radius $r_☾$ is also written in terms of the orbital elements like $r_☾ = a_☾(1 - e_☾^2)/(1 + e_☾ \cos f_☾)$, with $a_☾$ and $e_☾$ denoting the semimajor axis and eccentricity of the Moon's orbit about the Earth, respectively.

The lunar disturbing potential is about $\chi_☾ (n_☾/n)^2$ times smaller than the Keplerian, as checked in Eq. (8.21), and is of the order of one millionth for Galileo orbits. Besides, the effect of each consecutive term of the Legendre polynomial expansion of the lunar disturbing potential is, on average, $a/r_☾$ times smaller than the preceding one, which is 29600/384400 ≈ 0.077 for Galileo orbits. In that case, effects of the Legendre polynomial of degree three are, on average, multiplied by $a/r_☾ \approx 0.08$, and those of P_4 by $a^2/r_☾^2 \approx 0.006$. The latter is very small and, in consequence, we neglect the disturbing effects on Galileo orbits of terms of degree four and higher in the Legendre polynomial expansion of the Moon disturbing potential.

From this truncation model, the mean anomaly of the Moon is removed in closed form resorting to the differential relation between the true and mean anomalies of the Moon $a_☾^2 \eta_☾ \, d\ell_☾ = r_☾^2 \, df_☾$, which is analogous to Eq. (4.64). After the new averaging, the part of the Moon's long-term disturbing potential due to P_2, denoted $\langle \mathcal{V}_{☾,2} \rangle_{\ell,\ell_☾}$, is arranged in the form

$$\langle \mathcal{V}_{☾,2} \rangle_{\ell,\ell_☾} = \frac{\mu}{a} \chi_☾ \frac{n_☾^2}{n^2} \frac{1}{\eta_☾^3} \sum_{i=0}^{1} \sum_{j=-2}^{2} \sum_{k=-2}^{2} B_{2,i}(e) Q_{2,i,j}(I) \times \mathcal{E}_{2,j,k}(\varepsilon) \mathcal{J}_{2,k}(J) \cos(2i\omega + j\Omega + kN), \qquad (8.33)$$

in which $B_{2,0} = 6 + 9e^2$, $B_{2,1} = -15e^2$; the inclination polynomials $Q_{i,j,l}$ are listed in Table 8.3; $\mathcal{J}_{2,0} = 1 - \frac{3}{2}\sin^2 J$, $\mathcal{J}_{2,\pm1} = \sin 2J$, $\mathcal{J}_{2,\pm2} = \sin^2 J$, and the functions $\mathcal{E}_{2,j,k}$ of the obliquity of the ecliptic are given in Table 8.4; cf. [436]. In the tables, upper (resp. lower) signs in subindices match upper (resp. lower) signs in the corresponding functions. Long-period terms due to P_2 do not depend on the Moon's argument of the perigee γ, and $\langle \mathcal{V}_{☾,2} \rangle_{\ell,\ell_☾}$ comprises only 38 different trigonometric terms.

Table 8.3: Inclination polynomials $Q_{2,j,l}$ and $Q_{3,j,l}$.

l	$Q_{2,0,l}$	$Q_{2,1,l}$	$Q_{3,0,l}$	$Q_{3,1,l}$
0	$\frac{1}{48}(3c^2 - 1)$	$-\frac{1}{16}s^2$	$-\frac{15}{128}(5c^2 - 1)s$	$-\frac{175}{128}s^3$
±1	$\frac{1}{8}cs$	$\frac{1}{8}(c \pm 1)s$	$-\frac{15}{512}(c \pm 1)(15c^2 \mp 10c - 1)$	$-\frac{525}{512}(c \pm 1)s^2$
±2	$-\frac{1}{32}s^2$	$\frac{1}{32}(c \pm 1)^2$	$\frac{75}{256}(c \pm 1)(3c \mp 1)s$	$-\frac{525}{256}(c \pm 1)^2 s$
±3			$\frac{75}{512}(c \pm 1)s^2$	$\frac{175}{512}(c \pm 1)^3$

Table 8.4: Terms $\mathcal{E}_{2,j,k} = \mathcal{E}_{2,-j,-k}$ in the obliquity of the ecliptic in Eq. (8.33).

k	$j = 0$	$j = -1$	$j = -2$
0	$\frac{1}{2}(1 - 3\cos^2\varepsilon)$	$-\frac{1}{2}\cos\varepsilon \sin\varepsilon$	$\frac{1}{2}\sin^2\varepsilon$
±1	$\frac{3}{4}\cos\varepsilon \sin\varepsilon$	$\frac{1}{8}(1 \pm \cos\varepsilon)(1 \mp 2\cos\varepsilon)$	$\pm\frac{1}{4}(1 \pm \cos\varepsilon)\sin\varepsilon$
±2	$-\frac{3}{8}\sin^2\varepsilon$	$\pm\frac{1}{8}(1 \pm \cos\varepsilon)\sin\varepsilon$	$\frac{1}{8}(1 \pm \cos\varepsilon)^2$

Note that terms affected by the coefficient $\mathcal{J}_{2,0}$ are free from N, and may give rise to quasi-secular terms of argument $2\omega + \Omega$ [66, 519]. In the case of Galileo orbits the period is of the order of several thousands of years, as readily checked from Eqs. (5.24) and (5.26), thus contributing quasi-secular terms to the evolution of both the eccentricity and the inclination [443].

Proceeding analogously with the disturbing terms contributed by P_3, we obtain

$$\langle \mathcal{V}_{\rightmoon,3}\rangle_{\ell,\ell_\rightmoon} = \frac{\mu}{a}\chi_\rightmoon \frac{n_\rightmoon^2}{n^2}\frac{a}{a_\rightmoon}\frac{e_\rightmoon}{\eta_\rightmoon^5}\sum_{i=0}^{1} B_{3,i}(e)\sum_{j=-3}^{3} Q_{3,i,j}(I)\sum_{k=-3}^{3}\sum_{l=0}^{1}(2l-1)\mathcal{E}_{3,j,k}(\varepsilon)$$
$$\times \mathcal{J}_{3,(2l-1)k}(J)\cos[(2i+1)\omega + j\Omega + kN - (2l-1)\gamma], \tag{8.34}$$

with $B_{3,0} = 4e + 3e^3$, $B_{3,1} = e^3$, $Q_{3,i,j}$ are given in Table 8.3,

$$\mathcal{J}_{3,0} = -(1 - 5\cos^2 J)\sin J,$$
$$\mathcal{J}_{3,\pm1} = \pm(1 \mp \cos J)(1 \mp 10\cos J - 15\cos^2 J),$$
$$\mathcal{J}_{3,\pm2} = -(1 \mp \cos J)(1 \pm 3\cos J)\sin J,$$
$$\mathcal{J}_{3,\pm3} = \pm(1 \mp \cos J)\sin^2 J,$$

and terms $\mathcal{E}_{3,j,k}$ in the obliquity of the ecliptic are given in Table 8.5. Now, all 196 trigonometric terms comprising Eq. (8.34) depend on the argument of the perigee of the Moon's orbit relative to the ecliptic γ, as well as on the argument of the perigee of the satellite's orbit with respect to the equator ω.

Table 8.5: Coefficients $\mathcal{E}_{3,j,k} = \mathcal{E}_{3,-j,-k}$ in Eq. (8.34).

k	$j = 0$	$j = -1$
0	$\frac{3}{16}(3 - 5\cos^2\varepsilon)\cos\varepsilon$	$-\frac{3}{16}(1 - 5\cos^2\varepsilon)\sin\varepsilon$
±1	$\frac{3}{64}(1 - 5\cos^2\varepsilon)\sin\varepsilon$	$\pm\frac{1}{64}(1 \pm \cos\varepsilon)(1 \pm 10\cos\varepsilon - 15\cos^2\varepsilon)$
±2	$\frac{15}{32}\cos\varepsilon\sin^2\varepsilon$	$-\frac{5}{32}(1 \pm \cos\varepsilon)(1 \mp 3\cos\varepsilon)\sin\varepsilon$
±3	$\frac{15}{64}\sin^3\varepsilon$	$\pm\frac{15}{64}(1 \pm \cos\varepsilon)\sin^2\varepsilon$
k	$j = -2$	$j = -3$
0	$-\frac{3}{16}\cos\varepsilon\sin^2\varepsilon$	$-\frac{3}{16}\sin^3\varepsilon$
±1	$-\frac{1}{64}(1 \pm \cos\varepsilon)(1 \mp 3\cos\varepsilon)\sin\varepsilon$	$\pm\frac{3}{64}(1 \pm \cos\varepsilon)\sin^2\varepsilon$
±2	$\mp\frac{1}{32}(1 \pm \cos\varepsilon)^2(2 \mp 3\cos\varepsilon)$	$\frac{3}{32}(1 \pm \cos\varepsilon)^2\sin\varepsilon$
±3	$\frac{3}{64}(1 \pm \cos\varepsilon)^2\sin\varepsilon$	$\mp\frac{3}{64}(1 \pm \cos\varepsilon)^3$

The second averaging has made explicit the factor $e_\star/\eta_\star^5$ in the contribution of P_3. For a Galileo orbit, $e_\rightmoon/\eta_\rightmoon^5 \approx e_\rightmoon \sim a/a_\rightmoon$, making the perturbation of P_3 of the same order as P_4, which has been neglected. Nevertheless, the double-averaged P_3 holds the Moon's perigee dynamics, whereas we checked that those terms of the double-averaged P_4 depending on γ are multiplied by $e_\rightmoon^2$. Therefore, P_3 must not be neglected in corresponding long-term propagations because it yields observable qualitative effects related to the perigee dynamics of the Moon orbit.

The Cartesian coordinates of the Sun's apparent orbit about the Earth are

$$(x_\odot, y_\odot, z_\odot)^\tau = R_1(-\varepsilon)\, R_3(-\theta_\odot)\, (r_\odot, 0, 0)^\tau, \tag{8.35}$$

where $\theta_\odot$ is the Sun argument of the latitude. We neither find apsidal or nodal resonances of Galileo orbits with the mean motion of the Sun, and annual oscillations of the orbital elements are also removed in the study of long-term motion.

Now $\chi_\odot \approx 1$ in Eq. (8.21), showing that the disturbing potential of the Sun is roughly $(n_\odot/n)^2$ times smaller than the Keplerian—again, of the order of one millionth for Galileo orbits, yet smaller than in the case of the Moon. We only take the Legendre polynomial of the second degree in the solar disturbing potential because P_3 is scaled by an additional reduction factor $a/r_\odot \sim 10^{-4}$.

Because Eq. (8.35) is equivalent to Eq. (8.32) with $J = N = 0$, we take advantage of the previous computation of Eq. (8.33). Taking, besides, $\eta_\odot^3 = 1$ due to the small eccentricity of the Sun's apparent orbit about the Earth, we obtain

$$\langle \mathcal{V}_\odot \rangle_{\ell,\theta_\odot} = \frac{\mu}{a} \frac{n_\odot^2}{n^2} \sum_{i=0}^{1} B_{2,i}(e) \sum_{j=-2}^{2} Q_{2,i,j}(I) \mathcal{E}_{2,j,0}(\varepsilon) \cos(2i\omega + j\Omega), \tag{8.36}$$

which is made only of eight different trigonometric terms. Quasi-secular trigonometric terms of argument $2\omega + \Omega$, previously identified for Galileo orbits in the long-term disturbing potential of the Moon, are now more clearly identified.

8.3.2 Additional simplifications. Long-term Hamiltonian

Due to the smallness of the inclination of the Moon orbit with respect to the ecliptic, terms factored by $\sin^2 J$ in Eq. (8.33) are less than 20 times smaller than those factored by $\sin 2J$ and are further neglected in the investigation of the long-term dynamics. This simplification reduces the range of variation of the summation index k from ± 2 to ± 1 and makes $\mathcal{J}_{2,0} \approx 2$ in Eq. (8.33). Conversely, terms factored by $\sin 2J$, while also small, are retained in $\langle \mathcal{V}_{☾,2} \rangle_{\ell,\ell_☾}$ due to the important long-period effects related to the lunar node, which would be removed if these terms were also neglected. Of the same order as $\sin J$ is the eccentricity of the Moon, which is present in Eq. (8.33) only through the factor $\eta_☾^{-3} = 1 + \mathcal{O}(e_☾^2)$ and is also neglected. Hence,

$$\langle \mathcal{V}_{☾,2} \rangle_{\ell,\ell_☾} \approx \frac{\mu}{a} \chi_☾ \frac{n_☾^2}{n^2} \sum_{i=0}^{1} B_{2,i} \sum_{j=-2}^{2} Q_{2,i,j} \sum_{k=-1}^{1} \mathcal{E}_{2,j,k} \mathcal{J}_{2,k} \cos(2i\omega + j\Omega + kN), \tag{8.37}$$

which now comprises only 23 trigonometric terms.

Analogous simplifications are applied to Eq. (8.34), where, on account of the small eccentricity of the orbit of the Moon, we also neglect terms involving the product $e_☾ \sin J$. Recalling that $\cos J = 1 - \mathcal{O}(\sin^2 J)$, we immediately see that the only relevant Moon inclination function is $\mathcal{J}_{3,-1} \approx 8$. In consequence, the summation in k in Eq. (8.34) is replaced by a single term, yielding

$$\langle \mathcal{V}_{☾,3} \rangle_{\ell,\ell_☾} \approx \frac{\mu}{a} \chi_☾ \frac{n_☾^2}{n^2} \frac{a}{a_☾} e_☾ \sum_{i=0}^{1} B_{3,i}(e) \sum_{j=-3}^{3} Q_{3,i,j}(I) \sum_{l=0}^{1} 8(1-2l)\mathcal{E}_{3,j,2l-1}(\epsilon) \times \cos[(2i+1)\omega + j\Omega + (2l-1)(N+\gamma)], \tag{8.38}$$

which consists only of 28 trigonometric terms.

The important contribution of the Earth's oblateness must also be taken into account, at least for the first order of J_2 as given in Eq. (5.60). The effects of higher-order harmonics may be neglected for an orbit of the characteristics of Galileo except for those of J_3; cf. [183]. The latter is obtained from the last term of Eq. (6.42) particularized for $i = 3$ using Eq. (6.14). Thus, the long-term dynamics of Galileo orbits can be efficiently investigated with the radically simplified Hamiltonian

$$\widetilde{\mathcal{K}} = \widetilde{\mathcal{M}} + \frac{\mu}{a} J_3 \frac{R_\oplus^3}{p^3} \frac{3}{4} e\eta(5s^2 - 4)s \sin\omega + \langle \mathcal{V}_\odot \rangle_{\ell,\ell_\odot} + \langle \mathcal{V}_{☾,2} \rangle_{\ell,\ell_☾} + \langle \mathcal{V}_{☾,3} \rangle_{\ell,\ell_☾}, \tag{8.39}$$

in which $\widetilde{\mathcal{M}}$ is formally equal to Eq. (5.71), the double-averaged disturbing potential of the Sun $\langle \mathcal{V}_\odot \rangle_{\ell,\ell_\odot}$ is taken from Eq. (8.36), and that of the Moon $\langle \mathcal{V}_{☾,2} \rangle_{\ell,\ell_☾} + \langle \mathcal{V}_{☾,3} \rangle_{\ell,\ell_☾}$ from Eqs. (8.37) and (8.38). Recall that all the symbols are now functions of double-prime Delaunay variables.

Note that, while most of the simplifications carried out would apply also to other global navigation satellite orbits, the double-averaged Hamiltonian (8.39) may need to be complemented with tesseral resonant terms. In particular, as checked in Eq. (7.67), Hamiltonian terms related to the $J_{3,2}$ coefficient should be taken into account when dealing with GPS orbits [183].

Due to the low eccentricity of Galileo orbits, Hamilton equations derived from Eq. (8.39) should be reformulated in nonsingular variables, at least for the eccentricity. The usual set defined by the semi-equinoctial variables given projections of the eccentricity vector in the nodal frame $C = e\cos g$, $S = e\sin g$, previously defined in Eq. (5.74), the mean argument of the latitude $F = \ell + g$, and the Delaunay variables L, h, and H, is adequate for the current case. Since the mean anomaly of the satellite has been removed, L'' is constant, and, for the investigation of the long-term orbit evolution, it is enough to deal with the reduced system involving the variations of H, h, C, and S, which, after properly having been formulated in the C and S variables, are checked to be free from the eccentricity in denominators. They still may show trouble for almost equatorial orbits, but this is not of concern for Galileo orbits.

The numerical integration of the double-averaged flow is very efficient and allows for the fast exploration of the long-term behavior of Galileo orbits. For instance, Fig. 8.3 illustrates how the difference between the initial nodes of the orbiter and the Moon may introduce evident changes on the dynamics; cf. [443].

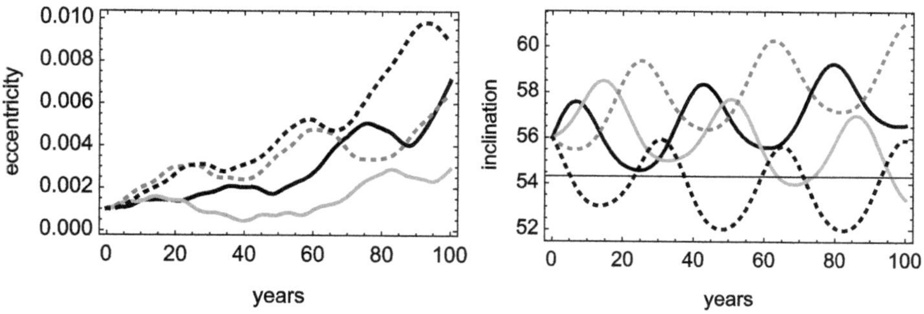

Figure 8.3: Galileo-type orbit evolution for $N_0 = \gamma_0 = 0$ and $h_0 = 60$ (black), 150 (gray), 240 (gray, dashed), and 330 deg (black, dashed curve).

8.4 Perturbations in the ecliptic frame

The orbit of an artificial satellite of the Earth under geopotential disturbances is naturally described by referring the orbit's ascending node and inclination to the equatorial plane. However, the effect of third-body perturbations is better understood when referring both said orbital parameters to the orbital plane of the third-body orbit and, in some cases, the analysis of the combined perturbations is simplified when referred to this plane [364, 450, 502]—or to the Laplace plane [384, 632] when both effects are comparable or the effects of additional bodies are taken into account [20, 578, 649].

Referring the coordinates of the satellite to other reference frame than the equatorial one does not complicate the orbital elements formulation, which is still given by Eq. (4.4), and only means that the argument of the perigee as well as the inclination and the longitude of the node have different values. On the contrary, when the coordinates of the Moon are referred to the ecliptic rather than to the equator, we avoid the final rotation $R_1(-\varepsilon)$ in Eq. (8.32), corresponding to the obliquity of the ecliptic, to give

$$(x_{☾}, y_{☾}, z_{☾})^\tau = R_3(-N)\,R_1(-J)\,R_3(-\theta_{☾})\,(r_{☾}, 0, 0)^\tau. \tag{8.40}$$

Therefore, the cosine of the elongation of the satellite with respect to the Moon as given by Eq. (8.5) will be simpler in the ecliptic frame formulation and, in consequence, the disturbing potential of the Moon is expected to be notably simplified. Obviously, the same happens to the coordinates of the Sun in Eq. (8.35), which when referred to the ecliptic simply turn into

$$(x_\odot, y_\odot, z_\odot)^\tau = R_3(-\theta_\odot)\,(r_\odot, 0, 0)^\tau. \tag{8.41}$$

If besides we assume that the Sun evolves in a circular orbit then $r_\odot = a_\odot$, and $\theta_\odot$ grows linearly with angular velocity $n_\odot$.

The Cartesian coordinates of the satellite in the ecliptic frame are

$$(x, y, z)^\tau = R_3(-h)\,R_1(-i)\,R_3(-\vartheta)\,(r, 0, 0)^\tau, \tag{8.42}$$

where $\vartheta = f + \omega_E$, ω_E is the argument of the perigee of the satellite referred to the ecliptic, i denotes orbit inclination with respect to the ecliptic, and $h = \Omega_E$ is the ecliptic longitude of the node.

The inclination of the orbit with respect to the equatorial plane I is easily obtained in terms of the inclination with respect to the ecliptic i. Indeed, let $\boldsymbol{n}$ and $\boldsymbol{b}_3$ be unit vectors in the direction perpendicular to the orbital plane and to the equator, respectively; then, by direct computation, $\cos I = \boldsymbol{b}_3 \cdot \boldsymbol{n}$. On the other hand, the components of $\boldsymbol{n}$ in the equatorial frame can be computed using the ecliptic as an intermediate plane. Thus, $\boldsymbol{n} = R_1(-\varepsilon)\, R_3(-\Omega_E)\, R_1(-i)\, (0,0,1)^\tau$, and hence

$$\cos I = \cos\varepsilon \cos i - \sin\varepsilon \sin i \cos \Omega_E. \tag{8.43}$$

Analogously, calling $\boldsymbol{s}_3$ the direction orthogonal to the ecliptic, $\cos i = \boldsymbol{s}_3 \cdot \boldsymbol{n}$. Computing the components of $\boldsymbol{n}$ in the ecliptic using the equator as an intermediate plane, $\boldsymbol{n} = R_1(\varepsilon)\, R_3(-\Omega)\, R_1(-I)\, (0,0,1)^\tau$. Therefore,

$$\cos i = \cos\varepsilon \cos I + \sin\varepsilon \sin I \cos \Omega. \tag{8.44}$$

8.4.1 The disturbing potential

We constrain ourselves to distances at which the oblateness and the third-body lunisolar perturbations are comparable, say between 5 and 9 Earth radii. At these altitudes long-period effects due to other zonal harmonics than J_2 are clearly of high order and we neglect them. In addition, for both Sun and Moon, we only take the P_2 Legendre polynomial into account in the expansion of the third-body disturbing potential.

We use the notation (ξ, η, ζ) to distinguish equatorial coordinates from the ecliptic ones (x, y, z) used in Eq. (8.42). They are related by a single rotation of amplitude given by the obliquity of the ecliptic $(\xi, \eta, \zeta)^\tau = R_1(-\varepsilon)(x, y, z)^\tau$. The oblateness disturbing potential is then obtained by neglecting the Keplerian term from Eq. (5.4), in which $\sin\varphi = \zeta/r$. Straightforward computations yield

$$\mathcal{V}_\oplus = -J_2 \frac{\mu}{2r} \frac{R_\oplus^2}{r^2} \left[1 - 3\left(\frac{z}{r} \cos\varepsilon + \frac{y}{r} \sin\varepsilon \right)^2 \right]. \tag{8.45}$$

Hereafter, we may abbreviate $s_\alpha \equiv \sin\alpha$, $c_\alpha \equiv \cos\alpha$ in the case of (almost) constant angles or when they are functions of the momenta.

If we now replace (x, y, z) from Eq. (8.42), we obtain

$$\begin{aligned}\mathcal{V}_\oplus = \frac{\mu}{2a} J_2 \frac{R_\oplus^2}{r^2} \frac{a}{r} \frac{1}{8} \{ &(6s_\varepsilon^2 - 4)(2 - 3s_i^2 + 3s_i^2 \cos 2\vartheta) + 12 c_\varepsilon s_\varepsilon s_i [(1 - c_i) \\ &\times \cos(h - 2\vartheta) + 2c_i \cos h - (1 + c_i)\cos(h + 2\vartheta)] - 3s_\varepsilon^2 [(1 - c_i)^2 \\ &\times \cos(2h - 2\vartheta) + 2s_i^2 \cos 2h + (1 + c_i)^2 \cos(2h + 2\vartheta)] \},\end{aligned} \tag{8.46}$$

thus adding four new periodic terms that carry the dependence on the node of the orbit in the ecliptic frame, with respect to the equatorial formulation of the J_2 disturbing potential in Eq. (5.15). The customary expression of $\mathcal{V}_\oplus$ in the equatorial plane is recovered when making $\varepsilon = 0$ (and hence $s_i = s$, $\vartheta = \theta$) in Eq. (8.46).

The solar disturbing potential is obtained from Eq. (8.6) with $\star \equiv \odot$, in which $\cos\psi_\odot$ is computed by replacing Eqs. (8.42) and (8.41) in Eq. (8.5). Limited to the contribution of P_2, we obtain

$$\begin{aligned}\mathcal{V}_\odot = &-\frac{\mu}{2a}\frac{n_\odot^2}{n^2}\frac{r^2}{a^2}\frac{1}{8}\big[4 - 6s_i^2 + 6s_i^2\cos 2\vartheta + 6s_i^2\cos 2(h-\theta_\odot) + 3(1-c_i)^2 \\ &\times \cos 2(h-\theta_\odot-\vartheta) + 3(c_i+1)^2\cos 2(h-\theta_\odot+\vartheta)\big].\end{aligned} \tag{8.47}$$

Proceeding analogously with the Moon ($\star \equiv ☾$), now using Eq. (8.40), we obtain

$$\begin{aligned}\mathcal{V}_☾ = &-\frac{\mu}{2a}\chi_☾\frac{n_☾^2}{n^2}\frac{r^2}{a^2}\frac{a_☾^3}{r_☾^3}\frac{1}{8}\{(2-6w^2)(2-3s_i^2+3s_i^2\cos 2\vartheta) + 6uv \\ &\times\big[(1-c_i)^2\sin(2h-2\vartheta) + 2s_i^2\sin 2h + (1+c_i)^2\sin(2h+2\vartheta)\big] \\ &+ 3(u^2-v^2)\big[(1-c_i)^2\cos(2h-2\vartheta) + (1+c_i)^2\cos(2h+2\vartheta) \\ &+ 2s_i^2\cos 2h\big] - 12uws_i\big[(1-c_i)\sin(h-2\vartheta) + 2c_i\sin h \\ &- (1+c_i)\sin(h+2\vartheta)\big] + 12vws_i\big[(1-c_i)\cos(h-2\vartheta) \\ &+ 2c_i\cos h - (1+c_i)\cos(h+2\vartheta)\big]\},\end{aligned} \tag{8.48}$$

where, for brevity, the components of the Moon direction vector in the ecliptic frame $(u, v, w) = (x_☾, y_☾, z_☾)/r_☾$, are not yet replaced.

8.4.2 Removing short-period effects

We use the Lie transforms method to find the transformation that removes the mean anomaly of the orbiter ℓ. To avoid time-related issues, we move to an extended phase space $(\ell, g, h, \ell_☾, \gamma, N, \theta_\odot, L, G, H, L_☾, G_☾, H_☾, \Theta_\odot)$ in which the symbols $L_☾$, $G_☾$, and $H_☾$ stand for the conjugate momenta to the mean anomaly, argument of the perigee, and longitude of the node of the Moon, respectively, which evolve with frequencies $n_☾$, n_γ, and n_N, respectively, while $\Theta_\odot$ is the conjugate momentum to the polar angle of the Sun $\theta_\odot$, which evolves with frequency $n_\odot$. Recall that $g = \omega_E$, and $h = \Omega_E$, are referred to the ecliptic frame.

In the previous assumptions that oblateness and third-body perturbations are of the same order, and using the extended phase space formulation to deal with the time dependency of the lunisolar ephemeris, we set up the usual perturbation Hamiltonian (2.30) with a formal small parameter ϵ,

$$\mathcal{H}_{0,0} = -\frac{\mu}{2a}, \tag{8.49}$$

$$\mathcal{H}_{1,0} = L_{☾} n_{☾} + G_{☾} n_{\gamma} + H_{☾} n_N + \Theta_{\odot} n_{\odot}, \tag{8.50}$$

$$\mathcal{H}_{2,0} = 2!(\mathcal{V}_{\oplus} + \mathcal{V}_{☾} + \mathcal{V}_{\odot}), \tag{8.51}$$

and $\mathcal{H}_{m,0} = 0$ for $m \geq 3$.

The homological equation (4.68) is solved in closed form of the eccentricity. This is achieved with the help of the differential relation (4.64) between the mean and true anomalies for geopotential perturbations, and using the differential relation (4.62) between the mean and eccentric anomalies for those related to third-body lunisolar perturbations.

Because $\widetilde{\mathcal{H}}_{0,1} = \mathcal{H}_{1,0}$ does not depend on ℓ, we choose $\mathcal{H}_{0,1} = \mathcal{H}_{1,0}$, and $\mathcal{W}_1 = 0$. At the second order $\mathcal{H}_{0,2} = \langle \mathcal{H}_{2,0} \rangle_\ell = 2!(\langle \mathcal{V}_\oplus \rangle_\ell + \langle \mathcal{V}_{☾} \rangle_\ell + \langle \mathcal{V}_\odot \rangle_\ell)$, and the homological equation splits into three different parts that are solved separately.

The part of $\mathcal{H}_{0,2}$ corresponding to the Earth's J_2 disturbing potential becomes $\langle \mathcal{V}_\oplus \rangle_\ell = \langle \mathcal{V}_\oplus r^2/(a^2\eta) \rangle_f$. That is,

$$\langle \mathcal{V}_\oplus \rangle_\ell = -\frac{\mu}{2a} J_2 \frac{R_\oplus^2}{p^2} \frac{\eta}{4} [(2 - 3s_\varepsilon^2)(2 - 3s_i^2) - 12 s_\varepsilon c_\varepsilon s_i c_i \cos h + 3 s_\varepsilon^2 s_i^2 \cos 2h], \tag{8.52}$$

which adds two new periodic terms with respect to the equatorial plane formulation in Eq. (5.60)—from which it could have been directly derived replacing $\cos I$ from Eq. (8.43). Corresponding short-period corrections are derived from the part $\mathcal{W}_\oplus$ of the generating function $\mathcal{W} = \mathcal{W}_\oplus + \mathcal{W}_{☾} + \mathcal{W}_\odot$. From Eq. (4.69), we obtain

$$\begin{aligned}\mathcal{W}_\oplus = L J_2 \frac{R_\oplus^2}{p^2} 8\eta \sum_{k=-2}^{2} (-1)^k \Bigg[& (f - \ell) A_2(e) Q_{2,0,k}(i) \mathcal{E}_{2,k,0}(\varepsilon) \cos kh \\ & + \sum_{l=0}^{1} \sum_{j=1}^{2l+1} \frac{1}{(-2)^l} A_j(e) Q_{2,l,k}(i) \mathcal{E}_{2,k,0}(\varepsilon) \sin(jf + 2lg + kh) \Bigg],\end{aligned}$$

where $A_1 = 3e$, $A_2 = 3$, $A_3 = e$, the inclination polynomials $Q_{2,l,k}$ are formally the same as those in Table 8.3, but now with the inclination referred to the ecliptic, and the functions of the obliquity of the ecliptic $\mathcal{E}_{2,k,0}$, are those of Table 8.4.

The part of the solar perturbation becomes $\langle \mathcal{V}_\odot \rangle_\ell = \langle \mathcal{V}_\odot (r/a) \rangle_u$. Namely,

$$\langle \mathcal{V}_\odot \rangle_\ell = -\frac{\mu}{a} \frac{n_\odot^2}{n^2} \sum_{l=0}^{1} b_l(e) \sum_{k=-1}^{1} (-1)^{k-l} Q_{2,l,2k}(i) \cos[2k(h - \theta_\odot) + 2lg], \tag{8.53}$$

with $b_0 = 6 + 9e^2$, $b_1 = 15e^2$, and inclination polynomials $Q_{2,l,2k}$ from Table 8.3. Now, the number of periodic terms is reduced from 15 in the equatorial frame to just 4. From Eq. (4.70) we obtain the solar part of the generating function

$$\mathcal{W}_\odot = L \frac{n_\odot^2}{n^2} \sum_{l=0}^{1} \sum_{j=-3}^{3} \sum_{k=-1}^{1} (-1)^k B_{2,l,j}(e) Q_{2,l,2k}(i) \sin[ju + 2k(h - \theta_\odot) + 2lg], \tag{8.54}$$

with the eccentricity polynomials $B_{2,l,j}$ from Table 8.6, and the inclination polynomials $Q_{2,l,2k}$ again from Table 8.3. The terms $B_{2,l,0}$ are dispensable, yet they prevent the appearance of hidden long-period terms in $\mathcal{W}_\odot$.

Table 8.6: Eccentricity polynomials $B_{2,l,j}(e)$ in Eqs. (8.54) and (11.5).

l	$j=0$	$j=\pm1$	$j=\pm2$	$j=\pm3$
0	$6+9e^2$	$\pm\frac{3}{4}e(8-3e^2)$	$\mp\frac{9}{4}e^2$	$\pm\frac{1}{4}e^3$
1	$-\frac{15}{2}e^2\eta$	$\mp\frac{15}{4}e(1\pm\eta)^2$	$\pm\frac{3}{4}(1\pm\eta)^2(3\mp2\eta)$	$\mp\frac{1}{4}e(1\pm\eta)^2$

Finally, $\langle\mathcal{V}_☾\rangle_\ell = \langle\mathcal{V}_☾(r/a)\rangle_u$ produces

$$\begin{aligned}\langle\mathcal{V}_☾\rangle_\ell = {}& \frac{\mu}{2a}\chi_☾\frac{n_☾^2}{n^2}\frac{a_☾^3}{r_☾^3}\frac{1}{16}\{[(2+3e^2)(1-3c_i^2)-15s_i^2e^2\cos 2g] \\ & \times(2-6w^2)+6uv[5e^2(1-c_i)^2\sin(2g-2h)-(4+6e^2)s_i^2 \\ & \times\sin 2h-5e^2(1+c_i)^2\sin(2g+2h)]+3(v^2-u^2) \\ & \times[5e^2(1-c_i)^2\cos(2g-2h)+(4+6e^2)s_i^2\cos 2h+5e^2(1+c_i)^2 \\ & \times\cos(2g+2h)]-12uws_i[5e^2(1-c_i)\sin(2g-h)-(4+6e^2) \\ & \times c_i\sin h+5e^2(1+c_i)\sin(2g+h)]-12vw[(4+6e^2)c_i\cos h \\ & +5e^2(1-c_i)\cos(2g-h)-5e^2(1+c_i)\cos(2g+h)]s_i\},\end{aligned} \tag{8.55}$$

which, as far as we keep the lunar ephemeris given by the direction vector $\hat{\boldsymbol{r}}_☾ = (u,v,w)$, remains with the same complexity as when using the equatorial frame formulation. The lunar term of the generating function is obtained, once more, from the homological equation in the form of Eq. (4.70). Namely,

$$\begin{aligned}\mathcal{W}_☾ = {}& L\frac{n_☾^2}{n^2}\chi_☾\frac{a_☾^3}{r_☾^3}\sum_{l=0}^{1}\sum_{j=-3}^{3}\sum_{k=-2}^{2}B_{2,l,j}(e)Q_{2,l,k}(i) \\ & \times[C_{2,k}^\star(\hat{\boldsymbol{r}}_☾)\cos(ju+2lg+kh)-S_{2,k}^\star(\hat{\boldsymbol{r}}_☾)\sin(ju+2lg+kh)],\end{aligned}$$

in which, in addition to the eccentricity and inclination polynomials given before, we need the third-body direction polynomials $C_{2,0}^\star=0$, $C_{2,\pm1}^\star=\pm uw$, $C_{2,\pm2}^\star=\pm2uv$, $S_{2,0}^\star=-1+3w^2$, $S_{2,\pm1}^\star=-vw$, $S_{2,\pm2}^\star=u^2-v^2$.

The amplitude of the periodic corrections of the transformation from original to prime variables that removes the mean anomaly of the satellite from the Hamiltonian is small, with oblateness, lunar, and solar terms multiplied by $J_2(R_\oplus/p)^2$, $\chi_☾(n_☾/n)^2$, and $(n_\odot/n)^2$, respectively.

After neglecting higher orders, the new Hamiltonian

$$\mathcal{K} = -\frac{\mu}{2a}+L_☾'n_☾+G_☾'n_\gamma+H_☾'n_N+\Theta_\odot'n_\odot+\langle\mathcal{V}_\oplus\rangle_\ell+\langle\mathcal{V}_☾\rangle_\ell+\langle\mathcal{V}_\odot\rangle_\ell, \tag{8.56}$$

which is obtained by replacing old by new variables in Eqs. (8.52), (8.53), and (8.55), is free from periodic effects related to the mean anomaly of the satellite. Therefore, L' (or the semimajor axis $a = \mu/L'^2$) is a formal integral that decouples the Hamiltonian flow.

8.4.3 Removing monthly and annual effects

The new Hamiltonian (8.56) has one degree of freedom less than the original one, but it still depends on two other angles as well as the time. However, as far as we are interested in the evolution of the system in time scales of tens of years, we can carry out a new Lie transformation to double-prime variables such that, up to the truncation order, the mean anomalies of the Sun and Moon are removed from the transformed Hamiltonian.

Like in §8.3, we choose $\mathcal{K}_{0,0} = -\mu/(2a) + \Theta'_\odot n_\odot + L'_☾ n_☾$, and hence the Lie derivative (2.49) involves partial differentiation with respect to the mean anomaly of the Moon and the polar angle of the Sun. Namely, $\mathcal{L}_0 = n_☾ \partial/\partial \ell'_☾ + n_\odot \partial/\partial \theta'_\odot$. In consequence, the first-order Hamiltonian term is $\mathcal{K}_{1,0} = G'_☾ n_\gamma + H'_☾ n_N$. Because it is free from angles, the first-order homological equation is trivially solved by choosing $\mathcal{K}_{0,1} = \mathcal{K}_{1,0}$, and assigning the null value to the first-order term of the generating function, $\mathcal{U}_1 = 0$.

The second-order term of the Hamiltonian is

$$\mathcal{K}_{2,0} = 2!(\langle \mathcal{V}_\oplus \rangle_\ell + \langle \mathcal{V}_☾ \rangle_\ell + \langle \mathcal{V}_\odot \rangle_\ell), \tag{8.57}$$

where now the components of the Moon direction in Eq. (8.55) are written in terms of the Moon orbital elements using Eq. (8.40). Since there is no coupling between lunar and solar terms at second order, the homological equation can be solved separately for each kind of perturbation.

For the Moon, we choose the new Hamiltonian term by averaging $\mathcal{K}_{2,0}$, or just $\langle \mathcal{V}_☾ \rangle_\ell$ in Eq. (8.57), over the mean anomaly of the Moon. The averaging is achieved in closed form of the eccentricity in the same way as we did in §8.3.1. We obtain

$$\begin{aligned}\langle \mathcal{V}_☾ \rangle_{\ell,\ell_☾} &= -\frac{\mu}{2a}\frac{n_☾^2}{n^2}\frac{\chi_☾}{\eta_☾^3}\frac{1}{32}\{(4-6s_J^2)[(2+3e^2)(2-3s_i^2)+15e^2s_i^2\cos 2g'] \\ &\quad + 12c_J s_J s_i[5e^2(1-c_i)\cos(\Delta-2g') + (4+6e^2)c_i\cos\Delta - 5e^2 \\ &\quad \times (1+c_i)\cos(\Delta+2g)] + 3s_J^2[(4+6e^2)s_i^2\cos 2\Delta + 5(1-c_i)^2 \\ &\quad \times e^2\cos(2\Delta-2g') + 5e^2(1+c_i)^2\cos(2\Delta+2g')]\},\end{aligned} \tag{8.58}$$

where we abbreviate $\Delta = h' - N' = \Omega_E - N'$. Like with the equatorial frame formulation, the argument of the perigee of the lunar orbit is removed in this process. Now, the simplification is radical and the length of the series representing the lunar disturbing potential is shortened from the 38 periodic terms in the equatorial frame formulation of Eq. (8.33) to just the eight explicitly shown in Eq. (8.58). That is, almost 80 % of the terms of the double-averaged lunar disturbing potential are removed by the expedient of referring it to the ecliptic frame, and an analogous shortening applies to the simplifications achieved when neglecting $s_J^2 \equiv \sin^2 J$. In this last case Eq. (8.58) is reduced to only five trigonometric terms, compared to the 23 trigonometric terms that remain in Eq. (8.37).

Corresponding periodic corrections, which are derived from the lunar generating function term computed using Eq. (4.69), have notably larger amplitude than in the previous Lie transformation that removed the mean anomaly of the satellite. Up to $\mathcal{O}(s_J)$, that is, neglecting terms $\mathcal{O}(s_J^2)$ and higher, among which those $s_J e_☾$ and $s_J(f_☾ - \ell_☾)$ are also included, we obtain

$$\begin{aligned}
\mathcal{U}_{2,☾} = L\chi_☾ \frac{n_☾}{n}\frac{1}{64}\{&15e^2(1-c_i)^2\sin(2\theta_☾ - 2\Delta + 2g) + 6(2+3e^2)s_i^2 \\
&\times \sin(2\theta_☾ - 2\Delta) + 15e^2(1+c_i)^2\sin(2\theta_☾ - 2\Delta - 2g) + (f_☾ - \ell_☾) \\
&\times [(4+6e^2)(4-6s_i^2) + 60e^2 s_i^2\cos 2g] - 6s_J s_i[5e^2(1-c_i) \\
&\times \sin(2\theta_☾ - \Delta + 2g) + (4+6e^2)c_i\sin(2\theta_☾ - \Delta) - 5e^2(1+c_i) \\
&\times \sin(2\theta_☾ - \Delta - 2g)] + e_☾(4+6e^2)[(4-6s_i^2)\sin f_☾ + 3s_i^2 \\
&\times \sin(f_☾ + 2\gamma - 2\Delta) + s_i^2\sin(3f_☾ + 2\gamma - 2\Delta)] + 5e_☾ e^2[6s_i^2 \\
&\times \sin(f_☾ + 2g) + 3(1-c_i)^2\sin(f_☾ + 2\gamma - 2\Delta + 2g) + (1-c_i)^2 \\
&\times \sin(3f_☾ + 2\gamma - 2\Delta + 2g) + 6s_i^2\sin(f_☾ - 2g) + 3(1+c_i)^2 \\
&\times \sin(f_☾ + 2\gamma - 2\Delta - 2g) + (1+c_i)^2\sin(3f_☾ + 2\gamma - 2\Delta - 2g)]\},
\end{aligned}$$

where $\theta_☾ = f_☾ + \gamma$ and we removed primes for brevity. Note that the first two rows accommodate the dominant terms of the lunar perturbation.

So two important benefits stem from the ecliptic frame formulation. Namely, the dramatic reduction of the number of perturbation terms, and the realization that Moon disturbing effects depend on the difference between the longitude of the nodes of the Moon and the satellite, and not on their particular values. Remarkably, trigonometric terms of argument $2\omega + \Omega$ vanish when the node is referred to the ecliptic.

As regards solar perturbations, we only keep in the new Hamiltonian those terms of Eq. (8.57) that are free from the Sun polar angle. That is, as follows from Eq. (8.53),

$$\langle \mathcal{V}_\odot \rangle_{\ell,\theta_\odot} = -\frac{\mu}{2a}\frac{n_\odot^2}{n^2}\frac{1}{8}[(2+3e^2)(2-3s_i^2) + 15e^2 s_i^2\cos 2g], \tag{8.59}$$

which reduces the number of trigonometric terms from eight in Eq. (8.36) to just two here. Periodic corrections related to solar perturbations are derived from the solar term of the generating function, which is also obtained from Eq. (4.69),

$$\begin{aligned}
\mathcal{U}_{2,\odot} = L\frac{n_\odot}{n}\frac{3}{64}[&5e^2(1+c_i)^2\sin(2\theta_\odot - 2h - 2g) \\
&+ (4+6e^2)s_i^2\sin(2\theta_\odot - 2h) + 5e^2(1-c_i)^2\sin(2\theta_\odot - 2h + 2g)],
\end{aligned}$$

which is proportional to $n_\odot/n$ and yields periodic corrections of even larger amplitude than those of the lunar case.

The remaining term $\langle\mathcal{V}_\oplus\rangle_\ell$ of the Hamiltonian is not affected by the new transformation and remains formally the same as Eq. (8.52), yet in the new variables.

8.4.4 Elimination of the Moon's longitude of the node

The new Hamiltonian $\mathcal{T} = \mathcal{T}(g'', h'', N'', G'', H'', H''_☾)$, which is obtained after changing prime by double-prime variables in the terms $\mathcal{K}_{0,i}$, $i = 0, 1, 2$, maintains the form of a perturbation problem $\mathcal{T} = \mathcal{T}_{0,0} + \epsilon\mathcal{T}_{1,0}$, with $\mathcal{T}_{0,0} = -\mu/(2a) + \Theta''_\odot n_\odot + L''_☾ n_☾ + H''_☾ n_N$, and $\mathcal{T}_{1,0} = \langle\mathcal{V}_\oplus\rangle_\ell + \langle\mathcal{V}_☾\rangle_{\ell,\ell_☾} + \langle\mathcal{V}_\odot\rangle_{\ell,\theta_\odot}$, yet it is still of three degrees of freedom.

Nevertheless, in time scales of several tens of years, it is meaningful to remove the longitude of the node of the Moon, whose period is about 18.6 years, by a new Lie transformation to triple-prime variables. Note that the validity of this new transformation will not apply to resonances with either the node or the perigee of the satellite, which, besides, are assumed to evolve in a much longer scale.

The Lie derivative now includes a partial derivation with respect to N, which is in fact the only variable with respect to which the homological equation needs to be solved. At first order we choose $\mathcal{T}_{0,1} = \langle\mathcal{V}_\oplus\rangle_\ell + \langle\mathcal{V}_☾\rangle_{\ell,\ell_☾,N} + \langle\mathcal{V}_\odot\rangle_{\ell,\theta_\odot}$, where

$$\langle\mathcal{V}_☾\rangle_{\ell,\ell_☾,N} = \frac{\mu}{2a}\frac{n_☾^2}{n^2}\frac{\chi_☾}{\eta_☾^3}(3s_J^2 - 2)\frac{1}{16}[(2 + 3e^2)(2 - 3s_i^2) + 15e^2s_i^2\cos 2g], \tag{8.60}$$

and compute the generating function

$$\begin{aligned}\mathcal{S} = L\chi_☾ \frac{n_☾/n}{n_N/n_☾}\frac{s_J}{\eta_☾^3}\frac{3}{128}\{&8c_Js_i[5e^2(1 - c_i)\sin(\Delta - 2g) + (4 + 6e^2)c_i \\ &\times \sin\Delta - 5e^2(1 + c_i)\sin(\Delta + 2g)] + s_J[5e^2(1 - c_i)^2\sin(2\Delta - 2g) \\ &+ (4 + 6e^2)s_i^2\sin 2\Delta + 5e^2(1 + c_i)^2\sin(2\Delta + 2g)]\}.\end{aligned}$$

Note that Eq. (8.60) is completely analogous to the double-averaged disturbing potential of the Sun in Eq. (8.59). This is not a surprise and simply shows that, save for the periodic corrections that materialize the Lie transformation, removing the node of the Moon is equivalent to neglecting the small inclination of the Moon's orbit over the ecliptic in Eq. (8.40). Therefore, in the long term, the lunisolar disturbing potential can be combined into a single third-body perturbation,

$$\langle\mathcal{V}_\star\rangle = -\frac{\mu}{2a}\frac{n_\star^2}{n^2}\frac{1}{8}[(2 + 3e^2)(2 - 3s_i^2) + 15e^2s_i^2\cos 2g], \tag{8.61}$$

in which

$$n_\star = n_\odot\sqrt{1 + \frac{n_☾^2}{n_\odot^2}\frac{\chi_☾}{\eta_☾^3}\left(1 - \frac{3}{2}s_J^2\right)} \approx 1.8n_\odot. \tag{8.62}$$

Since the extended phase space is no longer needed, we remove the terms $L''_{☾} n_{☾}$, $H''_{☾} n_N$, and $\Theta''_{\odot} n_{\odot}$ from the final Hamiltonian. After changing double- by triple-prime variables, it results in the two-degrees-of-freedom Hamiltonian

$$\mathcal{Q} = \mathcal{Q}(g''', h''', G''', H''') \equiv -\frac{\mu}{2a} + \langle \mathcal{V}_{\oplus} \rangle_{\ell} + \langle \mathcal{V}_{\star} \rangle, \tag{8.63}$$

with the oblateness perturbation $\langle \mathcal{V}_{\oplus} \rangle_{\ell}$ and the third-body perturbation $\langle \mathcal{V}_{\star} \rangle$ given in Eqs. (8.52) and (8.61), respectively.

8.5 Kudielka's balanced orbits

While the Hamiltonian flow stemming from Eq. (8.63) still can be seen as made of distorted ellipses, it is no longer a Hamiltonian perturbation problem. Indeed, the Keplerian is just a constant term and we do not have an analogous term depending only on momenta which a new Lie transformation could hinge on. Moving to a rotating frame could give remedy to this situation [542]. However, we do not pursue this approach here, which is delayed to the third part of this monograph, and only try to get some insight on the reduced problem by finding equilibria solutions of the flow in the (g''', h''', G''', H''') reduced phase space.

Simplifying notation, we will do without the primes in what follows. Thus, after neglecting constant terms and scaling Hamiltonian (8.63), we get the nondimensional Hamiltonian

$$\widetilde{\mathcal{Q}} = (2 + 3e^2)(2 - 3s_i^2) + 15e^2 s_i^2 \cos 2g + \frac{8}{3}\frac{\rho^2}{\eta^3}[3\cos^2 I(h) - 1], \tag{8.64}$$

where $\widetilde{\mathcal{Q}} = -8(n/n_{\star})^2 - 16\mathcal{Q}/(a^2 n_{\star}^2)$, $\cos I$ yields the dependence on $h = \Omega_E$ from (8.43), and

$$\rho = \left(\frac{3}{2} J_2\right)^{1/2} \frac{R_{\oplus}/a}{n_{\star}/n}, \tag{8.65}$$

with $n_{\star}$ given in Eq. (8.62). The parameter $\rho = \rho(a)$ is of order one for the range in which Hamiltonian (8.64) is representative of the dynamics, and, as illustrated in Fig. 8.4, is an indicator of the strength of the oblateness perturbation relative to the lunisolar disturbing effect. The coefficient $\frac{3}{2}$ is chosen trying to give the same weight when $\rho = 1$ to oblateness and lunisolar perturbations for the different arguments of the node and perigee as well as possible inclinations, on average. Still, different choices have been done in the case of Earth orbits [372, 581]. Other choices could be more adequate in third-body systems whose obliquity is different for that of the ecliptic [389].

The simple model in by Eq. (8.64) shows the existence of high-altitude and highly inclined orbits with fixed node and perigee, on average, yielding orbits with frozen eccentricity when the right ascension of the ascending node is 0 or π. These "balanced"

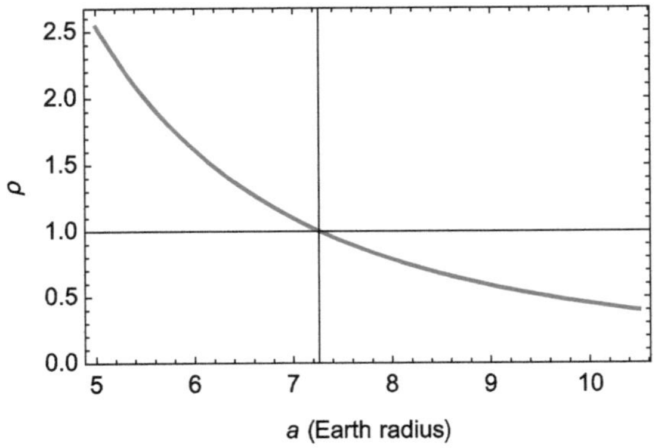

Figure 8.4: Ratio of oblateness lunisolar perturbations $\rho = \rho(a)$ in Eq. (8.65).

orbits survive with minor variations in the full, non-simplified dynamics [371]. Indeed, to avoid the singularities of the Delaunay variables for circular orbits, we replace the pair of Delaunay variables g, G by the semi-equinoctial elements C, S. Thus, taking into account that L is constant,

$$\frac{dC}{dt} = -\frac{\eta^2}{e}\frac{1}{G}\frac{dG}{dt}\cos\omega - e\sin\omega\frac{dg}{dt}, \tag{8.66}$$

$$\frac{dS}{dt} = -\frac{\eta^2}{e}\frac{1}{G}\frac{dG}{dt}\sin\omega + e\cos\omega\frac{dg}{dt}, \tag{8.67}$$

where the appearance of the eccentricity in denominators is canceled by corresponding factors appearing in the mean variation of the total angular momentum.

Therefore, from Hamilton equations, and in the time scale $t' = -\frac{1}{16}n_\star^2 a^2 t$ of Hamiltonian (8.64), we obtain

$$\begin{aligned}\frac{dh}{dt'} &= \frac{4}{np^2}\Big\{3\eta^3 c_i(1 - C^2 + 4S^2) + 2\rho^2 \\ &\quad \times \Big[(2 - 3s_\varepsilon^2)c_i - 2\frac{1 - 2c_i^2}{s_i}c_\varepsilon s_\varepsilon \cos h - s_\varepsilon^2 c_i \cos 2h\Big]\Big\},\end{aligned} \tag{8.68}$$

$$\frac{dH}{dt'} = -16\frac{\rho^2}{\eta^3}s_\varepsilon s_i(c_\varepsilon c_i - s_\varepsilon s_i\cos h)\sin h, \tag{8.69}$$

$$\frac{dC}{dt'} = \frac{4S}{np^2}\{3\eta^3[3S^2 - (3 - 5c_i^2)(1 - C^2)] + \rho^2 R\}, \tag{8.70}$$

$$\frac{dS}{dt'} = -\frac{4C}{np^2}\{3\eta^3[2(1 - C^2) + (3 - 5s_i^2)S^2] + \rho^2 R\}, \tag{8.71}$$

in which $c_i = H/(L\eta)$, $\eta = \sqrt{1 - C^2 - S^2}$, and $R = R(C, S, h, H)$ is

$$R = (1 - 3c_\varepsilon^2)(1 - 5c_i^2) + 4c_\varepsilon s_\varepsilon\frac{c_i}{s_i}(1 - 5s_i^2)\cos h + s_\varepsilon^2(3 - 5c_i^2)\cos 2h.$$

Note that, except for Eq. (8.69), the variation equations are singular for ecliptic orbits. This singularity is not essential and can be avoided using an alternative set of nonsingular variables.

8.5.1 The manifold of circular orbits

When $C = S = 0$ Eqs. (8.70) and (8.71) vanish and, therefore, circular orbits form an invariant manifold of Hamiltonian (8.64). Because of that, the Hamiltonian dynamics of this particular manifold $G = L$ can be obtained making $e = 0$ and $\eta = 1$ in Eq. (8.64). On that manifold,

$$\mathfrak{C} = 4 - 6s_i^2 + 4\rho^2\left[\left(s_\varepsilon^2 - \frac{2}{3}\right)(3s_i^2 - 2) - 4s_\varepsilon c_\varepsilon s_i c_i \cos h + s_\varepsilon^2 s_i^2 \cos 2h\right], \tag{8.72}$$

which is a one-degree-of-freedom Hamiltonian in the conjugate pair (h, H) depending on the dynamical parameter $G = L$.

Greater insight is obtained when the flow is described in terms of the average inclination rather than H. Thus, in the time new scale $t'' = 16t'/(na^2) = -(n_\star^2/n)t$, the reduced flow is obtained from the integration of the variations of $i = \arccos(H/G)$ and $h = \Omega_E$. On account of $\mathrm{d}i/\mathrm{d}t'' = -1/(Gs_i)\mathrm{d}H/\mathrm{d}t''$, from the Hamilton equations of Eq. (8.72), we obtain

$$\frac{\mathrm{d}i}{\mathrm{d}t''} = \rho^2(c_\varepsilon c_i - s_\varepsilon s_i \cos\Omega_E)s_\varepsilon \sin\Omega_E, \tag{8.73}$$

$$\frac{\mathrm{d}\Omega_E}{\mathrm{d}t''} = \frac{3}{4}c_i + \rho^2\left[c_\varepsilon^2 c_i - \left(c_\varepsilon \frac{1 - 2c_i^2}{s_i} + s_\varepsilon c_i \cos\Omega_E\right)s_\varepsilon \cos\Omega_E\right], \tag{8.74}$$

whose solution depends on special functions and, therefore, does not provide too much insight on the dynamics. Therefore, rather than pursuing the general solution of Eqs. (8.73)–(8.74) we search only for their equilibria.

Ignoring the case of ecliptic orbits, which cannot be studied in orbital elements, it is found that the variation of the inclination with respect to the ecliptic vanishes in three cases: $\Omega_E = 0$, $\Omega_E = \pi$, and $\cos\Omega_E = -\cot\varepsilon \cot i$. The inclination value required to get an equilibrium is then obtained by imposing a null variation of the argument of the node for each of these Ω_E particular values.

Thus, replacing $\cos\Omega_E = -\cot\varepsilon\cot i$ in the right side of Eq. (8.74), we find that it vanishes for $i = \frac{\pi}{2}$, which in turn requires that $\Omega_E = \pm\frac{1}{2}\pi$. Note that these particular values of the node and inclination are exactly the same as when referred to the equator.

Also, replacing $\Omega_E = 0$ in the right side of Eq. (8.74), the equilibrium condition becomes $\rho^2 = -\frac{3}{4}\sin 2i/\sin(2\varepsilon + 2i)$, from which

$$i = \frac{1}{2}\left(-\arctan\frac{4\rho^2 \sin 2\varepsilon}{3 + 4\rho^2\cos 2\varepsilon} + k\pi\right), \tag{8.75}$$

with k integer. Analogously, $\Omega_E = \pi$ results in $\rho^2 = \frac{3}{4}\sin 2i/\sin(2\varepsilon - 2i)$, from which

$$i = \frac{1}{2}\left(\arctan\frac{4\rho^2\sin 2\varepsilon}{3 + 4\rho^2\cos 2\varepsilon} + k\pi\right). \tag{8.76}$$

An illustration of the evolution of the inclination of this equilibrium for the different values of the ratio of oblateness lunisolar perturbations is depicted in Fig. 8.5. In the limit $\rho \to \infty$, in which the oblateness perturbation dominates the dynamics, if $\Omega_E = \pi$ then $i = \varepsilon$ ($k \bmod 2 = 0$), or $i = \frac{\pi}{2} + \varepsilon$ ($k \bmod 2 = 1$); if $\Omega_E = 0$, then $i = \pi - \varepsilon$ ($k \bmod 2 = 0$), or $i = \frac{\pi}{2} - \varepsilon$ ($k \bmod 2 = 1$). In the other limit $\rho \to 0$, in which the third body dominates, the inclination with respect to the ecliptic becomes polar in both cases for $k \bmod 2 = 1$; for $k \bmod 2 = 0$, $i = 0$ when $\Omega_E = \pi$, and $i = \pi$ when $\Omega_E = 0$.

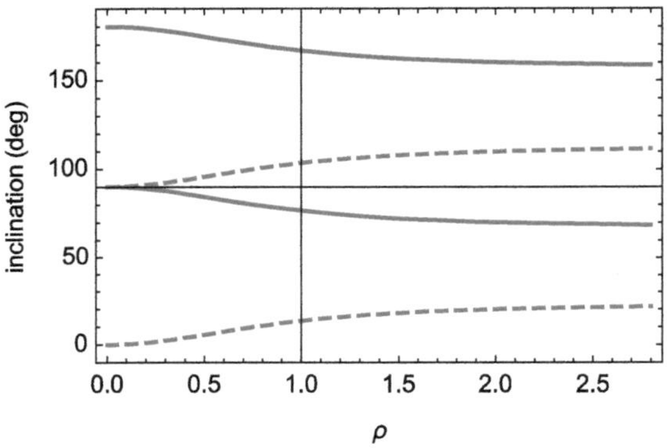

Figure 8.5: Ecliptic inclination of the equilibria at $\Omega_E = 0$ (full line) and $\Omega_E = \pi$ (dashed line) as a function of the ratio of oblateness lunisolar perturbations ρ.

When referred to the equator the node has the same values as Ω_E, but those of the inclination are $I = \varepsilon \pm i$ as follows from Eq. (8.43)—the plus sign (resp. minus) for $\Omega_E = 0$ (resp. $\Omega_E = \pi$), thus yielding some kind of a double vertical-horizontal reflection of Fig. 8.5.

On the other hand, the Hamiltonian flow is readily visualized without need of integrating Eqs. (8.73)–(8.74) by depicting contour plots of Eq. (8.72) for different values $\mathfrak{C}$ = constant. Again, instead of using the (h, H) representation, it is more descriptive to display the flow in the (Ω_E, i) plane, in which it is parameterized by ρ. An example for $\rho = 1$ is presented in Fig. 8.6, where we observe four fixed points of the elliptic type: two at $\Omega_E = 90°, 270°$, and $i = 90°$, and the other two at $\Omega_E = 0$, $i = 166.53°$, and $\Omega_E = 180°$, $i = 13.47°$. Besides, we note two fixed points of the hyperbolic type at $\Omega_E = 0$, $i = 76.53°$, and $\Omega_E = 180°$, $i = 103.47°$, which are joined by heteroclinic connections that separate rotation and libration regions. As observed in Fig. 8.6, orbits in these libration regions, as well as circulation orbits close to the boundary, change periodically from direct to retrograde inclinations and vice versa.

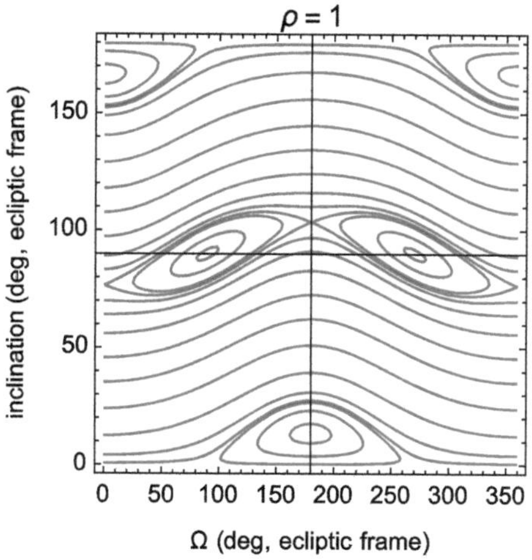

Figure 8.6: Contour plot of the circular-orbit Hamiltonian (8.72) for $a \approx 6.7$ Earth radius.

The stability character of the equilibria can be guessed from the graphic. Alternatively, the stability can be computed analytically from the usual linearization of the flow. Thus,

$$\begin{aligned}\frac{\mathrm{d}\delta i}{\mathrm{d}t'} &= \rho^2 s_\varepsilon [s_\varepsilon s_i + (c_\varepsilon c_i - 2 s_\varepsilon s_i \cos h)\cos h]\delta h \\ &\quad - \rho^2 s_\varepsilon (c_\varepsilon s_i + s_\varepsilon c_i \cos h)\sin h\,\delta i, \\ \frac{\mathrm{d}\delta h}{\mathrm{d}t'} &= \left\{-\frac{3 s_i}{4} - \rho^2\left[s_\varepsilon\left(c_\varepsilon \frac{c_i}{s_i^2} + 2c_\varepsilon c_i - s_\varepsilon s_i \cos h\right)\cos h + c_\varepsilon^2 s_i\right]\right\}\delta i \\ &\quad + \rho^2 \frac{s_\varepsilon}{s_i}[c_\varepsilon(1 - 2c_i^2) + 2 s_\varepsilon c_i s_i \cos h]\sin h\,\delta h.\end{aligned}$$

Then, when $\Omega_E = 90°$, $i = 90°$ we get the characteristic matrix

$$\begin{pmatrix} -\rho^2 c_\varepsilon s_\varepsilon & \rho^2 s_\varepsilon^2 \\ -\frac{3}{4} - \rho^2 c_\varepsilon^2 & \rho^2 c_\varepsilon s_\varepsilon \end{pmatrix} \tag{8.77}$$

whose eigenvalues $\pm\mathbf{i}\frac{\sqrt{3}}{2}\rho \sin\varepsilon$ are purely imaginary, and, therefore, this equilibrium is always stable on the invariant manifold of circular orbits. On the other hand, the stability in the out-of-manifold direction is computed from the remaining variational equations of C and S. In this case, we get the matrix

$$\begin{pmatrix} 0 & -\frac{9}{4} - \frac{1}{2}\rho^2 \\ -\frac{3}{2} + \frac{1}{2}\rho^2 & 0 \end{pmatrix}, \tag{8.78}$$

whose eigenvalues $\pm\frac{1}{2}(\frac{9}{2} + \rho^2)^{1/2}(3 - \rho^2)^{1/2}$ are real if $\rho < \sqrt{3} \approx 1.7$, a case in which these balanced orbits are unstable.

Proceeding analogously, when $\Omega_E = 0^\circ$, we get the eigenvalues

$$\pm\rho\frac{\sqrt{\sin\varepsilon\cos(\varepsilon+i)}}{4\sin i}\sqrt{4\rho^2[\sin(2\varepsilon+3i)-3\sin(2\varepsilon+i)]-9\sin i+3\sin 3i}.$$

For Earth's orbits, when i is given by Eq. (8.75) we find that the eigenvalues are purely imaginary for any value of ρ when $k = 0$, whereas they are real if $k = 1$. On the other hand, for $\Omega_E = \pi$, the eigenvalues are the same as above changing i by $-i$. Now, the inclination with respect to the ecliptic is given by Eq. (8.76), a case in which we find purely imaginary eigenvalues for $k = 0$ and real eigenvalues for $k = 1$, again, particularized for the Earth case.

In both stable cases we always find stability also in the out-of-manifold direction. On the contrary, in both unstable cases we find instability in the out-of-manifold direction only when ρ is smaller than $\rho \approx 1.7$.

8.5.2 The manifold of polar orbits orthogonal to the equinox

When $H = 0$ ($i = 90^\circ$) and $h = \pm\frac{\pi}{2}$, their variations in Eqs. (8.68) and (8.69) identically vanish. Then the set of polar orbits orthogonal to the equinox constitute also an invariant manifold $\mathfrak{P} = \mathfrak{P}(g, G)$ of Hamiltonian (8.64). That is,

$$\mathfrak{P} = -2 - 3e^2 - \frac{8\rho^2}{3\eta^3} + 15e^2\cos 2g. \tag{8.79}$$

The flow on this manifold is obtained from the reduced system $\mathrm{d}g/\mathrm{d}t' = \partial\mathfrak{P}/\partial G$, $\mathrm{d}G/\mathrm{d}t' = -\partial\mathfrak{P}/\partial g$, which is better written in the non-canonical semi-equinoctial variables

$$\frac{\mathrm{d}C}{\mathrm{d}t'} = -\frac{2\rho^2 + 9\eta^5}{4\eta^4}S, \qquad \frac{\mathrm{d}S}{\mathrm{d}t'} = \frac{\rho^2 - 3\eta^5}{2\eta^4}C, \tag{8.80}$$

where $\eta = \sqrt{1 - C^2 - S^2}$. It is easy to check that this invariant manifold accepts three different equilibria; namely, circular orbits ($C = S = 0$), and elliptic orbits with eccentricity $e = \sqrt{1 - 3^{-2/5}\rho^{4/5}}$ (derived from the condition $\rho^2 = 3\eta^5$) and argument of the perigee either 0 or π (derived from the condition $S = 0$). Thus, while circular orbits of this invariant manifold are always equilibria, the elliptic orbits with the line of apsides on the ecliptic exist only when $\rho < \sqrt{3}$. The evolution of the eccentricity of these balanced orbits with respect to the ratio of oblateness lunisolar perturbations is illustrated in Fig. 8.7.

The stability of the equilibria is obtained from the linearization of the flow given by Eq. (8.80). For elliptic orbits $S = 0$, $C = e$, $\rho^2 = 3\eta^5$, we obtain the characteristic matrix

$$\begin{pmatrix} 0 & -\frac{15}{4}\eta \\ \frac{15}{2}e^2\eta^{-1} & 0 \end{pmatrix},$$

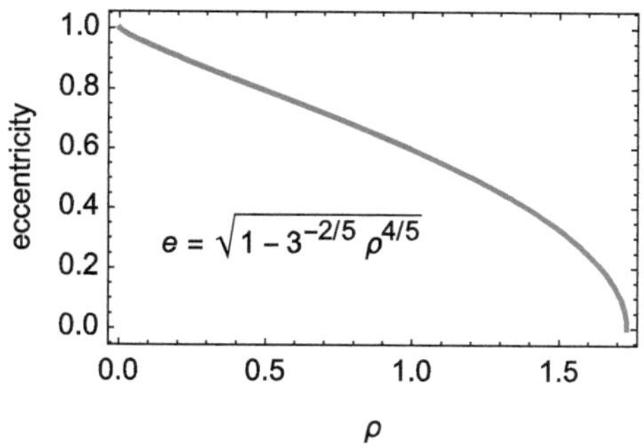

Figure 8.7: Eccentricity of balanced orbits in the plane orthogonal to the equinox as a function of the ratio of oblateness lunisolar perturbations ($\omega_E = 0, \pi$).

whose eigenvalues are purely imaginary $\pm\frac{15}{\sqrt{8}}\mathbf{i}\,e$, showing the stable character of the eccentric orbits on this manifold.

On the other hand, the case of circular orbits ($S = C = 0$, $\eta = 1$) with $\Omega_E = \pm\frac{\pi}{2}$ is one of the particular solutions of the invariant manifold of circular orbits that was already discussed in §8.5.1. The analysis carried out there applies here too with the simple expedient of interchanging the roles of the in- and out-of-manifold directions. Therefore, the eigenvalues are those of Eq. (8.78), which vanish when $\rho = \sqrt{3}$ and are purely imaginary for higher values of ρ. This behavior shows that elliptic balanced orbits bifurcate from circular in a change of stability of the latter, as illustrated in Fig. 8.8, which has been depicted by direct evaluation of the Hamiltonian (8.79) without the need of integrating the differential system (8.80).

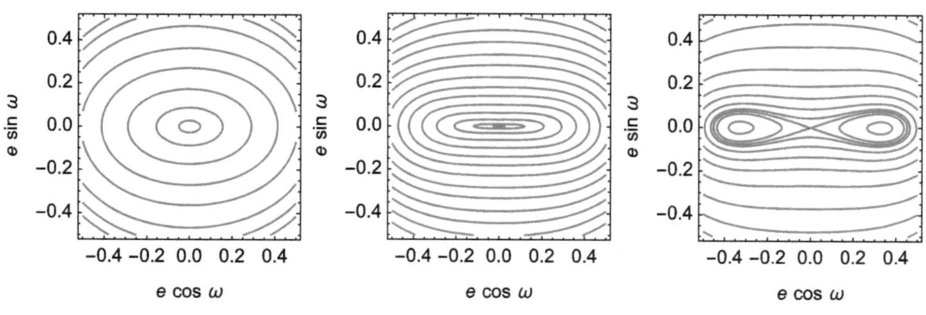

Figure 8.8: Balanced orbits in the plane orthogonal to the equinox from $\rho = \frac{5}{2}$ (left), $\rho = \sqrt{3}$ (center), and $\rho = \frac{3}{2}$ (right plot).

Finally, the eigenvalues of Eq. (8.77) show that circular orbits are always stable in the out-of-manifold direction—contrary to the weak stability obtained when the analysis is performed in the equatorial frame; cf. [372].

8.5.3 Other equilibria

For Earth orbits, an additional equilibrium of Eqs. (8.68)–(8.71) occurs when the perigee lies on the ecliptic ($S = 0, C = e$) in the direction of the equinox ($\Omega_E = 0$). Then Eqs. (8.69) and (8.70) identically vanish, whereas Eq. (8.68) yields the condition $\rho^2 = -\frac{3}{4}\eta^5 \sin 2i/\sin(2\varepsilon + 2i)$, which requires that either $i < 90^\circ$ and $\varepsilon > 90^\circ$, or $i > 90^\circ$ and $\epsilon < 90^\circ$. This equation in turn is plugged into Eq. (8.71) to give the equilibrium condition

$$0 = 2\sin 2i - \sin 2\varepsilon - 12\sin(2i + 2\varepsilon) + 5\sin(4i + 2\varepsilon). \tag{8.81}$$

To get this solution in the equatorial frame, like originally obtained in [372], we only need to replace $i = I - \varepsilon$, as obtained from Eq. (8.44) when $\Omega = \Omega_E = 0$. Both cases are shown superimposed in Fig. 8.9 for $0 \leq \varepsilon < \pi$. For the Earth the obliquity of the ecliptic is $\varepsilon = 23.4$, yielding $i = 139.42^\circ$ with respect to the ecliptic, and $I = 94.55^\circ$ with respect to the equator.

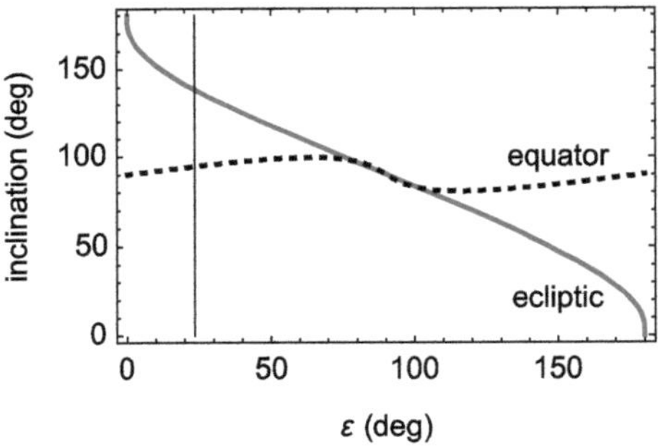

Figure 8.9: Inclination of balanced orbits with respect to the ecliptic (gray line, $\Omega_E = \omega_E = 0$) and to the equator (black-dashed line, $\Omega = \omega = 0$).

We do not find other balanced orbits for the Earth, yet they may be found for other planets; cf. [372]. In general, the flow must be integrated numerically from Eqs. (8.68)–(8.71).

9 Non-conservative effects

Some configurations of Earth's satellites are clearly affected by non-conservative effects that do not admit a Hamiltonian formulation. This is the case of the solar-radiation pressure (SRP) and the atmospheric drag. However, in different instances, the cannonball approximation may be a reasonable assumption when the aim is long-term orbit prediction. This simplification allows one to derive the SRP acceleration from a "potential" function with minor additional simplifications. Then the long-term effects of SRP are approached like a case of Hamiltonian perturbations.

The Hamiltonian approach to SRP perturbations also serves to illustrate how time-dependency issues can be avoided without need of moving to the extended phase space by the simple expedient of formulating the problem in a synodic ecliptic frame. It also shows that the vectorial formulation immediately discloses the integrability of the SRP evolution equations, trivialize its achievement, and directly characterizes the geometric nature of the motion [159, 480]. From the vectorial formulation and the standard differentiation of a vector in a moving frame, setting the problem in the inertial equatorial frame is straightforward. The latter includes the time dependency in the long-term Hamiltonian, and is the usual formulation when the radiation pressure is not the dominant perturbation.

On the contrary, the case of perturbations due to atmospheric drag, which are quite relevant in the propagation of satellite orbits with low perigee altitudes, cannot be approached in the Hamiltonian setting. Nevertheless, perturbation methods are not constrained to the case of Hamiltonian flows and widely apply to the case of vectorial flows, the method of Lie transforms not being an exception [282, 304, 333]. Since a detailed description of these methods falls outside the scope of this monograph, the dissipative effects of the atmospheric drag are formulated using the classical approach based on the Gauss variation equations [462]. The same solution would, of course, be obtained if approached by vectorial Lie transforms up to first-order effects. These non-Hamiltonian forces are treated like generalized forces in the usual mean-element orbit propagators [436].

9.1 Solar-radiation pressure

Objects irradiated by the light energy of the Sun experience a pressure whose magnitude is inversely proportional to the square of the distance to the Sun. Assuming that there is no re-radiation from Earth, the perturbing acceleration caused by solar-radiation pressure (SRP) is [506]

$$\boldsymbol{\alpha}_{\mathrm{srp}} = -P_\odot(A/m)(a_\odot/d_\odot)^2(\boldsymbol{e}_\odot\cdot\boldsymbol{n})\big[(1-\beta)\boldsymbol{e}_\odot + 2\beta(\boldsymbol{e}_\odot\cdot\boldsymbol{n})\,\boldsymbol{n}\big],$$

where $P_\odot$ is the solar-radiation pressure constant at one astronomical unit (AU), A/m is the area-to-mass ratio of the spacecraft, $d_\odot = \|\boldsymbol{r}_\odot - \boldsymbol{r}\|$ is the distance between the

https://doi.org/10.1515/9783110668513-009

Sun and the irradiated object, $a_\odot$ is the semi-major axis of the Sun's apparent orbit around Earth, $\boldsymbol{e}_\odot$ is the Sun direction from the satellite, $\boldsymbol{n}$ is the normal to the radiated surface, and β is the index of reflection ($0 < \beta < 1$).

For common artificial satellites this effect is small, and can be treated as a perturbation of the Keplerian motion. Still, the SRP effect on objects with high area-to-mass ratio may have great importance, to the extent of preventing the perturbation approach. We only deal with the first case, and approach the SRP problem by Lie transforms.

9.1.1 The disturbing SRP "potential"

For the sake of orbit long-term prediction using perturbation methods, common additional simplifications are that the solar flux is constant along the orbit of the satellite, which allows one to replace $d_\odot$ by the radius of the apparent orbit of the Sun around the Earth $r_\odot$, and that the parallax of the Sun is negligible, which allows one to replace $\boldsymbol{e}_\odot$ by the Sun geocentric direction $\hat{\boldsymbol{r}}_\odot$ [361]. If, besides, we assume that the solar panels remain oriented to the Sun or the satellite is a sphere (cannonball approximation), then $\boldsymbol{e}_\odot \equiv \boldsymbol{n}$, and hence $\boldsymbol{\alpha}_{\text{srp}} = -F_{\text{srp}}\hat{\boldsymbol{r}}_\odot$, whose magnitude is $F_{\text{srp}} = P_\odot(A/m)(a_\odot/r_\odot)^2(\beta+1) > 0$, which may further be taken as constant.

When these simplifications are allowed, in spite of SRP being a non-conservative effect, the acceleration it produces can be derived from a scalar function like $\boldsymbol{\alpha}_{\text{srp}} = -\nabla_{\boldsymbol{r}}\mathcal{S}$, where

$$\mathcal{S} = F_{\text{srp}}\hat{\boldsymbol{r}}_\odot \cdot \boldsymbol{r}. \tag{9.1}$$

Because $\mathcal{S}$ plays the role of a potential, the perturbation treatment of SRP can be approached by canonical perturbation theory.

For Earth satellite orbits, SRP perturbations are customarily referred to the equatorial frame. However, as discussed in §8.4 and §8.5 for third-body perturbations, the formulation is simpler and provides the highest insight when referred to the ecliptic, like it was done in early studies related to zodiacal light [549].

9.1.2 Hamiltonian short-period reduction in the synodic frame

The perturbed Keplerian flow stemming from Eq. (9.1) is of three degrees of freedom and, due to the dependence of $\hat{\boldsymbol{r}}_\odot$ on the ecliptic longitude of the Sun $\lambda_\odot$, time-dependent. Nevertheless, the explicit appearance of time is removed when choosing a synodic frame of reference rotating with the mean motion of the Sun $n_\odot$, in which, besides, the axis of abscissas is customarily aligned with the direction of radiation pressure [104]. Thus, we deal with a perturbation Hamiltonian (2.30) in which $\mathcal{H}_{0,0}$

is the Keplerian, and, using Delaunay variables, $\mathcal{H}_{1,0} = -n_\odot H + \mathcal{S}(\ell, g, h, L, G, H)$. The Coriolis term $-n_\odot H$ keeps the Hamiltonian structure of the dynamical model when the longitude of the node in the synodic frame $h = \Omega_E - \lambda_\odot$ is used. Now, $g = \omega_E$ and $c_i = H/G$.

Like in §8.1.2, the short-period effects are better isolated in the apsidal frame $(O, \hat{\boldsymbol{e}}, \hat{\boldsymbol{b}}, \boldsymbol{n})$ defined in §4.5.1, where we recall that $\boldsymbol{n}$ is the direction of the angular momentum vector $\boldsymbol{G} = G\boldsymbol{n}$, $\hat{\boldsymbol{e}}$ is the direction of the eccentricity vector $\boldsymbol{e} = e\hat{\boldsymbol{e}}$, and $\hat{\boldsymbol{b}} = \boldsymbol{n} \times \hat{\boldsymbol{e}}$. Therefore, replacing Eqs. (8.5) and (8.10) into Eq. (9.1), we write the disturbing potential like

$$\mathcal{H}_{1,0} = -n_\odot \boldsymbol{k}_\odot \cdot \boldsymbol{G} + F_{\mathrm{srp}} a\left[(\hat{\boldsymbol{e}} \cdot \hat{\boldsymbol{r}}_\odot)(\cos u - e) + (\hat{\boldsymbol{b}} \cdot \hat{\boldsymbol{r}}_\odot)\eta \sin u\right], \tag{9.2}$$

where $\boldsymbol{k}_\odot$ is the normal direction to the ecliptic. The components of $\boldsymbol{G}$ and $\boldsymbol{e}$ in the synodic frame, when required, are computed using Eqs. (4.52) and (4.53), respectively, where I is replaced by the inclination with respect to the ecliptic, and h is measured in the synodic frame.

The short-period effects due to SRP are removed from the perturbation Hamiltonian by carrying out a Lie transformation to prime variables. The Lie derivative is, once more, Eq. (4.66), and, like in the case of third-body perturbations, the homological equation is (4.70).

The term $\mathcal{H}_{0,1}$ is chosen such that it cancels the terms pertaining to the kernel out of Eq. (9.2). That is, $\mathcal{H}_{0,1} = \langle \widetilde{\mathcal{H}}_{0,1} \rangle_\ell = \langle \mathcal{H}_{1,0}(r/a) \rangle_u$, from which

$$\mathcal{H}_{0,1} = -n_\odot (\boldsymbol{k}_\odot \cdot \boldsymbol{G}) - \frac{3}{2} F_{\mathrm{srp}} a(\hat{\boldsymbol{r}}_\odot \cdot \boldsymbol{e}). \tag{9.3}$$

Then the homological equation is analogously solved to obtain

$$\mathcal{W}_1 = \frac{F_{\mathrm{srp}}}{4} \frac{a}{n} \{[(4 - 2e^2)\sin u - e \sin 2u]\hat{\boldsymbol{e}} - \eta(4\cos u - e\cos 2u)\hat{\boldsymbol{b}}\} \cdot \hat{\boldsymbol{r}}_\odot, \tag{9.4}$$

with misprints in [159] amended, which is used in the computation of the first row of Eq. (2.17) to obtain the short-period corrections of the transformation.

After truncation to first-order effects, the new Hamiltonian is obtained changing original by prime variables in Eq. (9.3). Disregarding the constant Keplerian term, the long-term Hamiltonian adopts the strikingly simple form

$$\mathcal{K}_1 = -n_\odot (L' \boldsymbol{\eta} \cdot \boldsymbol{k}_\odot) - n_{\mathrm{srp}} (L' \boldsymbol{e} \cdot \hat{\boldsymbol{r}}_\odot), \tag{9.5}$$

where we recall that $\boldsymbol{\eta} = \eta \boldsymbol{n}$, from Eq. (4.59), and $n_{\mathrm{srp}} = n_{\mathrm{srp}}(L')$ is the frequency

$$n_{\mathrm{srp}} = \frac{3}{2} F_{\mathrm{srp}} \frac{a}{L'}. \tag{9.6}$$

The coefficients of the reduced Hamiltonian (9.5) are intrinsic physical magnitudes representing the component of the angular momentum vector normal to the ecliptic, and the projection of the eccentricity vector (or the Laplace vector $\boldsymbol{A} = L'\boldsymbol{e}$) in the direction of the radiation source.

On account of $\hat{\boldsymbol{r}}_\odot \cdot \boldsymbol{e} = e(\cos g' \cos h' - c_i \sin g' \sin h')$, from Eq. (4.53), $\boldsymbol{k}_\odot \cdot \boldsymbol{\eta} = \eta c_i$, from Eq. (4.52), and L' is constant, the Hamiltonian (9.5) has two degrees of freedom.

9.1.3 Particular solutions

The Hamilton equations are reformulated like the variations of the semi-equinoctial variables in Eq. (8.66),

$$\frac{dC}{dt} = \frac{n_{\text{srp}}}{\eta}(1 - C^2)c_i \sin h', \qquad \frac{dS}{dt} = \frac{n_{\text{srp}}}{\eta}(\eta^2 \cos h' - CSc_i \sin h'),$$

and, like we did in §8.5.1, the longitude of the node,

$$\frac{dh}{dt} = -n_\odot + \frac{n_{\text{srp}}}{\eta} S \sin h',$$

and the inclination

$$\frac{di}{dt} = \frac{n_{\text{srp}}}{\eta} Cs_i \sin h',$$

which show that, on average, orbits initially lying on the plane orthogonal to the ecliptic ($i = \frac{\pi}{2}$), will remain in this plane when the initial node and perigee are equal and take the value $g' = h' = \pm\frac{\pi}{2}$ ($C = 0, S = e$), and the eccentricity of the orbit is $e = e^* \equiv n_\odot/(n_\odot^2 + n_{\text{srp}}^2)^{1/2}$.

On the other hand, the variation of the inclination vanishes for orbits initially resting on the ecliptic ($i = 0, \pi$), showing that, on average, planar, ecliptic orbits form an invariant manifold of Hamiltonian (9.5), in which $|H'| = G' = \Theta$ and $\hat{\boldsymbol{r}}_\odot \cdot \boldsymbol{e} = e \cos\theta$. Since (θ, Θ) are canonical conjugate variables, the dynamics on the coplanar manifold is derived from the one-degree-of-freedom Hamiltonian

$$\mathcal{K}_E = -n_\odot \Theta - n_{\text{srp}} L' e \cos\theta, \tag{9.7}$$

in which $e = \sqrt{1 - \Theta^2/L'^2}$, whose Hamilton equations

$$\frac{d\theta}{dt} = \frac{\partial \mathcal{K}_E}{\partial \Theta} = -n_\odot + n_{\text{srp}} \frac{\eta}{e} \cos\theta, \qquad \frac{d\Theta}{dt} = -\frac{\partial \mathcal{K}_E}{\partial \theta} = -n_{\text{srp}} Le \sin\theta,$$

show that $\theta = 0$, $e = n_{\text{srp}}/(n_\odot^2 + n_{\text{srp}}^2)^{1/2}$ is also an equilibrium of the averaged system. Orbits of this manifold are readily obtained solving θ from Eq. (9.7),

$$\cos\theta = -\frac{\eta + \mathcal{K}_E^*}{(n_{\text{srp}}/n_\odot)e}, \qquad \mathcal{K}_E^* = \frac{\mathcal{K}_E(\theta_0, \Theta_0)}{L' n_\odot}.$$

A sample phase space of Hamiltonian (9.7) scaled by $L'n_\odot$ is depicted in the left plot of Fig. 9.1 for the particular case $n_{\text{srp}}/n_\odot = \tan\frac{\pi}{6}$ [495]. The right plot provides the typical eccentricity-vector diagram of the same case. Dashed lines highlight the limit case for librational motion of the perigee. Between both dashed lines the perigee circulates about the 360 deg, while outside both dashed lines the orbits reach the limit eccentricity 1 before completing a cycle.

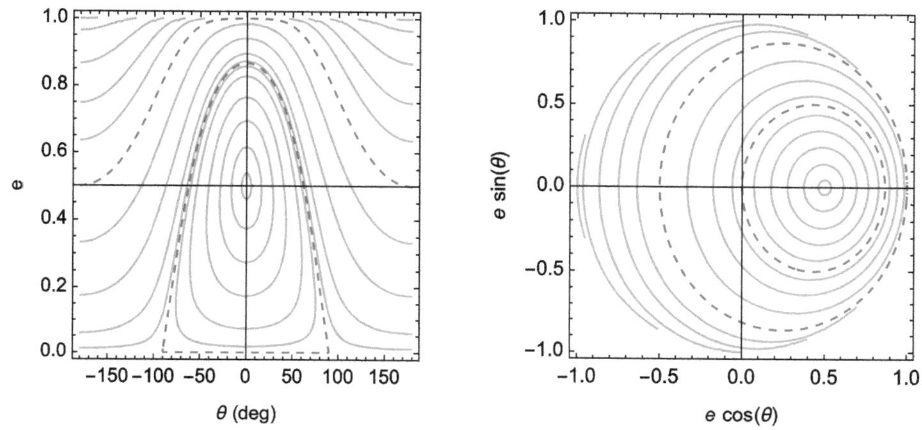

Figure 9.1: Contour plots $\mathcal{K}_E/(L'n_\odot)$ of the coplanar manifold for $n_\odot/n_{\text{srp}} = \sqrt{3}$ (after [495]).

The general solution of the coplanar manifold is delayed to the next section, in which the vectorial elements formulation of the variation equations will disclose that it is solved from a linear differential system with constant coefficients.

9.1.4 General solution in vectorial elements

The flow stemming from Hamiltonian (9.5) is conveniently solved in vectorial elements replacing $\mathcal{Q} = -\mathcal{K}_1/L'$ into Eqs. (4.60)–(4.61). Thus, $\nabla_{\boldsymbol{\eta}}\mathcal{Q} = n_\odot \boldsymbol{k}_\odot$, $\nabla_{\boldsymbol{e}}\mathcal{Q} = n_{\text{srp}}\hat{\boldsymbol{r}}_\odot$, and, therefore,

$$\frac{\mathrm{d}\boldsymbol{\eta}}{\mathrm{d}t} = n_\odot \boldsymbol{\eta} \times \boldsymbol{k}_\odot + n_{\text{srp}} \boldsymbol{e} \times \hat{\boldsymbol{r}}_\odot, \tag{9.8}$$

$$\frac{\mathrm{d}\boldsymbol{e}}{\mathrm{d}t} = n_\odot \boldsymbol{e} \times \boldsymbol{k}_\odot + n_{\text{srp}} \boldsymbol{\eta} \times \hat{\boldsymbol{r}}_\odot, \tag{9.9}$$

in which we identify two separate subsystems each of which is a linear differential system with constant coefficients, of trivial solution. The first one is[1]

$$\frac{\mathrm{d}\eta_1}{\mathrm{d}t} = n_\odot \eta_2, \quad \frac{\mathrm{d}\eta_2}{\mathrm{d}t} = n_{\text{srp}} e_3 - n_\odot \eta_1, \quad \frac{\mathrm{d}e_3}{\mathrm{d}t} = -n_{\text{srp}} \eta_2, \tag{9.10}$$

1 The equivalent variables $X = -\eta_1$, $Y = -\eta_2$, $Z = -e_3$ were used in [495].

which deals essentially with the motion of the node of the orbital plane on the ecliptic plane. The other differential subsystem,

$$\frac{de_1}{dt} = n_\odot e_2, \quad \frac{de_2}{dt} = n_{\text{srp}}\eta_3 - n_\odot e_1, \quad \frac{d\eta_3}{dt} = -n_{\text{srp}} e_2, \tag{9.11}$$

concerns basically the motion of the perigee in the orbital plane.

Alternatively to the standard solution in exponentials, both subsystems can be reduced by means of obvious integrals. Thus, by inspection of Eq. (9.10), we find

$$\eta_1 \frac{d\eta_1}{dt} + \eta_2 \frac{d\eta_2}{dt} + e_3 \frac{de_3}{dt} = 0,$$

which constrains the flow to the surface of a sphere of radius $\rho = \sqrt{\eta_1^2 + \eta_2^2 + e_3^2}$. Also by inspection of Eq. (9.10),

$$n_\odot \frac{de_3}{dt} + n_{\text{srp}} \frac{d\eta_1}{dt} = 0,$$

from which we obtain the integral $\chi = n_\odot e_3 + n_{\text{srp}}\eta_1$. The change of the physical parameters $(n_\odot, n_{\text{srp}}) \mapsto (q, \gamma)$

$$q = (n_\odot^2 + n_{\text{srp}}^2)^{1/2}, \quad \cos\gamma = n_\odot/q, \quad \sin\gamma = n_{\text{srp}}/q, \quad 0 \le \gamma \le \frac{\pi}{2},$$

is used to replace the integral χ by the scaled one $\zeta = \chi/q$. That is,

$$\zeta = e_3 \cos\gamma + \eta_1 \sin\gamma.$$

Then, on account of $\rho^2 = (\eta_1 \cos\gamma - e_3 \sin\gamma)^2 + \eta_2^2 + \zeta^2$, the choice of rectangular coordinates, $x = \eta_1 \cos\gamma - e_3 \sin\gamma$, $y = \eta_2$, $z = \zeta$, immediately shows that the flow is made of parallels $z = \zeta = \text{const.}$ on the sphere.

On the other hand, the flow is naturally described in spherical coordinates (ρ, λ, φ), from which $x = \rho\cos\varphi\cos\lambda$, $y = \rho\cos\varphi\sin\lambda$, $z = \rho\sin\varphi$. Then differentiation of x yields

$$\frac{d\eta_1}{dt}\cos\gamma - \frac{de_3}{dt}\sin\gamma = -\rho\cos\varphi\sin\lambda\frac{d\lambda}{dt} = -y\frac{d\lambda}{dt}.$$

If we now replace $y = \eta_2$ and take the variations of η_1 and e_3 from Eq. (9.10), we finally obtain $d\lambda/dt = -q$.

In summary, the flow stemming from the differential subsystem (9.10) evolves on the sphere with constant elevation φ and constant rate of precession of the azimuth λ given by minus the *gyration frequency* q.

The differential subsystem (9.11) is solved analogously. The obvious integrals are now $\rho^* = \sqrt{e_1^2 + e_2^2 + \eta_3^2}$, $\zeta^* = \eta_3 \cos\gamma + e_1 \sin\gamma$, and an analysis similar to the previous case shows that the motion also takes place along parallels in the new sphere with minus the gyration frequency. This degeneracy of the frequencies to a common value is broken when second-order effects of the perturbation approach are taken into account [480].

Note that the coplanar manifold previously discussed in §9.1.3 is here obtained from Eq. (9.11) in the particular case in which the radius of the first sphere collapses to zero ($\eta_1 = \eta_2 = e_3 = 0$).

9.1.5 Complete Hamiltonian reduction in action-angle variables

While the averaged problem has been correctly integrated up to the first order in the previous section, obtaining the action-angle variables of the solution might be convenient if higher orders of the normalization are pursued. The action-angle variables of Hamiltonian (9.5) are the Kramers–Deprit variables (ϕ, ψ, Φ, Ψ) [159, 368], which are given by the equations

$$\Phi = S_3 \cos y + S_1 \sin y, \tag{9.12}$$
$$\sin\phi = 2S_2(L^2 - 4\Phi^2)^{-1/2}, \tag{9.13}$$
$$\cos\phi = 2(S_1 \cos y - S_3 \sin y)(L^2 - 4\Phi^2)^{-1/2}, \tag{9.14}$$
$$\Psi = D_3 \cos y - D_1 \sin y, \tag{9.15}$$
$$\sin\psi = 2D_2(L^2 - 4\Psi^2)^{-1/2}, \tag{9.16}$$
$$\cos\psi = 2(D_1 \cos y + D_3 \sin y)(L^2 - 4\Psi^2)^{-1/2}, \tag{9.17}$$

where $S_i, D_i, i = 1, 2, 3$, are the components of the Moser elements defined in Eq. (4.54)—which are written in Delaunay variables (g, h, G, H) using Eqs. (4.52) and (4.53). The demonstration of the canonical character of this transformation can be found in [159].

Indeed, if we reformulate the reduced Hamiltonian (9.5) in the geometric variables, S_i, D_i,

$$\mathcal{K}_1 = -q[(S_3 + D_3)\cos y + (S_1 - D_1)\sin y],$$

in which the components of the Moser elements are replaced using Eqs. (9.12) and (9.15), we obtain

$$\mathcal{K}_1 = -q(\Phi + \Psi),$$

which only depends on the Kramers–Deprit actions, thereby disclosing the integrability of the reduced problem. The corresponding Hamilton equations directly show that the two angles evolve with the same gyration frequency $-q$.

The variables S_i, D_i, are readily solved from Eqs. (9.12)–(9.17), to obtain

$$S_1 = \Phi \sin y + \frac{1}{2}(L^2 - 4\Phi^2)^{1/2} \cos\phi \cos y,$$

$$S_2 = \frac{1}{2}(L^2 - 4\Phi^2)^{1/2} \sin\phi,$$
$$S_3 = \Phi \cos\gamma - \frac{1}{2}(L^2 - 4\Phi^2)^{1/2} \cos\phi \sin\gamma,$$
$$D_1 = \frac{1}{2}(L^2 - 4\Psi^2)^{1/2} \cos\psi \cos\gamma - \Psi \sin\gamma,$$
$$D_2 = \frac{1}{2}(L^2 - 4\Psi^2)^{1/2} \sin\psi,$$
$$D_3 = \frac{1}{2}(L^2 - 4\Psi^2)^{1/2} \cos\psi \sin\gamma + \Psi \cos\gamma.$$

9.1.6 Short-period reduction in the equatorial frame

On the other hand, in most cases the SRP is just a second-order perturbation among the different ones that affect artificial satellites. In that case, the removal of short-period terms is customarily done in the equatorial frame, to be propagated semi-analytically jointly with the other perturbations.

The synodic frame is a rotating frame with angular velocity $n_\odot \boldsymbol{k}_\odot$. Therefore, the variation equations in the equatorial (inertial) frame are obtained by simply adding to the variation equations (9.8) and (9.9) the terms $n_\odot \boldsymbol{k}_\odot \times \boldsymbol{h}$ and $n_\odot \boldsymbol{k}_\odot \times \boldsymbol{e}$, respectively, stemming from the derivative of a vector in a rotating frame. Thus, using Eq. (3.29),

$$\frac{\mathrm{d}\boldsymbol{h}}{\mathrm{d}t} = \frac{3aF_{\mathrm{srp}}}{2L'} \boldsymbol{e} \times \hat{\boldsymbol{r}}_\odot, \qquad \frac{\mathrm{d}\boldsymbol{e}}{\mathrm{d}t} = \frac{3aF_{\mathrm{srp}}}{2L'} \boldsymbol{h} \times \hat{\boldsymbol{r}}_\odot, \tag{9.18}$$

where n_{srp} has been replaced by its definition in Eq. (9.6). Recall that now the coordinates of $\hat{\boldsymbol{r}}_\odot$ in the apsidal frame involve the explicit appearance of time through the coordinates of the Sun. Namely,

$$\hat{\boldsymbol{r}}_\odot = R_3(\omega)\, R_1(I)\, R_3(\Omega)\, R_1(-\varepsilon)\, R_3(-\lambda_\odot)(1, 0, 0)^{\tau},$$

where ε is the obliquity of the ecliptic and we neglected the ecliptic latitude of the Sun because it is always less than 2 arcseconds [482].

Alternatively, the Hamilton equations are obtained directly from the averaged Hamiltonian (9.3) after removing the Coriolis term. Namely,

$$\begin{aligned}\mathcal{H}_{0,1} = -\frac{3}{8}F_{\mathrm{srp}}ae\{&2s_\varepsilon s[\cos(\omega - \lambda_\odot) - \cos(\omega + \lambda_\odot)] + (1 - c_\varepsilon)\\ &\times [(1 - c)\cos(\omega - \Omega - \lambda_\odot) + (1 + c)\cos(\omega + \Omega + \lambda_\odot)] + (1 + c_\varepsilon)\\ &\times [(1 + c)\cos(\omega + \Omega - \lambda_\odot) + (1 - c)\cos(\omega - \Omega + \lambda_\odot)]\},\end{aligned} \tag{9.19}$$

which now depends on time through the explicit appearance of the ecliptic longitude of the Sun. From the latter, Kozai's variations of the mean orbital elements [361] are readily derived.

Up to first-order effects, the transformation equations are derived from Eq. (2.17) using Eq. (9.4), which remains unaltered in the equatorial frame. That is,

$$\mathcal{W}_1 = \frac{aF_{\mathrm{srp}}}{32n} \sum_{l=-2}^{2} \sum_{k=-1}^{1} \sum_{j=0}^{1} B_l(e) Q_k(I) \mathcal{E}_{j,k}(\varepsilon) \sin[lu + \omega + k\Omega + (2j-1)\lambda_\odot],$$

with the eccentricity polynomials $B_0 = 0$, $B_{\pm 1} = \pm 2(1 \pm \eta)^2$, and $B_{\pm 2} = \mp e(1 \pm \eta)$, the inclination polynomials $Q_0 = 2s$ and $Q_{\pm 1} = 1 \pm c$, and the obliquity polynomials $\mathcal{E}_{0,0} = -\mathcal{E}_{1,0} = s_\varepsilon$ and $\mathcal{E}_{0,\pm 1} = \mathcal{E}_{1,\mp 1} = 1 \pm c_\varepsilon$.

9.2 Atmospheric drag

The magnitude of the atmospheric drag force depends on the local density of the atmosphere ρ and the cross-sectional area A of the spacecraft in the direction of motion. The drag force per unit of mass m is (see, for instance, [464, 506])

$$\boldsymbol{\alpha}_{\mathrm{drag}} = -\frac{1}{2} n_{\mathrm{d}} \boldsymbol{v}, \quad n_{\mathrm{d}} = (A/m) C_{\mathrm{drag}} \rho v > 0, \tag{9.20}$$

where $\boldsymbol{v}$ is velocity relative to the atmosphere, $v = \|\boldsymbol{v}\|$, and C_{drag} is the dimensionless drag coefficient. For a typical satellite $1.5 \leq C_{\mathrm{drag}} \leq 3.0$, yet the value may vary when calculated for the different constituents of the thermosphere, which have different densities, and hence variants of Eq. (9.20) are sometimes used for increased precision [214]. However, the complexity of accurately predicting the dependence of the atmospheric behavior on the solar flux and geomagnetic activity, makes the efforts in refining the drag coefficient value questionable for long-term predictions, for which the use of Eq. (9.20) seems adequate. Existing tables of drag coefficients for different simple shapes and range of temperatures [503] justify the use of the constant value $C_{\mathrm{drag}} = 2.2$ for typical satellites in low-Earth orbit [129].

9.2.1 Atmospheric density

The drag force in Eq. (9.20) depends on time through the atmospheric density at the satellite location ρ. The latter is a complex parameter that depends on the satellite altitude, the solar flux, the Earth magnetism, the time of day, and the geocentric longitude and latitude of the spacecraft. It must be computed from some density model, among which Roberts' analytical formulation [567], adapted to the Jacchia 1971 model [318], and the Harris–Priester model [265, 266] are popular ones. We only deal with the Harris–Priester model, which neglects the explicit dependence of semi-annual and seasonal latitude variations, but, after modifications in [464], it considers the diurnal density bulge and allows for reasonably accurate density computations. For the computation of ρ we follow the descriptions in [464].

The height of the satellite above the Earth's reference ellipsoid $h = r - \rho_\oplus$ is first computed, where $\rho_\oplus = R_\oplus(1 - f_\oplus)[1 - f_\oplus(2 - f_\oplus)\cos^2\varphi]^{-1/2}$ depends on latitude φ and Earth's flattening $f_\oplus \approx 1/300$. The atmospheric density $\rho(h)$ is then estimated from

$$\rho = \rho_m(h) + [\rho_M(h) - \rho_m(h)]\cos^n\left(\frac{1}{2}\Psi\right), \tag{9.21}$$

where ρ_m and ρ_M are the density coefficients at antapex and apex of the atmospheric diurnal bulge, respectively, Ψ is the angle encompassed by the direction of the apex of the diurnal bulge $\boldsymbol{q}$ and the satellite direction $\hat{\boldsymbol{r}} \equiv \boldsymbol{r}/r$, and the exponent n takes the value 2 for low-inclination orbits, and 6 for polar orbits.

The Harris–Priester density model replaces the minimum and maximum density values $\rho_m(h)$ and $\rho_M(h)$ in Eq. (9.21) by the interpolated values $(x = m, M)$

$$\rho_x(h) = \rho_x(h_i)\exp\left[\frac{h_i - h}{h_i - h_{i+1}}\log\frac{\rho_x(h_{i+1})}{\rho_x(h_i)}\right],$$

where $h_i \le h \le h_{i+1}$, and $\rho_m(h_i)$ and $\rho_M(h_i)$ are given in Table 9.1 [464]. Besides, the apex of the diurnal bulge is assumed to happen two hours later than the local noon. Thus,

$$\cos^n\left(\frac{1}{2}\Psi\right) = \left[\frac{1}{2}(1 + \cos\Psi)\right]^{n/2} = \left[\frac{1}{2}(1 + \boldsymbol{q}\cdot\hat{\boldsymbol{r}})\right]^{n/2},$$

where $\boldsymbol{q}$ is located about 30° east of the subsolar point—or delayed by 2 hours in right ascension from the Sun's direction, given by the its right ascension and declination $(\alpha_\odot, \delta_\odot)$. Then replacing the coordinates of $\boldsymbol{r}$ from Eq. (4.4), with $\Omega \equiv \nu$, $\boldsymbol{q} = R_3(-\alpha)\,R_2(\delta_\odot)(1,0,0)^\tau$, and $\alpha = \alpha_\odot + 30°$, we obtain

$$\begin{aligned}\boldsymbol{q}\cdot\hat{\boldsymbol{r}} &= s\sin\delta_\odot\sin\theta + \frac{1}{2}(c-1)\cos\delta_\odot[\sin\alpha\sin(\theta-\Omega) - \cos\alpha\cos(\theta-\Omega)]\\ &\quad + \frac{1}{2}(c+1)\cos\delta_\odot[\sin\alpha\sin(\theta+\Omega) + \cos\alpha\cos(\theta+\Omega)].\end{aligned} \tag{9.22}$$

9.2.2 Rotating atmosphere

As is customary, we assume that the atmosphere co-rotates with the Earth [506]. Then the velocity relative to this rotating atmosphere, which is needed in the computation of the drag acceleration in Eq. (9.20), is computed like the derivative of a vector in a rotating frame (3.29). Then, assuming that the rotation takes place in the direction orthogonal to the Earth's equator,

$$\boldsymbol{v} = \frac{\mathrm{d}\boldsymbol{r}}{\mathrm{d}t} - \omega_\oplus\boldsymbol{k}\times\boldsymbol{r}, \tag{9.23}$$

where $\omega_\oplus$ denotes the Earth's rotation rate and $\boldsymbol{k}$ is the unit vector in the z direction.

Table 9.1: Minimum and maximum density values ρ (g/km^3) for an altitude $100 \le h \le 1000$ km, and mean solar activity (Harris–Priester model, from [464]).

h	ρ_m	ρ_M	h	ρ_m	ρ_M	h	ρ_m	ρ_M
100	497400.0	497400.0	270	34.30	61.82	580	0.1092	0.7997
110	111289.1	111289.1	280	26.97	50.95	600	0.08070	0.6390
120	24900.0	24900.0	290	21.39	42.26	620	0.06012	0.5123
130	8377.0	8710.0	300	17.08	35.26	640	0.04519	0.4121
140	3899.0	4059.0	320	10.99	25.11	660	0.03430	0.3325
150	2122.0	2215.0	340	7.214	18.19	680	0.02632	0.2691
160	1263.0	1344.0	360	4.824	13.37	700	0.02043	0.2185
170	800.8	875.8	380	3.274	9.955	720	0.01607	0.1779
180	528.3	601.0	400	2.249	7.492	740	0.01281	0.1452
190	361.7	429.7	420	1.558	5.684	760	0.01036	0.1190
200	255.7	316.2	440	1.091	4.355	780	0.008496	0.09776
210	183.9	239.6	460	0.7701	3.362	800	0.007069	0.08059
220	134.1	185.3	480	0.5474	2.612	840	0.004680	0.05741
230	99.49	145.5	500	0.3916	2.042	880	0.003200	0.04210
240	74.88	115.7	520	0.2819	1.605	920	0.002210	0.03130
250	57.09	93.08	540	0.2042	1.267	960	0.001560	0.02360
260	44.03	75.55	560	0.1488	1.005	1000	0.001150	0.01810

Then the components of $\boldsymbol{v}$ in the radial, normal, and binormal direction are obtained by replacing $\boldsymbol{k} = R_3(\theta)\, R_1(I)(0,0,1)^\tau$ into Eq. (9.23). Using polar variables $\dot{r} = R$ and $r\dot{\theta} = \Theta/r$, and we get

$$\boldsymbol{v} = \begin{pmatrix} R \\ \Theta/r \\ 0 \end{pmatrix} + r\omega_\oplus \begin{pmatrix} 0 \\ -\cos I \\ \cos\theta \sin I \end{pmatrix} = \begin{pmatrix} R \\ -r\omega_\oplus \cos I + \Theta/r \\ r\omega_\oplus \sin I \cos\theta \end{pmatrix}, \tag{9.24}$$

from which, replacing $R = (\Theta/p) e \sin f$, from Eq. (4.30), and $\Theta = \sqrt{\mu p}$, from the first of Eq. (4.21), we obtain

$$v = \frac{na}{\eta}\left[1 + 2e\cos f + e^2 - 2\eta^3 \frac{\omega_\oplus}{n} c + \frac{r^2}{a^2}\eta^2 \frac{\omega_\oplus^2}{n^2}(1 - s^2 \sin^2\theta)\right]^{1/2}. \tag{9.25}$$

9.2.3 Gauss equations of variation

Substitution of Eqs. (9.24) and (9.25) in Eq. (9.20) provides the radial, normal, and binormal components of the drag acceleration $\boldsymbol{\alpha}_{\text{drag}}$, which are directly plugged into Gauss planetary equations (see [42], for instance), to obtain

$$\frac{\mathrm{d}a}{\mathrm{d}t} = -an_\mathrm{d}\frac{1}{\eta^2}\left(\frac{pR}{\Theta} e \sin f + \frac{p^2}{r^2} - \frac{\omega_\oplus p^2}{\Theta} c\right),$$

$$\frac{\mathrm{d}e}{\mathrm{d}t} = -\frac{1}{2}n_\mathrm{d}\left\{\frac{pR}{\Theta}\sin f + \left(1 - \frac{\omega_\oplus r^2}{\Theta} c\right)\left[\left(\frac{p}{r} + 1\right)\cos f + e\right]\right\},$$

$$\frac{dI}{dt} = -\frac{1}{2}n_d\frac{\omega_\oplus r^2}{\Theta}s\cos^2\theta,$$
$$\frac{d\Omega}{dt} = -\frac{1}{2}n_d\frac{\omega_\oplus r^2}{\Theta}\sin\theta\cos\theta,$$
$$\frac{d\omega}{dt} = -\frac{1}{2}n_d\frac{1}{e}\left[-\frac{pR}{\Theta}\cos f + \left(1-\frac{\omega_\oplus r^2}{\Theta}c\right)\left(\frac{p}{r}+1\right)\sin f\right] - c\frac{d\Omega}{dt},$$
$$\frac{dM}{dt} - n = n_d\eta\frac{r}{p}\frac{pR}{\Theta} - \eta\left(\frac{d\omega}{dt} + c\frac{d\Omega}{dt}\right).$$

If required, the corresponding variations of the Delaunay variables are computed using the differential relations $d\ell = dM$, $dg = d\omega$, $dh = d\Omega$, and

$$\frac{dL}{L} = \frac{1}{2}\frac{da}{a}, \qquad \frac{dG}{G} = \frac{dL}{L} - \frac{e}{\eta^2}de, \qquad \frac{dH}{H} = \frac{dG}{G} - \frac{s}{c}dI.$$

Then we replace $p/r = 1 + e\cos f$ and $pR/\Theta = e\sin f$, from Eqs. (4.29) and (4.30), respectively, and $\Theta = na^2\eta$, from Eq. (4.63). Hence

$$\frac{da}{dt} = -\frac{a}{\eta^2}n_d[(1-\delta)(1+2e\cos f) + e^2(1-\delta c\cos^2 f)], \tag{9.26}$$
$$\frac{de}{dt} = -n_d\left[\left(1-\frac{1}{2}\delta c\right)e + \left(1-\delta c-\frac{1}{2}\delta c e\cos f\right)\cos f\right], \tag{9.27}$$
$$\frac{dI}{dt} = -\frac{1}{2}n_d\delta s\cos^2\theta, \tag{9.28}$$
$$\frac{d\Omega}{dt} = -\frac{1}{2}n_d\delta\sin\theta\cos\theta, \tag{9.29}$$
$$\frac{d\omega}{dt} = -n_d\frac{1}{e}\sin f\left[1-\delta c\left(1+\frac{e}{2}\cos f\right)\right] - c\frac{d\Omega}{dt}, \tag{9.30}$$
$$\frac{dM}{dt} = n + n_d e\sin u - \eta\left(\frac{d\omega}{dt} + c\frac{d\Omega}{dt}\right), \tag{9.31}$$

in which we abbreviate $\delta = \delta(f) \equiv (\omega_\oplus/n)(r/p)^2\eta^3$. It is worth mentioning that Eqs. (9.26)–(9.31) are in agreement with the corresponding equations in [462] except for the last term in Eq. (9.30) which is missing in the equation for the variation of the argument of the perigee given in [462].

Orbit contraction due to atmospheric drag effects is now easily observed for the lower orbits, a case in which $\delta \approx (\omega_\oplus/n) \ll 1$ and can be neglected. In that case, $dI/dt = d\Omega/dt = 0$, and

$$\frac{da}{dt} = -an_d\frac{1}{\eta^2}(1+2e\cos f + e^2) = -an_d[1+2e\cos f + \mathcal{O}(e^2)],$$
$$\frac{de}{dt} = -n_d(e+\cos f),$$
$$\frac{dF}{dt} = n + n_d(e\sin u - \beta\sin f) = n + \frac{1}{2}n_d[e\sin f + \mathcal{O}(e^2)],$$

where $F = M + \omega$, $\beta = e/(1+\eta)$ was previously defined in Eq. (5.63), and Eq. (4.71) was used to express the eccentric anomaly in terms of the true one.

9.2.4 Perturbation equations

The removal of short-period effects from the Gauss equations is made here in the fashion of Kozai [360] up to first-order effects of the atmospheric drag perturbation.

Thus, we assume that the solution of Eqs. (9.26)–(9.31) takes the form $\xi_i(t) = \xi_i'(t) + \Delta_i(\xi_j'(t))$, $i,j = 1,2,\dots 6$, where $\Delta_i \equiv \Delta\xi_i'$ denotes the short-period terms of the solution while ξ_i' refers to the secular and long-period contributions. That is, since $\Delta_i(\xi_j')$ is purely periodic in the mean anomaly, $\xi_i' = \frac{1}{2\pi}\int_0^{2\pi} \xi_i \,\mathrm{d}M$.

The Gauss equations are equally decomposed into

$$\frac{\mathrm{d}\xi_i}{\mathrm{d}t} = \frac{\mathrm{d}\xi_i'}{\mathrm{d}t} + \frac{\mathrm{d}\Delta_i(\xi_j')}{\mathrm{d}t}, \tag{9.32}$$

where the variations of the prime (mean) elements are computed by averaging the right sides of Eqs. (9.26)–(9.31) over the mean anomaly. In particular, using the notation $\sigma = e\sin f$, $\kappa = e\cos f$ introduced in Eq. (5.21),

$$\frac{\mathrm{d}a'}{\mathrm{d}t} = -\frac{a'}{\eta'^2}\frac{1}{2\pi}\int_0^{2\pi} [(1+\kappa')^2(1-\delta c') + \sigma'^2] n_\mathrm{d}' \,\mathrm{d}M', \tag{9.33}$$

$$\frac{\mathrm{d}e'}{\mathrm{d}t} = -\frac{1}{2\pi}\frac{1}{e'}\int_0^{2\pi} \left[e'^2 + \kappa' - \delta' c'\left(\kappa'^2 + \kappa' + \frac{1}{2}\sigma'^2\right)\right] n_\mathrm{d}' \,\mathrm{d}M', \tag{9.34}$$

$$\frac{\mathrm{d}I'}{\mathrm{d}t} = -\frac{1}{2}s'\frac{1}{2\pi}\int_0^{2\pi} n_\mathrm{d}'\delta'\cos^2\theta' \,\mathrm{d}M', \tag{9.35}$$

$$\frac{\mathrm{d}\Omega'}{\mathrm{d}t} = -\frac{1}{2}\frac{1}{2\pi}\int_0^{2\pi} n_\mathrm{d}'\delta'\sin\theta'\cos\theta' \,\mathrm{d}M', \tag{9.36}$$

$$\frac{\mathrm{d}\omega'}{\mathrm{d}t} = -c'\frac{\mathrm{d}\Omega'}{\mathrm{d}t} - \frac{1}{2\pi e'^2}\int_0^{2\pi} \sigma'\left[1 - \delta' c'\left(1 + \frac{1}{2}\kappa'\right)\right] n_\mathrm{d}' \,\mathrm{d}M', \tag{9.37}$$

$$\frac{\mathrm{d}M'}{\mathrm{d}t} = n' + \frac{\eta'}{2\pi p'}\int_0^{2\pi} \sigma' r' n_\mathrm{d}' \,\mathrm{d}M' - \eta'\left(c'\frac{\mathrm{d}\Omega'}{\mathrm{d}t} + \frac{\mathrm{d}\omega'}{\mathrm{d}t}\right), \tag{9.38}$$

in which symbols with primes stand for functions of the prime (mean) orbital elements. As usual, alternative orbital elements can be used to avoid specific singularities.

Since both the relative velocity with respect to the rotating atmosphere, v, and the atmospheric density ρ are naturally expressed like functions of the true anomaly, as shown by Eqs. (9.25) and (9.21)–(9.22), respectively, it happens that $n_\mathrm{d} \equiv n_\mathrm{d}(f)$, as follows from the definition of n_d in Eq. (9.20). Therefore, Eqs. (9.33)–(9.38) are rather integrated in f than in $M \equiv \ell$ using the differential relation (4.64). Besides, due to the complicated representation of the atmospheric density, these quadratures are usually evaluated numerically [184, 635, 674].

It is worth mentioning that the use of some intermediary solution of the main problem, rather than the standard Keplerian solution, in the evaluation of the integrals has been claimed to provide additional benefits by accounting for the "drag–oblateness coupling". In particular, since the atmospheric density depends essentially on altitude, improvements are observed when using the osculating J_2-perturbed orbit in the computation of ρ [253, 674]. This is commonly done modifying the Keplerian radius with the correction $r_{0,1}$ given in §5.7.2, sometimes called an Izsak correction [317].[2] This improved value of the radius can also be applied to other elements than n'_d in the evaluation of the quadratures by simply replacing $\kappa = -1 + p/r$, as given in Eq. (5.42).

The short-period corrections $\Delta_i(\xi'_j)$ are computed as follows. First, their variations are solved from Eq. (9.32). Next, they are computed along the Keplerian flow, in which $\mathrm{d}t = (1/n)\mathrm{d}M$, by indefinite integration

$$\Delta_i(\xi'_j) = \int \left(\frac{\mathrm{d}\xi_i}{\mathrm{d}t} - \frac{\mathrm{d}\xi'_i}{\mathrm{d}t} \right) \mathrm{d}t, \tag{9.39}$$

replacing the variations of ξ_i by the right sides of Eqs. (9.26)–(9.31), and those of ξ'_i by the right sides of Eqs. (9.33)–(9.38).

The right side of Eq. (9.39) involves both prime and non-prime elements. Nevertheless, as far as $n_\mathrm{d} \ll n$, the difference between mean and osculating elements is small and, up to first-order effects of $\epsilon = n_\mathrm{d}/n$, the osculating orbital elements can be taken as prime variables in the integration. There is an exception, however, which occurs in the integration of the short-period corrections to the mean anomaly. In that case, the first-order corrections cannot be neglected in the first summand of Eq. (9.31). On the contrary, on account of $n = \sqrt{\mu/a^3}$, as derived from Eqs. (4.44) and (4.45), $n = n' + \Delta n' \approx n'(1 - \frac{3}{2}\Delta a'/a')$, which must be replaced in Eq. (9.31) in order to compute the correction $\Delta M'$ from Eq. (9.39).

In summary,

$$\Delta\xi = \frac{1}{n'} \int \left(\left.\frac{\mathrm{d}\xi}{\mathrm{d}t}\right|_{\substack{\xi=\xi' \\ M=M'}} - \frac{\mathrm{d}\xi'}{\mathrm{d}t} \right) \mathrm{d}M', \quad \xi \in (a, e, I, \Omega, \omega),$$

$$\Delta M = \frac{1}{n'} \int \left(\left.\frac{\mathrm{d}M}{\mathrm{d}t}\right|_{\substack{\xi=\xi' \\ M=M'}} - \frac{\mathrm{d}M'}{\mathrm{d}t} \right) \mathrm{d}M' - \frac{3}{2} \int \frac{\Delta a'}{a'} \,\mathrm{d}M',$$

where $\Delta a'$ must be replaced by the previously computed solution of the short-period correction to a.

2 Short-period corrections to the radius due to oblateness perturbations were, in fact, first given by Kozai [360].

Alternatively, the approximation of Eqs. (9.26)–(9.31) by means of Fourier series in the true anomaly in which the coefficients are computed numerically, allows for the analogous construction of the short-period corrections by means of Fourier series. Full details on this procedure can be found in [185, 253, 467].

Part III: Relative motion and perturbed non-Keplerian motion

10 The Hill problem

The investigation of the three-body dynamics was originally motivated by efforts made in providing accurate descriptions of the motion of the Moon [260]. The *restricted* approximation, in which the mass of the Moon is negligible, can be traced back to the lunar theories of Euler in 1772 [628]. Since then, the common trend in computing lunar theories was approaching the motion of the Moon as a perturbed Keplerian problem. Hill's departure from the tradition by computing the Moon's orbit as a variation of a particular solution of the three-body problem not only produced a breakthrough in the computation of lunar theories, but also introduced the use of the Jacobian constant to determine forbidden regions of motion of the massless body [292].

The Hill problem does not depend on essential physical parameters and, therefore, has wide applicability to different three-body systems beyond the Earth–Moon–Sun system. This chapter briefly discusses the derivation of the Hill problem equations as well as relevant facts of the model. As previously mentioned in §1.3.2, the Hill problem has much wider generality than being a mere limiting case of the restricted problem of three bodies. However, for our aim of computing perturbation solutions, we find it natural to adhere to the common practice of deriving the Hill problem from the restricted three-body problem.

10.1 The circular restricted three-body problem

The "restricted" three-body problem, or RTBP, studies the motion of a particle of negligible mass under the gravitational attraction of two point-mass bodies, called *primaries*, which are orbiting about their common center of mass under their mutual gravitational interaction.

Let m_1 and m_2 be the masses of the primaries. In the restricted approximation they evolve with Keplerian motion of constant angular momentum vector $\boldsymbol{G}$. We take the plane of motion of the primaries like the x–y plane, and the z axis in the direction of $\boldsymbol{G}$. Then, from Newton's gravitational law, the acceleration of the third body is given by

$$\frac{\mathrm{d}^2\boldsymbol{s}}{\mathrm{d}t^2} = -\frac{\mathcal{G}m_1}{\rho^3}\boldsymbol{\rho} - \frac{\mathcal{G}m_2}{r^3}\boldsymbol{r}, \tag{10.1}$$

where $\boldsymbol{s}$ is the position vector of the massless body with respect to the origin in an inertial or *sidereal* frame, $\boldsymbol{\rho}$ and $\boldsymbol{r}$ are the vectors joining m_1 and m_2 with the massless body, respectively, of moduli $\rho = \|\boldsymbol{\rho}\|$ and $r = \|\boldsymbol{r}\|$, and $\mathcal{G}$ is the gravitational constant. The acceleration is obtained like the gradient of the potential function,

$$U = \frac{\mathcal{G}m_1}{\rho} + \frac{\mathcal{G}m_2}{r}. \tag{10.2}$$

https://doi.org/10.1515/9783110668513-010

A sketch of the RTBP is depicted in Fig. 10.1, where, due to the preservation of the total linear momentum, the origin of the reference system O is chosen at the center of mass of m_1 and m_2, and m' is the projection of m in the orbital plane of the primaries.

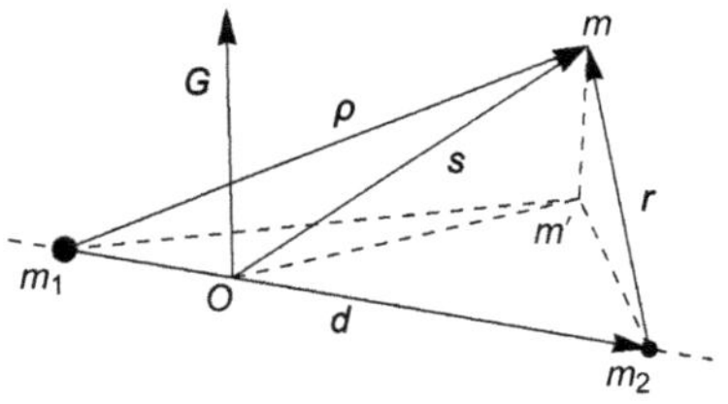

Figure 10.1: Geometry of the restricted three-body problem.

10.1.1 Synodic frame. The Jacobi integral

The RTBP is conveniently formulated in a rotating or *synodic* frame in which the direction of abscissas is defined by the vector $\boldsymbol{d}$ joining the primaries from m_1 to m_2, of modulus d, and the direction of ordinates completes and orthogonal frame with $\boldsymbol{d}$ and $\boldsymbol{G}$. From Eq. (4.2), the velocity vector $\boldsymbol{\omega} = \dot{\vartheta}\boldsymbol{G}/\|\boldsymbol{G}\| = \boldsymbol{G}/d^2$ has constant direction due to the Keplerian motion of the primaries, and $\dot{\vartheta} = \|\boldsymbol{G}\|/d^2$ is the system rotation rate.

Recalling from Eq. (3.29), the derivative of a vector in a moving frame, the derivative of $\boldsymbol{s}$ is

$$\frac{\mathrm{d}\boldsymbol{s}}{\mathrm{d}t} = \boldsymbol{s}' + \boldsymbol{\omega} \times \boldsymbol{s}, \tag{10.3}$$

where the prime symbol means differentiation in the rotating frame. We limit ourselves to the case in which $\boldsymbol{\omega}$ is constant. Then the orbit of the primaries is circular and the problem is called circular RTBP, or CRTBP in short. In that case,

$$\frac{\mathrm{d}^2\boldsymbol{s}}{\mathrm{d}t^2} = \boldsymbol{s}'' + 2\boldsymbol{\omega} \times \boldsymbol{s}' + \boldsymbol{\omega} \times (\boldsymbol{\omega} \times \boldsymbol{s}),$$

where $\boldsymbol{s} = \boldsymbol{r} + m_1\boldsymbol{d}/(m_1 + m_2)$, from the definition of the center of mass, and $\boldsymbol{d}$ is constant in the rotating frame. Hence,

$$\frac{\mathrm{d}^2\boldsymbol{s}}{\mathrm{d}t^2} = \boldsymbol{r}'' + 2\boldsymbol{\omega} \times \boldsymbol{r}' + \frac{m_1}{m_1 + m_2}\boldsymbol{\omega} \times (\boldsymbol{\omega} \times \boldsymbol{d}) + \boldsymbol{\omega} \times (\boldsymbol{\omega} \times \boldsymbol{r}). \tag{10.4}$$

Then, equating the right sides of Eqs. (10.4) and (10.1), we obtain

$$\boldsymbol{r}'' + 2\boldsymbol{\omega} \times \boldsymbol{r}' = -\boldsymbol{\omega} \times (\boldsymbol{\omega} \times \boldsymbol{r}) + \frac{m_1\dot{\vartheta}^2}{m_1 + m_2}\boldsymbol{d} - \frac{\mathcal{G}m_1}{\rho^3}\boldsymbol{\rho} - \frac{\mathcal{G}m_2}{r^3}\boldsymbol{r}, \tag{10.5}$$

both sides of which are to be multiplied in the scalar way by $\boldsymbol{r}'$. Recalling the triple product $\boldsymbol{r}' \cdot [\boldsymbol{\omega} \times (\boldsymbol{\omega} \times \boldsymbol{r})] = (\boldsymbol{r}' \times \boldsymbol{\omega}) \cdot (\boldsymbol{\omega} \times \boldsymbol{r})$, we obtain

$$\boldsymbol{r}' \cdot \boldsymbol{r}'' = (\boldsymbol{\omega} \times \boldsymbol{r}) \cdot (\boldsymbol{\omega} \times \boldsymbol{r}') + \boldsymbol{r}' \cdot \left(-\frac{\mathcal{G} m_1}{\rho^3} \boldsymbol{\rho} - \frac{\mathcal{G} m_2}{r^3} \boldsymbol{r} + \frac{m_1 \dot{\vartheta}^2}{m_1 + m_2} \boldsymbol{d} \right), \tag{10.6}$$

which is readily integrated to give the Jacobi integral

$$(\boldsymbol{\omega} \times \boldsymbol{r}) \cdot (\boldsymbol{\omega} \times \boldsymbol{r}) - 2V - \boldsymbol{r}' \cdot \boldsymbol{r}' = \widetilde{C}, \tag{10.7}$$

where, replacing $\boldsymbol{\rho} = \boldsymbol{d} + \boldsymbol{r}$,

$$V = -\frac{\mathcal{G} m_1}{\|\boldsymbol{d} + \boldsymbol{r}\|} - \frac{\mathcal{G} m_2}{\|\boldsymbol{r}\|} - \frac{m_1 \dot{\vartheta}^2}{m_1 + m_2} \boldsymbol{r} \cdot \boldsymbol{d}, \tag{10.8}$$

and $\widetilde{C}$ is an arbitrary constant which is known as the Jacobi constant.

10.1.2 Hamiltonian formulation

Equation (10.5) can be derived from the Lagrangian

$$\mathcal{L} = \frac{1}{2} \boldsymbol{r}' \cdot \boldsymbol{r}' + (\boldsymbol{\omega} \times \boldsymbol{r}) \cdot \boldsymbol{r}' + \frac{1}{2} (\boldsymbol{\omega} \times \boldsymbol{r}) \cdot (\boldsymbol{\omega} \times \boldsymbol{r}) - V.$$

The conjugate momentum to $\boldsymbol{r}$ is $\boldsymbol{R} = \partial \mathcal{L} / \partial \boldsymbol{r}' = \boldsymbol{r}' + \boldsymbol{\omega} \times \boldsymbol{r}$, which, by comparison with Eq. (10.3), is the velocity of the massless particle in the inertial frame. Then the usual construction of the Hamiltonian, $\mathcal{H} = \boldsymbol{R} \cdot \boldsymbol{r}' - \mathcal{L}$, yields

$$\mathcal{H} = \frac{1}{2} \boldsymbol{R} \cdot \boldsymbol{R} - (\boldsymbol{\omega} \times \boldsymbol{r}) \cdot \boldsymbol{R} + V. \tag{10.9}$$

Note that $\mathcal{H} = -\frac{1}{2}\widetilde{C}$.

In view of Eq. (10.8), Hamiltonian (10.9) depends on the physical parameters $\mathcal{G}$, m_1, m_2, $\dot{\vartheta}$, and d. However, a simple nondimensionalization shows that not all of them are essential. Indeed, the transformation to nondimensional variables $\boldsymbol{r} = d\boldsymbol{r}^*$, $\boldsymbol{R} = \dot{\vartheta} d \boldsymbol{R}^*$, is canonical with multiplier $\alpha = 1/(\dot{\vartheta} d^2)$, as checked from the differential form $\boldsymbol{R}^* \cdot \mathrm{d}\boldsymbol{r}^* = \alpha \boldsymbol{R} \cdot \mathrm{d}\boldsymbol{r}$ [243]. When Eq. (10.9) is written in the nondimensional variables we obtain $\mathcal{H}(\boldsymbol{r}(\boldsymbol{r}^*, \boldsymbol{R}^*), \boldsymbol{R}(\boldsymbol{r}^*, \boldsymbol{R}^*)) = \mathcal{K}(\boldsymbol{r}^*, \boldsymbol{R}^*)/\alpha$ [684]. Straightforward operations yield $\mathcal{K} = \dot{\vartheta} \mathcal{H}^*(\boldsymbol{r}^*, \boldsymbol{R}^*)$, with

$$\mathcal{H}^* = \frac{1}{2} \boldsymbol{R}^* \cdot \boldsymbol{R}^* - (\boldsymbol{k} \times \boldsymbol{r}^*) \cdot \boldsymbol{R}^* + V^*, \tag{10.10}$$

in which $\boldsymbol{k} = \boldsymbol{\omega}/\dot{\vartheta}$, and, taking into account that $\mathcal{G}(m_1 + m_2) = \dot{\vartheta}^2 d^3$ from the Keplerian motion of the primaries. The scaled potential $V^* = V/(\dot{\vartheta}^2 d^2)$ is

$$V^* = -\frac{1 - \mu}{\|\boldsymbol{i} + \boldsymbol{r}^*\|} - \frac{\mu}{\|\boldsymbol{r}^*\|} - (1 - \mu) \boldsymbol{r}^* \cdot \boldsymbol{i}, \tag{10.11}$$

where $\boldsymbol{i} = \boldsymbol{d}/d$, and $\mu = m_2/(m_1 + m_2)$ stands for the nondimensional mass of the smaller primary ($0 < \mu \le \frac{1}{2}$). Hence, $1 - \mu = m_1/(m_1 + m_2)$.

Therefore, in the new time scale $t' = \dot{\vartheta}t$, in which the period of the primaries' orbit is 2π, the flow of the CRTBP is derived from the nondimensional Hamiltonian (10.10) that only depends on the physical nondimensional parameter μ. Namely,

$$\frac{\mathrm{d}\boldsymbol{r}^*}{\mathrm{d}t'} = \boldsymbol{R}^* - \boldsymbol{k} \times \boldsymbol{r}^*, \qquad \frac{\mathrm{d}\boldsymbol{R}^*}{\mathrm{d}t'} = -\boldsymbol{k} \times \boldsymbol{R}^* - \nabla_{\boldsymbol{r}*} V^*. \tag{10.12}$$

10.1.3 Surfaces and curves of zero velocity

The Jacobi integral (10.7) is also conveniently written in the nondimensional coordinates, from which we solve the square of the nondimensional velocity in the rotating frame $v = \|\boldsymbol{r}'\|/(d\dot{\vartheta})$. Denoting $C = \widetilde{C}/(d\dot{\vartheta})^2$, we obtain

$$v^2 = (\boldsymbol{k} \times \boldsymbol{r}^*) \cdot (\boldsymbol{k} \times \boldsymbol{r}^*) + 2V^* - C, \tag{10.13}$$

which for each value of C defines a *zero-velocity surface*

$$S(\boldsymbol{r}^*; C) \equiv (\boldsymbol{k} \times \boldsymbol{r}^*) \cdot (\boldsymbol{k} \times \boldsymbol{r}^*) + 2V^* - C = 0. \tag{10.14}$$

The surfaces of zero velocity define regions of allowed motion for a particle moving with a certain energy. Indeed, points on one side of the surface will make $v^2 > 0$ in Eq. (10.13), whereas points on the other side of the surface would make $v^2 < 0$, which does not admit real solutions, thus precluding motion. An example of these kinds of surfaces is illustrated in Fig. 10.2 for $\mu = \frac{1}{8}$ and $C = 2.6$, where the motion only can happen in the region defined by the outside of the ring which is connected with the interior of the two-sphere-type surface. The representation has been limited to $z = \pm 0.5$ although the outer ring exists for any value of z.

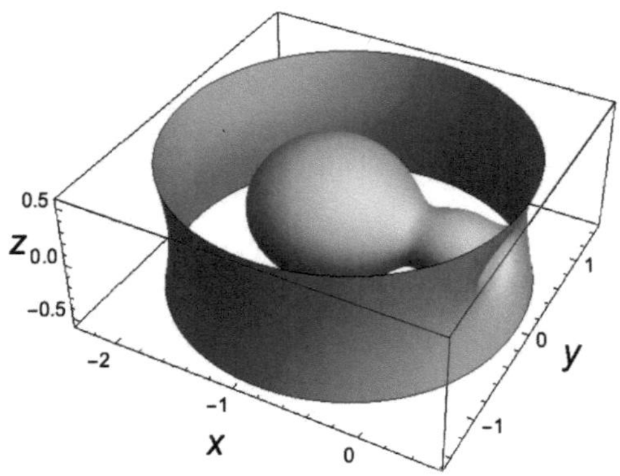

Figure 10.2: Surface $C = 2.6$ of zero velocity of the CRTBP for $\mu = \frac{1}{8}$.

Intersections of different surfaces of zero velocity with the coordinate planes produce zero-velocity *curves*, which are also used to show how the regions of allowed motion evolve with corresponding values of C. This is illustrated in Fig. 10.3 where the evolution from three disconnected regions to two (the transition represented by the dotted zero-velocity curve), and then to one (the transition represented by the black zero-velocity curve) for decreasing values of C is observed. Note that the connection between the different regimes occurs always at critical points on the x–y plane.

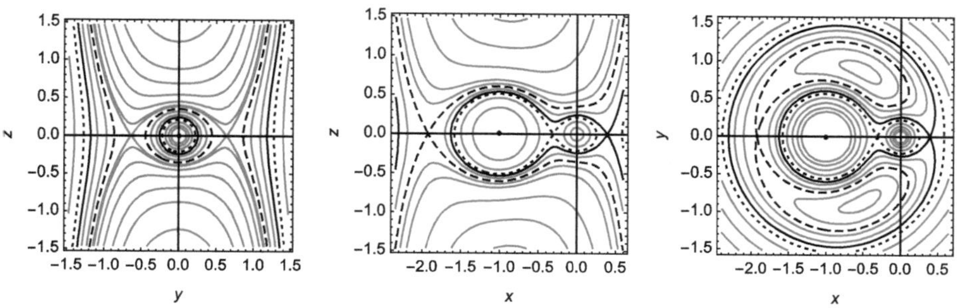

Figure 10.3: Cuts of different surfaces of zero velocity of the CRTBP ($\mu = \frac{1}{8}$, $0.5 < C < 5$) with the planes, left to right, $x = 0$, $y = 0$ and $z = 0$.

10.2 Hill's simplifying assumptions

For all practical effects, the CRTBP is not integrable. Alternatively, a wealth of information on the system's behavior is obtained by computing those solutions that may be predictable when time goes to infinity. These are equilibria, bounded regions, periodic orbits and other invariant manifolds of Eqs. (10.12). For instance, five equilibria resting on the plane of the primaries, the so-called Lagrangian or libration points, are guessed from Fig. 10.3. Details on particular solutions of the CRTBP can be found in classical treatments, for instance [514, 517, 628], and we do not discuss them. On the contrary, we focus on the specific case of motion always close to the smaller primary, for which useful analytical approximations to the dynamics are obtained by perturbations.

We take the smaller primary like the origin of the synodic frame, in which the coordinates of the massless body are denoted $\boldsymbol{r} = (x, y, z)$. Then $r = \|\boldsymbol{r}\|$ and

$$\rho = \|\boldsymbol{d} + \boldsymbol{r}\| = \sqrt{(d + x)^2 + y^2 + z^2} = d\sqrt{1 + 2\epsilon(x/r) + \epsilon^2},$$

with $\epsilon = r/d < 1$. Next, the inverse of ρ is expanded in powers of ϵ,

$$\frac{1}{\rho} = \frac{1}{d}\sum_{i\geq 0} \epsilon^i P_i(-x/r) = \frac{1}{d}\left[1 - \epsilon\frac{x}{r} + \sum_{i\geq 2} \epsilon^i P_i(-x/r)\right], \tag{10.15}$$

where the P_i are Legendre polynomials. Replacing Eq. (10.15) into Eq. (10.8), the potential is written in the form

$$V = -\frac{\mathcal{G}m_1}{d} - \frac{\mathcal{G}m_2}{r} - \dot{\vartheta}^2 d^2(1-\mu)\sum_{i\geq 2}\epsilon^i P_i(-x/r), \tag{10.16}$$

in which the term $\mathcal{G}m_1/d$ can be ignored because it does not affect the dynamics of the massless body.

When $r/d = \epsilon \ll 1$ Eq. (10.16) can be safely truncated to the lower orders. If, besides, we constrain to the case in which μ is of the order of ϵ or higher, then

$$V = -\frac{\mathcal{G}m_2}{r} + \frac{1}{2}\dot{\vartheta}^2(r^2 - 3x^2) + \mathcal{O}(\epsilon^3),$$

which, after replaced in Eq. (10.9) and neglecting higher orders, gives rise to the Hill problem Hamiltonian

$$\mathcal{H} = \frac{1}{2}(X^2 + Y^2 + Z^2) - \dot{\vartheta}(xY - yX) - \frac{\mathcal{G}m_2}{r} + \frac{1}{2}\dot{\vartheta}^2(r^2 - 3x^2). \tag{10.17}$$

For instance, the distance between Saturn and Enceladus is $d \approx 238000$ km. If neglecting $\mathcal{O}(10^{-6})$ were acceptable, then in the Hill problem approximation of neglecting $\mathcal{O}(\epsilon^3)$ we obtain $\epsilon \leq 10^{-2}$, which roughly means dealing with orbits about Enceladus with shorter semi-major axes than $a = 10^{-2}d = 2380$ km—or about 9 times Enceladus' equatorial radius of 256 km. Besides, since the mass of Enceladus is about 2×10^{-7} times the mass of Saturn, the Hill problem model fits quite well for describing the dynamics of these types of orbits.

That Eq. (10.17) does not depend on essential physical parameters is shown, like we did in §10.1.2, by choosing such units of length and time that $\dot{\vartheta} = 1$ and $\mathcal{G}m_2 = 1$. That is,

$$\mathcal{H} = \frac{1}{2}(X^2 + Y^2 + Z^2) - (xY - yX) + \frac{1}{2}(r^2 - 3x^2) - 1/r, \tag{10.18}$$

where, with abuse of notation, we used the same variables as in the previous case. Therefore, application of the Hill problem to different systems becomes a simpler matter, one of rescaling units of length and time. For instance, the gravitational parameter of Enceladus is $\mathcal{G}m_2 = 7.2095\,\mathrm{km}^3/\mathrm{s}^2$ and the rotation rate of the Saturn–Enceladus system is $\dot{\vartheta} = 5.30364\times 10^{-5}\,\mathrm{s}^{-1}$. Hence, the Hill problem units of length and time turn out to be 1368.52 km and 18 855 s, respectively.

The Hill problem scaling is illustrated in Fig. 10.4, where the equatorial radii of different bodies are presented in units of the Hill problem together with the Hill radius, to be defined later. Note that the small-mass assumption of the smaller primary of the Hill problem may not apply to the Earth–Moon system, whereas the assumption of a circular orbit of the primaries may not apply to the Sun–Mercury system [420].

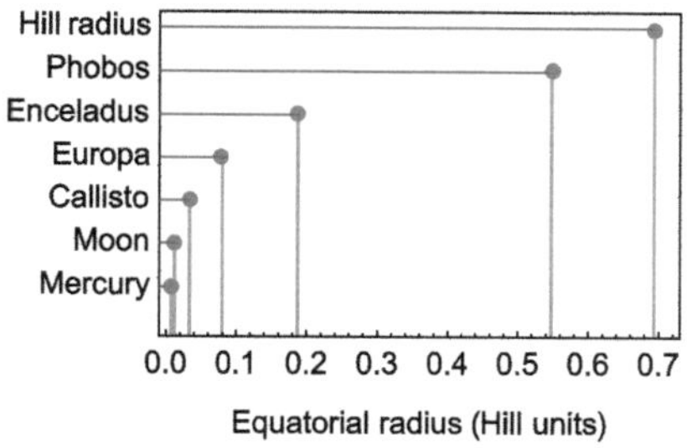

Figure 10.4: Equatorial radius of different bodies in the Hill problem scale.

On the other hand, physical dimensions may eventually be retained in what follows for their insight, as well as the help they provide in checking dimensions throughout the different analytical manipulations that are carried out.

In the Hill problem units, the zero-velocity surfaces are given by

$$3x^2 - z^2 + (2/r) - C = 0. \tag{10.19}$$

A sample zero-velocity surface is presented in Fig. 10.5, while Fig. 10.6 shows a sequence of the allowed regions of motion obtained from the projections of the zero-velocity surfaces in the (x, y) plane. The latter also shows the collinear libration points, which will be discussed in §10.2.1.

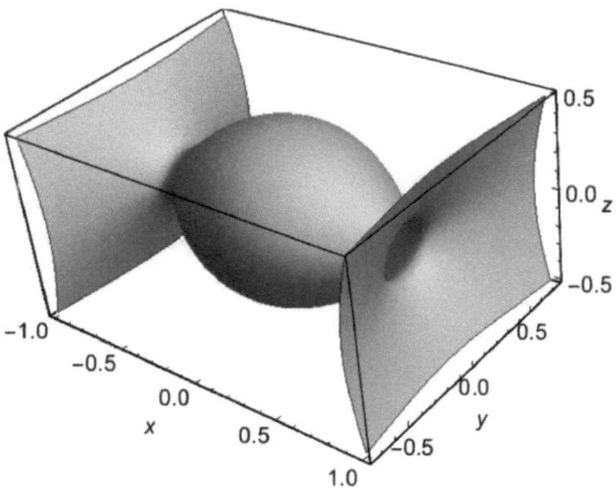

Figure 10.5: Surface $C = 4.24$ of zero velocity of the Hill problem.

Zero-velocity curves also serve to illustrate the dimension scaling of the Hill problem. Thus, the (x, y) zero-velocity curve corresponding to the value of the Jacobi constant of the equilibrium points, which determines the maximum size of the region of bounded motion about the primary, is depicted in Fig. 10.7 with the equatorial projections of different celestial bodies superimposed. In particular, Fig. 10.7 shows that regions of bounded motion about Phobos may not exist in the Hill problem approximation.

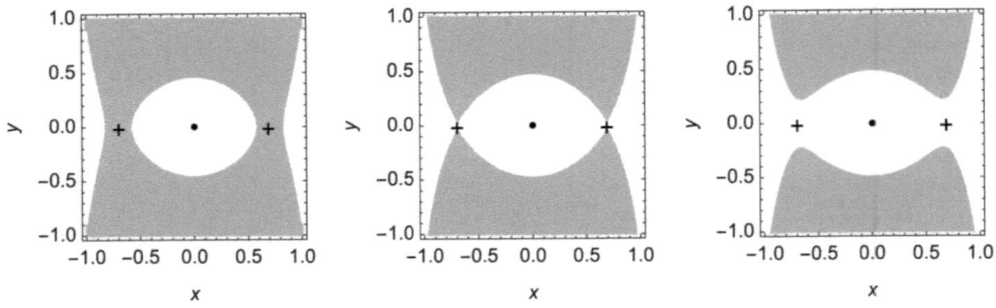

Figure 10.6: Allowed regions of motion (white areas) of the Hill problem. Left to right: $C = 4.5, 4.327$, and 4.2. Crosses mark the collinear equilibrium points.

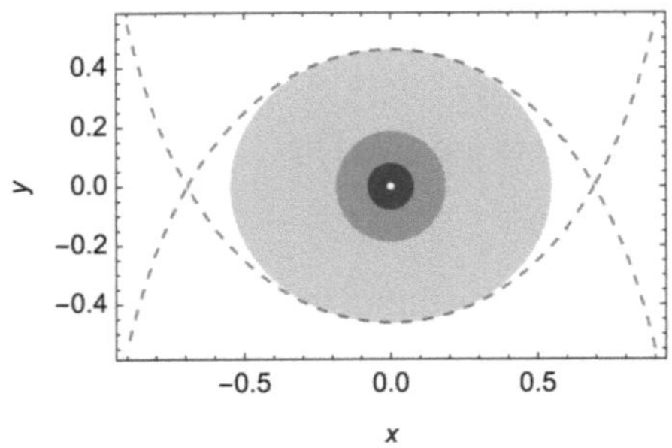

Figure 10.7: Sizes in units of the Hill problem of Phobos (pale gray), Enceladus (gray), Europa (dark gray), and the Moon (white disk). The dashed line is the zero-velocity curve $C = 3^{4/3}$ of the limit case of bounded motion.

10.2.1 Equilibria. Hill's sphere

The Hamiltonian flow stemming from Eq. (10.18) is given by the Hamilton equations

$$\dot{x} = X + y, \tag{10.20}$$

$$\dot{y} = Y - x, \tag{10.21}$$

$$\dot{z} = Z, \tag{10.22}$$

$$\dot{X} = -\frac{x}{r^3} + 2x + Y, \tag{10.23}$$

$$\dot{Y} = -\frac{y}{r^3} - y - X, \tag{10.24}$$

$$\dot{Z} = -\frac{z}{r^3} - z, \tag{10.25}$$

where over dots denote time differentiation in the Hill problem units. As follows from Eqs. (10.22) and (10.25), equilibria can occur only in the (x, y) plane. Besides, Eqs. (10.20) and (10.21) vanish when $X = -y$ and $Y = x$, respectively. For these values, Eq. (10.24) becomes zero only when $y = 0$, which in turn requires that $r^3 = \frac{1}{3}$ to make null Eq. (10.23). Therefore, the Hill problem only accepts two equilibria, which are collinear along the axis of abscissas with $x = \pm 3^{-1/3}$. The value of the Jacobi constant of the two symmetric equilibria is $C = 3^{4/3} \approx 4.32675$, as follows from evaluation of Eq. (10.19).

Then the coordinates of the collinear libration points in the phase space of the Hill problem are $\pm(r_{\mathrm{H}}, 0, 0, 0, r_{\mathrm{H}}, 0)$, where the distance

$$r_{\mathrm{H}} = 3^{-1/3}, \tag{10.26}$$

is called the Hill sphere radius, or Hill radius in short, and determines the distance in which the attraction of the central body m_2 dominates the dynamics.

The elliptic or hyperbolic character of the libration points is computed from the linear variations $(\delta x, \delta y, \delta z, \delta X, \delta Y, \delta Z)$ obtained from Eqs. (10.20)–(10.25) after particularization for the collinear libration points. They show that the equations $\delta\dot{z} = \delta Z$ and $\delta\dot{Z} = -4\delta z$ decouple from the rest of the system. They are readily integrated to yield

$$\begin{aligned} \delta z &= \delta z_0 \cos 2t + \frac{1}{2}\delta Z_0 \sin 2t, \\ \delta Z &= \delta Z_0 \cos 2t - 2\delta z_0 \sin 2t, \end{aligned} \tag{10.27}$$

thus giving rise to small harmonic oscillations of frequency

$$\nu = 2. \tag{10.28}$$

Therefore, the z axis is a stable direction in all cases. More precisely, the projection of the tangent flow in the (z, Z) plane is made of ellipses, so that, relative to this plane, the collinear equilibria are of the "center" type.

It remains to integrate the reduced linear system with constant coefficients

$$(\delta\dot{x}, \delta\dot{y}, \delta\dot{X}, \delta\dot{Y})^{\tau} = \mathbb{M}\,(\delta x, \delta y, \delta X, \delta Y)^{\tau}, \tag{10.29}$$

with

$$\mathbb{M} = \begin{pmatrix} 0 & 1 & 1 & 0 \\ -1 & 0 & 0 & 1 \\ 8 & 0 & 0 & 1 \\ 0 & -4 & -1 & 0 \end{pmatrix}, \tag{10.30}$$

and τ denotes transposition. The general solution is made of a linear combination of exponentials. Namely,

$$\delta_i = \sum_{j=1}^{4} A_{i,j}\, e^{\lambda_j t}, \tag{10.31}$$

where, for $i = 1, \ldots, 4$, δ_i stands for δx, δy, δX, and δY, respectively, $(A_{i,j})$ is a 4×4 matrix of arbitrary coefficients, and the characteristic exponents λ_j are the eigenvalues of $\mathbb{M}$. The latter are computed from the characteristic equation

$$\mathrm{Det}(\mathbb{M} - \mathbb{I}_4\,\lambda) \equiv \lambda^4 - 2\lambda^2 - 27 = 0, \tag{10.32}$$

which is biquadratic in λ. Therefore, the characteristic exponents will appear in opposite pairs. We find

$$\lambda_{1,2} = \pm w\,\mathbf{i}, \quad w = (2\sqrt{7} - 1)^{1/2}, \tag{10.33}$$

which are purely imaginary, thus giving rise to an elliptic direction—or a center-type component. However,

$$\lambda_{3,4} = \pm \lambda, \quad \lambda = (2\sqrt{7} + 1)^{1/2}, \tag{10.34}$$

are always real, thus giving rise to a hyperbolic direction—or a saddle-type component. Hence, the instability of the collinear equilibria is said to be of the center × center × saddle type.

10.2.2 Motion near the equilibrium points

The existence of periodic oscillations with small amplitude in the z-axis direction was already shown in Eq. (10.27). We will see now that, in spite of the saddle component on the collinear equilibria, particular instances of Eq. (10.31) give rise to small periodic oscillations in the (x, y) plane [514].

First of all, we note that the arbitrary coefficients $A_{i,j}$ in Eq. (10.31) are not completely independent; cf. [628]. Indeed, plugging Eq. (10.31) into both members of the linearized equations (10.29), we find that

$$\begin{aligned} A_{1,j} &= -A_{2,j}\lambda_j + A_{4,j}, \\ A_{2,j} &= A_{1,j}\lambda_j - A_{3,j}, \\ A_{3,j} &= -A_{4,j}\lambda_j - 4A_{2,j}, \\ A_{4,j} &= A_{3,j}\lambda_j - 8A_{1,j}. \end{aligned} \tag{10.35}$$

The elimination of the arbitrary constants between them leads to the constraint equation $\lambda_j^4 - 2\lambda_j^2 - 27 = 0$, which is automatically fulfilled because it is the same as the characteristic equation (10.32). Besides, solving the two first for $A_{4,j}$ and $A_{3,j}$, which are then put into the two last, we find

$$A_{2,j} = \alpha_j A_{1,j}, \quad \alpha_j = \frac{\lambda_j^2 - 9}{2\lambda_j}. \tag{10.36}$$

Analogous computations demonstrate that all the arbitrary constants of the general solution in Eq. (10.31) are expressed as functions of the four $A_{1,j}$.

Therefore, we can kill the undesired effect of the two real eigenvalues $\lambda_{3,4}$ by simply choosing such initial conditions that $A_{1,3} = A_{1,4} = 0$. The corresponding particular solution is a small ellipse in the (x, y) plane centered in the libration point. Indeed, plugging $\lambda_{1,2} = \pm\mathbf{i}\, w$, into Eq. (10.31) we get

$$\delta x = \delta x_0 \cos w\tau - (\delta y_0/\alpha) \sin wt, \tag{10.37}$$

$$\delta y = \alpha\delta x_0 \sin w\tau + \delta y_0 \cos wt, \tag{10.38}$$

where the initial conditions are related to the arbitrary constants by

$$\delta x_0 = A_{1,1} + A_{1,2}, \quad \delta y_0 = \mathbf{i}\,\alpha(A_{1,2} - A_{1,1}), \tag{10.39}$$

and, from Eq. (10.36),

$$\alpha = \mathbf{i}\,\alpha_1 = -\mathbf{i}\,\alpha_2 = -\frac{9 + w^2}{2w} = -\sqrt{2\sqrt{7} + 5} \approx -3.21. \tag{10.40}$$

The corresponding solutions $\delta X = \delta\dot{x} - \delta y$ and $\delta Y = \delta\dot{y} + \delta x$, as derived from Eqs. (10.20) and (10.21), respectively, are readily obtained after differentiation of Eqs. (10.37) and (10.38).

In this way we have found two different types of periodic motion around the collinear points: On the one hand, small vertical oscillations as given by Eq. (10.27) will happen when the initial conditions $(x_0, y_0, \dot{x}_0, \dot{y}_0)$ are chosen null. On the other hand, small planar oscillations as given by Eqs. (10.37)–(10.38), and their corresponding velocities, will exist when choosing the initial conditions $z_0 = \dot{z}_0 = 0$. Note that the lack of commensurability between the vertical and planar frequencies given by Eqs. (10.28) and (10.33), respectively, prevents three-dimensional periodic orbits consisting of the combination of both infinitesimal motions.

10.2.3 Basic families of periodic orbits

We remark that the validity of the analytical solutions in Eqs. (10.27) and (10.37)–(10.38) is limited to the case of small oscillations about the libration points because they were obtained from the linearization of the flow. For larger displacements from the libration points, the non-linearities of the Hill problem render these solutions only approximately periodic. However, true periodic solutions of Eqs. (10.20)–(10.25) exist, and can be obtained by the computation of differential corrections to the initial conditions and period. Besides, since the Hill problem is conservative, periodic orbits are grouped in families which may be computed for variations of the Jacobi constant using standard differential corrections algorithms. Full details on the topic can be consulted in specialized texts, as, for instance, [22, 449, 500, 609, 610].

Using this procedure, the families of the so-called Lyapunov planar and vertical periodic orbits are readily computed [176]. As an example, the family of Lyapunov planar orbits for decreasing values of the Jacobi constant is shown in Fig. 10.8. It is characterized by the periodic orbits' stability, which, in the linear approximation, is obtained from the eigenvalues of the state transition matrix at the end of one period. The six eigenvalues occur in reciprocal pairs, as a consequence of the Hamiltonian character, and one of the eigenvalues takes the trivial value 1 because we are dealing with periodic orbits. Therefore, only two stability indices are needed, which are obtained from the sum of the non-trivial reciprocal pairs. Since the orbits are planar, the "horizontal" stability index b_h in Fig. 10.8 is related to in-plane perturbations, whereas the "vertical" stability index b_v is related to perturbations in the out-of-plane direction. When both stability indices are real numbers in the range $(-2, 2)$ the corresponding periodic orbit is stable, whereas any other case yields instability. Critical values ± 2 of one of these indices show possible bifurcation orbits from which new families of periodic orbits could emerge. Note that the stability indices are represented in Fig. 10.8 in an inverse hyperbolic sine scale.

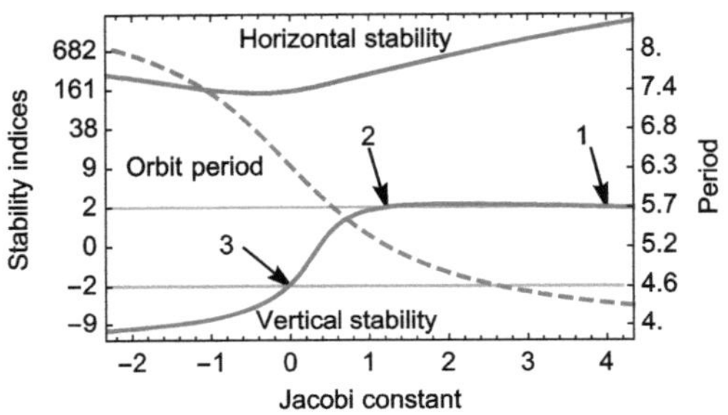

Figure 10.8: Family of planar Lyapunov orbits of the Hill problem. Arrows point to the vertical bifurcations at $C \approx 4, 1.2$ and 0 [396].

For the horizontal index to be always much higher than 2, planar Lyapunov orbits of the Hill problem are highly unstable. On the contrary, the vertical index takes critical values in different occasions, which are pointed to with arrows in Fig. 10.8, and labelled 1, 2, and 3. At point 1, two symmetric three-dimensional *halo* orbits emerge in a vertical bifurcation. At point 2, two new families of three-dimensional periodic orbits emerge, which will end in an orbit of the family of Lyapunov vertical orbits, thus constituting some kind of bridge between the two Lyapunov families. Both kind of orbits, halo orbits and orbits of the bridge family, will be discussed from the point of view of perturbation theory in Chapter 12. At point 3 a new bifurcation happens in which three-dimensional periodic orbits bifurcate with period doubling. For additional details on periodic orbits of the Hill problem, as well as other invariant manifolds, the reader is referred to [250].

11 Motion inside Hill's sphere

When the motion of the massless body takes place always inside Hill's sphere, the motion can be approached like a case of perturbed Keplerian motion. This is the case of typical science missions about planetary satellites, which require spacecraft orbits close to the surface of the satellite to take full advantage of the science instruments. The ratio between the orbital periods of the spacecraft and the natural satellite then becomes a small quantity that can be used for arranging the Hill problem Hamiltonian like a perturbation Hamiltonian in power series of a formal small parameter. Then, similarly to Garfinkel's procedure in §7.2, the mean anomaly and the longitude of the node in the rotating frame are stepwise eliminated from the three-degrees-of-freedom Hamiltonian by Lie transforms. While lower orders of the perturbation solution suffice for describing the qualitative features of the long-term dynamics in this particular region of the Hill problem, higher orders may be needed in mission-designing procedures for particular binary systems [419].

Third-body perturbations have the effect of destabilizing typical mapping orbits—which require a high-inclination and low-eccentricity design—in what is known as the *Kozai mechanism* [264, 311, 364, 365], also called the Lidov–Kozai resonance [450, 651]. The eccentricity–perigee dynamics is then crucial to the design of the nominal science orbit, and common strategies form maximizing lifetimes are based on the natural dynamics of the Kozai mechanism. For this aim, the accurate computation of the mean to osculating elements transformation becomes essential, yet simplified expressions are allowed in different orbital configurations.

On the other hand, the non-central mass distribution of the attracting body may have observable effects on the long-term dynamics, mitigating the instabilities that affect science orbits to some extent. Because common planetary satellites are in synchronous rotation with their orbital motion, the disturbing effect of the ellipsoidal figure of the central body is incorporated like a perturbation of the Hill problem Hamiltonian without augmenting the number of degrees of freedom, and hence is analogously approached by perturbations.

11.1 Perturbed Keplerian motion

The Hill problem Hamiltonian (10.17) can be written like $\mathcal{H} = \mathcal{H}_K + \mathcal{H}_C + \mathcal{H}_{3b}$, where $\mathcal{H}_K = \frac{1}{2}(X^2+Y^2+Z^2) - \mathcal{G}m_2/r$ is the Keplerian, $\mathcal{H}_C = -\dot{\vartheta}(xY - yX)$ stands for the Coriolis term, and $\mathcal{H}_{3b} = \frac{1}{2}\dot{\vartheta}^2(r^2 - 3x^2)$ accounts for third-body effects in the Hill problem approximation. Then, in those regions of phase space in which both the Coriolis effect and the third-body perturbations are small when compared to the Keplerian attraction, the Hill problem can be approached as a case of a perturbed Keplerian motion.

In the case in which $\mathcal{H}_{3b} \ll \mathcal{H}_C \ll \mathcal{H}_K$, one can take a formal small parameter that roughly indicates the smallness of the time varying ratio $\epsilon \sim \dot{\vartheta}/n$. Then, using Eq. (4.4),

https://doi.org/10.1515/9783110668513-011

the perturbation Hamiltonian (2.30) is arranged in the form

$$\mathcal{H}_{0,0} = -\frac{\mu}{2a}, \tag{11.1}$$

$$\mathcal{H}_{1,0} = -\dot{\vartheta}H = 2\mathcal{H}_{0,0}\frac{\dot{\vartheta}}{n}c\eta, \tag{11.2}$$

$$\begin{aligned}\mathcal{H}_{2,0} = \mathcal{H}_{0,0}\frac{\dot{\vartheta}^2}{n^2}\frac{r^2}{a^2}\frac{1}{4}\{2(2-3s^2+3s^2\cos 2\theta)+3[(1-c)^2\cos(2h-2\theta)\\ +2s^2\cos 2h+(1+c)^2\cos(2h+2\theta)]\},\end{aligned} \tag{11.3}$$

in which a, n, c, η, r, $\theta = f + \omega$, and $h = \Omega - \dot{\vartheta}t$ are the usual functions of the Delaunay canonical variables, and $\mathcal{H}_{m,0} = 0$ if $m \geq 3$. It is not a surprise to check that $\mathcal{H}_{2,0}$ is equivalent to twice Eq. (8.47) if replacing h by $h - \theta_{\odot}$; that is, when the latter is viewed in a frame rotating with the mean motion of the sun.

The long-term evolution of the flow is then investigated by removing the short-period terms from Eqs. (11.1)–(11.3) by perturbations based on Lie transforms. Due to the rotating-frame formulation, the longitude of the node in the rotating frame h evolves fast when compared with the slow motion of the argument of the periapsis. In consequence, a double-averaging procedure is appropriate. That is, short-period terms related to the mean anomaly are removed first by the usual Delaunay normalization, which is followed by a second Lie transformation that removes periodic terms related to h.

11.1.1 Short-period elimination

Thus, we apply the Lie transforms method to find the canonical transformation $(\ell, g, h, L, G, H) \mapsto (\ell', g', h', L', G', H', \epsilon)$ such that, up to some truncation order, removes the short-period effects from the Hill problem Hamiltonian in the prime variables. Since we are dealing with perturbed Keplerian motion, the Lie derivative is the one in Eq. (4.66) and the homological equation is solved by indefinite integration from Eq. (4.68), which is defined up to an arbitrary "constant" that will be used to guarantee the purely periodic character of the generating function.

On account of the fact that $\mathcal{H}_{1,0}$ does not depend on the mean anomaly, we choose $\mathcal{H}_{0,1} = \widetilde{\mathcal{H}}_{0,1} \equiv \mathcal{H}_{1,0}$, and $\mathcal{W}_1 = 0$. Therefore, there are no short-period corrections of the order of $\sim \dot{\vartheta}/n$. At second order we obtain $\widetilde{\mathcal{H}}_{0,2} = \mathcal{H}_{2,0}$ from Eq. (2.37), and the new Hamiltonian term is chosen by averaging $\mathcal{H}_{2,0}$ over the mean anomaly.

The averaging is made in closed form of the eccentricity with the help of the differential relation Eq. (4.62) between the mean and eccentric anomalies. Namely, $\mathcal{H}_{0,2} = \langle\mathcal{H}_{2,0}\rangle_{\ell} = \langle\mathcal{H}_{2,0}(r/a)\rangle_{u}$. We obtain

$$\mathcal{H}_{0,2} = -2!\frac{\mu}{a}\frac{\dot{\vartheta}^2}{n^2}\sum_{l=0}^{1} b_l(e)\sum_{k=-1}^{1}(-1)^{k-l}Q_{2,l,2k}(I)\cos(2kh+2lg), \tag{11.4}$$

which is twice Eq. (8.53). The term $\mathcal{W}_2$ is then computed from Eq. (4.70), to obtain

$$\mathcal{W}_2 = 2!L\frac{\dot{\vartheta}^2}{n^2}\sum_{l=0}^{1}\sum_{k=-1}^{1}\sum_{j=-3}^{3}(-1)^k B_{2,l,j}(e)Q_{2,l,2k}(I)\sin(ju+2kh+2lg), \tag{11.5}$$

which is twice Eq. (8.54), where the eccentricity polynomials $B_{2,l,j}$ are given in Table 8.6, and the inclination polynomials $Q_{2,l,2k}$ in Table 8.3. Derivation of short-period corrections from Eq. (11.5), which are of the order of $(\dot{\vartheta}/n)^2$, follows the standard sequence in Eq. (2.17).

After neglecting higher-order terms, we obtain the new Hamiltonian $\mathcal{K} = \mathcal{K}_{0,0} + \epsilon\mathcal{K}_{1,0} + \frac{1}{2}\epsilon^2\mathcal{K}_{2,0}$, in which the terms $\mathcal{K}_{i,0} = \mathcal{H}_{0,i}$ $(i = 0,1,2)$ are obtained by replacing the original by prime variables in Eqs. (11.1), (11.2), and (11.4), respectively.

11.1.2 Elimination of the node in the rotating frame

The single-averaged Hill problem Hamiltonian $\mathcal{K}$ is still of two degrees of freedom. In spite of different particular solutions having been reported in the literature [504, 505, 653, 655], as well as periodic orbits emanating from them [654], not unexpectedly, these known analytical solutions depend on special functions and do not provide much insight on the dynamics. On the other hand, the new Hamiltonian $\mathcal{K}$ still remains in the form of a perturbation Hamiltonian, in which, due to the rotating-frame formulation, the variable h is assumed to evolve faster than g. Therefore, the long-term motion can be explored after a new Lie transformation to double-prime variables that removes from the new Hamiltonian the periodic terms related to h''.

Like in Brouwer's elimination of the perigee in §6.3.1 or in the elimination of the longitude of the node from Garfinkel's tesseral Hamiltonian in §7.2.3, the Lie derivative (4.66) identically vanishes at each order, producing the shift between the orders by which the new Hamiltonian terms and the generating function terms get determined. The new homological equation becomes formally analogous to Eq. (6.47), where now

$$\mathcal{L}_1 = \frac{\partial\mathcal{K}_{1,0}}{\partial H'}\frac{\partial}{\partial h'} = -\dot{\vartheta}\frac{\partial}{\partial h'}, \tag{11.6}$$

and hence

$$\mathcal{V}_{m-1} = -\frac{1}{m\dot{\vartheta}}\int(\mathcal{K}_{0,m} - \widetilde{\mathcal{K}}_{0,m})\,\mathrm{d}h'. \tag{11.7}$$

Then $\mathcal{K}_{0,1} = \mathcal{K}_{1,0}$, and we select $\mathcal{K}_{0,2} = \langle\mathcal{K}_{2,0}\rangle_{h'}$. The only term from Eq. (11.4) that remains after averaging is the one with the summation index $k = 0$. In preparation of the computations following in §11.2, we arrange it in the general form

$$\mathcal{K}_{0,m} = -\frac{\mu}{2a}\frac{\dot{\vartheta}^m}{n^m}(c\eta)^{m^*}\sum_{l=0}^{m_0}\sum_{j=0}^{m_0-l}Q^*_{m,l,j}e^{2j}(es)^{2l}\cos 2lg, \tag{11.8}$$

where $m_0 = \lfloor \frac{1}{2}m \rfloor$ and $m^\star = m \bmod 2$, from the index convention in Eq. (6.5). For $m = 2$, we obtain the new inclination polynomials $Q^*_{2,0,0} = \frac{1}{2}(3c^2 - 1)$, $Q^*_{2,0,1} = \frac{3}{2}Q^*_{2,0,0}$, and $Q^*_{2,1,0} = \frac{15}{4}$.

The term $\mathcal{V}_1$ of the generating function is next computed from Eq. (11.7) with $m = 2$. We arrange the result like

$$\mathcal{V}_m = L'\eta^{(m-1)^\star} \frac{\dot{\vartheta}^m}{n^m} \sum_{l=-m'}^{m'} \sum_{k=1}^{m'} \sum_{j=0}^{m'-l'} Q^*_{m,l,k,j}(I) e^{2(j+l')} s^{2|l'-k|} \sin(2kh + 2lg), \tag{11.9}$$

with $l' = |l|$, $(m-1)^\star = (m-1) \bmod 2$, $m' = m_{-1} = \lfloor \frac{1}{2}(m+1) \rfloor$. At first order $m = 1$ and $Q^*_{1,\pm1,1,0} = \frac{15}{64}(c \pm 1)^2$, $Q^*_{1,0,1,0} = \frac{3}{16}$, and $Q^*_{1,0,1,1} = \frac{9}{32}$. The periodic corrections, which now are of the first order of $\dot{\vartheta}/n$, are computed as usual from Eq. (2.17).

The double-averaging procedure ends replacing prime by double-prime variables in the terms $\mathcal{K}_{0,i}$, $i = 0, 1, 2$, giving rise to the double-averaged Hamiltonian

$$\mathcal{Q} = -\frac{\mu}{2a}\left\{1 + 2\frac{\dot{\vartheta}}{n}c\eta - \frac{1}{4}\frac{\dot{\vartheta}^2}{n^2}[(3e^2 + 2)(3s^2 - 2) - 15e^2s^2\cos 2g'']\right\}, \tag{11.10}$$

in which both ℓ'' and h'' are cyclic variables. Therefore, $\mathcal{Q}$ is an integrable, one-degree-of-freedom Hamiltonian in (g'', G''), in which L'' and H'' are (formal) integrals. In particular, because $H'' = L''\sqrt{1-e^2}\cos I$, from Eq. (4.46), the constant value of the third component of the angular momentum vector creates a coupling between the eccentricity and the inclination. As shown in Fig. 11.1, this coupling leads, for given values L'' and H'', an increase of the eccentricity to yield a corresponding decrease of the inclination, and the highest eccentricities allowed for a given energy manifold $\mathcal{Q} = E$ of Eq. (11.10) correspond to orbits resting in the orbital plane of the primaries. A symmetric figure is obtained for the case of retrograde inclinations, whereas polar orbits ($H'' = 0$) remain polar. Like in Eq. (5.73), the dynamical parameters L'' and H'' in Fig. 11.1 are combined into a single parameter $y = c\eta = H''/L''$, which, on average, is the cosine of the inclination of the circular orbit for that case.

11.1.3 Third-body critical inclination. The Lidov–Kozai resonance

Neglecting constant terms in Eq. (11.10) does not affect the (g'', G'') dynamics. Then after scaling the time by $-\frac{1}{16}a^2\dot{\vartheta}^2$ we get the double-averaged Hamiltonian

$$\mathcal{Q}^* = \mathcal{Q}^*(g'', G''; y) \equiv (2 + 3e^2)(3c^2 - 1) + 15e^2(1 - c^2)\cos 2\omega, \tag{11.11}$$

where $c = y/\eta$. The Hamiltonian (11.11) is formally equivalent to the solar perturbation in Eq. (8.59), or to the combined lunisolar effects in Eq. (8.61), and can be integrated in terms of special functions [349, 350, 364].

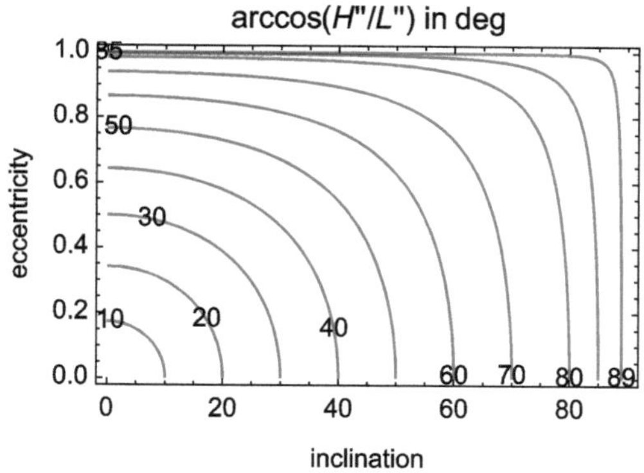

Figure 11.1: Eccentricity–inclination coupling in the double-averaged Hill problem.

Alternatively, as we did in §5.6.3, more insight in the long-term dynamics can be obtained from the graphic representation of the reduced phase space, which is done without need of integration by depicting contour plots of Eq. (11.11) for different values of y. An example is shown in Fig. 11.2, where the reduced phase space is presented in both the cylindrical (e, ω) representation and the planar eccentricity-vector representation. The situation found here is analogous to the bifurcation process that happens in the vicinity of the critical inclination of the satellite problem, previously discussed in §5.6. Indeed, circular orbits always remain circular, on average, and remain stable for the lower inclinations, as illustrated in the left column of Fig. 11.2. Eventually, their stability character changes to instability in a bifurcation process in which two stable elliptic orbits emerge with frozen argument of the periapsis at 90 and 270 deg (plots in the center column of Fig. 11.2). However, the second bifurcation from the circular orbits that happens in the artificial satellite problem, with the consequent return to stability (Fig. 5.4), does not occur now. Therefore, as shown in the plots in the right column of Fig. 11.2, the eccentricity of the orbits in the libration regions about the eccentric equilibria, as well as in the circulation region close to the separatrix, can vary by a large amount, contrary to the artificial satellite problem, with the consequent large variation of the inclination derived from the preservation of the third component of the angular momentum vector.

This phenomenon, which is known as the *Lidov–Kozai resonance*[1] [364, 450], has received a lot of attention in dynamical astronomy [485, 525, 607], but it also has important implications in mission designing of space probes.

Beyond the insights provided by the contour plot representation, the proper discussion of the equilibria should be done in variables on a compact manifold as those of Eq. (5.75). However, for the low eccentricities of science orbits about planetary satel-

1 A denomination that Vashkov'yak attributes to A. I. Neishtadt [651]. For a recent review on the topic and contributions by other authors, most notably H. von Zeipel, see [313].

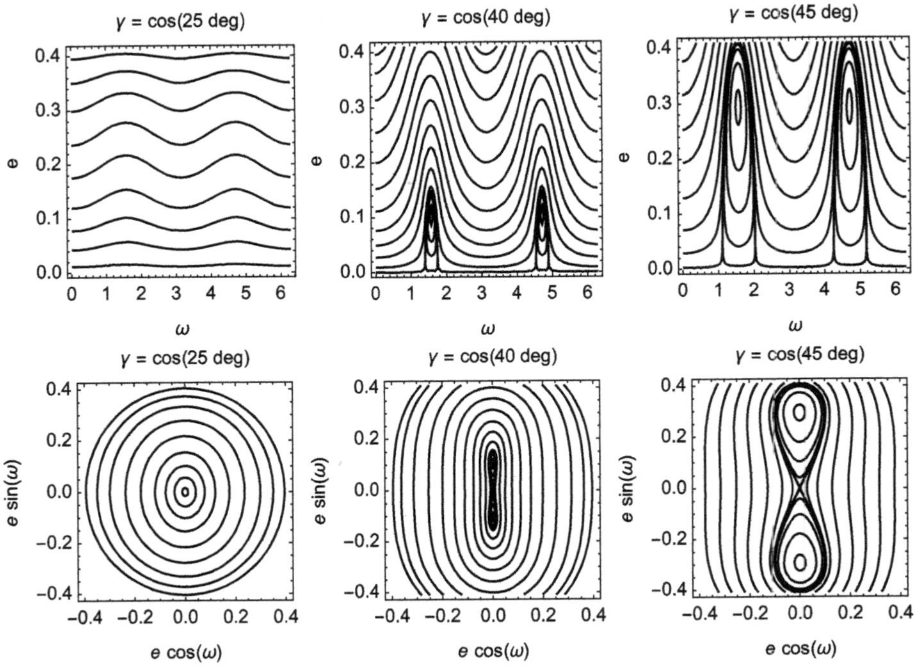

Figure 11.2: Contour plots of the double-averaged Hill problem Hamiltonian (11.11) for different values of γ. Compare with Figs. 5.3 and 5.4.

lites, in practice it is enough to use the components of the eccentricity vector in the orbital plane (C, S) defined in Eq. (5.74). Their average variation was given in Eq. (8.66), from which, replacing the Hamilton equations of the double-averaged Hamiltonian (11.11), we obtain

$$\frac{\mathrm{d}C}{\mathrm{d}t} = -12\frac{S}{L''}\left(3\eta - 5\gamma^2\frac{1-C^2}{\eta^3}\right), \tag{11.12}$$

$$\frac{\mathrm{d}S}{\mathrm{d}t} = -12\frac{C}{L''}\left(2\eta + 5\gamma^2\frac{S^2}{\eta^3}\right), \tag{11.13}$$

where $\eta = \eta(C, S) \equiv \sqrt{1 - C^2 - S^2}$.

The variation equations (11.12)–(11.13) show that circular orbits, $C = S = 0$, are always equilibria of the double-averaged flow. Besides, it is simple to check that Eq. (11.13) only vanishes when $C = 0$. Then the vanishing of Eq. (11.12) when $S \neq 0$ yields another equilibrium solution,

$$C = 0, \quad S = \pm(1 - \sqrt{5/3}\,|\gamma|)^{1/2}. \tag{11.14}$$

That is, if $|\gamma| < \sqrt{3/5}$ then frozen orbits with periapsis $\omega = \pm\frac{\pi}{2}$ exist whose eccentricity $e = |S|$ is given by Eq. (11.14). In the limit $\gamma = \sqrt{3/5}$ these equilibria bifurcate from circular orbits with $I_{\text{circular}} = \arccos\sqrt{3/5} \approx 39.23$ deg, which is sometimes called the third-body critical inclination.

11.2 Higher-order dynamics

While the lower-order truncation of the perturbation solution used is useful in disclosing the main facts of the qualitative dynamics, and in particular the Lidov–Kozai resonance, it establishes a symmetry of direct and retrograde orbits that is not part of the Hill problem. On the other hand, neglecting short-period terms of the order of $(\dot{\vartheta}/n)^3$ and higher may be insufficient for making reasonably accurate predictions. For instance, from the physical characteristics of the Sun–Mercury system, $(\dot{\vartheta}/n)^3 = \mathcal{O}(10^{-10})$ for a grazing orbit about Mercury, so the second-order truncation of the perturbation solution seems accurate enough. On the contrary, $(\dot{\vartheta}/n)^3 = \mathcal{O}(10^{-5})$ for grazing orbits about the Galilean moons, whereas in the case of Enceladus $(\dot{\vartheta}/n)^3 \approx 5\times10^{-4}$ only. The situation, of course, aggravates for higher altitudes of the orbits under investigation.

Higher orders of the perturbation approach provide more accurate predictions, but they also reveal additional details on the Hill problem dynamics. In particular, they show that the inclination at which the stability of the circular orbits changes varies with the semimajor axis of the orbit. Besides, the curve where eccentric orbits bifurcate from circular is no longer symmetric for direct and retrograde inclinations, a fact that had been already checked with numerical methods in the case of Europa [429].

11.2.1 Degeneracy at the third order

At third order, the known terms that enter the homological equation are those previously given in Eq. (6.88), which in the current case are limited to $\widetilde{\mathcal{H}}_{0,3} = 3\{\mathcal{H}_{0,1};\mathcal{W}_2\}$. After evaluation of the Poisson bracket, the term $\mathcal{H}_{0,3}$ is chosen by averaging the known terms over the mean anomaly. As before, the averaging is computed in closed form of the eccentricity, yielding[2]

$$\mathcal{H}_{0,3} = \langle \widetilde{\mathcal{H}}_{0,3}(r/a) \rangle_u = 0. \tag{11.15}$$

The term $\mathcal{W}_3$ is then computed from Eq. (4.70), to obtain

$$\mathcal{W}_3 = L\frac{\dot{\vartheta}^3}{n^3} \sum_{l=0}^{1} \sum_{k=-1}^{1} \sum_{j=-4}^{4} B_{3,l,j}(e) Q_{3,l,2k}(I) \sin(ju + 2kh + 2lg), \tag{11.16}$$

where $Q_{3,0,\pm2} = \pm\frac{1}{64}s^2$, $Q_{3,1,0} = 0$, $Q_{3,1,\pm2} = \mp\frac{1}{64}(c\pm1)^2$, which are no longer the same as the inclination polynomials of the equatorial frame formulation in Table 8.3, and the eccentricity polynomials $B_{3,l,j}$ are given Table 11.1.

2 Because $\mathcal{W}_2$ is purely periodic in the mean anomaly, as checked from Eq. (11.5), the term $\mathcal{H}_{0,3}$ is different from the one in Eq. (A.4) of [418]. Similar differences appear in higher orders.

Table 11.1: Eccentricity polynomials $B_{3,l,j}$ in Eq. (11.16).

j	$l=0$	$l=1$
0	$-9e^2(3e^2-16)$	$-9e^2(9\eta^2+13)$
±1	$9e(3\eta^2+13)$	$\mp 9e(\eta\pm1)(8\eta^2\pm\eta+13)$
±2	$-3e^2(4\eta^2+17)$	$-3(\eta\pm1)^2(8\eta^2\pm6\eta-17)$
±3	$11e^3$	$\pm e(\eta\pm1)^2(6\eta\mp11)$
±4	$-\frac{3}{4}e^4$	$\frac{3}{4}e^2(\eta\pm1)^2$

Because $\mathcal{K}_{3,0}=\mathcal{H}_{0,3}=0$ the new Hamiltonian in prime variables is the same as the second-order one, yet it is now accurate to $\mathcal{O}(\dot\vartheta^3/n^3)$. Therefore, we can compute a third-order approach of the Lie transformation that eliminates the node.

We first compute the known terms $\widetilde{\mathcal{K}}_{0,3}$ from Eq. (6.88) replacing $\mathcal{H}$ by $\mathcal{K}$. Then, like in §11.1.2, the longitude of the node is removed from the homological equation making $\mathcal{K}_{0,3}=\langle\widetilde{\mathcal{K}}_{0,3}\rangle_{h'}$. The term $\mathcal{K}_{0,3}$ takes the general form of Eq. (11.8) with $m=3$ and $Q^*_{3,0,0}=\frac{27}{16}s^2$, $Q^*_{3,0,1}=-\frac{27}{32}(17s^2-50)$, and $Q^*_{3,1,0}=\frac{405}{32}$. The homological equation (11.7) yields again Eq. (11.9) with $m=2$ and $Q^*_{2,\pm1,1,0}=\frac{45}{128}(c\pm1)^2(3c\mp2)$, $Q^*_{2,0,1,0}=-\frac{9}{32}c$, and $Q^*_{2,0,1,1}=\frac{153}{64}c$.

After neglecting constant terms, and scaling the double-averaged Hamiltonian analogously to Eq. (11.11), we obtain

$$\begin{aligned}\mathcal{Q}^* &= (2+3e^2)(2-3s^2)+15e^2s^2\cos 2g''\\ &\quad+\frac{9}{8}\varepsilon y[2s^2+(50-17s^2)e^2+15s^2e^2\cos 2g''],\end{aligned}\tag{11.17}$$

where we abbreviated $\varepsilon=\dot\vartheta/n$. The coefficient $y=H''/L''$ breaks the symmetry of direct and retrograde inclination orbits introduced by the early truncation to the second order, and the variation equations of the semi-equinoctial elements undergo concomitant modifications of order ε with respect to Eqs. (11.12)–(11.13). That is,

$$\frac{\mathrm{d}C}{\mathrm{d}t}=-12\frac{S}{L''}\left[3\eta\left(1-\frac{9}{8}y\varepsilon\right)-5y^2\frac{1-C^2}{\eta^3}\left(1+\frac{9}{8}y\varepsilon\right)\right],\tag{11.18}$$

$$\frac{\mathrm{d}S}{\mathrm{d}t}=-12\frac{C}{L''}\left[2\eta\left(1+\frac{9}{2}y\varepsilon\right)+5y^2\frac{S^2}{\eta^3}\left(1+\frac{9}{8}y\varepsilon\right)\right].\tag{11.19}$$

While these modifications do not affect the existence of circular orbits, $C=S=0$, which continue to be equilibria of the double-averaged problem, they slightly change the eccentricity of the eccentric equilibria. The latter still exist with the argument of the periapsis frozen at $g=\pm\frac{\pi}{2}$ (or $C=0$) for which Eq. (11.19) identically vanishes, and Eq. (11.18) vanishes too when

$$e=\left(1-|y|\sqrt{\frac{5}{3}}\sqrt{\frac{8+9y\varepsilon}{8-9y\varepsilon}}\right)^{1/2}\approx\left[1-|y|\sqrt{\frac{5}{3}}\left(1+\frac{9}{8}\varepsilon y\right)\right]^{1/2}.\tag{11.20}$$

In the limit $e = 0$, Eq. (11.20) gives the curves $y = y(\varepsilon)$ along which circular orbits change their stability in a bifurcation process where the eccentric orbits emerge. Instead of solving the cubic, it is simpler to give y in the implicit form

$$0 = 3 - 5y^2 - \frac{9}{8}\varepsilon(3 + 5y^2)y, \tag{11.21}$$

which shows the bifurcation lines as perturbations of the lines $3 - 5y^2 = 0$ of the third-body critical inclination. For direct-inclination orbits ($y > 0$) with $g = \pm\frac{\pi}{2}$, the bifurcation line starts from the critical direct inclination, which continuously increases for increasing values of ε. The bifurcation from retrograde circular orbits ($y < 0$) starts from the critical retrograde inclination, and approaches the equator for increasing values of ε. This curve ends in a retrograde equatorial orbit ($y = -1$) when $\varepsilon = 2/9 \approx 0.22$.

This order of approximation of the perturbation solution produces an additional root for $S = 0$. This value makes Eq. (11.18) vanish identically, whereas Eq. (11.19) only vanishes along the line $y\varepsilon = -\frac{2}{9}$. The nature of this bifurcation is degenerate, and orbits with the periapsis at either 0 or π and any eccentricity $0 < e < 1$ become frozen orbits along this line.

Bifurcation lines separate regions with different flows. They are illustrated in the left plot of Fig. 11.3 for the third-order double-averaged Hamiltonian. The flow in regions 1 and 3 is qualitatively the same as the flow in the left plot of Fig. 11.2, while the flow in region 2 of Fig. 11.3 is qualitatively the same as the flow in the center and right plots of Fig. 11.2. The right plot of Fig. 11.3 shows the (incorrect) flow provided by the third-order truncation of the perturbation solution in region 4.

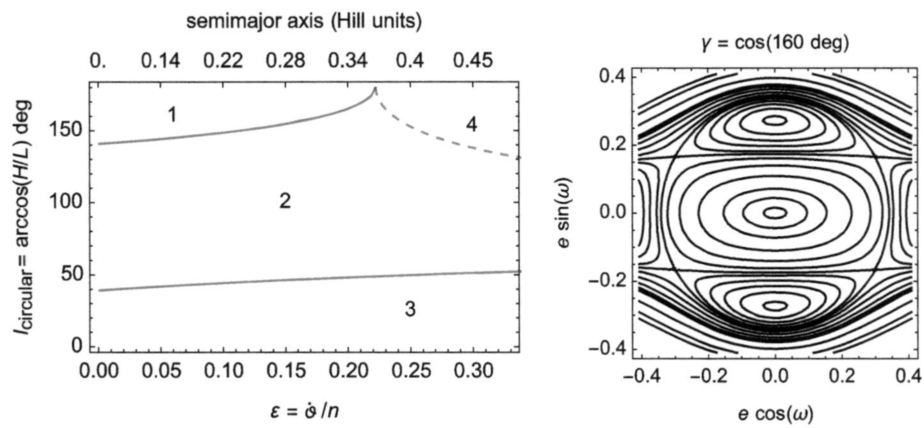

Figure 11.3: Left: Bifurcation lines of the third-order truncation of the Hill problem in the parameters plane. Right: illusory flow in region 4 for $\varepsilon = 0.25$.

The degeneracy of the third-order truncation indicates that we must proceed to higher orders in the normalization procedure. In general, the computation of higher orders will solve the issue, yet one must be aware that progressing to higher orders is not always a guarantee of success [168].

11.2.2 Fourth-order corrections

At fourth order, from the known terms in Eq. (6.89) we select $\mathcal{H}_{0,4}$ by averaging $\widetilde{\mathcal{H}}_{0,4}$ over the mean anomaly. We arrange it in the general form

$$\mathcal{H}_{0,m} = -\frac{\mu}{2a}\frac{\dot{\vartheta}^m}{n^m}\eta^{m^*}\sum_{l=0}^{m'}\sum_{k=-m'}^{m'}\sum_{j=0}^{m_0-l} Q_{m,l,k,j}(I)e^{2j+2l}\cos(2kh+2lg), \tag{11.22}$$

where $m^* = m \bmod 2$ and $m_0 = \lfloor \frac{1}{2}m \rfloor$, from Eq. (6.5), $m' = m_0 - m^*$, and the inclination polynomials $Q_{4,l,k,j}$ are given in Table 11.2. The fourth-order term of the generating function of the short-period elimination is then computed from Eq. (4.70), and is found to comprise 119 trigonometric terms.

Table 11.2: Inclination polynomials $Q_{4,l,k,j}$ in Eq. (11.22).

0,0,0 :	$-\frac{1}{32}(189c^4 + 846c^2 + 141)$	0,0,1 :	$\frac{27}{32}(209c^4 + 190c^2 + 377)$
0,±1,0 :	$\frac{9}{16}s^2(7s^2 - 34)$	0,±1,1 :	$-\frac{171}{16}s^2(11s^2 - 24)$
0,±2,0 :	$-\frac{63}{64}s^4$	0,±2,1 :	$\frac{1881}{64}s^4$
1,0,0 :	$-\frac{27}{32}s^2(195s^2 - 278)$	1,0,1 :	$\frac{9}{64}s^2(555s^2 - 686)$
1,±1,0 :	$\frac{27}{16}(c \pm 1)^2(65c^2 \mp 65c + 37)$	1,±1,1 :	$-\frac{9}{32}(c \pm 1)^2(185c^2 \mp 185c + 79)$
1,±2,0 :	$\frac{1755}{64}(c \pm 1)^2 s^2$	1,±2,1 :	$-\frac{1665}{128}(c \pm 1)^2 s^2$
2,0,0 :	$\frac{5535}{256}s^4$	0,0,2 :	$-\frac{189}{256}(143c^4 + 170c^2 + 239)$
2,±1,0 :	$\frac{1845}{128}(c \pm 1)^2 s^2$	0,±1,2 :	$\frac{63}{128}s^2(143s^2 - 342)$
2,±2,0 :	$\frac{1845}{512}(c \pm 1)^4$	0,±2,2 :	$-\frac{9009}{512}s^4$

Next, we change original variables by primes in Eq. (11.22) to compute $\mathcal{K}_{4,0} = \mathcal{H}_{0,4}$. The known terms at this order of the elimination of the node $\widetilde{\mathcal{K}}_{0,4}$ are those shown in Eq. (6.89), in which we now change $\mathcal{H}$ by $\mathcal{K}$. Then we select the new Hamiltonian term $\mathcal{K}_{0,4} = \langle \widetilde{\mathcal{K}}_{0,4} \rangle_{h'}$. Once more, it takes the form of Eq. (11.8), with $m = 4$ and corresponding inclination polynomials in Table 11.3. The homological equation (11.7) produces $\mathcal{V}_3$ in the form of Eq. (11.9) with $m = 3$ and the corresponding inclination polynomials of Table 11.4.

After neglecting constant terms and scaling the refined double-averaged Hamiltonian, we compute the equilibria from the Hamilton equations of the double-averaged flow. The condition $C = 0$, $e = |S|$ gives new values of the eccentric equilibria with the

Table 11.3: Fourth- and fifth-order inclination polynomials $Q^*_{m,l,j}$ in Eq. (11.8).

l,j	$m=4$	$m=5$
0,0	$\frac{3}{64}(9c^4-726c^2-67)$	$\frac{15}{1024}(351c^5-19666c^2-30349)$
0,1	$\frac{27}{32}(344c^4+253c^2+329)$	$\frac{15}{1024}(209763c^5+544816c^2+397761)$
0,2	$-\frac{27}{512}(5407c^4+2794c^2+2527)$	$-\frac{45}{8192}(1155789c^5+845358c^2+325853)$
1,0	$\frac{27}{32}(345c^2+68)$	$\frac{45}{1024}(70155c^2+104101)$
1,1	$-\frac{9}{128}(4035c^2+307)$	$-\frac{45}{2048}(338805c^2+168689)$
2,0	$\frac{945}{512}$	$-\frac{8932275}{8192}$

Table 11.4: Third- and fourth-order inclination polynomials $Q^*_{m,l,k,j}$ in Eq. (11.9).

l,k,j	$m=3$	$m=4$
±2,1,0	$-\frac{2385}{4096}(c\pm1)^2$	$-\frac{135}{65536}(c\pm1)^2(26730c\mp9601)$
±2,2,0	$\frac{7065}{32768}(c\pm1)^3$	$\frac{2025}{131072}(c\pm1)^4(939c\mp140)$
±1,1,0	$\frac{27}{512}(c\pm1)^2$ $\times(310c^2\mp175c+79)$	$\frac{15}{16384}(c\pm1)^2$ $\times(252639c^3\mp84861c^2+118641c\mp17939)$
±1,1,1	$-\frac{9}{1024}(c\pm1)^2$ $\times(2125c^2\mp1315c+563)$	$-\frac{15}{32768}(c\pm1)^2$ $\times(986418c^3\mp737271c^2+382002c\pm1361)$
±1,2,0	$\frac{405}{2048}(c\pm1)^2$	$\frac{135}{16384}(c\pm1)^2(387c\pm226)$
±1,2,1	$\frac{5445}{8192}(c\pm1)^2$	$\frac{135}{32768}(c\pm1)^2(13311c\mp1502)$
0,1,0	$\frac{9}{256}(13c^2-57)$	$-\frac{3}{4096}c(1539c^2+11197)$
0,1,1	$\frac{9}{256}(904c^2+215)$	$\frac{3}{8192}c(1269351c^2+546293)$
0,1,2	$-\frac{9}{2048}(8131c^2+1355)$	$-\frac{27}{16384}c(481869c^2+81992)$
0,2,0	$-\frac{261}{2048}$	$\frac{7047}{8192}c$
0,2,1	$\frac{333}{512}$	$\frac{38151}{8192}c$
0,2,2	$\frac{12627}{16384}$	$\frac{5342841}{65536}c$

periapsis at $\pm\frac{\pi}{2}$, which, in the limit case $e\to0$, provide the new bifurcation line of circular orbits

$$\sum_{j=0}^{m-2}\Gamma_j(y)\varepsilon^j=0, \tag{11.23}$$

with $\Gamma_0=3-5y^2$, $\Gamma_1=-\frac{9}{8}(5y^2+3)y$, and $\Gamma_2=-\frac{1}{64}(2070y^4-193y^2+783)$. The term Γ_2 not only introduces observable quantitative changes, but it also corrects the degeneracy of the third-order truncation. Now, bifurcations from retrograde circular orbits exist for all values of ε of interest, as shown in Fig. 11.4. Region 4 no longer exists and circular direct orbits can remain stable for higher inclinations than retrograde ones.

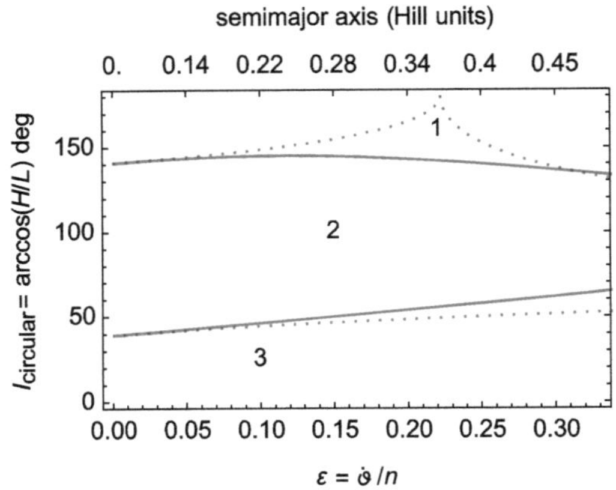

Figure 11.4: Bifurcation lines of circular orbits of the fourth-order truncation (full lines) superimposed to the analogous lines in Fig. 11.3 (dotted lines).

11.2.3 Higher-order refinements

Proceeding to higher orders of the Lie transforms method is standard, although the solution of non-trivial integrals can appear in the procedure. Strategies like those developed in [2, 324] may be of definitive help in that case.

Thus, at fifth order we get the new, single-averaged Hamiltonian term in the general form of Eq. (11.22) with $m = 5$ and the inclination polynomials of Table 11.5. It is next written in the prime variables to obtain $\mathcal{K}_{5,0} = \mathcal{H}_{0,5}$. The Fourier series that provides the fifth-order (purely periodic) term of the generating function is found to comprise 132 trigonometric terms.

Table 11.5: Inclination polynomials $Q_{5,l,k,j}$ in Eq. (11.22).

$0,0,0: -\frac{15}{2}c(53c^2+44)$	$0,\pm1,0: -\frac{795}{4}cs^2$	$1,0,0: \frac{4545}{4}cs^2$	$1,0,1: -\frac{765}{4}cs^2$
$0,0,1: \frac{75}{8}c(379c^2+555)$	$0,\pm1,1: \frac{28425}{16}cs^2$	$1,\pm1,0: \frac{1515}{4}(c\pm1)^2(3c\mp2)$	
$0,0,2: -\frac{225}{8}c(45c^2+58)$	$0,\pm1,2: -\frac{10125}{16}cs^2$	$1,\pm1,1: -\frac{255}{4}(c\pm1)^2(3c\mp2)$	

The node is then removed after computing the known terms $\widetilde{\mathcal{K}}_{0,5}$ from Deprit's recursion (2.15). The new, double-averaged Hamiltonian term $\mathcal{K}_{0,5} = \langle\widetilde{\mathcal{K}}_{0,5}\rangle_{h'}$ takes the form of Eq. (11.8) with the inclination polynomials $Q^*_{5,l,j}$ given in Table 11.3. The homological equation (11.7) is solved, again, in the form of Eq. (11.9), with $m = 4$ and the corresponding inclination polynomials $Q^*_{m,l,k,j}$ of Table 11.4.

The bifurcation line from circular orbits is given by Eq. (11.23) with $m = 5$. The new term $\Gamma_3 = -\frac{1}{1024}y(14243 + 104498y^2 + 70155y^4)$ prevents the bifurcation of circular retrograde inclination orbits to exist for moderate values of ε, and the degeneracy of equilibria with $\omega = 0, \pi$ occurs at the new bifurcation line

$$\frac{1}{128}\gamma(8819\gamma^2 + 14793)\varepsilon^3 + \frac{1}{64}(1469\gamma^2 + 1191)\varepsilon^2 + 9\gamma\varepsilon + 2 = 0,$$

which is a refinement of the line $9\gamma\varepsilon + 2 = 0$ found at third order. The situation is again amended at the next order, $m = 6$, for which this last bifurcation line ceases to exist, and the bifurcation from circular orbits in Eq. (11.23) is complemented with the term

$$\Gamma_4 = -\frac{1}{12288}(3086640\gamma^6 + 4388089\gamma^4 + 2133651\gamma^2 + 102040).$$

Recall that the differences between the coefficients Γ_3 and Γ_4 from analogous ones in [419] are a consequence of the purely periodic nature of the generating function of the short-period elimination computed here.

The sixth-order terms of the single- and double-averaged Hamiltonians still adhere to the general form of Eqs. (11.22) and (11.8), respectively. The former comprises 58 terms and corresponding inclination polynomials are not provided to avoid long listings. The terms $Q^*_{6,l,j}$ of the latter are provided in Table 11.6.

Table 11.6: Inclination polynomials $Q^*_{6,l,j}$ in Eq. (11.8).

0,0 :	$-\frac{15}{2048}(1737c^6 + 254540c^4 + 1402545c^2 + 268346)$
0,1 :	$\frac{15}{2048}(9265131c^6 + 24834958c^4 + 21356085c^2 + 4723182)$
0,2 :	$-\frac{15}{16384}(303051753c^6 + 346139368c^4 + 190389621c^2 + 23680002)$
0,3 :	$\frac{45}{16384}(83046117c^6 + 47146752c^4 + 24371729c^2 + 4316018)$
1,0 :	$\frac{45}{1024}(1543320c^4 + 1877096c^2 + 685157)$
1,1 :	$-\frac{15}{4096}(88656795c^4 + 59142109c^2 - 838246)$
1,2 :	$\frac{45}{16384}(103612815c^4 + 51973518c^2 + 3613243)$
2,0 :	$-\frac{75}{16384}(10323423c^2 - 351202)$
2,1 :	$\frac{135}{16384}(7081485c^2 + 404026)$
3,0 :	$\frac{30290625}{16384}$

The higher-order refinements of the perturbation solution further increase the maximum inclination of the region where circular direct-inclination orbits remain stable. This is illustrated in Fig. 11.5, in which the bifurcation line of the sixth-order solution (gray curve) is shown jointly with the fourth-order bifurcation line (dashed curve) and the classical third-body critical inclination (dotted horizontal line). This fact would allow one to find an almost circular stable nominal orbit with $I \approx 60°$—contrary to the 39.23° limit of the classical result—for a prospective science orbit about Enceladus with a semimajor axis about twice the equatorial radius of Enceladus (of ~ 256 km), which would provide global coverage for the science mission [96, 585]. When expressed in

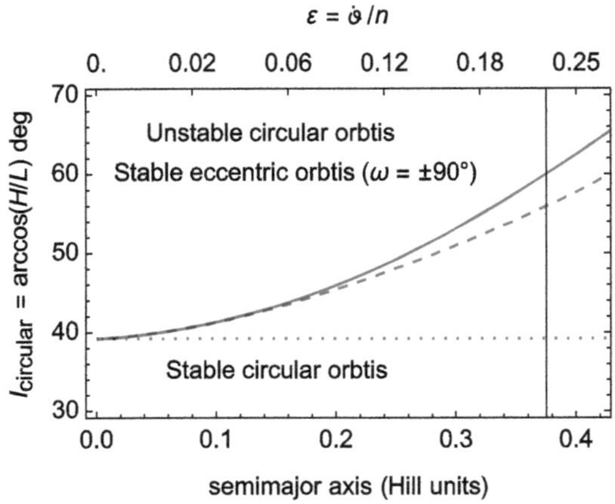

Figure 11.5: Departure of the bifurcation lines of the sixth- (full line) and fourth-order solutions (dashed) from the third-body critical inclination (dotted horizontal line). The vertical line marks 2 times the equatorial radius of Enceladus (after [419]).

units of the Hill problem, this semimajor axis is $a \sim 0.375$, and is marked with a vertical line in Fig. 11.5 to highlight how the sixth-order solution modifies the fourth-order prediction by about 4 degrees.

Using the values in §10.2, we find that, for a science orbit about Enceladus with $a = 512\,\text{km}$, the approximation provided by the Hill problem neglects terms $(a/d)^3 = \mathcal{O}(10^{-8})$. Besides, $n = (\mathcal{G}m_2/a^3)^{1/2} \approx 2 \times 10^{-4}\text{s}^{-1}$ and, since $\dot{\vartheta} = 5.3 \times 10^{-5}\text{s}^{-1}$, a sixth-order perturbation solution neglects $\mathcal{O}(\dot{\vartheta}/n)^7 = \mathcal{O}(10^{-5})$, which provides an acceptable approximation to the Hill problem dynamics.

11.3 The case of planetary satellites

While the dynamics of planetary satellites is generally well described by the Hill problem equations, typical science orbits evolve close to the surface of the planetary satellite, thus making the mass-point approximation invalid. In this case the primary at the origin is replaced in first approximation by a triaxial ellipsoid. In addition, planetary satellites commonly orbit their mother planets in synchronous rotation with their mean motion, and the axial tilt is very small and can be neglected. When that is the case, the perturbed Hill problem remains invariant in the rotating frame.

The Hill problem Hamiltonian (11.1)–(11.3) is then supplemented with additional terms that take the non-centralities of the gravitational potential into account [365, 451]. The more relevant terms of this perturbation are those related to the oblateness and dynamical ellipticity, yet the influence of a possible latitudinal asymmetry is sometimes considered [389, 421, 548]. On the other hand, most planetary satellites fit to the hypothesis of hydrostatic equilibrium, which permits one to reduce the number of physical parameters of the perturbation model taking $C_{2,2} = \frac{3}{10}J_2$ and $J_3 = 0$ [88].

Assuming that the principal axis of inertia of the triaxial ellipsoid are aligned with the rotating frame, the part of disturbing potential due to the non-central gravitation of the smaller mass body is obtained from Eq. (5.2),

$$V_\otimes = -\frac{\mu}{r}\frac{R_\otimes^2}{r^2}\left[\frac{1}{2}J_2\left(1-3\frac{z^2}{r^2}\right)+3C_{2,2}\frac{x^2-y^2}{r^2}\right],$$

where the subindex $_\otimes$ is used to denote the planetary satellite. The Cartesian coordinates are expressed in polar-nodal variables using Eq. (4.4) in which the angle ν is replaced by the longitude of the node in the rotating frame h. That is,

$$\begin{aligned}V_\otimes = {}& -\frac{1}{4}\frac{\mu}{r}\frac{R_\otimes^2}{r^2}\{J_2(2-3s^2+3s^2\cos 2\theta)+3C_{2,2}\\ &\times[(1-c)^2\cos(2h-2\theta)+2s^2\cos 2h+(1+c)^2\cos(2h+2\theta)]\}\end{aligned}$$

A common assumption is that the disturbing effects due to the ellipsoidal figure of the natural satellite are of the same order as the planetary third-body perturbations.[3] Therefore, the second-order term of the usual perturbation Hamiltonian (2.30) comprises both effects. Thus, $\mathcal{H}_{0,0}$ and $\mathcal{H}_{1,0}$ are given by Eqs. (11.1) and (11.2), respectively, while $\mathcal{H}_{2,0} = \mathcal{H}_{2,0,\mathrm{Hill}} + \mathcal{H}_{2,0,\otimes}$, with $\mathcal{H}_{2,0,\mathrm{Hill}}$ given by Eq. (11.3), and, applying the hydrostatic equilibrium condition to the disturbing potential $V_\otimes$,

$$\begin{aligned}\mathcal{H}_{2,0,\otimes} = {}& \mathcal{H}_{0,0}\frac{\dot\vartheta^2}{n^2}\beta^2\frac{a^3}{r^3}\bigg\{2-3s^2+3s^2\cos 2\theta+\frac{9}{10}[(1-c)^2\cos 2(h-\theta)\\ &+2s^2\cos 2h+(1+c)^2\cos 2(h+\theta)]\bigg\},\end{aligned} \tag{11.24}$$

where the function

$$\beta = J_2^{1/2}\frac{R_\otimes/a}{\dot\vartheta/n} \tag{11.25}$$

provides an indication on how the strength of the oblateness perturbation relative to the third-body gravitational pull varies with the orbiter's semimajor axis. When expressed in Delaunay variables β depends only on the Delaunay action L, and its derivative with respect to L can be written in the form $\partial\beta/\partial L = -5\beta/L$.

Like in the Hill problem alone, to study the long-term orbital behavior of planetary satellites we eliminate the short-period effects by means of two consecutive Lie transformations. The first one removes the short-period effects related to the mean anomaly of the orbiter, and the second the remaining periodic effects related to the longitude of the node in the rotating frame.

---

3 This would not be the case of the science orbit about Enceladus discussed previously, where the ellipsoidal figure perturbations would be of higher order than the third-body effects.

11.3.1 Elimination of the mean anomaly

Like in the Hill problem alone in §11.1.1, we make $\mathcal{H}_{0,1} = \mathcal{H}_{1,0}$, $\mathcal{W}_1 = 0$ and, in consequence, there are not short-period corrections of order $\dot{\vartheta}/n$.

At the second order the known terms are simply $\widetilde{\mathcal{H}}_{0,2} = \mathcal{H}_{2,0}$, from which the term $\mathcal{H}_{0,2}$ is chosen by averaging. That is, $\mathcal{H}_{0,2} = \langle\mathcal{H}_{2,0}\rangle_\ell \equiv \langle\mathcal{H}_{2,0,\mathrm{Hill}}\rangle_\ell + \langle\mathcal{H}_{2,0,\otimes}\rangle_\ell$, where the term $\langle\mathcal{H}_{2,0,\mathrm{Hill}}\rangle_\ell$ has been computed in Eq. (11.4). The averaging $\langle\mathcal{H}_{2,0,\otimes}\rangle_\ell = \langle\mathcal{H}_{2,0,\otimes}\, r/(a^2\eta)\rangle_f$ is standard. We obtain

$$\mathcal{H}_{0,2,\otimes} = \mathcal{H}_{0,0}\frac{\dot{\vartheta}^2}{n^2}\frac{\beta^2}{\eta^3}\left(2 - 3s^2 + \frac{9}{5}s^2\cos 2h\right). \tag{11.26}$$

Now, making $\mathcal{W}_2 = \mathcal{W}_{2,\mathrm{Hill}} + \mathcal{W}_{2,\otimes}$ the homological equation Eq. (4.67) is decomposed into two parts that are readily solved in closed form of the eccentricity using Eqs. (4.70) and (4.69), respectively. The term $\mathcal{W}_{2,\mathrm{Hill}}$ is the same as $\mathcal{W}_2$ in Eq. (11.5), whereas

$$\begin{aligned}\mathcal{W}_{2,\otimes} = -L\frac{\dot{\vartheta}^2}{n^2}\frac{\beta^2}{\eta^3}\Bigg\{&\phi\left(1 - \frac{3}{2}s^2 + \frac{9}{10}s^2\cos 2h\right)\\ &+\sum_{l=0}^{1}\sum_{k=-1}^{1}\sum_{j=0}^{2l+1}\left[Q_{l,k}B_{l,j}\sin(jf + 2kh + 2lg)\right]\Bigg\},\end{aligned} \tag{11.27}$$

where $Q_{1,\pm1} = \frac{3}{10}(c \pm 1)^2$, $Q_{1,0} = s^2$, $Q_{0,\pm1} = \frac{9}{20}s^2$, and $Q_{0,0} = \frac{1}{2}(3c^2 - 1)$; and $B_{0,1} = e$, $B_{1,0} = \frac{1}{4}(2\eta + 1)(1 - \eta)/(1 + \eta)$, $B_{1,1} = \frac{3}{4}e$, $B_{1,2} = \frac{3}{4}$, and $B_{1,3} = \frac{1}{4}e$. Note that, different from [389], the generating function is free from long-period terms. That is, $\langle\mathcal{W}_2\rangle_\ell = 0$. Besides, a slightly different arrangement of the summations from the one chosen in [389] has been made for better efficiency.

At third order, the known terms of the homological equation are $\widetilde{\mathcal{H}}_{0,3} = 3\{\mathcal{H}_{1,0};\mathcal{W}_2\}$, like in the Hill problem alone, from which the new Hamiltonian term $\mathcal{H}_{0,3} = \langle\widetilde{\mathcal{H}}_{0,3}\rangle_\ell = 0$ is computed. The vanishing of $\mathcal{H}_{0,3}$ was in fact expected from the analogous result in Eq. (11.15) in view of the non-central gravity perturbations due to the elipsoidal figure having been downgraded to the same order as the third-body planetary perturbations.

Then, to the third order of the small parameter, the procedure ends by replacing original by prime variables in the terms $\mathcal{H}_{0,i}$, $(i = 0,\ldots,2)$, of the new Hamiltonian. After neglecting terms of $\mathcal{O}(\dot{\vartheta}^4/n^4)$ and higher, the partially reduced Hamiltonian in prime variables is

$$\mathcal{K} = \mathcal{K}_{0,0} + \mathcal{K}_{1,0} + \frac{1}{2}\mathcal{K}_{2,0} + \frac{1}{3!}\mathcal{K}_{3,0}, \tag{11.28}$$

where $\mathcal{K}_{0,0}$ is the Keplerian, $\mathcal{K}_{1,0}$ is the Coriolis term, $\mathcal{K}_{2,0} = \mathcal{H}_{0,2}$ is given by the sum of Eqs. (11.4) and (11.26), $\mathcal{K}_{3,0} = 0$, and prime variables replace the original ones. From

Hamilton equations, we note that the variation of H',

$$\frac{\mathrm{d}H'}{\mathrm{d}t} = -nL'\frac{3}{16}\frac{\dot\vartheta^2}{n^2}\Big\{2\Big(2+3e^2+\frac{12\beta^2}{5\eta^3}\Big)s^2\sin 2h' + 5e^2[(c+1)^2\sin(2g'+2h') - (c-1)^2\sin(2g'-2h')]\Big\},$$

vanishes in the case of polar orbits when $h' = k\frac{\pi}{2}$, with k integer. Hence, polar orbits either collinear with or orthogonal to the primaries' line of apsides remain polar, on average.

Up to the order of $\dot\vartheta^2/n^2$, the short-period corrections of the Lie transformation from prime to original variables are computed following the sequence in Eq. (2.17).

11.3.2 Elimination of the longitude of the node

Like in §11.1.2, the formulation in the rotating frame makes h' evolve fast when compared to the rate of variation of g'. Therefore, we further remove from Hamiltonian (11.28) the periodic terms related to the longitude of the node by means of a new Lie transformation, from prime to double-prime variables, in which the Lie derivative and the homological equation are given by Eqs. (11.6) and (11.7), respectively.

Thus, $\mathcal{K}_{0,1} = \mathcal{K}_{1,0}$ while $\mathcal{V}_1$ remains undetermined. At second order, the new Hamiltonian term is chosen by averaging $\widetilde{\mathcal{K}}_{0,2} = \mathcal{K}_{2,0}$ over h'. That is,

$$\mathcal{K}_{0,2} = -\frac{\mu}{2a}\frac{\dot\vartheta^2}{n^2}\frac{1}{4}\Big[(2-3s^2)\Big(2+3e^2+4\frac{\beta^2}{\eta^3}\Big) + 15e^2s^2\cos 2g'\Big],$$

which allows the computation to be made of $\mathcal{V}_1$ from Eq. (11.7). We obtain

$$\mathcal{V}_1 = \frac{3}{64}L'\frac{\dot\vartheta}{n}\Big\{2\Big(2+3e^2+\frac{12}{5}\frac{\beta^2}{\eta^3}\Big)s^2\sin 2h' + 5e^2 \times[(1+c)^2\sin(2g'+2h') - (1-c)^2\sin(2g'-2h')]\Big\}, \tag{11.29}$$

and fill Deprit's triangle computing the intermediate term $\mathcal{K}_{1,1} = \mathcal{K}_{0,2} - \{\mathcal{K}_{1,0}, \mathcal{V}_1\}$.

At third order, the known terms are $\widetilde{\mathcal{K}}_{0,3} = \{\mathcal{K}_{0,2} + \mathcal{K}_{1,1} + \mathcal{K}_{2,0}, \mathcal{V}_1\}$, from which we choose $\mathcal{K}_{0,3} = \langle\widetilde{\mathcal{K}}_{0,3}\rangle_{h'}$. That is,

$$\mathcal{K}_{0,3} = -\frac{\mu}{2a}\frac{\dot\vartheta^3}{n^3}\frac{27}{32}c\eta\Big[2\Big(2+3e^2+\frac{6\beta^2}{5\eta^3}\Big)\frac{6\beta^2}{5\eta^5}s^2 + 2s^2 - (17s^2-50)e^2 + 15\Big(1+\frac{6\beta^2}{5\eta^5}\Big)s^2e^2\cos 2g'\Big],$$

and hence, from Eq. (11.7),

$$\mathcal{V}_2 = -G' \frac{\dot{\vartheta}^2}{n^2} \frac{15}{128} \left\{ \frac{4}{5} \left[\frac{3}{2}(2 - 17e^2) + \left(2 + 3e^2 + \frac{3\beta^2}{2\eta^3} \right) \frac{24\beta^2}{5\eta^5} \right] cs^2 \sin 2h' \right.$$
$$+ (1-c)^2 \left[6 + 9c + (7 + 7c - 5c^2) \frac{6\beta^2}{5\eta^5} \right] e^2 \sin(2g' - 2h')$$
$$\left. + (1+c)^2 \left[6 - 9c + (7 - 7c - 5c^2) \frac{6\beta^2}{5\eta^5} \right] e^2 \sin(2g' + 2h') \right\}. \tag{11.30}$$

From the latter, the derivation of medium-period corrections of $\mathcal{O}(\dot{\vartheta}^2/n^2)$ is straightforward from Eq. (2.17).

Up to third order of $\dot{\vartheta}/n$, the procedure ends replacing prime by double-prime variables in the terms $\mathcal{K}_{0,i}$, $i = 0, 1, 2, 3$.

11.3.3 Reduced phase space in the parameters plane

The double-averaged Hamiltonian is of one degree of freedom in (g'', G'') in which both H'' and $L'' = L'$ are formal integrals. To study the dynamics in the reduced phase space we neglect the constant terms and scale the double-averaged Hamiltonian by $-\frac{1}{16}a^2\dot{\vartheta}^2$, to get

$$\mathcal{H} = -(3s^2 - 2)\left(2 + 3e^2 + 4\frac{\beta^2}{\eta^3} \right) + 15\left[1 + \varepsilon \frac{9}{8} \gamma \left(1 + \frac{6\beta^2}{5\eta^5} \right) \right] s^2 e^2 \cos 2\omega$$
$$+ \varepsilon \frac{9}{8} \gamma \left[2s^2 - (17s^2 - 50)e^2 + 2s^2 \left(2 + 3e^2 + \frac{6\beta^2}{5\eta^3} \right) \frac{6\beta^2}{5\eta^5} \right], \tag{11.31}$$

where $s^2 = 1 - \gamma^2/\eta^2$, $e^2 = 1 - \eta^2$, $\eta = G''/L''$, $\varepsilon = \dot{\vartheta}/n(L'')$, and $\gamma = H''/L''$. Therefore, the reduced flow depends on three parameters, namely ε, γ, and β.

Disregarding the case of rectilinear orbits the reduced phase space is the sphere. A detailed discussion of the reduced flow in the variables on the sphere given in Eq. (5.75) can be found in [433, 592]. Here, we limit our discussions to the more relevant facts that may affect the design of science orbits. Like in previous sections of this chapter, we rely on the usual eccentricity-vector diagram representation. Analogously to Eqs. (11.12)–(11.13), we obtain the variation equations,

$$\frac{\mathrm{d}C}{\mathrm{d}t} = -3\frac{S}{L}\left[12\eta - 20(1 - C^2)\frac{\gamma^2}{\eta^3} + 4\left(1 - 5\frac{\gamma^2}{\eta^2} \right) \frac{\beta^2}{\eta^4} \right]$$
$$+ \frac{27}{10}\frac{S}{L}\varepsilon\frac{\gamma}{\eta}\left\{ 25(1 - C^2)\frac{\gamma^2}{\eta^2} + 15\eta^2 + 12\left(\frac{4}{5} - \frac{\gamma^2}{\eta^2} \right)\frac{\beta^4}{\eta^8} \right.$$
$$\left. + \left[2 + 123C^2 - 27S^2 - 5(2 + 33C^2 - 9S^2)\frac{\gamma^2}{\eta^2} \right] \frac{\beta^2}{2\eta^5} \right\}, \tag{11.32}$$

$$\frac{\mathrm{d}S}{\mathrm{d}t} = -3\frac{C}{L}\left[8\eta + 20S^2\frac{\gamma^2}{\eta^3} - 4\left(1 - 5\frac{\gamma^2}{\eta^2}\right)\frac{\beta^2}{\eta^4}\right]$$
$$- \frac{27}{10}\frac{C}{L}\varepsilon\frac{\gamma}{\eta}\left\{25S^2\frac{\gamma^2}{\eta^3} + 40\eta^2 + 12\left(\frac{4}{5} - \frac{\gamma^2}{\eta^2}\right)\frac{\beta^4}{\eta^8}\right.$$
$$\left. + \left[62 + 63C^2 - 87S^2 - 35(2 + 3C^2 - 3S^2)\frac{\gamma^2}{\eta^2}\right]\frac{\beta^2}{2\eta^5}\right\}. \tag{11.33}$$

On the other hand, as will be illustrated in §11.3.4, for the orbits of interest higher-order terms only introduce quantitative variations with respect to the second order. Then the discussion of the relative equilibria is made in the simpler approximation, which is obtained after neglecting the terms factored by ε in Eqs. (11.32) and (11.33). Still, the refinements provided by third-order terms of the long-term Hamiltonian (11.31) definitely improve the computation of particular solutions, and will be taken into account in §11.4.

Then, in the second-order approximation, the equilibria are computed from

$$0 = -3\frac{S}{L}\left[12\eta - 20(1 - C^2)\frac{\gamma^2}{\eta^3} + 4\left(1 - 5\frac{\gamma^2}{\eta^2}\right)\frac{\beta^2}{\eta^4}\right], \tag{11.34}$$

$$0 = -3\frac{C}{L}\left[8\eta + 20S^2\frac{\gamma^2}{\eta^3} - 4\left(1 - 5\frac{\gamma^2}{\eta^2}\right)\frac{\beta^2}{\eta^4}\right]. \tag{11.35}$$

In particular, circular orbits, $C = S = 0$, are immediately identified as equilibria of the double-averaged phase space. Besides, the condition $S = 0$ makes Eq. (11.34) vanish identically, and turns Eq. (11.35) into a polynomial equation in η,

$$2\eta^7 - \beta^2\eta^2 + 5\beta^2\gamma^2 = 0, \tag{11.36}$$

whose solution will provide the eccentricity of elliptic orbits with frozen perigee at $g = 0, \pi$. Elliptic orbits also exist with frozen perigee at $g = \pm\frac{\pi}{2}$, whose eccentricity is given by the solution of the polynomial equation

$$3\eta^7 - 5\gamma^2\eta^3 + \beta^2\eta^2 - 5\beta^2\gamma^2 = 0, \tag{11.37}$$

which is obtained by making $C = 0$ in Eq. (11.34).

Bifurcations of these eccentric frozen orbits from circular orbits happen along the lines obtained by making $\eta \to 1$ in Eqs. (11.37) and (11.36). We obtain

$$\gamma^2 = \frac{1}{5}\left(1 + \frac{2}{1 + \beta^2}\right), \quad \gamma^2 = \frac{1}{5}\left(1 - \frac{2}{\beta^2}\right), \tag{11.38}$$

which provide symmetric bifurcation lines $\gamma = \gamma(\beta)$ for direct and retrograde inclinations in the parameters (β, γ) plane. The latter shows the limit $\beta \geq \sqrt{2}$ for the existence of eccentric frozen orbits with $g = 0, \pi$, while the former yields the critical inclination value of the Hill problem $\gamma^2 = \frac{3}{5}$ when $\beta = 0$. Both kinds of bifurcation lines tend to the value $y^2 \to \frac{1}{5}$ when $\beta \to \infty$, corresponding to the critical inclination of the artificial satellite problem discussed in §5.6.

Elimination of η between Eq. (11.36) and its partial derivative with respect to η, namely $7\eta^5 - \beta^2 = 0$, yields the line of double roots

$$y = \pm\beta^{2/5}/7^{7/10}, \tag{11.39}$$

in the parameters plane (β, y), where new stable and unstable eccentric frozen orbits appear in a saddle-node bifurcation with $g = 0, \pi$. Since $y^2 \le 1$, this line exists only for $0 \le \beta \le 7^{7/4}$. However, after been plugged into Eq. (11.36), it only provides roots that make dynamical sense, namely $0 < \eta \le 1$, when $\beta \le \sqrt{7}$. For increasing values of y, the stable equilibria migrate along the lines $g'' = 0, \pi$, towards the lower eccentricities, whereas the unstable equilibrium do that along the same line to the higher eccentricities. The latter always get the critical value 1 when $y = 0$, whereas the former may collapse to circular orbits if $\beta \ge \sqrt{2}$. Besides, the eccentricity with which these bifurcated orbits stem reduces for increasing values of β until the lines given by the second of Eq. (11.38) and Eq. (11.39) osculate at $\beta = \sqrt{7}$, where $y = 1/\beta$ or $I_{\text{circular}} = 67.8°$, in which case they become circular and saddle-node bifurcations no longer exist. This is illustrated in Fig. 11.6 for different values of β, including the critical cases $\beta = \sqrt{2}$ and $\beta = \sqrt{7}$.

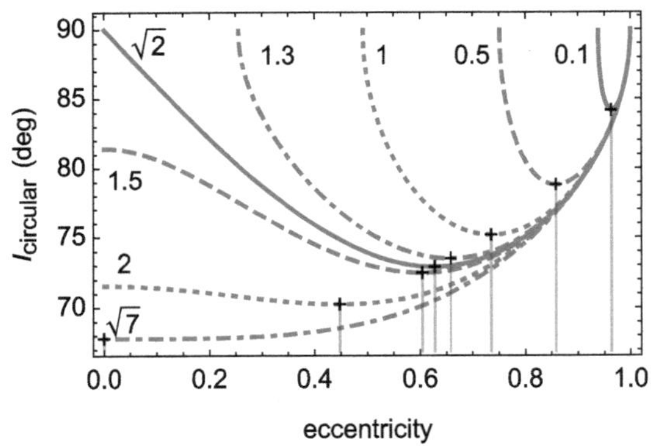

Figure 11.6: Eccentric frozen orbits with $g'' = 0, \pi$, from Eq. (11.36), stemming in a saddle-node bifurcation from points marked "+" for $\beta = 0.1, \ldots \sqrt{2}, \ldots \sqrt{7}$.

The bifurcation lines described by Eqs. (11.38) and (11.39) are illustrated in Fig. 11.7, in which the range of β is constrained to the region of interest. In the second-order approximation, retrograde inclination orbits are symmetric to direct ones and are not displayed. Typical representations of the reduced phase space by means of eccentricity-vector diagrams in the different regions of the parameters plane (β, y), with numbers 1, 2, 3, 4 in Fig. 11.7, are illustrated in Figs. 11.8 and 11.9.

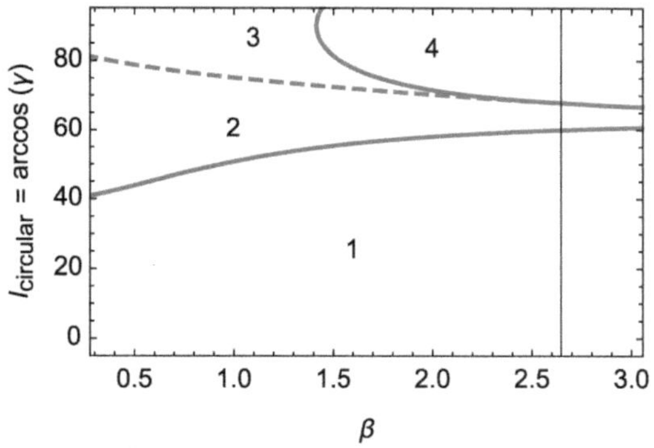

Figure 11.7: Bifurcation lines of circular (full lines) and eccentric orbits (dashed line) that determine regions of the reduced phase with different number of equilibria. The vertical line marks the value $\beta = \sqrt{7}$ where the dashed line ends.

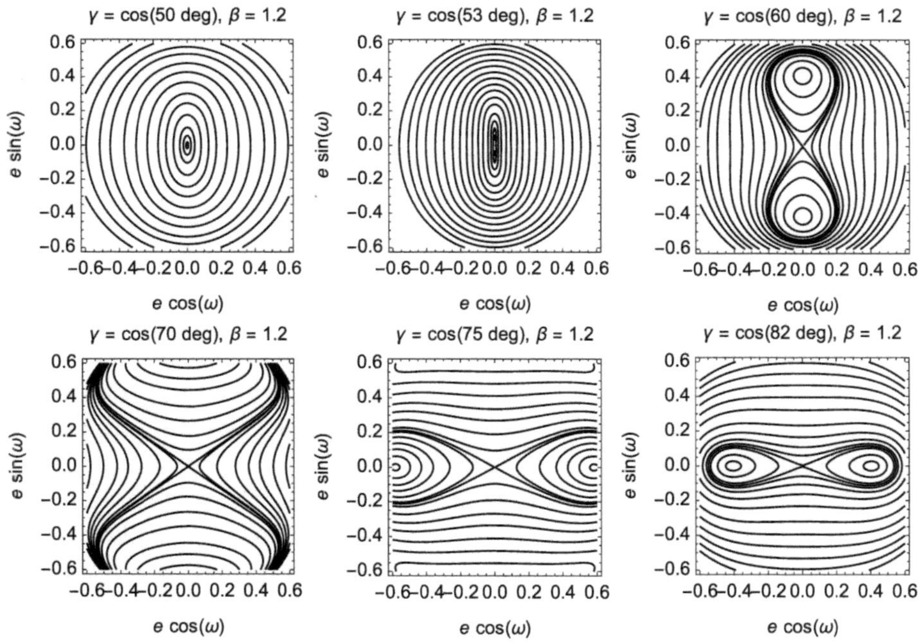

Figure 11.8: Typical phase spaces in the regions shown in Fig. 11.7 when $\beta < \sqrt{2}$.

In region 1 circular orbits enjoy stability and are the only frozen orbits of interest (left plots in the first row of Figs. 11.8 and 11.9). In region 2 the bifurcation line given by the first equation of Eq. (11.38)—the lower full line in Fig. 11.7—has already been crossed, and circular orbits became unstable while two new stable eccentric orbits exist with the periapsis frozen at $g = \pm\frac{\pi}{2}$ that may enjoy moderate eccentricities (center and right plots in the first row of Figs. 11.8 and 11.9). The eccentricities of the stable elliptic

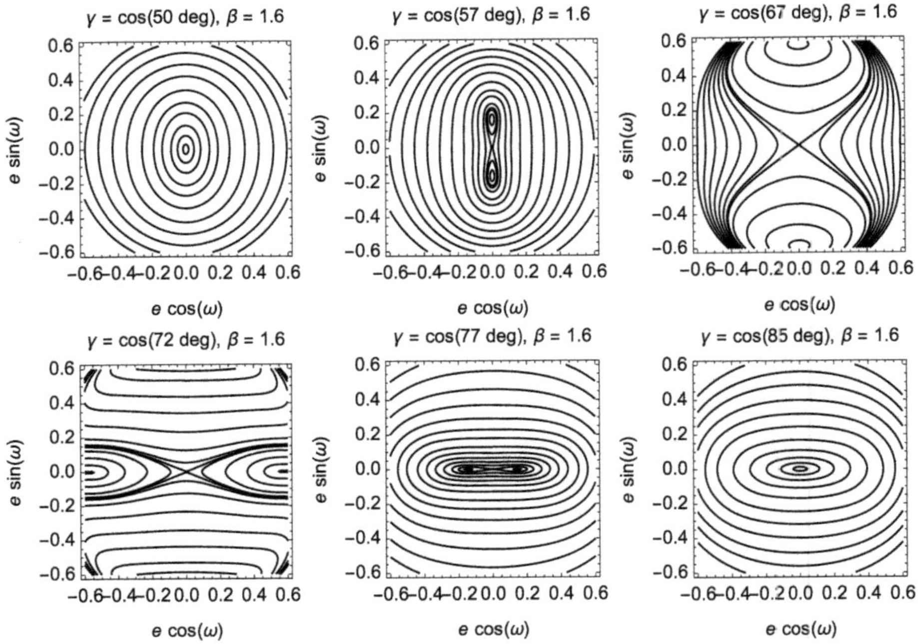

Figure 11.9: Typical phase spaces in the regions shown in Fig. 11.7 when $\beta > \sqrt{2}$.

frozen orbits grow when the absolute value of the dynamical parameter $y = \cos I_{\text{circular}}$ diminishes, soon reaching high values.

In the region numbered 3 the saddle-node bifurcations have happened at the crossing of the bifurcation line in Eq. (11.39)—the dashed line in Fig. 11.7. In addition to the unstable circular orbits and the highly eccentric frozen orbits with $g = \pm\frac{\pi}{2}$, two stable eccentric orbits exist with the periapsis frozen at $g = 0, \pi$, as well as two unstable ones with higher eccentricities (left plots in the second row of Figs. 11.8, and 11.9). Eventually, for some value I_{circular} the unstable circular orbits and the unstable eccentric frozen orbits with $\omega = 0, \pi$ get the same energy, thus producing global changes in the flow through a saddle connection; see [433]. The homoclinic passing through the circular orbits changes from an 8-shaped trajectory, surrounding the eccentric stable frozen orbits with periapses $\omega = \pm\frac{\pi}{2}$, into an ∞-shaped trajectory, which surrounds the eccentric stable frozen orbits with periapses $\omega = 0, \pi$ (center plots in the second row of Figs. 11.8, and 11.9). For values $\beta < \sqrt{2}$ there are no more regions in the parameters plane, so the flow remains qualitatively the same for higher values of the inclination of the circular orbits, with the only quantitative effect of reducing the eccentricity of the frozen orbits with $\omega = 0, \pi$, and increasing the eccentricity of the stable frozen orbits with $\omega = \pm\frac{\pi}{2}$ (right plot in the second row of Fig. 11.8). On the contrary, one more region exists in the parameter plane when $\beta > \sqrt{2}$. Indeed, for increasing values of I_{circular}, the eccentric frozen orbits with $\omega = 0, \pi$ approach the bifurcation line provided by the second equation of Eq. (11.38), which they reach eventually collapsing into a circular orbit in a pitchfork bifurcation. In consequence, circular orbits are stable in region 4 (right plot in the second row of Fig. 11.9).

In summary, for small values of β the dynamics is mostly dominated by the planetary perturbations, and the Lidov–Kozai resonance is the most relevant feature, analogously to the case of the Hill problem alone; cf. Fig. 11.2. On the contrary, for higher values of β the non-centralities of the planetary satellite dominate, yielding analogous dynamics to the main problem of the artificial satellite, in which high-inclination circular orbits are stable; cf. Fig. 5.5. Still, while bifurcations happen always from circular orbits, like in the main problem, the sequence is not the same, as shown by simple inspection of Figs. 11.10 and 5.4.

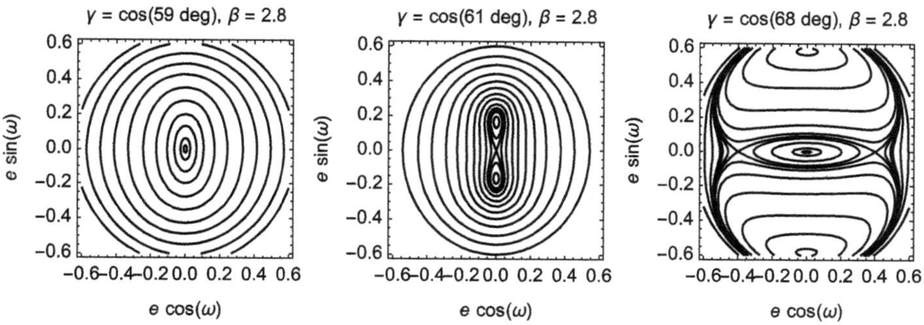

Figure 11.10: Typical phase spaces in the regions of Fig. 11.7 when $\beta > \sqrt{7}$.

11.3.4 Third-order effects. The space of parameters

The symmetry of direct and retrograde inclination orbits is broken when the third order of the perturbation theory is taken into account. Now, the equilibria are computed from Eqs. (11.32)–(11.33), which show that circular orbits remain as equilibria, as well as eccentric orbits with $g = \pm\frac{\pi}{2}$, and $g = 0, \pi$. The bifurcation lines of the latter from circular orbits are obtained making $\eta \to 1$ in the corresponding equilibria equations. We obtain

$$\begin{aligned} 0 &= 4(\beta^2 + 3) - 20(\beta^2 + 1)\gamma^2 \\ &\quad - \frac{9}{50}\varepsilon\gamma[48\beta^4 + 5\beta^2 + 75 - 5(12\beta^4 + 5\beta^2 - 25)\gamma^2], \end{aligned} \tag{11.40}$$

$$\begin{aligned} 0 &= 8 - 4\beta^2 + 20\beta^2\gamma^2 \\ &\quad + \frac{9}{50}\varepsilon\gamma[48\beta^4 + 155\beta^2 + 200 - 5(12\beta^2 + 35)\beta^2\gamma^2], \end{aligned} \tag{11.41}$$

which are modifications of order $\dot{\vartheta}/n$ with respect to Eq. (11.38) involving cubic equations in γ. As regards the saddle-node bifurcation, the third-order refinement of the bifurcation line in Eq. (11.39) can be computed from the elimination of η between Eq. (11.41) and its partial derivative with respect to η.

Now, the relative strength of third-body and non-central gravity perturbations cannot be combined into a single parameter, and the bifurcation lines of frozen orbits turn into bifurcation surfaces $\mathcal{S}(\gamma, \beta, \dot{\vartheta}/n) = 0$ [433]. For the orbits of interest in mission designing about planetary satellites we do not find qualitative changes in the flow. This is illustrated in Fig. 11.11, where the bifurcation surfaces of the third-order theory are depicted in the range $0 < \dot{\vartheta}/n \leq 0.1$. As expected from the higher orders of the Hill problem alone discussed in §11.2, the bifurcation surfaces bend towards the higher inclinations for increasing values of $\dot{\vartheta}/n$.

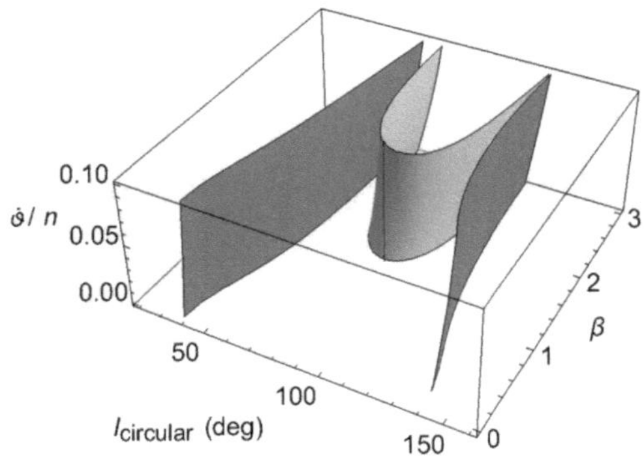

Figure 11.11: Bifurcation surfaces of circular orbits $\mathcal{S}(\gamma, \beta, \dot{\vartheta}/n) = 0$ (after [433]).

11.4 Application. Computation of the science orbit

Science missions around planetary satellites commonly require low-eccentricity orbits with low altitude and high inclination, which, from the previous discussion, are unstable when the third-body perturbation dominates the dynamics—case $\beta < \sqrt{2}$. This fact was illustrated in Fig. 11.8, in which low-eccentricity frozen orbits are shown to be stable only for the lower inclinations (upper-left plot). Due to the unstable character of low-eccentricity frozen orbits with high inclinations, a small perturbation would make the eccentricity of the science orbit grow fast, leading to the orbiter, if uncontrolled, make impact on the surface of the planetary satellite in short times.

A common strategy for maximizing orbit lifetime with minimum control consist in designing tours on the stable–unstable manifold of a reference frozen orbit [547].[4] An example is presented below for the physical characteristics of the Jovian

4 These kinds of tours were originally devised in the non-averaged restricted three-body problem dynamics [248]. In particular, they are efficiently designed based on the stable and unstable manifolds associated to repeat ground-track orbits of the planetary-satellite problem [247, 427].

moon Europa. Namely, $\mu = 3202.7\,\mathrm{km}^3/\mathrm{s}^2$, $R_\otimes = 1565\,\mathrm{km}$, $J_2 = 4.355 \times 10^{-4}$, and $\dot{\vartheta} = 2.05 \times 10^{-5}\,\mathrm{s}^{-1}$; cf. [34].

Given a nominal frozen orbit with $a = a_0$, $e = e_0$, $I = I_0$, $\omega = \omega_0$, the procedure starts by computing the dynamical parameters $\varepsilon = \dot{\vartheta}/n_0$, where $n_0 = (\mu/a_0^3)^{1/2}$ from Eqs. (4.44) and (4.45), $\gamma = (1 - e_0^2)^{1/2} \cos I_0$, and $\beta = \beta(a_0)$ from Eq. (11.25). Next, the value $\mathcal{H}(e_0, \omega_0; \varepsilon, \beta, \gamma) = E_0$, which is the "energy" of the unstable frozen orbit as well as its associated stable and unstable manifolds, is evaluated from Eq. (11.31) with $c = \gamma\eta$. In particular, for a nominal *circular* frozen orbit,

$$E_0 = (3\gamma^2 - 1)(2 + 4\beta^2) + \frac{9}{4}\varepsilon\gamma(1 - \gamma^2)\left[1 + \left(2 + \frac{6}{5}\beta^2\right)\frac{6}{5}\beta^2\right]. \tag{11.42}$$

Then, for a maximum allowed eccentricity e_M derived from the mission requirements, the corresponding argument of the periapsis ω_S of one of the two stable manifolds is computed from Eq. (11.31) by solving $\mathcal{H}(e_M, \omega_S; \varepsilon, \beta, \gamma) = E_0$ for ω_S. In this way, starting from the initial conditions corresponding to $(a_0, e_M, I_0, \omega_S)$ the natural dynamics will drive the orbit over the stable manifold towards the nominal frozen orbit. The orbit will remain on the unstable equilibrium position for some time, until it eventually departs from the frozen orbit along one of its two unstable manifolds. A maneuver is needed when the eccentricity reaches the value e_M in order to come back to the starting point of the tour.

For instance, for a nominal circular orbit 100 km above the surface of Europa ($a_0 = 1665\,\mathrm{km}$), we compute $n_0 = 8.32984 \times 10^{-4}$, from which $\beta = 0.797034$ and $\varepsilon = \dot{\vartheta}/n_0 = 0.0246103$. If, besides, we choose $\gamma = \cos 85°$, Eq. (11.42) yields $E_0 = -4.4227$. The maximum eccentricity of non-impact orbits corresponds to a grazing perigee $a_0(1-e) = R_\otimes$, from which the conservative limit $e_{\mathrm{impact}} = 0.058$ is used for estimating orbit lifetime. Besides, we consider a maximum eccentricity $e_M = 0.01$ acceptable in this example.

The reduced phase space of the double-averaged flow is illustrated with the eccentricity-vector diagram of Fig. 11.12, in which the stable and unstable manifolds of the circular frozen orbit are highlighted with a thick-dashed line in the left plot. The right plot focus on the lower eccentricities, highlighting the eccentricity leading to impact e_{impact} with a dotted big circle, and the prospective eccentricity limit for the mission orbit e_M with a black small circle.

Then the initial eccentricity is set to $e_0 = e_M = 0.01$ and the argument of the periapsis ω_S is solved from Eq. (11.31) for $\mathcal{H} = E_0$, to get $\omega_S = 147.744°$. The longitude of the node in the rotating frame and the mean anomaly are not part of the double-averaged perturbed Hill problem and their initial values remain arbitrary. We choose both as vanishing at the initial time. Then, we propagate the corresponding initial conditions in the original, non-averaged flow. The left plot of Fig. 11.13 depicts the evolution of the instantaneous eccentricity vector superimposed to the double-averaged flow. It shows that the orbit first moves close to the stable manifold towards the lower eccentricities, within which proximity it remains for sometime. Eventually, the unstable manifold dynamics makes that the eccentricity starts to increase, and it continuously grows until

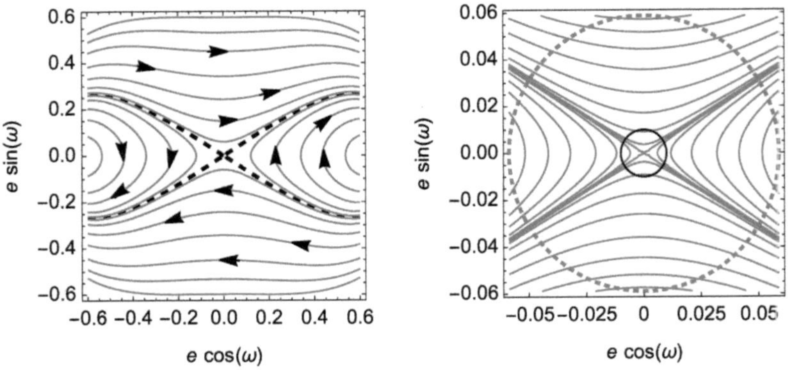

Figure 11.12: Averaged flow in the parameters plane $a = 1665$ km, $I_{circular} = 85°$.

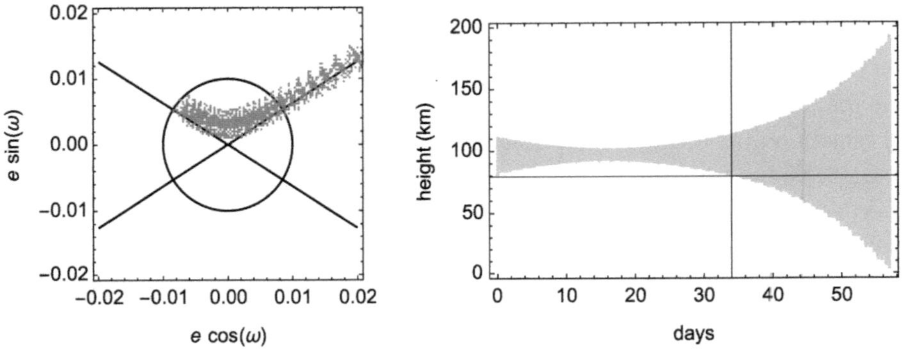

Figure 11.13: Eccentricity vector (left) and altitude (right) of the Europa orbit with $a = 1665$ km, $I = 85°$, and $h_0 = 0$.

the orbiter impacts the surface of Europa. The right plot of Fig. 11.13 shows that the impact happens after ≈ 57 days, but the orbit only remains below the maximum allowed eccentricity $e_M = 0.01$ for about one month.

On the other hand, if we repeat the computations arbitrarily choosing the initial longitude of the node $h = \frac{\pi}{2}$ instead of 0, then the orbit takes now a different tour and evolves much closer to the stable manifold of the double-averaged dynamics, as shown in the left plot of Fig. 11.14. The right plot of Fig. 11.14 presents the radius evolution superimposed to the previous case, showing that orbit lifetime is extended up to about three months, and the time between prospective maneuvers is now doubled.

However, the choice of the initial longitude of the node in the double-averaged perturbed Hill problem model is arbitrary. The important lifetime differences stem from the fact that we are using mean elements as if they were osculating. And in this particular example the (disregarded) periodic corrections needed for converting the double-prime argument of the periapsis—the more sensitive variable to travel the stable manifold of the frozen orbit—into the osculating one are certainly much smaller when $h'' = \frac{\pi}{2}$ than when $h'' = 0$. However, this might not be the general case.

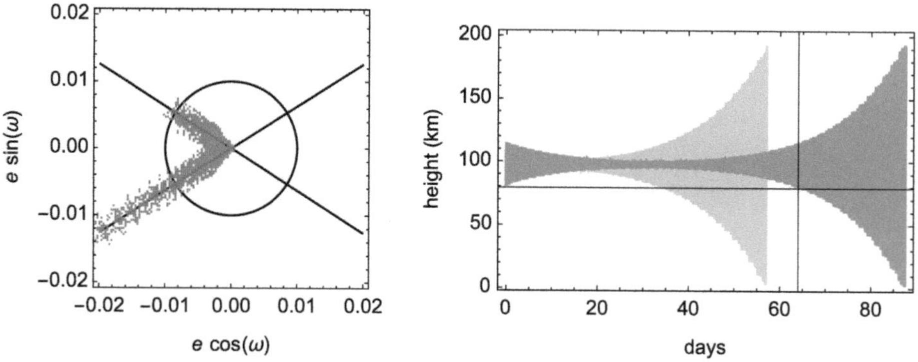

Figure 11.14: Same as Fig. 11.13 when $h_0 = \frac{\pi}{2}$. The height over the surface of Europa (in dark grey) is superimposed to the case $h_0 = 0$ (light gray).

In order to guarantee that an arbitrary choice of h'' allows one to get as close as possible to the stable manifold, thus yielding a reasonably long lifetime, the computation of the transformation mean to osculating variables is mandatory in the computation of the initial conditions. On the other hand, it is well known that an unstable equilibrium obtained after a normalization process can be viewed only like some kind of "landmark above the epicenter of a zone of chaotic dislocations" [168]. In consequence, even after recovering the periodic effects removed in the averaging, one rarely will be close enough to a maximum-lifetime solution [547].

11.4.1 Mean to osculating transformation

The periodic corrections of the mean to osculating transformation are computed from the usual sequence given in Eq. (2.17). First of all, we compute the transformation from double-prime to prime variables. Denoting ξ any canonical or non-canonical variable, it takes the form

$$\xi' = \xi'' + (\dot{\vartheta}/n)\Delta\xi'' + \frac{1}{2}(\dot{\vartheta}/n)^2\delta\xi'',$$

where $(\dot{\vartheta}/n)\Delta\xi'' = \{\xi'', \mathcal{V}_1\}$, $(\dot{\vartheta}/n)^2\delta\xi'' = \{\xi''; \mathcal{V}_2\} + \{\Delta\xi'', \mathcal{V}_1\}$, and the terms $\mathcal{V}_1$ and $\mathcal{V}_2$ of the generating function are given in Eqs. (11.29) and (11.30), respectively.

The corresponding first-order corrections to the classical orbital elements are $\Delta a'' = 0$, and

$$\Delta e'' = \frac{15}{32}e\eta[(1+c)^2\cos(2g+2h) - (1-c)^2\cos(2g-2h)], \tag{11.43}$$

$$\begin{aligned}\Delta I'' = \frac{15}{32}\frac{s}{\eta}\bigg[&(1+c)e^2\cos(2g+2h) + (1-c)e^2\cos(2g-2h)\\ &+\frac{2}{5}\left(2+3e^2+\frac{12\beta^2}{5\eta^3}\right)\cos 2h\bigg],\end{aligned} \tag{11.44}$$

$$\Delta h'' = \frac{15}{32}\frac{1}{\eta}\left[(1-c)e^2\sin(2g-2h) + (1+c)e^2\sin(2g+2h)\right.$$
$$\left. - \frac{2}{5}c\left(2 + 3e^2 + \frac{12\beta^2}{5\eta^3}\right)\sin 2h\right], \tag{11.45}$$

$$\Delta g'' = \frac{15}{32\eta}\left\{(c-1)(c-\eta^2)\sin(2g-2h) - (1+c)(c+\eta^2)\sin(2g-2h)\right.$$
$$\left. + \frac{2}{5}\left[3e^2 + (5c^2-3)\left(1 + \frac{6\beta^2}{5\eta^3}\right)\right]\sin 2h\right\}, \tag{11.46}$$

$$\Delta \ell'' = \frac{15}{32}\left\{(1+e^2)\left[(1+c)^2\sin(2g+2h) - (1-c)^2\sin(2g-2h)\right]\right.$$
$$\left. + \frac{2}{5}\left(7 + 3e^2 + \frac{42\beta^2}{5\eta^3}\right)s^2\sin 2h\right\}, \tag{11.47}$$

where the double-prime notation is omitted for brevity in the right side of the equations. The second-order corrections are made of much longer series. However, these higher-order corrections are not too relevant for mission-design purposes and we can safely ignore them.

For its part, the transformation from prime to original variables takes the form $\xi = \xi' + \frac{1}{2}(\dot{\vartheta}/n')^2\delta\xi'$, in which $(\dot{\vartheta}/n')^2\delta\xi' = \{\xi', \mathcal{W}_2\}$ and $\mathcal{W}_2$ is given in Eq. (11.27). Now, in spite of these corrections being of second order, we cannot completely neglect them. Indeed, the short-period correction to the argument of the periapsis, which is a fundamental design parameter in our case, $(\dot{\vartheta}/n')^2\delta g' = \{g', \mathcal{W}_2\} = \partial\mathcal{W}_2/\partial G'$ involves differentiation of the eccentricity with respect to the total angular momentum $\partial e/\partial G = -\eta^2/(eG)$. Therefore, the appearance of the eccentricity in denominators magnifies the effect of the short-period correction to the argument of the perigee, which can no longer be considered of higher order.

An analogous situation occurs with the short-period correction to the mean anomaly, yet an accurate value of ℓ is not of worry for mission-design purposes. Precisely because of that, the trigonometric terms of the series defining the short-period corrections, whose arguments involve both the eccentric and true anomalies and the argument of the periapsis and the longitude of the node in the rotating frame can be radically abbreviated. Indeed, the mean anomaly is absent for the corrections from double-prime to prime variables, so one can always assume that the initial value ℓ'' is chosen in such a way that $\ell' = 0$, and, therefore, both the true and the eccentric anomalies also vanish in the prime variables. In addition, because e_M is relatively small, the second-order correction $\delta g'$ is further simplified by expanding it in powers of the eccentricity. We obtain the manageable form

$$\delta g' = \sum_{j=-1}^{1}\sum_{k\geq 0}\Gamma_{k,j}(\beta,\gamma)e^{k-1}\sin(2g' + 2jh'), \tag{11.48}$$

where, up to $k = 1$,

$$\Gamma_{0,\pm 1} = \frac{1}{10}(3\beta^2 - 10)(\gamma \pm 1)^2,$$

$$\Gamma_{0,0} = (2-\beta^2)(\gamma^2-1),$$
$$\Gamma_{1,\pm1} = \frac{3}{80}(\gamma\pm1)[3\beta^2(11\gamma\pm7)-10(\gamma\pm2)],$$
$$\Gamma_{1,0} = -\frac{3}{8}[\beta^2(11\gamma^2-7)-2\gamma^2+4].$$

Needless to say that the appearance of divisions by the eccentricity makes the use of Eq. (11.48) inadequate for the computation of short-period corrections of almost circular orbits. This is not the current case, however, in which, quite on the contrary, the eccentricity is chosen to be non-negligible to initiate the tour over the stable manifold of the circular frozen orbit. Nevertheless, if desired, computation of the corrections in nonsingular variables is straightforward [389].

11.4.2 Mapping orbits

To be consequent with these truncations, analogous simplifications are carried out in Eqs. (11.43)–(11.47) neglecting terms $\mathcal{O}(e^2)$, and thus making $\eta = 1$. Besides, Eq. (11.47) can be completely ignored, as we already did with the correction $\delta\ell'$, and the tests can be constrained to the single case in which $\ell_0 = 0$. Moreover, on account of the high inclinations of mapping orbits, in addition to the low eccentricities, we can assume $\gamma^2 \sim e \sim \dot\vartheta/n$ and further simplify the periodic corrections to first-order effects. Also, as we already did with the mean anomaly, we may fix the initial longitude of the node in the rotating frame to some particular value. If we make this value zero, we obtain the extremely simple first-order corrections [389]

$$\Delta I = \frac{\dot\vartheta}{n}\frac{3}{8}\left(1+\frac{6}{5}\beta^2\right)\sin I,$$
$$\Delta g = -\frac{\dot\vartheta}{n}\left[\frac{15}{8}\cos I + \frac{\dot\vartheta/n}{e}\left(2-\frac{4}{5}\beta^2\right)\right]\sin 2g,$$

where quantities in the right sides are functions of the double-prime variables. That is, a good set of initial conditions to travel the stable manifold of the (unstable) circular frozen orbit is obtained directly from the double-averaged dynamics by choosing $a = a''$, $e = e''$, $I = I'' + \Delta I$, $g = g'' + \Delta g$, and $h = \ell = 0$.

When we apply these corrections to the previous example with $h = 0$, we obtain $\Delta I = 0.0162023$ and $\Delta g = 0.0851895$. That is, we have a correction of about 1° to the initial inclination and of almost 5° to the initial argument of the periapsis. When we propagate the corrected initial conditions we find that, on average, the orbit moves more closely to the stable manifold of the frozen orbit, extending lifetime to about five months, in four of which the eccentricity remains within operational limits. This is illustrated in Fig. 11.15, where the evolution of the radius is superimposed to the case in which the periodic corrections are not applied.

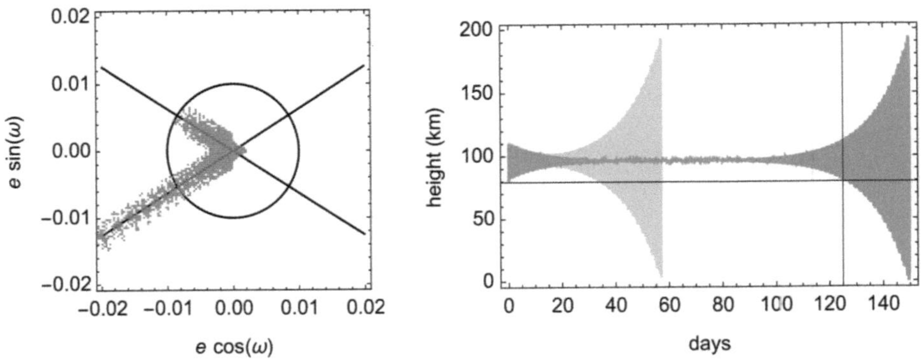

Figure 11.15: Lifetime improvement of the Europa orbiter after correcting the initial conditions of the double-averaged flow with periodic terms.

Due to the unstable dynamics, no major improvements are expected from a refinement of the periodic corrections taking the full expressions into account, yet they may result from a higher-order computation of the long-term dynamics. On the other hand, longer lifetimes can be obtained making minor adjustments to the initial mean eccentricity. Indeed, simply choosing $e_0'' = 0.0099$, which yields changes in ω_S of the order of 10^{-7}, lifetime is extended for almost one additional month. Moreover, after a few trials, we found that choosing a mean eccentricity $e = 0.0098738223$ extends the orbit's lifetime to more than one year, in agreement with analogous results in [427] based on repeat ground-track periodic orbit design.

The different trials made are summarized in Fig. 11.16, where numbers in the axis of abscissas indicate time between prospective maneuvers in each particular case. We remark that the final numerical refinement of the mean eccentricity is clearly different from, and apparently more effective than, the final refinement of the osculating argument of the periapsis advocated by other authors; cf. [547].

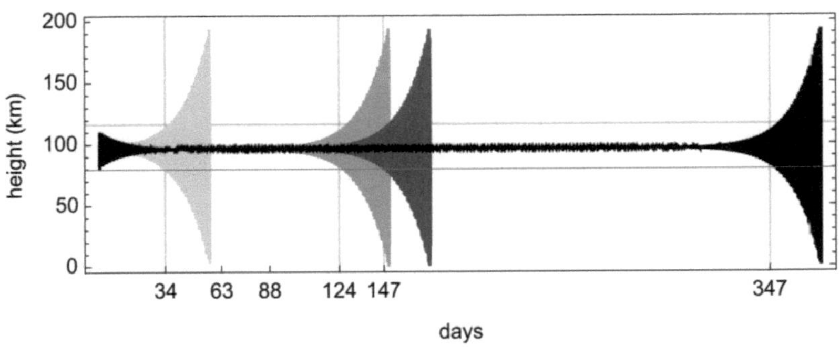

Figure 11.16: Altitude evolution of the Europa orbit with $a = 1665$ km under the different corrections to the mean elements.

12 Motion about the libration points

Periodic orbits about the libration points of the Hill problem are known to exist beyond the linearized dynamics discussed in §10.2.2 [515]. As an alternative to the numerical approach outlined in §10.2.3, the nonlinear effects that determine the dynamics of the main families of periodic orbits far from the equilibria can be approached analytically by Hamiltonian perturbations. The perturbation solution in this chapter is constructed by a chain of canonical transformations that reduces the Hill problem dynamics in the vicinity of the libration points to a one-degree-of-freedom Hamiltonian [396, 424].

The sequence of canonical transformations starts by a translation of the origin to the libration point, which sets the Hill problem Hamiltonian in a form amenable to perturbation treatment. Next, a linear transformation decouples the zeroth-order term of the Hamiltonian into a hyperbolic term and two harmonic oscillators of close frequencies. The latter are conveniently rewritten in the form of a perturbed *elliptic* oscillator after standard detuning [283].

The traditional way is then to carry out the reduction to the center manifold in order to eliminate the hyperbolic components of the motion by means of a Lie transformation. After that, the reduced dynamics can be approached, without constraint to any energy level, with standard tools of nonlinear dynamics like the computation of periodic orbits. Alternatively, for energy levels close enough to that of the libration points, a final reduction of the Hill problem Hamiltonian to a one-degree-of-freedom Hamiltonian can be achieved by an additional Lie transformation that removes the short-period terms. On the other hand, the reduction to the center manifold and the following removal of short-period terms can be combined into a single Lie transformation that is effectively approached in complex variables, whose practicality in dealing with perturbed harmonic motion was already illustrated in §3.2.7.

Integral manifolds of the reduced dynamics turn out to be spheres, on which the flow is advantageously described using the Hopf coordinates [300]. The equilibria of the normalized Hamiltonian flow clearly reveal the main families of periodic orbits originated from the libration points dynamics. Namely, we have the families of planar and vertical Lyapunov orbits, the family of halo orbits, and the two-lane bridge of periodic orbits that connects the families of planar and vertical Lyapunov orbits. The use of Deprit's Lissajous variables [160] then makes easier the analytical computation of each of these particular orbits.

12.1 Perturbation solution

The known symmetries of the Hill problem allow one to reduce the study of the dynamics about the libration points to just one of them, say $(r_{\mathrm{H}}, 0, 0)$, where the Hill radius $r_{\mathrm{H}} = 3^{-1/3}$ was previously defined in Eq. (10.26).

https://doi.org/10.1515/9783110668513-012

To study the dynamics about the libration point, the origin is first translated to $(r_{\mathrm{H}}, 0, 0)$ making the transformation $(\boldsymbol{x}', \boldsymbol{X}') \mapsto (\boldsymbol{x}, \boldsymbol{X})$, with $\boldsymbol{x} = (x, y, z)$ and $\boldsymbol{X} = (X, Y, Z)$, given by

$$x' = r_{\mathrm{H}} + x, \quad y' = y, \quad z' = z, \quad X' = X, \quad Y' = Y + r_{\mathrm{H}}, \quad Z' = Z. \tag{12.1}$$

This transformation is canonical and turns Eq. (10.18) into the new Hamiltonian $\mathcal{H}' = \mathcal{H}(\boldsymbol{x}(\boldsymbol{x}', \boldsymbol{X}'), \boldsymbol{X}(\boldsymbol{x}', \boldsymbol{X}'))$ given by

$$\mathcal{H}' = \frac{1}{2}(X'^2 + Y'^2 + Z'^2) - (x'Y' - y'X') + \frac{1}{2}(y'^2 + z'^2) - x'^2 - \frac{x'}{r_{\mathrm{H}}^2} - \frac{1}{r}, \tag{12.2}$$

where now $r = [(x' + r_{\mathrm{H}})^2 + y'^2 + z'^2]^{1/2}$. That is,

$$r = r_{\mathrm{H}} \sqrt{1 - 2(r'/r_{\mathrm{H}}) \cos \psi + (r'/r_{\mathrm{H}})^2},$$

with $\cos \psi = -x'/r'$, and $r' = \sqrt{x'^2 + y'^2 + z'^2}$ is the distance to the massless body from the libration point.

When the ratio r'/r_{H} is small the inverse of the radius in Eq. (12.2) can be replaced by the usual expansion in Legendre polynomials $P_n(\cos \psi)$, yielding

$$\mathcal{H}' = \sum_{n \geq 0} \frac{\epsilon^n}{n!} \mathcal{H}_n, \tag{12.3}$$

where ϵ is a formal small parameter that indicates the magnitude of the ratio r'/r_{H}, the zeroth-order term

$$\mathcal{H}_0 = \frac{1}{2}(X'^2 + Y'^2) - (x'Y' - X'y') + 2(y'^2 - 2x'^2) + \frac{1}{2}(Z'^2 + 4z'^2), \tag{12.4}$$

is integrable due to its quadratic character, and, for $n \geq 1$, the perturbation terms

$$\mathcal{H}_n = -\frac{1}{r_{\mathrm{H}}} \left(\frac{r'}{r_{\mathrm{H}}} \right)^{n+2} P_{n+2}(\cos \psi) \tag{12.5}$$

are homogeneous polynomials of degree $n + 2$, with monomials

$$M_{\boldsymbol{\kappa}} = q_{\boldsymbol{\kappa}} x'^{m_1} X'^{m_2} y'^{m_3} Y'^{m_4} z'^{m_5} Z'^{m_6}, \quad \boldsymbol{\kappa} = (m_1, m_2, m_3, m_4, m_5, m_6), \tag{12.6}$$

where m_i, $i = 1, \ldots 6$, are positive integers and $q_{\boldsymbol{\kappa}}$ are numeric coefficients.

12.1.1 The center manifold

The linearized dynamics about the libration points was already discussed in §10.2.1, where it was shown that the motion in the z-axis direction decouples from the rest of the flow and is made of harmonic oscillations of frequency $\nu = 2$, given by Eq. (10.27). This behavior is now directly observed from Eq. (12.4), where the last summand matches the Hamiltonian of a simple harmonic oscillator $\mathcal{H}_0 = \frac{1}{2}(Z^2 + \nu^2 z^2)$. The reduced linear flow

$$(\dot{x}', \dot{y}', \dot{X}', \dot{Y}')^\tau = \mathbb{M}(x', y', X', Y')^\tau, \tag{12.7}$$

with $\mathbb{M}$ given in Eq. (10.30) and τ denoting transposition, remains coupled and gives rise to elliptic and hyperbolic components stemming from the characteristic exponents in Eqs. (10.33) and (10.34), respectively.

On the other hand, the Hamiltonian in separate variables

$$\mathcal{K}_0 = \lambda x_1 X_1 + \frac{1}{2}(Y_1^2 + \omega^2 y_1^2) + \frac{1}{2}(Z_1^2 + \nu^2 z_1^2), \tag{12.8}$$

with λ, w, and ν, given by Eqs. (10.33), (10.34), and (10.28), respectively, enjoys the same dynamical behavior in the subindex-1 variables as Eq. (12.4). Disregarding oscillations in the z-axis direction, the flow stemming from Eq. (12.8) is

$$\begin{pmatrix} \dot{x}_1 \\ \dot{y}_1 \\ \dot{X}_1 \\ \dot{Y}_1 \end{pmatrix} = \mathbb{M}_1 \begin{pmatrix} x_1 \\ y_1 \\ X_1 \\ Y_1 \end{pmatrix}, \quad \text{with } \mathbb{M}_1 = \begin{pmatrix} \lambda & 0 & 0 & 0 \\ 0 & 0 & 0 & 1 \\ 0 & 0 & -\lambda & 0 \\ 0 & -\omega^2 & 0 & 0 \end{pmatrix}. \tag{12.9}$$

Then the question of whether a canonical transformation $(\boldsymbol{x}', \boldsymbol{X}') \to (\boldsymbol{x}_1, \boldsymbol{X}_1)$ exists such that it transforms $\mathcal{H}_0$ into $\mathcal{K}_0$ emerges naturally. The transformation certainly exists and is linear

$$(x', y', X', Y')^\tau = \mathbb{A}(x_1, y_1, X_1, Y_1)^\tau. \tag{12.10}$$

The matrix of constant coefficients $\mathbb{A} = \mathbb{A}(a_{i,j})$ is solved from the linear system obtained by replacing Eq. (12.10) and its time derivative into the right and left sides of (12.7), respectively. Then Eq. (12.9) is in turn plugged into the left side of (12.7), to finally obtain $\mathbb{A}\mathbb{M}_1 = \mathbb{M}\mathbb{A}$, which is an underdetermined linear system that gives rise to a family of solutions depending on two parameters [424].

Different diagonalizing transformations may be used depending on the context [161, 696], but the traditional computation of the matrix $\mathbb{A}$ is based on the eigenvector decomposition of $\mathbb{M}$ [329]. When this decomposition is applied to Eq. (10.30), we obtain

$$\mathbb{A} = \begin{pmatrix} 2\lambda/\sigma & 0 & -2\lambda/\sigma & 2/\upsilon \\ (\lambda^2 - 9)/\sigma & -(\omega^2 + 9)/\upsilon & (\lambda^2 - 9)/\sigma & 0 \\ (\lambda^2 + 9)/\sigma & (9 - \omega^2)/\upsilon & (\lambda^2 + 9)/\sigma & 0 \\ \lambda(\lambda^2 - 7)/\sigma & 0 & \lambda(7 - \lambda^2)/\sigma & -(\omega^2 + 7)/\upsilon \end{pmatrix}, \tag{12.11}$$

with $\sigma = 4\sqrt{\lambda(2\lambda^2 - 9)}$, $\upsilon = 2\sqrt{2(2\omega^2 + 9)}$.

As already discussed in §10.2.2, the removal of the hyperbolic direction is achieved in the linearized dynamics by a simple choice of initial conditions. This is immediately noted in Hamiltonian (12.8), which admits the integrals $J_x = x_1 X_1$, $J_y = Y_1^2 + \omega^2 y_1^2$, and $J_z = Z_1^2 + \nu^2 z_1^2$, defining corresponding invariant manifolds of the linearized motion for given particular values of each integral. In particular the saddle component is removed when the motion is constrained to the manifold $J_x = 0$, which, therefore, is of the center × center type. Therefore, the manifold $J_x = 0$ is called the center manifold.

When the transformation defined by Eqs. (12.10)–(12.11) is applied to all the terms of the perturbation Hamiltonian (12.3), we obtain

$$\mathcal{K} = \sum_{n\geq 0} \frac{\epsilon^n}{n!} \mathcal{K}_n(\boldsymbol{x}_1, \boldsymbol{X}_1), \tag{12.12}$$

in which J_x is no longer an integral. However, we will see that the existence of an integral encapsulating the unstable components of the motion, as well as the concomitant center manifold, can be extended to the nonlinear terms of the transformed Hamiltonian (12.12).

The procedure for extending the integral J_x to the nonlinear terms of Eq. (12.12) is called the *reduction to the center manifold*, and consists in finding a canonical transformation $(\boldsymbol{x}_1, \boldsymbol{X}_1) \mapsto (\boldsymbol{x}_2, \boldsymbol{X}_2)$ that converts Eq. (12.12) into a normal form in the variables with subindex 2, such that all the monomials (12.6) with $m_1 \neq m_2$ are removed from the Hamiltonian [237]; cf. [250].

The normalization is efficiently approached by Lie transforms and, after truncation to a certain order $n = \tilde{n}$, yields the normalized Hamiltonian

$$\mathcal{K}(\boldsymbol{x}_1(\boldsymbol{x}_2, \boldsymbol{X}_2), \boldsymbol{X}_1(\boldsymbol{x}_2, \boldsymbol{X}_2)) = \sum_{n\geq 0}^{\tilde{n}} \frac{\epsilon^n}{n!} \mathcal{K}_n(y_2, z_2, Y_2, Z_2; J'). \tag{12.13}$$

That $J' = x_2 X_2$ is a (formal) integral of the truncated Hamiltonian is checked from $\mathrm{d}J'/\mathrm{d}t = \{J', \mathcal{K}\} = 0$. The Hamiltonian (12.13) is, then, in the required normal form, and, therefore, can be particularized for the center manifold $J' = 0$.

12.1.2 Homological equation in complex variables

On the other hand, the structure of the center manifold is effectively obtained from a normal form that captures both the hyperbolic dynamics normal to the center manifold and the elliptic dynamics on the center manifold. Due to the polynomial nature of the perturbation, the normalization is best addressed in complex variables

$$y_1 = \frac{v + \mathbf{i}V}{\sqrt{2\omega}}, \quad z_1 = \frac{w + \mathbf{i}W}{\sqrt{2\nu}}, \quad Y_1 = \sqrt{\frac{\omega}{2}}(V + \mathbf{i}v), \quad Z_1 = \sqrt{\frac{\nu}{2}}(W + \mathbf{i}w), \tag{12.14}$$

by which we cast Eq. (12.18) into the form $\mathcal{K}_0 = \lambda uU + \mathbf{i}\omega vV + \mathbf{i}\nu wW$, in which $x_1 = u$, $X_1 = U$, whereas the perturbation terms remain as monomials of the type of Eq. (12.6) also in the complex variables.

The Lie derivative (2.49) is

$$\mathcal{L}_0 = \lambda\left(u\frac{\mathrm{d}}{\mathrm{d}u} - U\frac{\mathrm{d}}{\mathrm{d}U}\right) + \mathbf{i}\omega\left(v\frac{\mathrm{d}}{\mathrm{d}v} - V\frac{\mathrm{d}}{\mathrm{d}V}\right) + \mathbf{i}\nu\left(w\frac{\mathrm{d}}{\mathrm{d}w} - W\frac{\mathrm{d}}{\mathrm{d}W}\right), \tag{12.15}$$

and to convert Hamiltonian (12.12) in complex variables into the normal form, we must remove the terms that do not pertain to the kernel $\mathcal{L}_0(M_{\boldsymbol{\kappa}^*}) = 0$. When the Lie derivative (12.15) is applied to the general monomial in Eq. (12.6) we obtain

$$\mathcal{L}_0(M_{\boldsymbol{\kappa}}) = [\lambda(m_1 - m_4) + \mathbf{i}\omega(m_2 - m_5) + \mathbf{i}\nu(m_3 - m_6)]M_{\boldsymbol{\kappa}},$$

which shows that the kernel is characterized by those monomials $M_{\boldsymbol{\kappa}^*}$ such that $m_1 = m_4$, $m_2 = m_5$, and $m_3 = m_6$. In a Hamiltonian made of such kinds of monomials, $J = uU$, $L_1 = \mathbf{i}vV$, and $L_2 = \mathbf{i}wW$, remain constant.

Then, for a term of the image of the Lie derivative $M_{\boldsymbol{\kappa}} \neq M_{\boldsymbol{\kappa}^*}$, the homological equation in complex variables is solved without need of integration by noting that

$$\mathcal{L}_0^{-1}(M_{\boldsymbol{\kappa}}) = \frac{M_{\boldsymbol{\kappa}}}{\lambda(m_1 - m_4) + \mathbf{i}\omega(m_2 - m_5) + \mathbf{i}\nu(m_3 - m_6)}. \tag{12.16}$$

However, the transformation derived from a generating function obtained from Eq. (12.16) is valid only for non-resonant motion. Small divisors that may occur in the manifold J, in which $m_1 = m_4$, when $(m_2 - m_5)/(m_6 - m_3) \approx \nu/\omega$ will prevent convergence of the perturbation solution close to resonances, a case that needs specific treatment. Of particular interest is the 1:1 resonance, which begets relevant solutions of the libration point dynamics and gives rise to the family of halo orbits. Rather than using one solution for the non-resonant case and another one for the 1:1 resonance [103], both cases are effectively encompassed in a single analytical solution [396, 564].

12.1.3 Detuning. The perturbed elliptic oscillator

To cope with the 1:1-resonant dynamics, the unperturbed frequency of the oscillations in the z direction is written in the form $\nu = \omega\sqrt{1-\delta}$, where, on account of the particular values of ν in Eq. (10.28) and ω in Eq. (10.33),

$$\delta = \frac{1}{27}(23 - 8\sqrt{7}) \approx 0.068,$$

is a "detuning" parameter [283, 657] whose value is about one tenth of the Hill radius $r_{\mathrm{H}} = 3^{-1/3}$.

The frequency of the oscillations of the unperturbed problem in the z direction is then detuned by moving the term $-\frac{1}{2}\delta\omega^2 z_1^2$ to the perturbation part. That is, Eq. (12.12) is rearranged in the form

$$\mathcal{K} = \sum_{n\geq 0} \frac{\epsilon^n}{n!}\mathcal{K}_n^*, \tag{12.17}$$

in which the zeroth-order term is

$$\mathcal{K}_0^* = \lambda x_1 X_1 + \frac{1}{2}(Y_1^2 + \omega^2 y_1^2) + \frac{1}{2}(Z_1^2 + \omega^2 z_1^2), \tag{12.18}$$

consisting of the Hamiltonian of an elliptic oscillator in 1:1 resonance in addition to the hyperbolic part. The first-order term is $\mathcal{K}_1^* = \mathcal{K}_1 - \frac{1}{2}\delta\omega^2 z_1^2$, and the remaining terms stay unaltered; that is $\mathcal{K}_n^* = \mathcal{K}_n$ $(n > 1)$.

The transformation to complex variables is next applied by particularizing Eq. (12.14) to the case $\nu = \omega$. In this way, Eq. (12.18) turns into

$$\mathcal{K}_0^* = \lambda uU + \mathbf{i}\omega(vV + wW). \tag{12.19}$$

Accordingly, the Lie derivative is

$$\mathcal{L}_0 = \lambda\left(u\frac{\mathrm{d}}{\mathrm{d}u} - U\frac{\mathrm{d}}{\mathrm{d}U}\right) + \mathbf{i}\omega\left(v\frac{\mathrm{d}}{\mathrm{d}v} - V\frac{\mathrm{d}}{\mathrm{d}V} + w\frac{\mathrm{d}}{\mathrm{d}w} - W\frac{\mathrm{d}}{\mathrm{d}W}\right), \tag{12.20}$$

from which

$$\mathcal{L}_0(M_{\boldsymbol{\kappa}}) = [\lambda(m_1 - m_4) + \mathbf{i}\omega(m_2 - m_5 + m_3 - m_6)]M_{\boldsymbol{\kappa}} \tag{12.21}$$

shows that, now, monomials $M_{\boldsymbol{\kappa}^*}$ belonging to the kernel of the Lie derivative are characterized by $m_1 = m_4$ and $m_2 + m_3 = m_5 + m_6 = \frac{1}{2}\sigma$, where $\sigma = m_2 + m_3 + m_5 + m_6$ is even. The solution of the homological equation is simple analogously to Eq. (12.16), which is replaced by

$$\mathcal{L}_0^{-1}(M_{\boldsymbol{\kappa}}) = \frac{M_{\boldsymbol{\kappa}}}{\lambda(m_1 - m_4) + \mathbf{i}\omega(m_2 - m_5 + m_3 - m_6)}. \tag{12.22}$$

In particular, the monomials uU, vV, wW, vW, and wV belong to the kernel of the Lie derivative (12.20). Moreover, because m_2 (resp. m_6) is either smaller than, or equal to, or greater than m_5 (resp. m_3), for a term of the kernel we find

- $m_2 = m_5$: $v^{m_2}w^{m_3}V^{m_5}W^{m_6} = (vV)^{m_2}(wW)^{m_3}$
- $m_2 = m_5 + m_0$: $v^{m_2}w^{m_3}V^{m_5}W^{m_6} = (vV)^{m_5}(wW)^{m_3}(vW)^{m_0}$
- $m_5 = m_2 + m_0$: $v^{m_2}w^{m_3}V^{m_5}W^{m_6} = (vV)^{m_2}(wW)^{m_6}(wV)^{m_0}$

Therefore, the kernel of the Lie derivative (12.20) is generated precisely by these five types of monomials [136, 160]. However, while terms $J = uU$ of the kernel are easily identified with the conjugate momentum to the hyperbolic coordinate $\psi = \log\sqrt{u/U}$, the other four types of monomials in complex variables are not immediately identified with momenta conjugated to some either explicit or ignorable canonical variable.

Thing are much clearer when using the canonical set of Deprit's Lissajous variables (ℓ, g, L, G), defined by the transformation from Cartesian variables [160],

$$y = s\cos(g+\ell) - d\cos(g-\ell), \tag{12.23}$$

$$z = s\sin(g+\ell) - d\sin(g-\ell), \tag{12.24}$$

$$Y = -\omega[s\sin(g+\ell) + d\sin(g-\ell)], \tag{12.25}$$

$$Z = \omega[s\cos(g+\ell) + d\cos(g-\ell)], \tag{12.26}$$

where $s \equiv s(L, G; \omega)$ and $d \equiv d(L, G; \omega)$ are the state functions

$$s = \sqrt{\frac{L+G}{2\omega}}, \quad d = \sqrt{\frac{L-G}{2\omega}}. \tag{12.27}$$

At a given time, Deprit's Lissajous variables characterize a pair of ellipses. The relations $G = \omega ab$, $L = \frac{1}{2}\omega(a^2 + b^2)$, define the semimajor a and semiminor axis $|b|$ (resp. ωa and $\omega|b|$) of an ellipse in the y, z plane (resp. Y, Z) centered at the origin. The angle g defines the direction of the semiminor axis with respect to the y axis, whereas the angle ℓ measures the elliptic anomaly from the semiminor axis [160].

The transformation from Deprit's Lissajous variables to complex variables is readily obtained from Eq. (12.14), with v replaced by ω, and Eqs. (12.23)–(12.26). Hence,

$$v = \sqrt{\tfrac{1}{2}\omega}[(s-d)\cos g + \mathbf{i}(d+s)\sin g](\cos\ell + \mathbf{i}\sin\ell),$$

$$w = \sqrt{\tfrac{1}{2}\omega}[(s-d)\sin g - \mathbf{i}(d+s)\cos g](\cos\ell + \mathbf{i}\sin\ell),$$

$$\mathbf{i}V = \sqrt{\tfrac{1}{2}\omega}[(s-d)\cos g - \mathbf{i}(d+s)\sin g](\cos\ell - \mathbf{i}\sin\ell),$$

$$\mathbf{i}W = \sqrt{\tfrac{1}{2}\omega}[(s-d)\sin g + \mathbf{i}(d+s)\cos g](\cos\ell - \mathbf{i}\sin\ell),$$

from which we immediately check that the monomials vV, wW, vW, and wV are free from the elliptic anomaly ℓ. In consequence, the kernel of the Lie derivative remains of one degree of freedom in the pair of conjugate variables (g, G), whereas

$$J \equiv uU, \quad L \equiv \mathbf{i}(vV + wW), \tag{12.28}$$

are the two integrals to be normalized by Lie transforms, and, after normalization, will parameterize the reduced flow, as checked in Eq. (12.19).

Moreover, after normalization, for a given manifold L = constant, the reduced (g, G) dynamics in the center manifold $J = 0$ takes place on a two-dimensional sphere

[134, 373]. This fact is immediately disclosed when using the Hopf coordinates [300]

$$I_1 = \frac{1}{2}\mathbf{i}(wW - vV), \quad I_2 = -\frac{1}{2}\mathbf{i}(vW + wV), \quad I_3 = \frac{1}{2}(vW - wV), \tag{12.29}$$

which are linked by the constraint

$$I_0^2 = I_1^2 + I_2^2 + I_3^2 = \frac{1}{4}L^2. \tag{12.30}$$

12.1.4 Hamiltonian normalization

Normalization in complex variables is achieved through a single Lie transformation whose construction is straightforward because it only involves a polynomial algebra. That is, after rewriting the detuned Hamiltonian (12.17) in complex variables using Eqs. (12.14) with $\nu = \omega$, we obtain

$$\mathcal{K} = \sum_{n\geq 0} \frac{\epsilon^n}{n!}\mathcal{K}^*_{n,0}(u, v, w, U, V, W). \tag{12.31}$$

Then a Lie transformation $(u, v, w, U, V, W) \mapsto (u', v', w', U', V', W'; \epsilon)$ is constructed based on the properties of the fundamental equation (12.21) to normalize the Hamiltonian (12.31) up to some truncation order. After normalization, the quantities J and L in Eq. (12.28), which are now written in prime variables, become formal integrals.

Dropping the prime notation for brevity, the first few terms of the normalized Hamiltonian

$$\mathcal{N} = \sum_{n\geq 0} \frac{\epsilon^n}{n!}\mathcal{K}^*_{0,n}(v, w, V, W; J) \tag{12.32}$$

are, cf. [424],

$$\begin{aligned}
\mathcal{K}^*_{0,0} &= \mathbf{i}\omega(vV + wW),\\
\mathcal{K}^*_{0,1} &= -\frac{1}{2}\mathbf{i}\omega\delta wW,\\
\mathcal{K}^*_{0,2} &= -\frac{1}{4}\mathbf{i}\omega\delta^2 wW + r_{\mathrm{H}}\bigg[-\mathbf{i}\lambda\omega\left(\frac{2994}{4837}vV + \frac{349\omega^2 + 3737}{9674}wW\right)J\\
&\quad - \frac{4683\omega^2 + 7263}{38696}J^2 + \frac{4683\omega^2 + 2103}{38696}v^2V^2 + \frac{484\omega^2 - 193}{4146}w^2W^2\\
&\quad - \frac{23\omega^2 - 89}{42}vVwW + \frac{15235\omega^2 - 41269}{38696}(v^2W^2 + V^2w^2)\bigg],\\
\mathcal{K}^*_{0,3} &= -\frac{3}{8}\mathbf{i}\delta^3\omega wW + \delta r_{\mathrm{H}}\bigg[\mathbf{i}\frac{331467\omega^2 - 2387193}{13369468}\lambda\omega JwW\\
&\quad - \frac{109\omega^2 - 619}{84}vVwW + \frac{12286157\omega^2 - 60125147}{13369468}(v^2W^2 + V^2w^2)\\
&\quad + \frac{515150\omega^2 - 1475408}{1432443}w^2W^2\bigg].
\end{aligned}$$

12.2 Reduced dynamics in the center manifold

The hyperbolic dynamics is removed by constraining the motion to the center manifold $J = 0$. Besides, Eq. (12.32) is readily expressed in the Hopf variables based on the relations $wW = -\mathbf{i}(I_1 + I_0)$, $vV = \mathbf{i}(I_1 - I_0)$, $vW = \mathbf{i}I_2 + I_3$, and $Vw = \mathbf{i}I_2 - I_3$, which are readily obtained from Eqs. (12.29) and (12.30). The variation equations of the reduced Hamiltonian flow are readily derived in the Hopf variables taking into account that $\{I_1, I_2\} = I_3$, $\{I_2, I_3\} = I_1$, $\{I_3, I_1\} = I_2$. Namely,

$$\dot{I}_i = \frac{\mathrm{d}I_i}{\mathrm{d}t} = \{I_i; \mathcal{N}\} = \sum_{j=1}^{3} \{I_i; I_j\} \frac{\partial \mathcal{N}}{\partial I_j}, \quad i = 1, 2, 3.$$

Thus, $\mathcal{N}_{0,0} = \omega L$ is constant and, at the first order,

$$\mathcal{N}_{0,1} = -\frac{1}{4}\delta\omega L - \frac{1}{2}\delta\omega I_1. \tag{12.33}$$

Hence, $\dot{I}_1 = 0$, $\dot{I}_2 = \frac{1}{2}\delta\omega I_3$, $\dot{I}_3 = -\frac{1}{2}\delta\omega I_2$, whose solution yields circular motion in the (I_2, I_3) plane, of radius $\frac{1}{2}(L^2 - 4I_1^2)^{1/2}$ and frequency $\frac{1}{2}\delta\omega$. That is, orbits of the reduced dynamics transit parallel circles $I_1 = \text{const.}$ on the sphere, which reduce to stable equilibria at the poles $I_1 = \pm\frac{1}{2}L$. As we will further detail in §12.2.3, these equilibria agree with the small oscillations described in §10.2.2. However, the descriptions provided by the linearized Hamiltonian (12.33) constrain to energy values very close to the energy of the libration points.

At second order we find

$$\mathcal{N}_{0,2} = 2!\left[\frac{1}{4}\delta\mathcal{N}_{0,1} - k_1 L^2 + k_2 L I_1 - k_3(I_1^2 - I_2^2) + k_4 I_3^2\right], \tag{12.34}$$

in which the numeric coefficients

$$\begin{aligned}
k_1 &= \frac{1}{12}(1097 + 110641\omega^2)k_0,\\
k_2 &= \frac{2}{3}(11713 + 497\omega^2)k_0,\\
k_3 &= (841 - 79\omega^2)k_0,\\
k_4 &= -(329311 - 121801\omega^2)k_0,
\end{aligned}$$

with $k_0 = r_{\mathrm{H}}/154784 \approx 4.5 \times 10^{-6}$, are strictly positive irrational numbers ($k_1 \approx 0.18$, $k_2 \approx 0.041$, $k_3 \approx 0.002$, $k_4 \approx 0.87$). After neglecting constant terms, the reduced flow is derived from the non-trivial second-order Hamiltonian

$$\mathcal{I} = (k_2 L - \omega\delta^*)I_1 + k_3(I_2^2 - I_1^2) + k_4 I_3^2, \tag{12.35}$$

where we abbreviated $\delta^* = \frac{1}{2}\delta + \frac{1}{8}\delta^2 \approx 0.03454$. Strikingly, Eq. (12.35) is a particular case of the Zhukovsky–Volterra Hamiltonian describing the motion of a free gyrostat [41, 672, 700].

12.2.1 Visualization of the reduced flow

The orbits $O(I_1, I_2, I_3)$ on the sphere result from the intersection of the two-dimensional surfaces materialized by the sphere (12.30) and the Hamiltonian (12.35), and can be obtained without need of integrating the differential equations of the flow. Indeed, for given values of the formal integral L and the "energy" $\mathcal{I} = E$, the orbit is obtained like $O(I_1, L, E)$, by eliminating I_3 (resp. I_2) between Eqs. (12.30) and (12.35). We readily obtain

$$I_2^2 = -\frac{(k_3 + k_4)I_1^2 + (\delta^* \omega - k_2 L)I_1 + E - \frac{1}{4}L^2 k_4}{k_4 - k_3},$$

$$I_3^2 = \frac{2k_3 I_1^2 + (\delta^* \omega - k_2 L)I_1 + E - \frac{1}{4}L^2 k_3}{k_4 - k_3},$$

where, for each pair (E, L), the value of I_1 must be constrained to the subset of the closed interval $[-\frac{1}{2}L, \frac{1}{2}L]$ where both I_2 and I_3 take real values.

Figure 12.1 was constructed this way and illustrates the qualitative changes undergone by the reduced flow for increasing values of the dynamical parameter L. For clarity, the radius of the different spheres is normalized to 1. Thus, for small values of L we find two fixed points of the elliptic type at $I_1 = \pm\frac{1}{2}L$, and circulatory motion around them (first row of Fig. 12.1). Increasing the value of L turns the fixed point $(-\frac{1}{2}L, 0, 0)$ into hyperbolic in a bifurcation process in which two additional fixed points of the elliptic type appear in the *meridian* $I_2 = 0$ (second row of Fig. 12.1). Further increasing L displaces the elliptic points $(I_1, 0, \pm I_3)$ towards the "north" and "south" poles of the sphere (third row of Fig. 12.1). Eventually, the fixed point $(-\frac{1}{2}L, 0, 0)$ becomes elliptic again in a new bifurcation process, in which two fixed points of the hyperbolic type appear at the "equator" $I_3 = 0$ of the sphere (fourth row of Fig. 12.1). For still higher values of L these symmetric points $(I_1, \pm I_2, 0)$ migrate along the equator until they finally collapse at $(\frac{1}{2}L, 0, 0)$ which then changes its type to hyperbolic (fifth row of Fig. 12.1). No further qualitative changes of the reduced flow are observed for increasing values of L.

12.2.2 Equilibria and bifurcations. Analytical computation

The flow stemming from Hamiltonian (12.35) is obtained from the integration of

$$\dot{I}_1 = -2(k_4 - k_3)I_2 I_3, \tag{12.36}$$

$$\dot{I}_2 = 2(k_4 + k_3)I_1 I_3 + (\delta^* \omega - k_2 L)I_3, \tag{12.37}$$

$$\dot{I}_3 = 4k_3 I_1 I_2 - (\delta^* \omega - k_2 L)I_2. \tag{12.38}$$

Remarkably, this differential system becomes analogous to the Euler equations for the free rigid body rotation if $L = \tilde{L} = \delta^* \omega / k_2$, a case in which the closed-form solution is

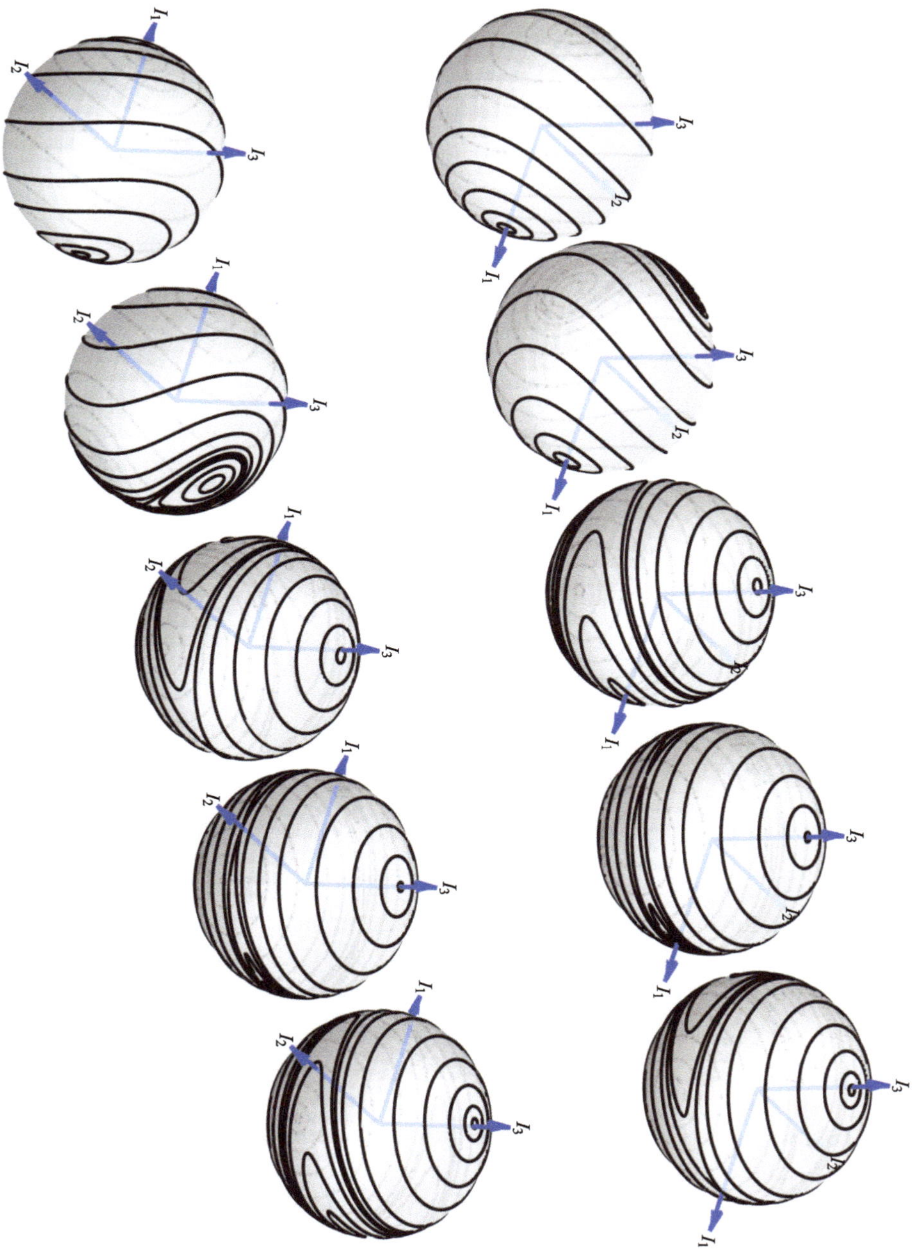

Figure 12.1: From top to bottom, opposite views (left/right col.) of the bifurcation sequence of the Hamiltonian flow (12.35) for increasing values of L. Reprinted by permission from Springer [396].

obtained in terms of Jacobi elliptic functions. Rather than pursuing the general solution, we are satisfied with computing the equilibria previously visualized, as well as the parameter values at which the bifurcations take place.

The equilibria are in full agreement with what was anticipated by depicting the flow. Indeed, that the points

$$E_{\pm 1} = \left(\pm\frac{1}{2}L, 0, 0\right) \tag{12.39}$$

are always equilibria results from the vanishing of Eqs. (12.36)–(12.38) when $I_2 = I_3 = 0$. When $I_2 = 0$ and $I_3 \neq 0$ Eqs. (12.36) and (12.38) identically vanish, but for Eq. (12.37) to vanish it is required that

$$I_1 = I_{1,\text{halo}} \equiv \frac{k_2 L - \delta^* \omega}{2(k_3 + k_4)} \Rightarrow |I_3| = I_{3,\text{halo}} \equiv \frac{1}{2}\sqrt{L^2 - 4I_{1,\text{halo}}^2}.$$

If $L > \delta^* \omega/(k_2 + k_3 + k_4) \approx 0.0768606$ then $|I_3| > 0$ and the two symmetric equilibria

$$E_{\pm 2} = (I_{1,\text{halo}}, 0, \pm I_{3,\text{halo}}), \tag{12.40}$$

also exist. Finally, Eqs. (12.36) and (12.37) vanish identically when $I_3 = 0$, but to make null Eq. (12.38) when $I_2 \neq 0$ we need

$$I_1 = I_{1,\text{Bridge}} \equiv \frac{k_2 L - \delta^* \omega}{4k_3} \Rightarrow |I_2| = I_{2,\text{Bridge}} \equiv \frac{1}{2}\sqrt{L^2 - 4I_{1,\text{Bridge}}^2},$$

which in turn requires that $\delta^* \omega/(k_2 + 2k_3) < L < \delta^* \omega/(k_2 - 2k_3)$, a case in which also exist the two symmetric equilibria

$$E_{\pm 3} = (I_{1,\text{Bridge}}, \pm I_{2,\text{Bridge}}, 0). \tag{12.41}$$

It is simple to check that the bifurcation at $L = \delta^* \omega/(k_2 + 2k_3)$ originates from E_{-1}, whereas the bifurcation at $L = \delta^* \omega/(k_2 - 2k_3)$ stems from E_1.

The usual linearization of the flow in Eqs. (12.36)–(12.38) could be used to show that the stability of the equilibria is in agreement with the elliptic or hyperbolic character of the fixed points visualized in the previous section.

12.2.3 Partner orbits of the equilibria in the center manifold

The equilibria of the reduced phase space correspond to particular orbits of the Hill problem whose dynamical characteristics are easily identified with the help of Deprit's Lissajous canonical variables. To do that, the Hopf variables are written in terms of the Lissajous variables defined in Eqs. (12.23)–(12.27). Namely,

$$I_0 = \frac{1}{2}L, \quad I_1 = \omega s d \cos 2g, \quad I_2 = \omega s d \sin 2g, \quad I_3 = \frac{1}{2}G, \tag{12.42}$$

which provide other evidence of the removal of short-period effects related to the elliptic anomaly ℓ carried out in the normalization process.

Thus, in the case of the equilibria $E_{\pm1}$ and $E_{\pm3}$, given by Eqs. (12.39) and (12.41), respectively, Eq. (12.42) shows that $G = \omega ab = 0$. That is, on average, either a or $|b|$, the semi-axes of the Lissajous ellipse, vanishes, thus yielding rectilinear oscillations on the center manifold. Because $g = 0$ for E_1, the harmonic oscillations are in the z_1 direction, as follows from Eqs. (12.23)–(12.26). On the contrary, $g = \frac{\pi}{2}$ for E_{-1}, and the oscillations take place in the y_1 direction. In the case of the two equilibria $E_{\pm3}$, the constant value $g \neq 0$ is determined from the components I_1 and I_2 in Eq. (12.41), and the oscillations take place on the (y_1, z_1) plane. Finally, in the case of $E_{\pm2}$, given in Eq. (12.40), because $G \neq 0$ both semi-axes have nonzero values, thus yielding, on average, elliptic motion on the center manifold. The elliptic oscillations take place in the (y_1, z_1) plane, and the area of the ellipse $\pi a|b| = \pi|G|/\omega$ depends on the value $|G| = 2I_{3,\text{halo}}$.

To recover the original dynamics in Cartesian variables, we need first to come back from the normalized to the subindex-1 variables, and then to use the linear transformation in Eq. (12.10). Then one finds that the equilibria of the reduced dynamics on the sphere match the familiar periodic orbits of the Hill problem about the libration points. Specifically, the Lyapunov planar orbits are the partners of the E_{-1} equilibria, and the Lyapunov vertical orbits are the partners of E_1. The two branches of halo orbits correspond to the $E_{\pm2}$ equilibria, whereas $E_{\pm3}$ match the two-lane bridge of periodic orbits that link the Lyapunov planar orbits with the vertical ones.

Due to the truncation order of the perturbation approach, the analytical solution will not provide exact periodic orbits of the Hill problem. However, the initial conditions provided by the perturbation solution are readily improved with the help of a differential-corrections algorithm to get the desired periodic orbit. Some of these algorithms need to be fed with an approximation of the period T, which can be estimated from the normalized variables from the rate of variation of the eccentric anomaly $\mathrm{d}\ell/\mathrm{d}t = 2\pi/T$. After writing Eq. (12.35) in Deprit's Lissajous variables using Eq. (12.42), the latter is obtained from Hamilton equations. In particular, if we limit to first-order effects, then $\mathcal{N} = \omega L - \frac{1}{4}\delta\omega(L + 2\omega sd\cos 2g)$, from which

$$\frac{\mathrm{d}\ell}{\mathrm{d}t} = \omega - \frac{1}{4}\delta\omega\left(1 + \frac{1}{\sqrt{1 - G^2/L^2}}\cos 2g\right), \tag{12.43}$$

where the needed values g, G, and L, are computed from Eq. (12.42) after replacing the Hopf variables by the values taken at the corresponding equilibria.

For the second-order truncation of the analytical solution hitherto discussed this procedure is successful only for energy values close to the energy of the libration point equilibrium (small values of L), which, therefore, exclude the computation of halo orbits as well as orbits of the bridge family. However, the validity of the perturbation solution is extended with the straightforward computation of higher orders, as it was already anticipated with the explicit presentation of the term $\mathcal{K}_{0,3}^*$ of the Hamiltonian (12.32).

12.3 Higher orders

Extension of the Hamiltonian normalization in complex variables (12.31) to higher orders is just a matter of mechanizing operations [424]. After reformulation in the Hopf variables, the higher-order normalized Hamiltonian takes the form

$$\mathcal{I} = \omega L + \sum_{n\geq 1} \frac{1}{n!}\mathcal{I}_n(I_1, I_2, I_3; L). \tag{12.44}$$

This normalized polynomial involves powers of the Hopf coordinates higher than 2, and, therefore, the orbits of the reduced flow cannot be solved explicitly as was previously done in §12.2.1. Alternatively, the flow is efficiently visualized by depicting contour plots of $\mathcal{I}(I_1, I_2, I_3; L) = E$. In particular, the flow is rendered in real time with the technique of painting Hamiltonians previously used in Fig. 5.6. This is illustrated in Fig. 12.2 for a seventh-order truncation of Eq. (12.44).

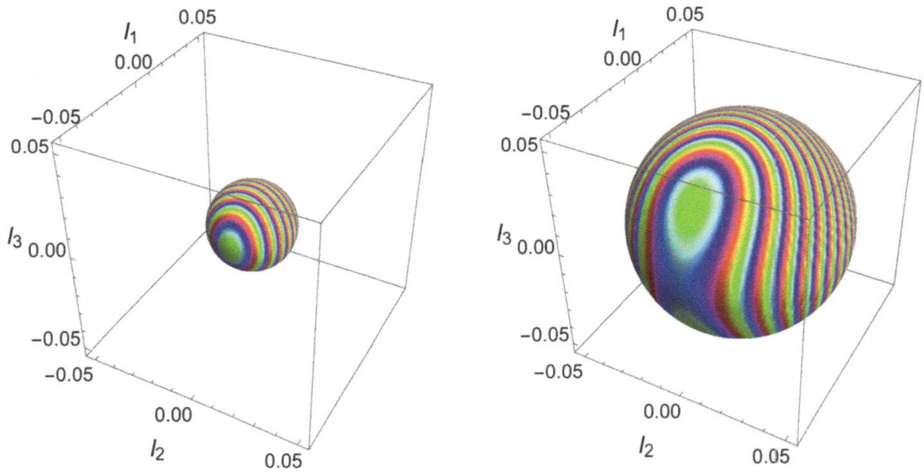

Figure 12.2: Painting the phase space of the Hill problem about the libration points to show the bifurcation of halo orbits from Lyapunov planar orbits $(-\frac{1}{2}L, 0, 0)$.

On the other hand, in order to use the analytical solution for the computation of libration-point orbits far away from the libration points, an accurate computation of the different equilibria is required. The flow stemming from the Hamiltonian (12.44) is now obtained from the differential system

$$\dot{I}_1 = I_2 I_3 \sum_{n\geq 0} F_{1,n}(I_1, I_2, I_3; L), \tag{12.45}$$

$$\dot{I}_2 = I_3 \sum_{n\geq 0} F_{2,n}(I_1, I_2, I_3; L), \tag{12.46}$$

$$\dot{I}_3 = I_2 \sum_{n\geq 0} F_{3,n}(I_1, I_2, I_3; L), \tag{12.47}$$

which still provides the same solutions as Eq. (12.39) for the Lyapunov orbits. Halo orbits $(I_1, 0, \pm I_3)$ also exist, but now the values I_1 and hence $I_3 = \frac{1}{2}\sqrt{L^2 - 4I_1^2}$, as results from Eq. (12.30), are different from those in Eq. (12.40), and must be numerically computed from the implicit equation in I_1,

$$\sum_{n\geq 0} F_{2,n}(I_1, 0, I_3(I_1; L); L) = 0, \tag{12.48}$$

which makes Eq. (12.46) vanish. The value of L in which the halo orbits bifurcate from the point $(-\frac{1}{2}L, 0, 0)$ is now given by the root $\sum_{n\geq 0} F_{2,n}(\frac{1}{2}L, 0, 0; L) = 0$. Finally, the analog equilibria $(I_1, \pm I_2, 0)$ to those in Eq. (12.41) are computed by solving the implicit equation in I_2,

$$\sum_{n\geq 0} F_{3,n}(I_1, I_2(I_1; L), 0; L) = 0, \tag{12.49}$$

where, from Eq. (12.30), $I_2 = \frac{1}{2}\sqrt{L^2 - 4I_1^2}$. The bifurcation and termination points of the two-lane bridge of periodic orbits linking planar and vertical Lyapunov orbits are obtained from the computation of the roots of the implicit equation $\sum_{n\geq 0} F_{3,n}(\frac{1}{2}L, 0, 0; L) = 0$.

The precision of the solution generally increases with the truncation order of the perturbation theory. However, one should keep in mind the risks inherent to a futile escalation in accuracy. Indeed, if the computation of the perturbation solution is carried out using integer arithmetic, the size of the integer numbers grows notably from one order to the next one. For instance, at the 11th order of the perturbation approach we found that the integer numbers to handle may involve more than 100 digits. This fact makes memory allocation an issue that notably slows the computation of higher orders, whose computational burden grows exponentially [424]. Computations are radically expedited working in real floating-point arithmetics, yet the propagation of truncation errors due to the physical length of the computer's registers must be carefully traced in this case [100, 328]. In particular, the transformation from complex to real variables produces some complex residuals, whose magnitude grows with the order of the perturbation theory at a faster rate than the growing of the real terms. Because this rate notably increases when the computations are extended beyond the order 15, the 16th order is recommended as a practical limit of applicability of the perturbation solution [424].

The construction of a higher-order analytical solution follows the steps outlined above. That is, for the selected value of the dynamical parameter L—which is roughly related with the size of the orbit—the equilibrium in the reduced phase space of the desired orbit (Lyapunov planar or vertical, halo, or pertaining to the bridge) is computed first. It may require one to solve Eq. (12.48) for a halo orbit, or Eq. (12.49) in the case of an orbit of the bridge. Then the corresponding Lissajous variables are computed and translated into complex (prime) variables for a discrete number of values $\ell \in [0, 2\pi)$. The (direct) Lie transformation of the normalization is then applied to recover the original dynamics in the subindex-1 variables. Finally, application of the linear transformation (12.10) will provide the original dynamics in Cartesian coordinates relative to the libration point.

The accuracy of the analytical perturbation solution thus computed is assessed by checking the periodicity $\Delta = \max |\xi_i(T) - \xi_i(0)|$, $\xi_i \in (x, y, z, X, Y, Z)$, of the orbit obtained by numerical propagation of one of the points provided by the analytical solution. This procedure is illustrated below with some examples for different truncations of the higher-order solution, which we borrow from [424]. A systematic scan showing the efficiency of the perturbation solution in an ample vicinity of the libration points of the Hill problem can be found in [424].

12.3.1 Lyapunov orbits

Lyapunov orbits exist always for energies above that of the libration points. Low-order truncations of the perturbation solution provide good approximations to the actual dynamics of orbits of moderate amplitude, yet higher-order truncations are required when the Lyapunov orbits grow in size.

Thus, for $L = 0.01$ we found that the orbit predicted by the fourth-order truncation matches its partner vertical Lyapunov orbit, but only at the precision of the graphics, and the periodicity error is $\Delta \approx 10^{-6}$ in Hill problem units. Periodicity improves gradually with the truncation order of the perturbation solution, becoming smaller than 10^{-10} at the 11th order. Analogous results are found for the Lyapunov planar orbit with $L = 0.01$, for which the 11th-order truncation yields $\Delta < 10^{-11}$. Errors slightly improve when using higher-order truncations, yet the improvements become negligible beyond order 13 for both Lyapunov orbits.

As expected, higher-order truncations are needed when the Lyapunov orbits evolve farther away from the libration points. Thus, for instance, when $L = 0.2$—that is, more than one order of magnitude larger than in the previous example—a truncation order between seven and nine is needed for the analytical perturbation solution to mimic the behavior of the actual orbit at the precision of the graphics. The accuracy of the 15th-order truncation of the analytical solution is $\Delta = \mathcal{O}(10^{-6})$ in both types of Lyapunov orbits, and no relevant improvements are found when extending the perturbation solution to higher orders. Lyapunov orbits provided by the analytical solution for $L = 0.2$ are shown with gray dots in Fig. 12.3 superimposed to their respective partner true periodic orbits (full lines). These plots are depicted in units of the Hill problem, and the origin is the libration point.

12.3.2 Resonant orbits

The analytical perturbation solution predicts the bifurcation of halo orbits from Lyapunov planar orbits when $L = 0.0781869$. On the other hand, because their size is

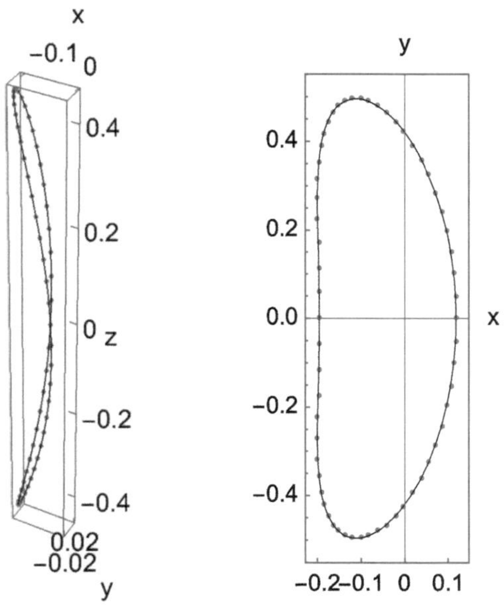

Figure 12.3: Analytical vertical and planar Lyapunov orbits for $L = 0.2$ (dots) superimposed to their partner, numerically integrated, periodic orbits (full lines).

large, we need higher-order truncations of the perturbation solution to mimic the actual dynamics. A sample resonant halo orbit for $L = 0.2$ is shown in the left plot of Fig. 12.4. In this particular example, a ninth-order truncation of the analytical solution yields a periodicity error of the order of one thousandth. The quality of the analytical solution gradually improves with higher-order truncations, yet we did not find improvements beyond the 17th-order truncation, in which case the periodicity is $\Delta = \mathcal{O}(10^{-7})$.

Periodic orbits of the two-lane bridge linking both kinds of Lyapunov orbits only exist for large values of L. Therefore, they may fall out of the range of validity of the Legendre polynomials expansion carried out in Eq. (12.5). In spite of that, we have already shown that their existence is correctly predicted by the second-order truncation of the reduced dynamics, yet their exactness notably varies depending on the truncation used. For instance, while the 14th-order truncation finds the bifurcation from a planar Lyapunov orbit at $L = 0.834458$, and the termination in a vertical Lyapunov orbit at $L = 1.0632$, these values non-negligibly vary when using different approximations. We checked that the 14th-order truncation correctly predicts the existence of an orbit of the bridge family for $L = 0.9$, as shown in the right plot of Fig. 12.4. Still, the periodicity error is gross, $\Delta = \mathcal{O}(10^{-2})$, and does not improve with higher-order truncations. On the other hand, the initial conditions provided by the analytical solution are amenable of being improved through differential corrections to converge to the true periodic orbit.

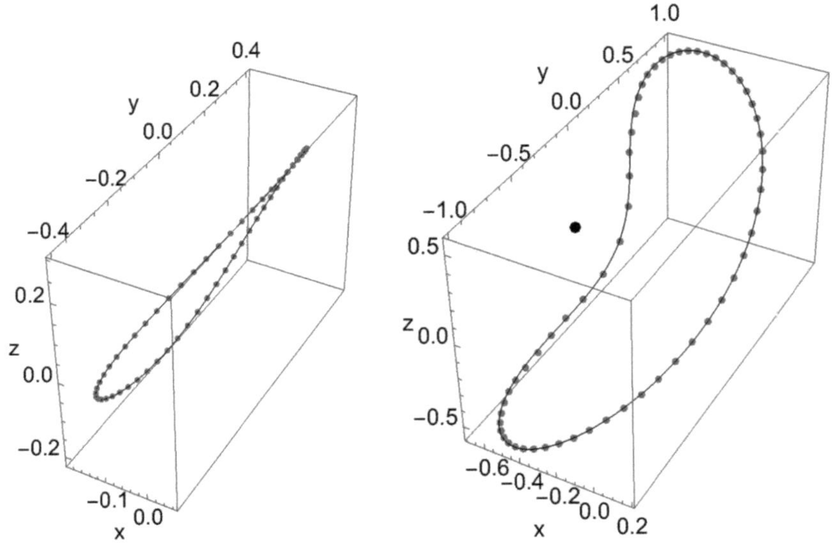

Figure 12.4: Resonant analytical orbits (gray dots) superimposed to their partner numerically integrated periodic orbits (full lines). Left: halo orbit for $L = 0.2$. Right: orbit of the bridge family for $L = 0.9$, in which the primary is depicted with a black spot. After [424].

13 Quasi-satellite orbits

The Hill problem Hamiltonian consists of quadratic terms except for the one representing the non-negligible interaction between the primary of smaller mass and the orbiter, which otherwise will both evolve with pure Keplerian motion about the primary of bigger mass. Because quadratic Hamiltonians give rise to linear dynamics, which is integrable, when this last term is small the Hill problem admits a natural perturbation arrangement in which the zeroth-order Hamiltonian comprises the quadratic terms and the nonlinear interaction is taken like the perturbation. In particular, this perturbation Hamiltonian may be representative of the coorbital dynamics of the lighter primary and the orbiter about the heavier primary.

As mentioned in the Introduction, conspicuous solutions of the coorbital motion are the so-called quasi-satellite orbits, which exist out of the Hill sphere and are useful in different astrodynamics applications. Because of that, a full description of the computation of analytical perturbation solutions is provided here, yet limited to the planar case. Like in previous chapters we hold formally the (superfluous) mass parameter and synodic rotation rate of the Hill problem due to the insight they provide on the dynamics, but also because they permit the systematic check of dimensions as an additional test to be applied to the analytical expressions obtained at each step of the perturbation procedure.

On the one hand, the usefulness of perturbation solutions for mission designing of artificial satellite missions is clearly illustrated in this chapter with the identification of orbit design parameters that are derived from the formal integrals obtained after the complete Hamiltonian reduction. On the other hand, the difficulties that may be found in the computation of the average dynamics of restricted three-body models are clearly manifested by the appearance of non-trivial integrals depending on special functions. The latter not only make the closed-form solution difficult, but neither admit a representation by fast-convergent series in this specific case.

13.1 Planar case. Epicyclic coordinates

In the planar case $z = Z = 0$, and the Hill problem Hamiltonian in Eq. (10.17) is now arranged in the form

$$\mathcal{H} = \mathcal{H}_0(x, y, X, Y; \dot{\vartheta}) - \mu/r, \tag{13.1}$$

where $r = \sqrt{x^2 + y^2}$ and

$$\mathcal{H}_0 = \frac{1}{2}(X + \dot{\vartheta}y)^2 + \frac{1}{2}(Y - \dot{\vartheta}x)^2 - \frac{3}{2}\dot{\vartheta}^2x^2. \tag{13.2}$$

The Hamilton equations of $\mathcal{H}_0$ give rise to a linear differential system, whose solution is straightforward [47, 113]. Alternatively, the Hamiltonian flow stemming from

https://doi.org/10.1515/9783110668513-013

Eq. (13.2) is readily integrated after performing the transformation from Cartesian to epicyclic variables $(x, y, X, Y) \mapsto (\phi, q, \Phi, Q; \dot{\vartheta}, k)$,[1] given by

$$\begin{aligned} x &= a\xi + b\sin\phi, \\ y &= a\eta + a\cos\phi, \\ X &= -2B\eta - B\cos\phi, \\ Y &= -B\xi - B\sin\phi, \end{aligned} \tag{13.3}$$

where

$$a = 2b, \quad b = (2\Phi/\dot{\vartheta})^{1/2}, \tag{13.4}$$

have units of length,

$$B = \dot{\vartheta} b = (2\dot{\vartheta}\Phi)^{1/2}, \tag{13.5}$$

has units of length divided by time,

$$\xi = \xi(Q, \Phi) \equiv \frac{1}{2k}\frac{Q}{B}, \quad \eta = \eta(q, \Phi) \equiv k\frac{q}{b}, \tag{13.6}$$

are nondimensional versions of Q and q, respectively, and k is a scaling factor.

The first two equations of Eq. (13.3) assign to each point of the orbit (x, y) a *reference* ellipse $(x - a\xi)^2/b^2 + (y - a\eta)^2/a^2 = 1$, where the ratio of semimajor axis a, which is in the y-axis direction, to semiminor axis b is always 2. That is, the eccentricity of the reference ellipse is $e = \sqrt{3/4}$. The coordinates of the center of the reference ellipse are

$$x_{\text{center}} = a\xi = \frac{1}{k\dot{\vartheta}}Q, \quad y_{\text{center}} = a\eta = 2kq, \tag{13.7}$$

and the orbit is traveled clockwise. Similarly, the last two equations of Eq. (13.3) assign to each point of the hodograph (X, Y) a circle $(X + 2B\eta)^2 + (Y + B\xi)^2 = B^2$, of radius B centered at $X_{\text{center}} = -2B\eta = -2k\dot{\vartheta}q$, $Y_{\text{center}} = -B\xi = -\frac{1}{2}Q/k$. The hodograph is traveled counterclockwise.

The transformation given by Eqs. (13.3)–(13.6) is canonical and yields the complete reduction of Hamiltonian (13.2) in epicyclic variables. Namely,

$$\mathcal{H}_0 = \mathcal{H}_0(-, -, \Phi, Q) \equiv \dot{\vartheta}\Phi - \frac{3}{8}(Q/k)^2, \tag{13.8}$$

which discloses the ignorable character of ϕ and q, and hence reveals that Φ and Q are the two needed independent integrals of the quadratic Hamiltonian (13.2).

---

1 Classically, the transformation to epicyclic variables is derived from the traditional solution of the Hamilton–Jacobi equation of the quadratic Hamiltonian (13.2) [257, 336, 529].

The flow stemming from $\mathcal{H}_0$ is obtained from the trivial integration of the Hamilton equations. That is,

$$\phi = \phi_0 + \dot{\vartheta}t, \quad q = q_0 - \frac{3}{4k^2}Qt,$$

which show that the phase of the reference ellipse ϕ evolves periodically whereas the coordinate of its center grows unbounded in the y-axis direction, as follows from Eq. (13.7). Solving Eq. (13.3) for the epicyclic variables, the constant values of Φ and Q are obtained from the initial conditions (x_0, y_0, X_0, Y_0), like

$$\Phi = \frac{1}{2\dot{\vartheta}}[(X_0 + \dot{\vartheta}y_0)^2 + (2Y_0 + \dot{\vartheta}x_0)^2], \quad Q = 2k(Y_0 + \dot{\vartheta}x_0).$$

Therefore, choosing such initial conditions that $Y_0 = -\dot{\vartheta}x_0$ makes $Q = 0$ and hence yields periodic motion (a fixed ellipse for the orbit and a fixed circle for the hodograph). Conversely, q, and hence η in Eq. (13.6), grows linearly when $Q \neq 0$. That is, initial conditions such that $Y_0 \neq -\dot{\vartheta}x_0$ give rise to drifting ellipses in the y-axis direction with corresponding drifting hodographs in the X-axis direction, as follows from Eq. (13.3).

However, perturbations of the integrable motion due to the gravitational attraction of the central body, which are given by the term $-\mu/r$ of Eq. (13.1), produce important qualitative changes in the solutions. Indeed, for a wide range of initial conditions, the drift of the ellipse is turned into slow oscillations of its center about the origin. The oscillatory behavior gives rise to the so-called quasi-satellite orbits, in which the orbiter remains relatively close to the primary at the origin yet clearly out of the Hill sphere of influence [46, 276, 524].

When the transformation to epicyclic variables (13.3) is applied to the radius, we obtain

$$r = a\sqrt{\Delta^2 + \xi \sin\phi + 2\eta\cos\phi + \xi^2 + \eta^2}, \tag{13.9}$$

in which the term

$$\Delta = \sqrt{1 - \frac{3}{4}\sin^2\phi} \tag{13.10}$$

allows one to foresee the appearance of elliptic functions in the perturbation solution.

Finally, replacing Eqs. (13.8) and (13.9) into Eq. (13.1), we obtain the Hamiltonian of the planar Hill problem in epicyclic variables. Namely,

$$\mathcal{H} = \dot{\vartheta}\Phi\left(1 - 3\xi^2 - \frac{\gamma}{\sqrt{\Delta^2 + \xi s + 2\eta c + \xi^2 + \eta^2}}\right), \tag{13.11}$$

in which we abbreviated $c \equiv \cos\phi$ and $s \equiv \sin\phi$ for the sake of conciseness, while the nondimensional function $\gamma \equiv \gamma(\Phi)$ stands for

$$\gamma = \frac{\mu}{a\dot{\vartheta}\Phi} = \frac{\mu\dot{\vartheta}}{B^3} = \frac{\mu\dot{\vartheta}}{(2\dot{\vartheta}\Phi)^{3/2}}. \tag{13.12}$$

For simplicity, we fix the scale factor of the transformation to epicyclic variables to $k = \sqrt{3/4}$, which turns $q = q_0 - Qt$ for the unperturbed problem. Then we make use of the Lie transforms method to compute a perturbation solution to the quasi-satellite orbits problem by the complete Hamiltonian reduction of Eq. (13.11), up to some truncation order. As usual, the reduction is split into the elimination of short-period terms, which is carried out first, and the consequent removal of long-period terms.

13.2 Elimination of short-period effects

First of all, Eq. (13.11) is arranged in the form of the usual perturbation Hamiltonian (2.30), in which ϵ is a formal small parameter. Then we compute the Lie transformation $(\phi, q, \Phi, Q) \mapsto (\phi', q', \Phi', Q'; \epsilon)$ that, up to some truncation order, makes ϕ cyclic in the transformed Hamiltonian in the prime variables.

Instead of choosing the quadratic Hamiltonian (13.8) like the integrable part of the perturbation Hamiltonian, we rather make the choice $\mathcal{H}_{0,0} = \dot{\vartheta}\Phi$, which is a harmonic oscillator of frequency $\dot{\vartheta}$, cf. §2.2.2, in this way implicitly assuming that the phase of orbiter in the reference ellipse evolves much faster than its center. Then the Lie derivative (2.49) is simply $\mathcal{L}_0 \equiv \{\ ;\dot{\vartheta}\Phi\} = \dot{\vartheta}\partial/\partial\phi$, showing that the kernel of the Lie operator is made of terms that are free from ϕ. The homological equation (2.48) is then solved by indefinite integration

$$\mathcal{W}_m = \frac{1}{\dot{\vartheta}} \int (\widetilde{\mathcal{H}}_{0,m} - \mathcal{H}_{0,m}) \mathrm{d}\phi. \tag{13.13}$$

Currently lacking a strategy for computing the perturbation solution in closed form, the perturbation term $-\mu/r$, whose integration would require the use of sophisticated special functions [47], is expanded under the assumptions that the third body and the central gravitation exert perturbations of the same order [401]. These assumptions are materialized taking

$$\eta = \mathcal{O}(\epsilon), \quad \xi = \mathcal{O}(\epsilon^2), \quad \gamma = \mathcal{O}(\epsilon^4). \tag{13.14}$$

Integration of terms related with the function Δ defined in Eq. (13.10) relies on the use of special functions. This fact makes the identification of the kernel and image of the Lie operator difficult, and the gradual coupling of the different orders of the perturbation approach establishes de facto a limit up to which the elimination of the fast angle ϕ can be carried out [401].

With the assumptions in Eq. (13.14), the perturbation terms of Eq. (13.11) are $\mathcal{H}_{1,0} = \mathcal{H}_{2,0} = \mathcal{H}_{3,0} = 0$, $\mathcal{H}_{4,0} = 4!\frac{1}{2}B^2(-3\xi^2 - \gamma/\Delta)$, and, up to ϵ^{10},

$$\mathcal{H}_{i,0} = i!\frac{1}{2}B^2 \frac{\gamma}{\Delta^{2i-7}} (c\eta)^{i^*} \sum_{j=1}^{i_2} t_{i^*+2j-2, i_2-j}(s)\eta^{2j-2}\xi^{i_2-j}, \tag{13.15}$$

where $i^* = i \bmod 2$ and $i_2 = \lfloor\frac{1}{2}(i-2)\rfloor$, from Eq. (6.5), $t_{1,0} = 1$, and the remaining coefficients $t_{k,l}$ are given in Table 13.1.

Table 13.1: Coefficients $t_{k,l}$ in Eq. (13.15).

$2,0: \frac{9}{8}s^2 - 1$	$0,3: -\frac{189}{512}s^9 + \frac{459}{256}s^7 - \frac{207}{64}s^5 + \frac{41}{16}s^3 - \frac{3}{4}s$
$0,1: \frac{1}{2}s - \frac{3}{8}s^3$	$2,2: \frac{297}{64}s^8 - \frac{4761}{256}s^6 + \frac{849}{32}s^4 - \frac{249}{16}s^2 + 3$
$3,0: 1 - \frac{11}{8}s^2$	$4,1: -\frac{7965}{1024}s^7 + \frac{6075}{256}s^5 - \frac{375}{16}s^3 + \frac{15}{2}s$
$1,1: \frac{9}{8}s^3 - \frac{3}{2}s$	$6,0: \frac{3309}{1024}s^6 - \frac{909}{128}s^4 + \frac{39}{8}s^2 - 1$
$4,0: -\frac{227}{128}s^4 + \frac{11}{4}s^2 - 1$	$1,2: \frac{27}{16}s^6 - \frac{171}{32}s^4 + \frac{21}{4}s^2 - \frac{3}{2}$
$2,1: \frac{153}{64}s^5 - \frac{87}{16}s^3 + 3s$	$3,1: -\frac{285}{64}s^5 + \frac{155}{16}s^3 - 5s$
$0,2: -\frac{27}{64}s^6 + \frac{45}{32}s^4 - \frac{3}{2}s^2 + \frac{1}{2}$	$5,0: \frac{303}{128}s^4 - \frac{13}{4}s^2 + 1$

13.2.1 Lower orders

On account of the Hamiltonian arrangement made, we trivially find $\mathcal{H}_{0,1} = \mathcal{H}_{0,2} = \mathcal{H}_{0,3} = 0$, and hence $\mathcal{W}_1 = \mathcal{W}_2 = \mathcal{W}_3 = 0$. Besides, since $\widetilde{\mathcal{H}}_{0,i} = \mathcal{H}_{i,0}$ for $i = 4,\ldots,7$, corresponding terms $\mathcal{H}_{0,i}$ are chosen by simply averaging the terms $\mathcal{H}_{i,0}$ of the original Hamiltonian over ϕ. That is, at fourth order,

$$\mathcal{H}_{0,4} = \langle\widetilde{\mathcal{H}}_{0,4}\rangle_\phi = -4!\frac{1}{2}B^2\left(3\xi^2 + \gamma\frac{1}{2\pi}\int_0^{2\pi}\frac{1}{\Delta}\mathrm{d}\phi\right) = -4!B^2\left(\frac{3}{2}\xi^2 + \gamma\tilde{K}\right),$$

where we use the notation $\tilde{K} \equiv K(k^2)/\pi$. Analogously, we find

$$\mathcal{H}_{0,5} = 0, \quad \mathcal{H}_{0,6} = 6!B^2\frac{2}{3}\gamma(\tilde{E} - \tilde{K})\eta^2, \quad \mathcal{H}_{0,7} = 0,$$

where $\tilde{E} \equiv E(k^2)/\pi$. Homologous terms of the generating function are then computed from Eq. (13.13), to obtain

$$\begin{aligned}
\mathcal{W}_4 &= 4!\Phi\gamma F^*,\\
\mathcal{W}_5 &= 5!\Phi\frac{1}{\Delta}\gamma\eta s,\\
\mathcal{W}_6 &= 6!\Phi\gamma\left\{\frac{2}{3}\left[F^* - E^* - \frac{1}{\Delta}\left(\frac{1}{\Delta^2} + 1\right)\frac{3}{4}sc\right]\eta^2 - 2\frac{\xi}{\Delta}c\right\},\\
\mathcal{W}_7 &= 7!\Phi\frac{2}{3}\gamma\eta\left[\frac{\eta^2}{32\Delta^5}(21s^5 - 70s^3 + 48s) - \left(\frac{1}{\Delta^3} - 8\tilde{E}\right)\xi\right],
\end{aligned}$$

where $F^* \equiv 2\tilde{K}\phi - F(\phi\,|\,k^2)$ and $E^* \equiv 2\tilde{E}\phi - E(\phi\,|\,k^2)$ are periodic functions of ϕ with period π. We remark that an integration constant has been added to the solution of Eq. (13.13) in order to guarantee that $\mathcal{W}$ is free from short-period effects. Namely $\langle \mathcal{W}_i \rangle_\phi = 0$, $i = 4, \ldots 7$.

Short-period direct and inverse corrections vanish at the first three orders. At the fourth,

$$\phi_{0,4} = -4!\frac{1}{2}\gamma F^*, \quad q_{0,4} = 0, \quad \Phi_{0,4} = 4!\Phi\gamma\left(\frac{1}{\Delta} - 2\tilde{K}\right), \quad Q_{0,4} = 0.$$

At fifth order,

$$\phi_{0,5} = -5!\gamma\frac{\eta}{\Delta}s, \quad q_{0,5} = 0, \quad \Phi_{0,5} = -5!\Phi\gamma\frac{\eta}{\Delta^3}c, \quad Q_{0,5} = -5!B\frac{k}{2}\frac{\gamma}{\Delta}s,$$

while the sixth-order corrections are

$$\begin{aligned}
\phi_{0,6} &= 6!\gamma\left\{\frac{c}{\Delta}\left[\frac{3}{4}\left(\frac{1}{\Delta^2}+1\right)\eta^2 s + 2\xi\right] + \eta^2[E^* - F^*]\right\},\\
q_{0,6} &= 6!b\frac{\gamma}{2k}\frac{1}{\Delta}\cos\phi,\\
\Phi_{0,6} &= 6!\Phi\gamma\left\{\frac{4}{3}\eta^2(\tilde{E} - \tilde{K}) + \frac{1}{2\Delta^3}\left[\left(3 - \frac{1}{\Delta^2}\right)\eta^2 - \xi s\right]\right\},\\
Q_{0,6} &= 6!B\frac{\gamma}{2k}\eta\left[\frac{3}{4\Delta}\left(\frac{1}{\Delta^2}+1\right)cs + E^* - F^*\right],
\end{aligned}$$

and, at seventh order,

$$\begin{aligned}
\phi_{0,7} &= 7!\gamma\eta\left[\left(\frac{1}{\Delta^3} - 8\tilde{E}\right)\xi + \frac{1}{9}\frac{s}{\Delta}\left(\frac{3}{\Delta^4} - \frac{7}{\Delta^2} - 14\right)\eta^2\right],\\
q_{0,7} &= -7!b\frac{\gamma}{6k}\eta\left(\frac{1}{\Delta^3} - 8\tilde{E}\right),\\
\Phi_{0,7} &= 7!\Phi\frac{\gamma\eta}{2\Delta^5}\left[\frac{1}{3}\left(\frac{5}{\Delta^2} - 11\right)\eta^2 + 3\xi s\right]c,\\
Q_{0,7} &= 7!B\frac{k}{3}\gamma\left[\left(\frac{1}{\Delta^3} - 8\tilde{E}\right)\xi + \frac{1}{4}\frac{s}{\Delta}\left(\frac{3}{\Delta^4} - \frac{7}{\Delta^2} - 14\right)\eta^2\right],
\end{aligned}$$

all of them to be evaluated using the prime variables. On the other hand $\mathcal{V}_i = -\mathcal{W}_i$, $i = 4, \ldots, 7$, and the inverse corrections are just the opposite of the direct ones, which now must be evaluated in the original variables.

13.2.2 Higher orders

The coupling of different Hamiltonian terms starts at the eight order, where $\widetilde{\mathcal{H}}_{0,8} = 35\{\mathcal{H}_{0,4} + \mathcal{H}_{4,0}, \mathcal{W}_4\} + \mathcal{H}_{8,0}$. The choice $\mathcal{H}_{0,8} = \langle \widetilde{\mathcal{H}}_{0,8} \rangle_\phi$ produces

$$\mathcal{H}_{0,8} = 8!B^2\gamma\left[\gamma(1 - 2\tilde{K}^2) + \frac{1}{9}(14\tilde{E} - 11\tilde{K})\eta^4 + \frac{\tilde{K} - 4\tilde{E}}{k^2}\xi^2\right].$$

Then, from Eq. (13.13),

$$\mathcal{W}_8 = 8!\Phi\frac{2}{3}y\left\{\frac{3}{4}y\left[\left(\tilde{K}+\frac{1}{2\Delta}\right)F^* - P^*\right] + \frac{1}{12}(11F^* - 14E^*)\eta^4 + (4E^* - F^*)\xi^2 + \frac{1}{\Delta}\left[\frac{1}{16}\left(\frac{5}{\Delta^6} - \frac{8}{\Delta^4} - \frac{11}{\Delta^2} - 14\right)\eta^4 s + \frac{1}{2}\left(\frac{3}{\Delta^4} - \frac{1}{\Delta^2} - 8\right)\eta^2\xi + \left(\frac{3}{4\Delta^2}+3\right)\xi^2 s\right]c\right\},$$

where $P^* \equiv 2\phi - \Pi(\frac{3}{4};\phi\,|\,0)$ is periodic with period π. We remark that the equality

$$\Pi\left(\frac{3}{4};\phi\,|\,0\right) = \int_0^{\phi}\frac{d\psi}{1-\frac{3}{4}\sin^2\psi} = 2\arctan\left(\frac{1}{2}\tan\phi\right),$$

is only true in the open interval $\phi \in (-\frac{\pi}{2},\frac{\pi}{2})$. Indeed, at $\phi = \frac{\pi}{2}$ (resp. $\phi = -\frac{\pi}{2}$) the incomplete elliptic integral of the third kind evaluates to π (resp. $-\pi$). On the contrary, the arc tangent function remains indeterminate at these values. Therefore, since ϕ may grow unbounded, the use of the former should be preferred for evaluation purposes, which, besides, avoids the demodulation that would be required if using the arc tangent function.

The eighth-order terms of the direct corrections are

$$\phi_{0,8} = 8!y\left\{y\left[P^* - \frac{1}{4}\left(6\tilde{K}+\frac{1}{\Delta}\right)F^*\right] + \frac{5}{36}(14E^* - 11F^*)\eta^4 + (F^* - 4E^*)\xi^2 + \frac{1}{\Delta}\left[\frac{2}{3}\left(8+\frac{1}{\Delta^2}-\frac{3}{\Delta^4}\right)\eta^2\xi - \left(3+\frac{3}{4\Delta^2}\right)\xi^2 s + \frac{5}{48}\left(14+\frac{11}{\Delta^2}+\frac{8}{\Delta^4}-\frac{5}{\Delta^6}\right)\eta^4 s\right]c\right\},$$

$$q_{0,8} = 8!b\frac{y}{k}\left\{\frac{4E^* - F^*}{3}\xi + \frac{c}{4\Delta}\left[\left(\frac{1}{\Delta^4}-\frac{1}{3\Delta^2}-\frac{8}{3}\right)\eta^2 + \left(\frac{1}{\Delta^2}+4\right)\xi s\right]\right\},$$

$$\Phi_{0,8} = 8!\Phi y\left\{y\left[1-\frac{1}{2\Delta^2}+\tilde{K}\left(\frac{1}{\Delta}-2\tilde{K}\right) - \frac{3}{8\Delta^3}F^* sc\right] + \frac{\eta^4}{9}\left[\frac{1}{8\Delta^5}\left(\frac{35}{\Delta^4}-\frac{190}{\Delta^2}+227\right)+14\tilde{E}-11\tilde{K}\right] + \xi^2\left[\frac{1}{\Delta^3}\left(\frac{1}{2\Delta^2}-1\right)+\frac{\tilde{K}-4\tilde{E}}{k^2}\right] + \frac{\eta^2\xi}{4\Delta^5}\left(\frac{5}{\Delta^2}-17\right)s\right\},$$

$$Q_{0,8} = 8!B\frac{k}{9}y\left\{(14E^* - 11F^*)\eta^2 + \frac{3}{\Delta}\left[\xi\left(8+\frac{1}{\Delta^2}-\frac{3}{\Delta^4}\right) + \eta^2\left(\frac{7}{2}+\frac{11}{4\Delta^2}+\frac{2}{\Delta^4}-\frac{5}{4\Delta^6}\right)s\right]c\right\}\eta,$$

which must be evaluated in the prime variables.

While the eighth-order term of the inverse generating function is still the opposite of the direct one, $\mathcal{V}_8 = -\mathcal{W}_8$, the coupling with the fourth-order short-period corrections happens at this order, making that the inverse and direct short-period corrections are no longer the same. Still, the inverse corrections of q and Q are $\tilde{q}_{0,8} = -q_{0,8}$ and $\tilde{Q}_{0,8} = -Q_{0,8}$, but $\tilde{\phi}_{0,8}$ and $\tilde{\Phi}_{0,8}$ must be complemented with additional terms. Namely,

$$\tilde{\phi}_{0,8} = -\phi_{0,8} + 8!y^2\left(\frac{1}{2\Delta} - \tilde{E}\right),$$
$$\tilde{\Phi}_{0,8} = -\Phi_{0,8} + 8!y^2\Phi\left[2\tilde{K}\left(\frac{1}{\Delta} - \tilde{K}\right) - \frac{1}{8\Delta^2}\left(4 + \frac{3}{\Delta}F^* sc\right)\right],$$

where the right sides of these equations are evaluated in original variables.

At ninth order, Deprit's recursion (2.15) yields

$$\widetilde{\mathcal{H}}_{0,9} = 56\{\mathcal{H}_{0,4}, \mathcal{W}_5\} + 70\{\mathcal{H}_{4,0}, \mathcal{W}_5\} + 56\{\mathcal{H}_{5,0}, \mathcal{W}_4\} + \mathcal{H}_{9,0},$$

from which $\mathcal{H}_{0,9} = \langle\widetilde{\mathcal{H}}_{0,9}\rangle_\phi = 0$. Next, from Eq. (13.13),

$$\begin{aligned}\mathcal{W}_9 = 9!\Phi y\bigg\{&\frac{1}{2\Delta}\left(\frac{7}{36\Delta^8} - \frac{17}{18\Delta^6} + \frac{11}{20\Delta^4} + \frac{11}{15\Delta^2} + \frac{22}{15}\right)\eta^5 s + \frac{2\eta\xi^2}{3\Delta}\\ &\times\left(\frac{3}{4\Delta^4} - \frac{1}{\Delta^2} - 2\right)s + \left[\frac{\eta^3}{9\Delta^5}\left(\frac{5}{\Delta^2} - 19\right) - k\log 8(\Delta + kc)^2\right]\xi\\ &+ \frac{2}{9}y\eta\left[\frac{1}{\Delta}\left(4\tilde{K} - \frac{7}{8\Delta}\right)s - \frac{F^*}{\Delta^3}c + \frac{9}{4}k\log\frac{1-ks}{1+ks}\right]\bigg\},\end{aligned}$$

and hence

$$\begin{aligned}\phi_{0,9} = 9!y\bigg\{&\frac{y\eta}{\Delta}\left[s\left(\frac{11}{8\Delta} - 4\tilde{K}\right) + \frac{cF^*}{2\Delta^2}\right] + \frac{\eta\xi^2}{3\Delta}\left(8 + \frac{4}{\Delta^2} - \frac{3}{\Delta^4}\right)s\\ &+ \xi\left[k\log 8(\Delta + kc)^2 + \frac{5\eta^3}{18\Delta^5}\left(19 - \frac{5}{\Delta^2}\right)\right] + \frac{5}{4}ky\eta\\ &\times\log\left(\frac{1+ks}{1-ks}\right) - \frac{\eta^5 s}{4\Delta}\left(\frac{7}{6\Delta^8} - \frac{17}{3\Delta^6} + \frac{33}{10\Delta^4} + \frac{22}{5\Delta^2} + \frac{44}{5}\right)\bigg\},\end{aligned}$$

$$q_{0,9} = 9!\frac{by}{4k}\left\{\frac{\eta^3}{9\Delta^5}\left(\frac{5}{\Delta^2} - 19\right) + \frac{\eta\xi s}{3\Delta}\left(\frac{3}{\Delta^4} - \frac{4}{\Delta^2} - 8\right) - k\log 8(\Delta + kc)^2\right\},$$

$$\begin{aligned}\Phi_{0,9} = 9!\Phi y\bigg\{&\frac{3\xi}{4\Delta}\left[\frac{5\eta^3}{9\Delta^6}\left(19 - \frac{7}{\Delta^2}\right)c - 2\right]s + \frac{y\eta}{2\Delta^2}\left[\left(2 - \frac{4\tilde{K}}{\Delta} + \frac{1}{\Delta^2}\right)c - \frac{F^*}{\Delta}\right.\\ &\left.\times\left(\frac{k^2}{\Delta^2} - 2\right)s\right] + \frac{c\eta}{\Delta^5}\left[\left(4 - \frac{5}{2\Delta^2}\right)\xi^2 - \frac{\eta^4}{8\Delta^2}\left(\frac{7}{\Delta^4} - \frac{98}{3\Delta^2} + \frac{101}{3}\right)\right]\bigg\},\end{aligned}$$

$$\begin{aligned}Q_{0,9} = 9!B\frac{k}{9}y\bigg\{&\frac{y}{2\Delta}\left[s\left(\frac{3}{4\Delta} - 2\tilde{K}\right) + \frac{cF^*}{2\Delta^2}\right] + \frac{\xi^2 s}{3\Delta}\left(2 - \frac{3}{4\Delta^4} + \frac{1}{\Delta^2}\right)\\ &+ \frac{\eta^2\xi}{6\Delta^5}\left(19 - \frac{5}{\Delta^2}\right) + \frac{k}{4}y\log\frac{1+ks}{1-ks}\\ &- \frac{\eta^4 s}{2\Delta}\left(\frac{35}{72\Delta^8} - \frac{85}{36\Delta^6} + \frac{11}{8\Delta^4} + \frac{11}{6\Delta^2} + \frac{11}{3}\right)\bigg\}.\end{aligned}$$

Now, the first time for the perturbation arrangement chosen, $\mathcal{V}_9 \neq -\mathcal{W}_9$. More precisely,

$$\mathcal{V}_9 = -\mathcal{W}_9 - 9!\Phi\frac{y^2\eta}{18\Delta}\left[\left(\frac{2}{\Delta} - 4\tilde{K}\right)s + \left(1 + \frac{3}{4\Delta^2}s^2\right)F^*c\right],$$

and the inverse short-period corrections are $\tilde{q}_{0,9} = -q_{0,9}$ and

$$\tilde{\phi}_{0,9} = -\phi_{0,9} + 9!\frac{y^2\eta}{36\Delta}\left[74\left(\frac{1}{\Delta} - 2\tilde{K}\right)s + \frac{1}{\Delta^2}F^*c\right],$$

$$\tilde{\Phi}_{0,9} = -\Phi_{0,9} + 9!\frac{\Phi y^2\eta}{18\Delta^2}\left[\left(14 - \frac{58}{\Delta}\tilde{K} + \frac{15}{\Delta^2}\right)c + \frac{11}{\Delta}\left(2 - \frac{3}{4\Delta^2}\right)F^*s\right],$$

$$\tilde{Q}_{0,9} = -Q_{0,9} + 9!B\frac{11k}{18\Delta}y^2\left[\left(\frac{1}{\Delta} - 2\tilde{K}\right)s + \frac{1}{2\Delta^2}F^*c\right],$$

whose right sides are evaluated in the original variables.

13.2.3 Additional Hamiltonian terms

At tenth order, Deprit's recursion (2.15) leads to

$$\widetilde{\mathcal{H}}_{0,10} = 84\{\mathcal{H}_{0,4}, \mathcal{W}_6\} + 126\{\mathcal{H}_{4,0}, \mathcal{W}_6\} + 126\{\mathcal{H}_{5,0}, \mathcal{W}_5\}$$
$$+ 126\{\mathcal{H}_{0,6}, \mathcal{W}_4\} + 84\{\mathcal{H}_{6,0}, \mathcal{W}_4\} + \mathcal{H}_{10,0},$$

from which $\mathcal{H}_{0,10}$ is chosen by averaging. Standard computations yield

$$\mathcal{H}_{0,10} = \frac{10!}{3}B^2\eta^2y\left[3y + 8\tilde{K}(\tilde{E} - \tilde{K})y + \frac{71\tilde{E} - 50\tilde{K}}{27}\eta^4 + (8\tilde{K} - 20\tilde{E})\xi^2\right].$$

The computation of the tenth-order term of the generating function addresses the indefinite integration of the incomplete elliptical integrals of the first and second kinds. Lacking a closed-form solution, we only apply the short-period corrections up to the ninth order. However, for the particular Hamiltonian arrangement chosen, the term $\mathcal{W}_{10}$ does not play a role in the computation of the next three orders of Deprit's triangle and allows us to obtain the next three Hamiltonian terms [398],

$$\mathcal{H}_{0,11} = -\frac{11!}{4}B^2y^2,$$

$$\mathcal{H}_{0,12} = \frac{12!}{2}B^2y\left\{\left[\frac{4}{3} + 5(\tilde{K} - \tilde{K}^3 - \tilde{E})\right]y^2 + \frac{4}{9}(\tilde{K} - 16\tilde{E})\xi^4\right.$$
$$+ \frac{1}{3}[(9 - 30\tilde{K}^2 - 8\tilde{E}^2 + 44\tilde{E}\tilde{K})\eta^4 + 8(3 + 2\tilde{K}^2 - 8\tilde{E}\tilde{K})\xi^2]y$$
$$\left.+ \frac{1}{324}(644\tilde{E} - 425\tilde{K})\eta^8 - \frac{10}{27}(74\tilde{E} - 35\tilde{K})\eta^4\xi^2\right\},$$

$$\mathcal{H}_{0,13} = -13!B^2y\eta\left\{\frac{8y^2}{13} + \frac{y}{36}[64\tilde{E}(\tilde{K} - \tilde{E}) + 15]\eta + \frac{2}{3}(\tilde{K} + 26\tilde{E})\eta\xi^2\right\}.$$

13.2.4 Long-period Hamiltonian

Once ϕ is removed up to $\mathcal{O}(\epsilon^{13})$, the new Hamiltonian terms are reformulated in the prime variables. We obtain the new Hamiltonian

$$\mathcal{H} = \dot{\vartheta}\Phi' - \frac{1}{2}(Q'^2 + \Omega^2 q'^2) + \mathcal{P}(-, q', \Phi', Q'), \tag{13.16}$$

where $\Omega = \Omega(\Phi')$ is the *libration frequency*

$$\Omega = \dot{\vartheta}\sqrt{\tilde{K} - \tilde{E}}\sqrt{y}, \tag{13.17}$$

which is obtained replacing Φ by Φ' in the computation of y from Eq. (13.12). Note that $\Omega/\dot{\vartheta} = \mathcal{O}(\epsilon^2)$ due to the assumption $y = \mathcal{O}(\epsilon^4)$ in Eq. (13.14).

The disturbing term $\mathcal{P}$ of the Hamiltonian (13.16) is conveniently arranged in the form

$$\mathcal{P} = \dot{\vartheta}\Phi' \sum_{i=0}^{2} \sum_{j=0}^{2-i} \sum_{n=0}^{4-2i-2j} p_{i,j,n} y^{i+1} \xi^{2j} \eta^{2n}, \tag{13.18}$$

where ξ and η are now obtained replacing q, Q, and Φ in Eq. (13.6) by q', Q', and Φ', respectively. The numeric coefficients $p_{i,j,n}$ are listed in Table 13.2.

Table 13.2: Coefficients $p_{i,j,n}$ in Eq. (13.18) (after [403]).

0,0,0 : $-2\tilde{K}$	0,1,2 : $-\frac{10}{27}(74\tilde{E} - 35\tilde{K})$
0,0,1 : 0	0,2,0 : $\frac{4}{9}(\tilde{K} - 16\tilde{E})$
0,0,2 : $\frac{1}{9}(14\tilde{E} - 11\tilde{K})$	1,0,0 : $\frac{1}{2} - 2\tilde{K}^2$
0,0,3 : $\frac{2}{81}(71\tilde{E} - 50\tilde{K})$	1,0,1 : $\frac{7}{6} + \frac{16}{9}(2\tilde{E}^2 - 3\tilde{K}^2 + \tilde{E}\tilde{K})$
0,0,4 : $\frac{1}{324}(644\tilde{E} - 425\tilde{K})$	1,1,0 : $8 + \frac{16}{3}(\tilde{K} - 4\tilde{E})\tilde{K}$
0,1,0 : $\frac{4}{3}(\tilde{K} - 4\tilde{E})$	1,0,2 : $3 - \frac{2}{3}(4\tilde{E}^2 + 15\tilde{K}^2 - 22\tilde{E}\tilde{K})$
0,1,1 : $4(\tilde{K} - 12\tilde{E})$	2,0,0 : $\frac{4}{3} + 5(\tilde{K} - \tilde{K}^3 - \tilde{E})$

As a result of the averaging, ϕ' is ignorable in Eq. (13.16), and hence Φ' is a dynamical parameter that remains constant along the motion. That is, on average, the motion of the center of the reference ellipse decouples from the ϕ' motion.

13.3 The nature of the long-term solution

The most relevant qualitative aspects of the quasi-satellite orbits dynamics are provided by the first two terms in the right side of Eq. (13.16), which brings the Hamiltonian into the form of two coupled harmonic oscillators. Indeed, after neglecting $\mathcal{P}$

from Eq. (13.16), we get the simplified Hamiltonian

$$\mathcal{H}_0 = \dot{\vartheta}\Phi' - \frac{1}{2}(Q'^2 + \Omega^2 q'^2), \tag{13.19}$$

which shows that the center of the reference ellipse moves with a simple harmonic motion of (constant) frequency Ω with respect to the primary of lesser mass. On the other hand, the phase ϕ' of the orbiter in the reference ellipse evolves with unperturbed frequency $\dot{\vartheta}$, yet the dependence of Ω on Φ' has the effect of modulating this phase with long-period variations due to the motion of the center of the reference ellipse.

The uncoupled motion of the center of the reference ellipse is readily solved after applying the harmonic transformation, which we already used in Eq. (2.40). That is, $(q', Q') \mapsto (\theta, \Theta; \Omega)$

$$q' = \sqrt{2\Theta/\Omega}\sin\theta, \quad Q' = \sqrt{2\Omega\Theta}\cos\theta, \tag{13.20}$$

which turns Eq. (13.19) into $\mathcal{H}_0 = \dot{\vartheta}\Phi' - \Omega(\Phi')\Theta$, thus showing that both Θ and Φ' are integrals of the simplified problem. The trivial integration of Hamilton equations shows that, on average, the motion combines the linear evolution of the phase of the center of the reference ellipse with respect to the origin $\theta = \theta_0 - \Omega(\Phi')t$, with the linear evolution of the phase of the reference ellipse $\phi' = \phi'_0 + [\dot{\vartheta} - \Theta\Omega(\Phi')]t$, which grows at a slightly modified rate with respect to the unperturbed motion. The initial phase and the value of Θ are computed from the initial conditions q'_0 and Q'_0 using the inverse transformation of Eq. (13.20). Namely,

$$\Theta = \frac{1}{2\Omega}(Q'^2 + \Omega^2 q'^2), \tag{13.21}$$

$$\sin\theta = \frac{\Omega q'}{\sqrt{Q'^2 + \Omega^2 q'^2}}, \quad \cos\theta = \frac{Q'}{\sqrt{Q'^2 + \Omega^2 q'^2}}. \tag{13.22}$$

A higher-order approximation that is also solvable in closed form is obtained by keeping explicit some additional terms of the long-term Hamiltonian (13.16). Indeed, taking additional terms from Eq. (13.18), we write

$$\mathcal{H} = \dot{\vartheta}^*\Phi' - f_1\left[\frac{1}{2}(Q'^2 + w^2 q'^2) + \beta^2 w^2 q'^4\right], \tag{13.23}$$

in which $\dot{\vartheta}^* = \dot{\vartheta}(1 + \sum_{i=0}^{2} p_{i,0,0}\gamma^{i+1})$ with the numeric coefficients $p_{i,0,0}$ of Table 13.2, whereas β and w are (constant) functions of Φ' defined by

$$\beta^2 = \frac{\dot{\vartheta}}{\Phi'}\frac{11\tilde{K} - 14\tilde{E}}{64(\tilde{K} - \tilde{E})}, \quad w^2 = \frac{\Omega^2}{f_1}, \quad \text{with } f_1 = 1 + \frac{4(10\tilde{E} - \tilde{K})}{9(\tilde{K} - \tilde{E})}\frac{\Omega^2}{\dot{\vartheta}^2}.$$

The reduced (q', Q') flow stemming from Hamiltonian (13.23) is immediately identified with the motion of a Duffing oscillator without forcing term [179, 255, 491] in a time scale proportional to f_1. However, the closed-form solution of the Duffing oscillator involves Jacobi elliptic functions, thus not leading to the desired insight. Therefore, Eq. (13.19) still provides a more enlightening integrable zeroth-order Hamiltonian than Eq. (13.23).

On the other hand, the use of special functions can be avoided using the Lindstedt series approach [459, 516, 526], which, besides, allows for the inclusion of higher-order effects of the long-period Hamiltonian (13.16). These kinds of solutions provide results of comparable accuracy to the semi-analytical integration that is customarily approached for the computation of quasi-satellite orbits with large librations [401, 403, 404, 456].

13.4 Complete Hamiltonian reduction

Alternatively to the Lindstedt series solution in [403], a higher-order perturbation solution can be approached by computing a new Lie transformation that removes the long-period terms from the mean-element Hamiltonian (13.16). In this way the new Hamiltonian is completely reduced, up to some truncation order, and, therefore, is trivially integrated.

When constraining to the order of ϵ^{10} used in the construction of the high-order Lindstedt series solution obtained in [403], and removing the prime notation for brevity, the mean-variable Hamiltonian (13.16) is

$$\mathcal{H} = \dot{\vartheta}\Phi\left\{1 - 3\xi^2 + \sum_{i=0}^{1}\sum_{j=0}^{1-i}\sum_{n=0}^{3-2i-2j} p^*_{i,j,n}\gamma^{i+1}\xi^{2j}\eta^{2n}\right\}, \tag{13.24}$$

with coefficients $p^*_{i,j,n} = p_{i,j,n}$ of Table 13.2 save for $p^*_{0,0,1} = -\frac{4}{3}(4\tilde{E} - \tilde{K})$.

13.4.1 Extended harmonic transformation

The transformation that removes the long-period terms from the mean-element Hamiltonian (13.24) is more naturally approached in the harmonic variables given in Eqs. (13.21) and (13.22). However, in view of Eqs. (13.17) and (13.12), the libration frequency Ω is not a physical parameter but depends on Φ'. Then the pair (ϕ, Φ) must also be transformed to make the transformation canonical in the whole phase space. This is achieved with the extended harmonic transformation

$$\Psi = \Phi, \tag{13.25}$$

$$\psi = \phi + \frac{\Gamma'}{2\Gamma}qQ, \tag{13.26}$$

$$\Theta = \frac{1}{2\Gamma}(Q^2 + \Gamma^2 q^2), \tag{13.27}$$

$$\sin\theta = \frac{\Gamma q}{\sqrt{Q^2+\Gamma^2 q^2}}, \qquad \cos\theta = \frac{Q}{\sqrt{Q^2+\Gamma^2 q^2}}, \tag{13.28}$$

in which Γ denotes an arbitrary function of Φ, and $\Gamma' \equiv \partial\Gamma/\partial\Phi$.

That the transformation given by Eqs. (13.25)–(13.28) is canonical is guaranteed after checking that the differential form $W \equiv \Psi\,\mathrm{d}\psi + \Theta\,\mathrm{d}\theta - Q\,\mathrm{d}q - \Phi\,\mathrm{d}\phi$ is an exact differential (see [243], for instance). Indeed, we readily check that $W \equiv \mathrm{d}(\Psi\psi - \Phi\phi - \frac{1}{2}\Theta\sin 2\theta)$. The inverse transformation is

$$\Phi = \Psi, \quad \phi = \psi - \frac{1}{2}(\Gamma'/\Gamma)\Theta\sin 2\theta, \quad Q = \sqrt{2\Theta\Gamma}\cos\theta, \quad q = \sqrt{2\Theta/\Gamma}\sin\theta,$$

where, because $\Phi = \Psi$, now $\Gamma \equiv \Gamma(\Psi)$ and $\Gamma' \equiv \partial\Gamma/\partial\Psi$.

In the current case, $\Gamma = \Omega(\Phi)$ is given by Eq. (13.17). Therefore, Eq. (13.26) turns into $\psi = \phi - \frac{3}{8}qQ/\Phi$, whereas Eqs. (13.27)–(13.28) are now replaced by Eqs. (13.21)–(13.22). Besides, the inverse transformation is replaced by

$$\Phi = \Psi, \quad \phi = \psi + \frac{3}{8}(\Theta/\Psi)\sin 2\theta, \quad Q = \sqrt{2\Theta\Omega}\cos\theta, \quad q = \sqrt{2\Theta/\Omega}\sin\theta. \tag{13.29}$$

After applying the new transformation of variables to the long-term Hamiltonian (13.24), it is rearranged in the form

$$\mathcal{H} = \dot{\vartheta}\Psi \sum_{m=0}^{2} \frac{\varepsilon^m}{m!}\mathcal{H}_{m,0}(\theta,\Theta;\Psi), \tag{13.30}$$

where ε is a new formal small parameter, and the new Hamiltonian terms are

$$\begin{aligned}
\mathcal{H}_{0,0} &= 1 + \kappa_{2,0}\frac{\Omega^2}{\dot{\vartheta}^2} + \kappa_{4,0}\frac{\Omega^4}{\dot{\vartheta}^4} + \kappa_{6,0}\frac{\Omega^6}{\dot{\vartheta}^6} + \kappa_{0,1}\frac{\Theta}{\Psi},\\
\mathcal{H}_{1,0} &= \kappa_{3,1}\frac{\Omega^3}{\dot{\vartheta}^3}\frac{\Theta}{\Psi}\cos^2\theta + \sigma_{3,1}\frac{\Omega^3}{\dot{\vartheta}^3}\frac{\Theta}{\Psi}\sin^2\theta + \sigma_{0,2}\frac{\Theta^2}{\Psi^2}\sin^4\theta,\\
\mathcal{H}_{2,0} &= \kappa_{2,2}\frac{\Omega^2}{\dot{\vartheta}^2}\frac{\Theta^2}{\Psi^2}\sin^2\theta\cos^2\theta + \sigma_{-1,3}\frac{\Theta^3/\Psi^3}{\Omega/\dot{\vartheta}}\sin^6\theta,
\end{aligned}$$

where $\Omega = \Omega(\Psi) \equiv \mu^{1/2}(\tilde{K} - \tilde{E})^{1/2}\dot{\vartheta}^{3/4}(2\Psi)^{-3/4}$, from Eq. (13.12), and the numeric coefficients κ and σ are given in Table 13.3. The integrable term $\mathcal{H}_{0,0}$ consists of the terms of the Hamiltonian (13.30) that are free from θ, independently of their smallness. Besides, $\mathcal{H}_{1,0}$ and $\mathcal{H}_{2,0}$ are homogeneous under the assumption $\Theta/\Psi = \mathcal{O}(\Omega^3/\dot{\vartheta}^3)$, which is the reason for the different arrangement from [404].

13.4.2 Secular Hamiltonian

A new Lie transformation $(\psi,\theta,\Psi,\Theta) \mapsto (\psi',\theta',\Psi',\Theta';\varepsilon)$ from mean to secular variables is applied to remove the remaining long-period effects from Eq. (13.30). Because

Table 13.3: Numeric coefficients in Hamiltonian (13.30).

$\kappa_{2,0} = -\frac{2\tilde{K}}{\tilde{K}-\tilde{E}}$	$\kappa_{4,0} = \frac{1}{2}\frac{1-4\tilde{K}^2}{(\tilde{K}-\tilde{E})^2}$	$\kappa_{6,0} = \frac{4+15(\tilde{K}-\tilde{K}^3-\tilde{E})}{3(\tilde{K}-\tilde{E})^3}$
$\kappa_{0,1} = -1$	$\kappa_{3,1} = \frac{4(\tilde{K}-4\tilde{E})}{9(\tilde{K}-\tilde{E})}$	$\kappa_{2,2} = 2\frac{\tilde{K}-12\tilde{E}}{\tilde{K}-\tilde{E}}$
$\sigma_{-1,3} = \frac{71\tilde{E}-50\tilde{K}}{48(\tilde{K}-\tilde{E})}$	$\sigma_{0,2} = \frac{14\tilde{E}-11\tilde{K}}{16(\tilde{K}-\tilde{E})}$	$\sigma_{3,1} = \frac{21-32(\tilde{K}-\tilde{E})(3\tilde{K}+2\tilde{E})}{24(\tilde{K}-\tilde{E})^2}$

ψ is a cyclic variable, Ψ is an integral of the mean-element Hamiltonian that is not affected by the transformation. That is, $\Psi' = \Psi$, and, in consequence, Ω remains unaltered. Besides, the Lie derivative can be simplified to just $\mathcal{L}_0 \equiv \{\ ; -\Omega\Theta\} = -\Omega\frac{\partial}{\partial\theta}$. Then the homological equation (2.48) is solved by indefinite integration

$$\mathcal{V}_m = \frac{1}{\Omega}\int(\mathcal{H}_{0,m} - \widetilde{\mathcal{H}}_{0,m})\mathrm{d}\theta.$$

The standard application of the Lie transforms method, up to the second order of ε, yields the secular Hamiltonian

$$\mathcal{H} = \mathcal{H}(-,-,\Psi',\Theta') \equiv \dot{\vartheta}\Psi'\left[1 - \sum_{i=0}^{3}\left(\frac{\Theta'/\Psi'}{\Omega/\dot{\vartheta}}\right)^{i}\sum_{j=0}^{3-i} h_{i,j}\left(\frac{\Omega}{\dot{\vartheta}}\right)^{2j+2}\right], \tag{13.31}$$

where the coefficients $h_{i,k}$ are given in Table 13.4. Note that $h_{0,2}$ and $h_{1,1}$ are different from the corresponding values in [404] because of the slightly different Hamiltonian arrangement made here.

Table 13.4: Coefficients $h_{i,j}$ ($h_{0,3} = 0$) in Eq. (13.31).

$0,0 : \frac{2\tilde{K}}{\tilde{K}-\tilde{E}}$	$0,1 : \frac{4\tilde{K}^2-1}{2(\tilde{K}-\tilde{E})^2}$	$0,2 : \frac{15(\tilde{K}^3-\tilde{K}+\tilde{E})-4}{3(\tilde{K}-\tilde{E})^3}$
$1,0 : 1$	$1,1 : \frac{64(\tilde{K}-\tilde{E})(4\tilde{K}+5\tilde{E})-63}{144(\tilde{K}-\tilde{E})^2}$	$1,2 : -\frac{1}{2}(\frac{4}{9}\frac{\tilde{K}-4\tilde{E}}{\tilde{K}-\tilde{E}} + h_{1,1})^2$
$2,0 : \frac{3}{4}\frac{11\tilde{K}-14\tilde{E}}{32(\tilde{K}-\tilde{E})}$	$2,1 : \frac{42h_{2,0}-61\tilde{K}^2+137\tilde{K}\tilde{E}-58\tilde{E}^2}{48(\tilde{K}-\tilde{E})^2}$	$3,0 : \frac{1829\tilde{K}^2-3652\tilde{K}\tilde{E}+1364\tilde{E}^2}{49152(\tilde{K}-\tilde{E})^2}$

The Hamiltonian (13.31) is completely reduced to a function of the new momenta Ψ' and Θ', which, therefore, are constant, whereas ψ' and θ' grow linearly with time. That is,

$$\psi' = \psi'_0 + n_\psi t, \quad \theta' = \theta'_0 + n_\theta t, \tag{13.32}$$

where, on account of $\partial\Omega/\partial\Psi' = -k^2\Omega/\Psi'$, the constant, secular frequencies $n_\psi = \partial\mathcal{H}/\partial\Psi'$ and $n_\theta = \partial\mathcal{H}/\partial\Theta'$ are

$$n_\theta = -\Omega\sum_{i=1}^{3} i\left(\frac{\Theta'/\Psi'}{\Omega/\dot{\vartheta}}\right)^{i-1}\sum_{j=0}^{3-i} h_{i,j}\left(\frac{\Omega}{\dot{\vartheta}}\right)^{2j}, \tag{13.33}$$

$$n_\psi = \dot{\vartheta}\left[1 + \frac{1}{4}\sum_{i=0}^{3}\left(\frac{\Theta'/\Psi'}{\Omega/\dot{\vartheta}}\right)^i \sum_{j=0}^{3-i}(2+i+6j)h_{i,j}\left(\frac{\Omega}{\dot{\vartheta}}\right)^{2j+2}\right]. \tag{13.34}$$

13.4.3 Long-period corrections

The generating function $\mathcal{V} = -\Theta' \sum_{j\geq 0}(\varepsilon^j/j!)\mathcal{V}_{j+1}$ of the transformation to prime variables has been obtained up to the second order of ε. At first order,

$$\mathcal{V}_1 = \left[\sqrt{-\frac{1}{2}h_{1,2}}\frac{\Omega^2}{\dot{\vartheta}^2} + \frac{1}{6}h_{2,0}\frac{\Theta/\Psi}{\Omega/\dot{\vartheta}}(4 - \cos 2\theta)\right]\sin 2\theta,$$

from which we obtain the long-period corrections

$$\begin{aligned}
\psi_{0,1} &= \frac{\Theta}{\Psi}\left[\frac{1}{24}h_{2,0}\frac{\Theta/\Psi}{\Omega/\dot{\vartheta}}(4 - \cos 2\theta) + \frac{3}{2}\chi\frac{\Omega^2}{\dot{\vartheta}^2}\right]\sin 2\theta \\
\theta_{0,1} &= -\left[\frac{1}{3}h_{2,0}\frac{\Theta/\Psi}{\Omega/\dot{\vartheta}}(4 - \cos 2\theta) + \chi\frac{\Omega^2}{\dot{\vartheta}^2}\right]\sin 2\theta \\
\Theta_{0,1} &= \Theta\left[\frac{1}{3}h_{2,0}\frac{\Theta/\Psi}{\Omega/\dot{\vartheta}}(4\cos 2\theta - \cos 4\theta) + 2\chi\frac{\Omega^2}{\dot{\vartheta}^2}\cos 2\theta\right],
\end{aligned}$$

with $\chi \equiv \frac{2}{9}(5\tilde{K} - \tilde{E})(\tilde{K} - \tilde{E})^{-1} - \frac{7}{32}(\tilde{K} - \tilde{E})^{-2}$. As usual, the right members are evaluated in prime variables for direct corrections, and in original variables for inverse corrections.

At second order,

$$\begin{aligned}
\mathcal{V}_2 = &\left(v_{2,2}\frac{\Omega^4}{\dot{\vartheta}^4} + v_{2,0}\frac{\Theta^2/\Psi^2}{\Omega^2/\dot{\vartheta}^2} + v_{2,1}\frac{\Theta}{\Psi}\frac{\Omega}{\dot{\vartheta}}\right)\sin 2\theta \\
&+ \left(v_{4,1}\frac{\Theta}{\Psi}\frac{\Omega}{\dot{\vartheta}} + v_{4,0}\frac{\Theta^2/\Psi^2}{\Omega^2/\dot{\vartheta}^2}\right)\sin 4\theta + v_{6,0}\frac{\Theta^2/\Psi^2}{\Omega^2/\dot{\vartheta}^2}\sin 6\theta,
\end{aligned}$$

with the numeric coefficients $v_{i,j}$ of Table 13.5. For our aims, second-order corrections are only relevant in the computation of the secular constants of the theory. In consequence, we only provide the inverse corrections. Namely,

$$\begin{aligned}
\psi_{0,2} = &\frac{\Theta}{\Psi}\kappa_{1,6,0}\frac{\Theta^2/\Psi^2}{\Omega^2/\dot{\vartheta}^2}\sin 6\theta + \frac{\Theta}{\Psi}\left(\kappa_{1,4,0}\frac{\Theta^2/\Psi^2}{\Omega^2/\dot{\vartheta}^2} + \kappa_{1,4,1}\frac{\Theta}{\Psi}\frac{\Omega}{\dot{\vartheta}}\right)\sin 4\theta \\
&+ \frac{\Theta}{\Psi}\left(\kappa_{1,2,0}\frac{\Theta^2/\Psi^2}{\Omega^2/\dot{\vartheta}^2} + \kappa_{1,2,2}\frac{\Omega^4}{\dot{\vartheta}^4} + \kappa_{1,2,1}\frac{\Theta}{\Psi}\frac{\Omega}{\dot{\vartheta}}\right)\sin 2\theta, \\
\theta_{0,2} = &\left(\kappa_{2,2,0}\frac{\Theta^2/\Psi^2}{\Omega^2/\dot{\vartheta}^2} + \kappa_{2,2,1}\frac{\Theta}{\Psi}\frac{\Omega}{\dot{\vartheta}} + \kappa_{2,2,2}\frac{\Omega^4}{\dot{\vartheta}^4}\right)\sin 2\theta \\
&+ \left(\kappa_{2,4,0}\frac{\Theta^2/\Psi^2}{\Omega^2/\dot{\vartheta}^2} + \kappa_{2,4,1}\frac{\Theta}{\Psi}\frac{\Omega}{\dot{\vartheta}} + \kappa_{2,4,2}\frac{\Omega^4}{\dot{\vartheta}^4}\right)\sin 4\theta
\end{aligned}$$

Table 13.5: Coefficients $v_{i,j}$ of the generating function term $\mathcal{V}_2$.

2,0 :	$\frac{7\tilde{K}^2+484\tilde{K}\tilde{E}-788\tilde{E}^2}{16384(\tilde{K}-\tilde{E})^2}$	2,1 :	$h_{2,0}\frac{1449-64(\tilde{K}-\tilde{E})(107\tilde{K}+55\tilde{E})}{864(\tilde{E}-\tilde{K})^2}$
4,0 :	$-\frac{548\tilde{E}^2-1012\tilde{K}\tilde{E}+437\tilde{K}^2}{16384(\tilde{K}-\tilde{E})^2}$	4,1 :	$-\frac{42h_{2,0}+502\tilde{E}^2-467\tilde{K}\tilde{E}-8\tilde{K}^2}{576(\tilde{K}-\tilde{E})^2}$
6,0 :	$\frac{1163\tilde{K}^2-2860\tilde{E}\tilde{K}+1724\tilde{E}^2}{147456(\tilde{K}-\tilde{E})^2}$	2,2 :	$\frac{63-64(\tilde{K}-\tilde{E})(4\tilde{K}+5\tilde{E})}{144(\tilde{K}-\tilde{E})^2}\sqrt{-2h_{1,2}}$

$$+\left(\kappa_{2,6,1}\frac{\Theta}{\Psi}\frac{\Omega}{\dot{\vartheta}}+\kappa_{2,6,0}\frac{\Theta^2/\Psi^2}{\Omega^2/\dot{\vartheta}^2}\right)\sin 6\theta+\kappa_{2,8,0}\frac{\Theta^2/\Psi^2}{\Omega^2/\dot{\vartheta}^2}\sin 8\theta,$$

$$\Theta_{0,2}=\Theta\Bigg[\kappa_{3,0,0}\frac{\Theta^2/\Psi^2}{\Omega^2/\dot{\vartheta}^2}+\kappa_{3,0,2}\frac{\Omega^4}{\dot{\vartheta}^4}+\kappa_{3,0,1}\frac{\Theta}{\Psi}\frac{\Omega}{\dot{\vartheta}}$$

$$+\left(\kappa_{3,2,2}\frac{\Omega^4}{\dot{\vartheta}^4}+\kappa_{3,2,1}\frac{\Theta}{\Psi}\frac{\Omega}{\dot{\vartheta}}+\kappa_{3,2,0}\frac{\Theta^2/\Psi^2}{\Omega^2/\dot{\vartheta}^2}\right)\cos 2\theta$$

$$+\left(\kappa_{3,4,1}\frac{\Theta}{\Psi}\frac{\Omega}{\dot{\vartheta}}+\kappa_{3,4,0}\frac{\Theta^2/\Psi^2}{\Omega^2/\dot{\vartheta}^2}\right)\cos 4\theta+\kappa_{3,6,0}\frac{\Theta^2/\Psi^2}{\Omega^2/\dot{\vartheta}^2}\cos 6\theta\Bigg],$$

which must be evaluated in original variables, and whose numeric coefficients $\kappa_{i,j,k}$ are given in Table 13.6.

Table 13.6: Coefficients $\kappa_{i,j,k}$ of the inverse second-order long-period corrections.

1,6,0 :	$-\frac{1724\tilde{E}^2-2860\tilde{E}\tilde{K}+1163\tilde{K}^2}{294912(\tilde{E}-\tilde{K})^2}$	1,2,2 :	$\frac{[64(\tilde{E}-\tilde{K})(5\tilde{E}+4\tilde{K})+63][64(\tilde{E}-\tilde{K})(\tilde{E}+5\tilde{K})+63]}{6912(\tilde{E}-\tilde{K})^4}$
1,4,0 :	$\frac{548\tilde{E}^2-1012\tilde{E}\tilde{K}+437\tilde{K}^2}{32768(\tilde{E}-\tilde{K})^2}$	1,4,1 :	$\frac{1218\tilde{E}^2-893\tilde{E}\tilde{K}-202\tilde{K}^2}{768(\tilde{E}-\tilde{K})^2}+\frac{119(14\tilde{E}-11\tilde{K})}{16384(\tilde{E}-\tilde{K})^3}$
1,2,0 :	$\frac{788\tilde{E}^2-484\tilde{E}\tilde{K}-7\tilde{K}^2}{32768(\tilde{E}-\tilde{K})^2}$	1,2,1 :	$-(14\tilde{E}-11\tilde{K})\frac{64(\tilde{E}-\tilde{K})(65\tilde{E}+129\tilde{K})+1743}{24576(\tilde{E}-\tilde{K})^3}$
2,2,0 :	$-\frac{1084\tilde{E}^2-572\tilde{E}\tilde{K}-71\tilde{K}^2}{8192(\tilde{E}-\tilde{K})^2}$	2,2,1 :	$\frac{(14\tilde{E}-11\tilde{K})}{(\tilde{E}-\tilde{K})}\frac{64(\tilde{E}-\tilde{K})(109\tilde{E}+209\tilde{K})+2835}{36864(\tilde{E}-\tilde{K})^2}$
2,4,0 :	$-\frac{76\tilde{E}^2-572\tilde{E}\tilde{K}+343\tilde{K}^2}{16384(\tilde{E}-\tilde{K})^2}$	2,4,1 :	$-\frac{64(\tilde{E}-\tilde{K})(530\tilde{E}^2-118\tilde{K}^2-349\tilde{E}\tilde{K})+189(14\tilde{E}-11\tilde{K})}{18432(\tilde{E}-\tilde{K})^3}$
2,4,2 :	$\frac{(64(\tilde{E}-\tilde{K})(\tilde{E}+5\tilde{K})+63)^2}{82944(\tilde{E}-\tilde{K})^4}$	2,2,2 :	$-[\frac{4}{9}\frac{5\tilde{E}-5\tilde{K}+9\tilde{K}}{\tilde{E}-\tilde{K}}+\frac{7}{16(\tilde{E}-\tilde{K})^2}][\frac{4}{9}\frac{\tilde{E}-\tilde{K}+6\tilde{K}}{\tilde{E}-\tilde{K}}+\frac{7}{16(\tilde{E}-\tilde{K})^2}]$
2,6,0 :	$-\frac{20\tilde{E}^2+44\tilde{E}\tilde{K}-37\tilde{K}^2}{24576(\tilde{E}-\tilde{K})^2}$	2,6,1 :	$\frac{(14\tilde{E}-11\tilde{K})(64(\tilde{E}-\tilde{K})(\tilde{E}+5\tilde{K})+63)}{36864(\tilde{E}-\tilde{K})^3}$
2,8,0 :	$\frac{(14\tilde{E}-11\tilde{K})^2}{65536(\tilde{E}-\tilde{K})^2}$	3,6,0 :	$-\frac{71\tilde{E}-50\tilde{K}}{1536(\tilde{E}-\tilde{K})}$
3,4,0 :	$\frac{548\tilde{E}^2-1012\tilde{E}\tilde{K}+437\tilde{K}^2}{4096(\tilde{E}-\tilde{K})^2}$	3,4,1 :	$\frac{64(\tilde{E}-\tilde{K})(502\tilde{E}^2-8\tilde{K}^2-467\tilde{E}\tilde{K})+63(14\tilde{E}-11\tilde{K})}{9216(\tilde{E}-\tilde{K})^3}$
3,2,0 :	$-\frac{122\tilde{E}^2-286\tilde{E}\tilde{K}+137\tilde{K}^2}{1024(\tilde{E}-\tilde{K})^2}$	3,2,1 :	$-(14\tilde{E}-11\tilde{K})\frac{64(\tilde{E}-\tilde{K})(13\tilde{E}+23\tilde{K})+315}{4608(\tilde{E}-\tilde{K})^3}$
3,0,0 :	$\frac{17(14\tilde{E}-11\tilde{K})^2}{8192(\tilde{E}-\tilde{K})^2}$	3,0,1 :	$-(14\tilde{E}-11\tilde{K})\frac{64(\tilde{E}-\tilde{K})(\tilde{E}+5\tilde{K})+63}{1536(\tilde{E}-\tilde{K})^3}$
3,0,2 :	$\frac{(64(\tilde{E}-\tilde{K})(\tilde{E}+5\tilde{K})+63)^2}{20736(\tilde{E}-\tilde{K})^4}$	3,2,2 :	$\frac{[64(\tilde{E}-\tilde{K})(5\tilde{E}+4\tilde{K})+63][64(\tilde{E}-\tilde{K})(\tilde{E}+5\tilde{K})+63]}{10368(\tilde{E}-\tilde{K})^4}$

13.4.4 Orbit design parameters

The formal integrals Ψ' and Θ' are related to the dimension of the reference ellipse, and to the amplitude of the oscillations of the center of this ellipse about the origin, respectively. Therefore, these dynamical parameters provide control over the size of the orbit and how close the orbiter can approach to the origin, and play an important role like orbit design parameters [398, 456].

Indeed, when the two last equations of Eq. (13.29) are plugged into Eq. (13.7), the coordinates of the center of the reference ellipse in the new variables become

$$x_{\text{center}} = \frac{1}{k}\frac{\Omega}{\dot{\vartheta}} M \cos(\theta_0' + n_\theta t), \tag{13.35}$$

$$y_{\text{center}} = 2kM \sin(\theta_0' + n_\theta t), \tag{13.36}$$

in which Θ', Ψ', and hence $\Omega \equiv \Omega(\Psi')$ are constant. So it is the amplitude of the oscillations

$$M = \sqrt{2\Theta'/\Omega}, \tag{13.37}$$

and the minimum distance with which the orbit will approach the primary of smaller mass in the y-axis direction,

$$y_{\min} = a - 2kM, \tag{13.38}$$

which corresponds to the maximum elongation of y_{center} in that direction, and this happens each time that $\theta_0' + n_\theta t = (2m-1)\frac{\pi}{2}$, with m integer.

Then an analytical orbit with given semimajor axis of the reference ellipse a that approaches the origin a minimum distance $y_{\min}$ is computed as follows. First, compute $M = (a - y_{\min})/(2k)$ from Eq. (13.38). Next, compute $\Psi' = \Phi$ from Eq. (13.4), and $\Omega = [\dot{\vartheta}(\tilde{K} - \tilde{E})/(\Psi' a)]^{1/2}$, from Eqs. (13.17) and (13.12). Finally, from Eq. (13.37),

$$\Theta' = \frac{1}{2}M^2\Omega. \tag{13.39}$$

While the choice of the initial conditions ψ_0' and θ_0' is arbitrary, their careful selection may help in the design of particular periodic orbits, as will be shown in brief.

13.4.5 Periodic orbits

The ratio $R = R(\Psi', \Theta') \equiv n_\psi/n_\theta$ computed from Eqs. (13.33) and (13.34) provides another design parameter useful in periodic orbit design. Indeed, for rational values of R the orbit will be periodic in the secular $(\psi', \theta', \Psi', \Theta')$ variables. However, the periodicity requirement will be generally in contradiction with the chosen values of a and

$y_{\min}$, which will result in R being real as opposite to rational. Nevertheless, the desired periodicity will normally be achieved with minor modifications of the parameters that define a and $y_{\min}$. Indeed, from Eq. (13.4) $a = a(\Psi')$, and from Eqs. (13.39) and (13.38) $\Theta = \Theta(\Psi'; y_{\min})$. Then, keeping R (rational) and $y_{\min}$ fixed, a new value of a is computed by solving Ψ' from the implicit equation $R(\Psi'; \Theta'(\Psi'; y_{\min})) - R = 0$. The new value of Ψ', which is readily obtained with a root-finding procedure, in turn provides a new value of a that now yields the desired commensurability.

The periodicity of the orbit in the secular variables yields a concomitant periodicity in the mean variables. This is due to the fact that the transformation from secular to mean variables only depends on the secular variables, which have been made periodic by design. Analogously, the transformation from mean to osculating variables only depends on mean elements, which, we have just seen, are periodic with the same period of the secular solution. However, due to the unavoidable truncation of the analytical solution, the numerical propagation of the initial conditions of a periodic orbit computed in such a way will produce an orbit of the planar Hill problem that is only approximately periodic. Improvement of the initial conditions to obtain a true periodic quasi-satellite orbit of the Hill problem with the desired characteristics will be achieved, in general, with the standard computation of differential corrections, which are efficiently integrated in Cartesian coordinates [422, 426].

It is worth mentioning also that the arbitrariness in the selection of the initial secular values of ψ' and θ' may compromise the convergence of the differential corrections procedure. This is the case of multiple periodic orbits with close tracks. Indeed, when the initial conditions and period predicted by the perturbation solution are not close enough to the desired periodic orbit, the differential correction algorithm will probably derail to a different, close, periodic orbit with stronger stability characteristics. On the other hand, as will be illustrated in §13.5, one can take advantage of this fact to compute 1:1-resonant periodic orbits in spite of them falling out of the convergence domain of the analytical solution.

For instance, disregarding primes for brevity, if the chosen initial conditions are $\psi = 0$ and $\theta = \pi/2$, application of the transformations in Eq. (13.29) yields $\phi = \psi = 0$, $\Phi = \Psi = \frac{1}{8}\dot{\vartheta}a^2$, $q = \sqrt{2\Theta/\Omega} = \frac{1}{2k}(a - y_{\min})$, $Q = 0$, from which $\eta = \eta_{\max} = 1 - y_{\min}/a$, $\xi = 0$, from Eq. (13.6). Then application of the transformation (13.3) would result in the Cartesian variables $x = 0$, $y = 2a - y_{\min}$, $X = -3B + \dot{\vartheta}y_{\min}$, $Y = 0$, corresponding to an *exterior* end of the quasi-satellite orbit (both for the orbit and the hodograph), a region in which consecutive tracks of the solution get close to each other. This fact may cause difficulty in the convergence of the differential corrections to the proper orbit. The same happens when choosing the initial conditions $\psi = \pi$, $\theta = -\pi/2$, which yield $x = 0$, $y = -2a + y_{\min}$, $X = 3B - \dot{\vartheta}y_{\min}$, $Y = 0$. The initial conditions $\psi = 0$, $\theta = -\pi/2$ or $\psi = \pi$, $\theta = \pi/2$, yield *interior* ends of the solution $x = 0$, $y = \pm y_{\min}$, $X = \pm B \mp \dot{\vartheta}y_{\min}$, $Y = 0$, which experience exactly the same problems.

On the other hand, it commonly happens that two consecutive tracks of a quasi-satellite orbit become more distant from each other along the y axis when the center of the corresponding reference ellipse is closer to the origin than in the exterior or interior parts of the orbit. This fact makes the differential corrections algorithm find more favorable conditions for the fast convergence to the desired periodic orbit when the initial conditions are chosen from a reference ellipse whose center is close to the origin. As a rule, choosing $\psi = 0$, $\theta = 0$, which correspond to the Cartesian variables of the planar Hill problem $x = \xi a$, $y = a$, $X = -B$, $Y = -\xi B$, where $\xi = (\Omega/\dot{\vartheta})(1-y_{\min}/a)/(2k^2) \ll 1$, will produce satisfactory results, yielding convergence to the desired periodic orbit with the expected quadratic rate theoretically predicted for differential correction algorithms.

Finally, it is worth noting that the applicability of the perturbation solution is restricted by construction to the case of slow librations of the reference ellipse when compared to the fast motion of the orbiter, $\Omega \ll \dot{\vartheta}$, hence excluding the useful case of 1:1 periodic orbits. Still, the perturbation solution can be used in practice to find this kind of strong resonant motion. Thus, for orbits with design parameters $a = y_{\min}$ the amplitude of the librations will be very small, and, in consequence, consecutive tracks of the orbit will be very close to each other in any place along the quasi-satellite orbit. This fact, which will make it almost impossible for the differential corrections algorithm to converge to the designed multiple-period periodic orbit, will usually let it converge to the dominant, strongly stable, 1:1 resonant orbit. This practical use of the analytical solution will be illustrated in the following examples.

13.5 Examples

To illustrate how the perturbation solution can be used in the design of quasi-satellite orbits we provide two examples. The first one deals with the challenging problem of quasi-satellite orbits with large amplitude libration. The second example shows how the perturbation solution can be effectively used in the design of a classical distant retrograde orbit under the 1:1 resonance condition.

13.5.1 Large amplitude libration

The parameters $a = 9$ and $y_{\min} = 2.5$, in units of the Hill problem, are chosen for the design of a quasi-satellite orbit with large amplitude libration. Then, from the sequence described in §13.4.4, we obtain the approximate values: $M = 3.75278$, $\Psi = 10.125$, $\Omega = 0.0574682$, and $\Theta' = 0.404672$. These values are set into Eqs. (13.33) and (13.34) to obtain the secular frequencies $n_\theta = -0.0734688$, and $n_\psi = 1.00971$, respectively. These frequencies are close to a 55:4 resonance but they are not exactly commensurable, yielding the ratio $R_0 = 13.7434$ between the libration period and orbital period.

To get an exact commensurability $R = 14$, the closest integer value to R_0, we modify sequentially the initial value of Ψ using the secant method. That is,

$$\Psi = \Psi_n + \frac{\Psi_{n+1} - \Psi_n}{R_{n+1} - R_n}(R - R_n). \tag{13.40}$$

After five iterations we found that the value $\Psi = 10.412533967155952$ produces the desired commensurability to the computer's numerical precision, corresponding to a slightly longer semimajor axis of the reference ellipse, as computed from Eq. (13.4) replacing Φ by Ψ, which is only 1.4 % longer than the original value. The new libration frequency, which is about 2 % slower than the original one, is computed from Eq. (13.17).

Next the new value $\Theta' = 0.41188510904143977$ is computed from Eq. (13.39), and the new orbital and libration periods

$$T_O = \frac{2\pi}{n_\psi} = 6.22511673870462, \quad T_L = \frac{2\pi}{|n_\theta|} = 87.1516343418646, \tag{13.41}$$

are computed from the corresponding secular frequencies in Eqs. (13.34) and (13.33), respectively. As expected, T_L is now exactly 14 times longer than T_O. For the improved values, Eq. (13.32) produces a quasi-satellite orbit that is periodic after the libration period.

The Cartesian coordinates of the 14:1-resonant quasi-satellite orbit, which are computed from Eqs. (13.29) and (13.3), are depicted in the left plot of Fig. 13.1. To show the agreement of the computed orbit with the required design ($a = 9.1269$, $y_{\min} = 2.5$) we displaced the x axis from the origin to the value $y = 2.5$. The right plot of Fig. 13.1 shows the evolution of the center of the reference ellipse scaled by the semimajor axis a—the nondimensional variables ξ and η in Eq. (13.7). Note that ξ and η are displayed in different scales for visualization purposes.

The long-period corrections in §13.4.3 provide the orbit in mean elements, which is also periodic with the same periodicity, and the short-period corrections also yield a periodic quasi-satellite orbit of the perturbation solution in osculating elements. As shown in the left plot of Fig. 13.2, the amplitude of the oscillations of the center of the reference ellipse is shorter both in mean and osculating elements in the y-axis direction and slightly longer in the x-axis direction than the amplitude in secular variables. Besides, this trajectory is much more involved in the osculating elements because it is modulated by the orbital period due to the coupling with the phase of the ellipse. The long-period corrections clearly modify also the orbit as shown in the right plot of Fig. 13.2. The short-period corrections introduce corrections of lesser amplitude and the orbit in osculating elements is not displayed.

On the other hand, due to the truncation of the perturbation solution, if we compute initial conditions from the osculating elements obtained from the perturbation solution, and propagate them in the original planar Hill problem equations, we do not

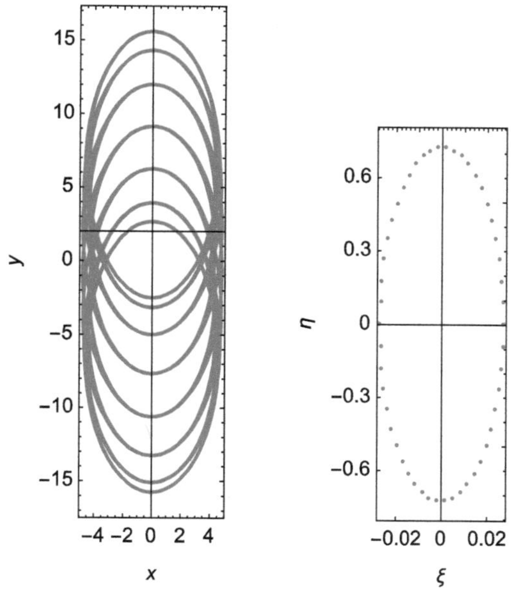

Figure 13.1: Left: Secular periodic orbit after 14 revolutions. Right: trajectory of the guiding center of the reference ellipse.

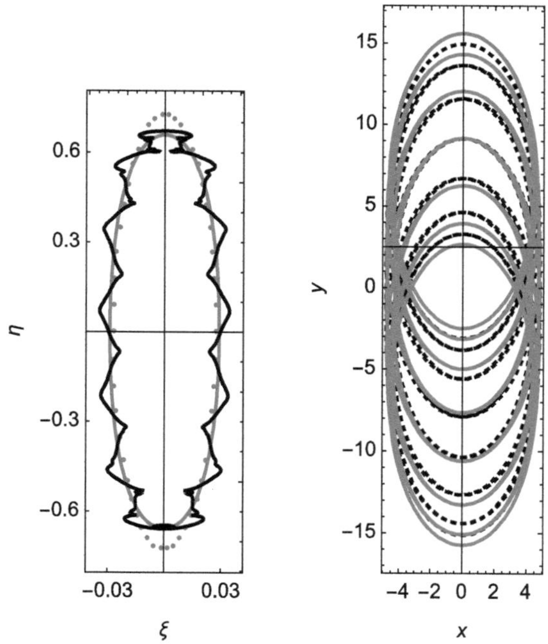

Figure 13.2: Left: Guiding center of the reference ellipse in osculating (black line), mean (gray curve) and secular elements (dots). Right: mean-element orbit (black dashed line) superimposed to the secular orbit in the left plot Fig. 13.1 (gray curve).

find an exactly periodic orbit. However, as discussed in §13.4.5, these initial conditions are readily improved by means of differential corrections to give a true periodic quasi-satellite orbit of the planar Hill problem with similar characteristics to those aimed by the design parameters. This is illustrated in Fig. 13.3, where the osculating solution predicted analytically and the improved solution computed using differential corrections are shown superimposed. We finally remark that convergence to a true periodic orbit with the desired characteristics is usually feasible differentially correcting the secular initial conditions and period. In this way, the need of conversion from secular to mean and to osculating elements is avoided. In this last case, the new periodic orbit may be slightly different from the one obtained from the osculating solution.

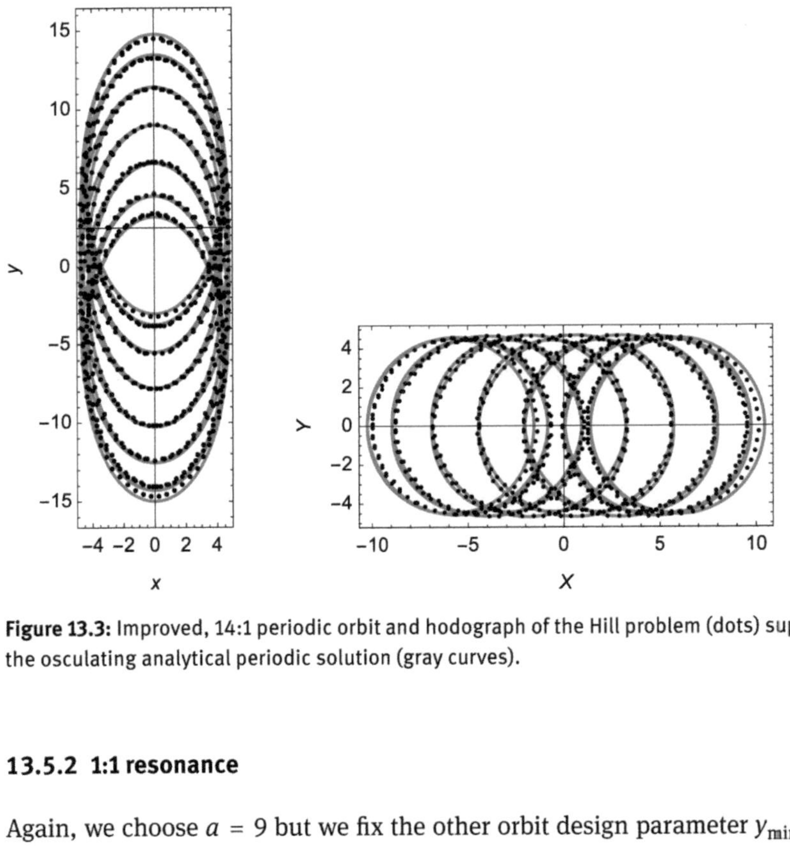

Figure 13.3: Improved, 14:1 periodic orbit and hodograph of the Hill problem (dots) superimposed to the osculating analytical periodic solution (gray curves).

13.5.2 1:1 resonance

Again, we choose $a = 9$ but we fix the other orbit design parameter $y_{\min} = a$, trying to get minimum librations of the reference ellipse. The new choice yields the same values as before except for $M = 0$ and $\Theta' = 0$. For these particular values Eqs. (13.33)–(13.34) take a simpler form, and turn the computation of the secular frequencies into the abridged formulas

$$n_\theta = -\Omega \sum_{j=0}^{2} h_{1,j}\left(\frac{\Omega}{\dot{\vartheta}}\right)^{2j}, \quad n_\psi = \dot{\vartheta}\left[1 + \frac{1}{2}\sum_{j=0}^{3}(1+3j)h_{0,j}\left(\frac{\Omega}{\dot{\vartheta}}\right)^{2j+2}\right]. \tag{13.42}$$

We obtain $n_\theta = -0.0578611$, $n_\psi = 1.00763$, and hence $R_0 = 17.4147$. Then a refined value $\Psi = 9.805769504452774$ is obtained by iteration of Eq. (13.40) with $R = 17$. Next, Eq. (13.39) yields $\Theta' = 2.0067075011 \times 10^{-4}$, and Eq. (13.41) $T_0 = 6.233222965910465$. Finally, the secular, periodic orbit is computed using Eq. (13.32), which is evaluated along the libration period $T_\mathrm{L} = 17 \times T_0$.

The propagation of initial conditions of this secular orbit in the original planar Hill problem does not yield a true periodic orbit. On the other hand, we failed in improving them with differential corrections to converge to a true periodic 17:1 resonant orbit of the planar Hill problem, as expected from the close proximity of the different tracks. However, as discussed in §13.4.5, when we require the differential corrections algorithm to find a periodic orbit with the orbital period, contrary to the libration one, the algorithm shows fast convergence to the desired 1:1-resonant orbit. The accuracy of the analytical prediction is illustrated in Fig. 13.4, where the secular and numeric orbits are shown superimposed—in the left plot for the orbit and in the right plot for the center of the reference ellipse.

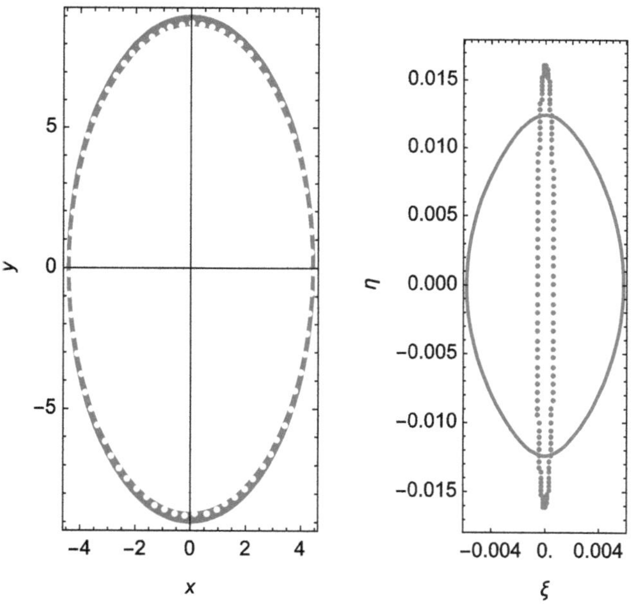

Figure 13.4: Left: secular periodic orbit after 17 revolutions (full line) superimposed to the true 1:1-periodic orbit (white dots). Right: corresponding trajectories for the center of the reference ellipse (dots represent the secular solution).

Bibliography

[1] R. Abraham and J. E. Marsden. *Foundations of Mechanics*. Addison-Wesley 2nd edition, 1987.

[2] M. K. M. Ahmed. On the normalization of perturbed Keplerian systems. *The Astronomical Journal*, 107:1900–1903, May 1994.

[3] J. Aiello. Numerical investigation of mapping orbits about Jupiter's icy moons. In B. G. Williams, L. A. D'Amario, K. C. Howell and F. R. Hoots, editors, *Astrodynamics 2005*, P.O. Box 28130, San Diego, California 92198, volume 123 of *Advances in the Astronautical Sciences*, pages 1965–1980. American Astronautical Society, Univelt, Inc., 2006.

[4] É. L. Akim and Z. P. Vlasova. Model of the moon's gravitational field based on the motion of the artificial lunar satellites Luna-10, 12, 14, 19, and 22. *Doklady Akademii Nauk SSSR*, 235:38–41, July 1977.

[5] Ye. P. Aksenov, Ye. A. Grebenikov, and V. G. Demin. General solution of the problem of the motion of an artificial satellite in the normal field of the earth's attraction. *Planetary and Space Science*, 9(8):491–498, August 1962.

[6] K. Aksnes. On the dynamical theory of a near-earth satellite, I. *Astrophysica Norvegica*, 10:69–77, August 1965.

[7] K. Aksnes. A Second-Order Artificial Satellite Theory Based on an Intermediate Orbit. *The Astronomical Journal*, 75:1066, November 1970.

[8] K. Aksnes. A note on 'The main problem of satellite theory for small eccentricities, by A. Deprit and A. Rom, 1970'. *Celestial Mechanics*, 4(1):119–121, September 1971.

[9] K. Aksnes. On the Use of the Hill Variables in Artificial Satellite Theory. *Astronomy and Astrophysics*, 17(1):70–75, February 1972.

[10] K. Aksnes. Short-period and long-period perturbations of a spherical satellite due to direct solar radiation. *Celestial Mechanics*, 13:89–104, February 1976.

[11] E. M. Alessi, G. Gómez, and J. J. Masdemont. Leaving the moon by means of invariant manifolds of libration point orbits. *Communications in Nonlinear Science and Numerical Simulation*, 14(12):4153–4167, 2009.

[12] E. M. Alessi, G. Schettino, A. Rossi, and G. B. Valsecchi. Natural highways for end-of-life solutions in the LEO region. *Celestial Mechanics and Dynamical Astronomy*, 130(5):34, May 2018.

[13] E. M. Alessi, G. Schettino, A. Rossi, and G. B. Valsecchi. Solar radiation pressure resonances in Low Earth Orbits. *Monthly Notices of the Royal Astronomical Society*, 473(2):2407–2414, January 2018.

[14] K. T. Alfriend and S. L. Coffey. Elimination of the perigee in the satellite problem. *Celestial Mechanics*, 32(2):163–172, February 1984.

[15] K. T. Alfriend, R. Dasenbrock, H. Pickard, and A. Deprit. The extended phase space formulation of the Vinti problem. *Celestial Mechanics*, 16(4):441–458, December 1977.

[16] K. T. Alfriend, S. R. Vadali, and H. Schaub. Formation Flying Satellites: Control by an Astrodynamicist. *Celestial Mechanics and Dynamical Astronomy*, 81:57–62, 2001.

[17] R. R. Allan. Satellite orbit perturbations due to radiation pressure and luni-solar forces. *The Quarterly Journal of Mechanics and Applied Mathematics*, 15(3):283–301, August 1962.

[18] R. R. Allan. Perturbations of a geostationary satellite by the longitude-dependent terms in the Earth's gravitational field. *Planetary and Space Science*, 11:1325–1334, November 1963.

[19] R. R. Allan. Resonance effects due to the longitude dependence of the gravitational field of a rotating primary. *Planetary and Space Science*, 15:53–76, January 1967.

[20] R. R. Allan and G. E. Cook. The Long-Period Motion of the Plane of a Distant Circular Orbit. *Proceedings of the Royal Society of London Series A*, 280:97–109, July 1964.

[21] R. R. Allan and G. N. Ward. Planetary equations in terms of vectorial elements. *Proceedings of the Cambridge Philosophical Society*, 59(3):669, January 1963.

https://doi.org/10.1515/9783110668513-014

[22] E. L. Allgower and K. Georg. *Numerical Continuation Methods: An Introduction*, volume 13 of *Springer Series in Computational Mathematics*. Springer-Verlag, Berlin Heidelberg, 1st edition, 1990.

[23] D. Amato, C. Bombardelli, G. Baù, V. Morand, and A. J. Rosengren. Non-averaged regularized formulations as an alternative to semi-analytical orbit propagation methods. *Celestial Mechanics and Dynamical Astronomy*, 131(5):21, May 2019.

[24] M-H. Andoyer. *Cours de mécanique céleste. Tome I*. Gauthier-Villars et cie, Paris, 1923-26.

[25] L. Anselmo and C. Pardini. Orbital Evolution of Geosynchronous Objects with High Area-To-Mass Ratios. In D. Danesy, editor, *4th European Conference on Space Debris*, volume 587 of *ESA Special Publication*, pages 279–284, August 2005.

[26] L. Anselmo and C. Pardini. Dynamical evolution of high area-to-mass ratio debris released into GPS orbits. *Advances in Space Research*, 43(10):1491–1508, 2009.

[27] R. Armellin and J. F. San-Juan. Optimal Earth's reentry disposal of the Galileo constellation. *Advances in Space Research*, 61:1097–1120, February 2018.

[28] R. Armellin, J. F. San-Juan, and M. Lara. End-of-life disposal of high elliptical orbit missions: The case of INTEGRAL. *Advances in Space Research*, 56(3):479–493, 2015. Advances in Asteroid and Space Debris Science and Technology, Part 1.

[29] V. I. Arnol'd. Proof of a theorem of A. N. Kolmogorov on the invariance of quasi-periodic motions under small perturbations of the Hamiltonian. *Russian Mathematical Surveys*, 18(5):9–36, October 1963.

[30] V. I. Arnol'd. *Mathematical Methods of Classical Mechanics*, volume 60 of *Graduate Texts in Mathematics*. Springer-Verlag, New York, 2nd edition, 1989.

[31] J. L. Arsenault, K. C. Ford, and P. E. Koskela. Orbit determination using analytic partial derivatives of perturbed motion. *AIAA Journal*, 8:4–12, 1970.

[32] J. A. Atchison and M. A. Peck. A passive, sun-pointing, millimeter-scale solar sail. *Acta Astronautica*, 67:108–121, July 2010.

[33] M. El-S. Awad. Analytical solution to the perturbed J_2 motion of Artificial Satellite in terms of Euler parameters. *Earth Moon and Planets*, 69(1):1–12, January 1995.

[34] F. Bagenal, T. E. Dowling and W. B. McKinnon, editors. *Jupiter: The Planet, Satellites and Magnetosphere*, pages xii, 732. *Cambridge Planetary Science*. Cambridge University Press, Cambridge, United Kingdom, 2004.

[35] N. Baresi, L. Dell'Elce, J. Cardoso dos Santos, and Y. Kawakatsu. Long-term evolution of mid-altitude quasi-satellite orbits. *Nonlinear Dynamics*, 99:2743–2763, January 2020.

[36] Y. V. Barkin. Unperturbed Chandler Motion and Perturbation Theory of the Rotation Motion of Deformable Celestial Bodies. *Astronomical and Astrophysical Transactions*, 17(3):179–219, 1998.

[37] R. B. Barrar. Some remarks on the motion of a satellite of an oblate planet. *The Astronomical Journal*, 66:11, February 1961.

[38] R. B. Barrar. Addendum to "Some remarks on the motion of a satellite of an oblate planet". *The Astronomical Journal*, 67:105, February 1962.

[39] R. Barrio and J. F. Palacián. Lie Transforms for Ordinary Differential Equations: Taking Advantage of the Hamiltonian Form of Terms of the Perturbation. *International Journal for Numerical Methods in Engineering*, 40:2289–2300, June 1997.

[40] R. Barrio and J. F. Palacián. High-order averaging of eccentric artificial satellites perturbed by the Earth's potential and air-drag terms. *Proceedings of the Royal Society of London Series A*, 459:1517–1534, June 2003.

[41] I. Basak. Explicit solution of the Zhukovski-Volterra gyrostat. *Regular and Chaotic Dynamics*, 14(2):223–236, April 2009.

[42] R. H. Battin. *An Introduction to the Mathematics and Methods of Astrodynamics*. American Institute of Aeronautics and Astronautics, Reston, VA, 1999.

[43] V. V. Beletskii. *Motion of an Artificial Satellite About Its Center of Mass*. Mechanics of Space Flight. Israel Program for Scientific Translations, S. Monson, Jerusalem, 1966. NASA TT F-429, Translated from Russian by Z. Lerman.

[44] D. Belorizky. Application pratique des méthodes de M. Sundman à un cas particulier du problème des trois corps. *Bulletin Astronomique Belgrade*, 6:417–434, 1930.

[45] S. Belyanin and P. Gurfil. Semianalytical study of geosynchronous orbits about a precessing oblate earth under lunisolar gravitation and tesseral resonance. *Journal of the Astronautical Sciences*, 57:517–543, March 2009.

[46] D. Benest. Effects of the Mass Ratio on the Existence of Retrograde Satellites in the Circular Plane Restricted Problem. *Astronomy and Astrophysics*, 32:39, April 1974.

[47] D. Benest. Libration effects for retrograde satellites in the restricted three-body problem. I - Circular plane Hill's case. *Celestial Mechanics*, 13:203–215, March 1976.

[48] A. F. Bertachini de Almeida Prado. Third-Body Perturbation in Orbits Around Natural Satellites. *Journal of Guidance Control Dynamics*, 26(1):33–40, January 2003.

[49] J. L. Bertaux and J. E. Blamont. Interpretation of Ogo 5 Lyman alpha measurements in the upper geocorona. *Journal of Geophysical Research*, 78(1):80, March January 1973.

[50] C. Bezrouk and J. S. Parker. Long term evolution of distant retrograde orbits in the Earth-Moon system. *Astrophysics and Space Science*, 362:176, September 2017.

[51] A. D. Biria and R. P. Russell. A satellite relative motion model including J_2 and J_3 via Vinti's intermediary. *Celestial Mechanics and Dynamical Astronomy*, 130:23, March 2018.

[52] L. Blitzer, E. M. Boughton, G. Kang, and R. M. Page. Effect of Ellipticity of the Equator on 24-Hour Nearly Circular Satellite Orbits. *Journal of Geophysical Research*, 67:329–335, January 1962.

[53] L. Blitzer, G. Kang, and J. B. McGuire. The Perturbed Motion of 24-Hour Satellites Due to Equatorial Ellipticity. *Journal of Geophysical Research*, 68:950–952, February 1963.

[54] D. Boccaletti and G. Pucacco. *Theory of orbits. Volume 2: Perturbative and geometrical methods. Astronomy and Astrophysics Library*. Springer-Verlag, Berlin Heidelberg New York, 1st edition, 2002.

[55] K. Bohlin. Zur Frage der Convergenz der Reihenentwickelungen in der Störungstheorie. *Astronomische Nachrichten*, 121:17, March 1889.

[56] F. Boigey. Une transformation canonique de Mathieu dans l'espace des phases du mouvement d'un solide mobile autour d'un point fixe. *Comptes rendus de l'Académie des Sciences de Paris, Série A Sciences Mathématiques*, 272:1115–1118, 1971.

[57] N. L. Bonavito, S. Watson, and H. Walden. An Accuracy and Speed Comparison of the Vinti and Brouwer Orbit Prediction Methods. Technical Report NASA TN D-5203, Goddard Space Flight Center, Greenbelt, Maryland, May 1969.

[58] V. R. Bond. An Analytical Singularity-Free Solution to the J_2 Perturbation Problem. Technical Report NASA-TM-58221; JSC-13128, NASA Johnson Space Center, July 1979.

[59] N. Borderies and P.-Y. Longaretti. A New Treatment of the Albedo Radiation Pressure in the Case of a Uniform Albedo and of a Spherical Satellite. *Celestial Mechanics and Dynamical Astronomy*, 49(1):69–98, March 1990.

[60] G. H. Born, J. L. Mitchell, and G. A. Heyler. Design of the GEOSAT Exact Repeat Mission. *Johns Hopkins APL Technical Digest*, 8(2):260–266, 1987.

[61] M. Born. *The Mechanics of the Atom*. G. Bells and Sons, Ltd, London, 1927.

[62] J. V. Breakwell and J. V. Brown. The 'halo' family of 3-dimensional periodic orbits in the earth-moon restricted 3-body problem. *Celestial Mechanics*, 20:389–404, November 1979.

[63] J. V. Breakwell and J. Vagners. On Error Bounds and Initialization in Satellite Orbit Theories. *Celestial Mechanics*, 2:253–264, June 1970.

[64] S. Ł. Breiter. Lunisolar Apsidal Resonances at low Satellite Orbits. *Celestial Mechanics and Dynamical Astronomy*, 74:253–274, August 1999.

[65] S. Ł. Breiter. Lunisolar Resonances Revisited. *Celestial Mechanics and Dynamical Astronomy*, 81(1–2):81–91, 2001.

[66] S. Ł. Breiter. On the coupling of lunisolar resonances for Earth satellite orbits. *Celestial Mechanics and Dynamical Astronomy*, 80:1–20, July 2001.

[67] A. J. Brizard. Jacobi zeta function and action-angle coordinates for the pendulum. *Communications in Nonlinear Science and Numerical Simulations*, 18:511–518, March 2013.

[68] R. A. Broucke. Periodic orbits in the restricted three-body problem with earth-moon masses. NASA Technical Report 32-1168, Jet Propulsion Laboratory, Pasadena, February 1968.

[69] R. A. Broucke. How to Assemble a Keplerian Processor. *Celestial Mechanics*, 2:9–20, March 1970.

[70] R. A. Broucke. Numerical integration of periodic orbits in the main problem of artificial satellite theory. *Celestial Mechanics and Dynamical Astronomy*, 58(2):99–123, February 1994.

[71] R. A. Broucke. Long-Term Third-Body Effects via Double Averaging. *Journal of Guidance Control Dynamics*, 26(1):27–32, January 2003.

[72] R. A. Broucke and P. J. Cefola. On the Equinoctial Orbit Elements. *Celestial Mechanics*, 5(3):303–310, May 1972.

[73] R. A. Broucke and P. J. Cefola. A Note on the Relations between True and Eccentric Anomalies in the Two-Body Problem. *Celestial Mechanics*, 7:388–389, April 1973.

[74] D. Brouwer. The motion of a particle with negligible mass under the gravitational attraction of a spheroid. *The Astronomical Journal*, 51:223–231, February 1946.

[75] D. Brouwer. Solution of the problem of artificial satellite theory without drag. *The Astronomical Journal*, 64:378–397, November 1959.

[76] D. Brouwer. Analytical study of resonance caused by solar radiation pressure. In M. Roy, editor, *Dynamics of Satellites/Dynamique des Satellites, IUTAM Symposia (International Union of Theoretical and Applied Mechanics)*, pages 34–39. Springer, Berlin, Heidelberg, 1963.

[77] D. Brouwer. The problem of the Kirkwood gaps in the asteroid belt. *The Astronomical Journal*, 68:152–158, April 1963.

[78] D. Brouwer and G. M. Clemence. *Methods of Celestial Mechanics*. Academic Press, New York and London, 1961.

[79] D. Brouwer and G.-I. Hori. Theoretical evaluation of atmospheric drag effects in the motion of an artificial satellite. *The Astronomical Journal*, 66:193–225, June 1961.

[80] E. W. Brown. *An introductory treatise on the lunar theory*. The University Press, Cambridge, 1896.

[81] S. Bruinsma, P. Exertier, G. Métris, and J. Bardina. Semi-analytical theory of mean orbital motion: A new tool for computing ephemerides. In T.-D. Guyenne, editor, *Space Flight Dynamics*, volume 403 of *ESA Special Publication*, pages 289–294, August 1997.

[82] E. Brumberg and T. Fukushima. Expansions of elliptic motion based on elliptic function theory. *Celestial Mechanics and Dynamical Astronomy*, 60:69–89, September 1994.

[83] V. A. Brumberg. *Analytical Techniques of Celestial Mechanics*. Springer-Verlag, Berlin Heidelberg, 1st edition, 1995.

[84] A. D. Bruno. *The restricted 3-body problem: Plane periodic orbits*, volume 17 of *De Gruyter Expositions in Mathematics*. Walter de Gruyter, Berlin, New York, 1994. English translation of the original Russian edition (Nauka, Moscow, 1990).

[85] R. W. Bryant. The effect of solar radiation pressure on the motion of an artificial satellite. *The Astronomical Journal*, 66:430, October 1961.

[86] J. A. Burns, P. L. Lamy, and S. Soter. Radiation forces on small particles in the solar system. *Icarus*, 40(1):1–48, October 1979.

[87] C. Burrau and E. Strömgren. Numerische Untersuchungen über eine Klasse einfach periodischer retrograder Bahnen im problème restreint. *Astronomische Nachrichten*, 202(12):185, March 1916.

[88] M. Burša. Figure and Dynamic Parameters of Synchronously Orbiting Satellites in the Solar System. *Bulletin of the Astronomical Institutes of Czechoslovakia*, 40:125–130, March 1989.
[89] P. F. Byrd and M. Friedman. *Handbook of Elliptic Integrals for Engineers and Physicists*. Springer-Verlag, Berlin, Heidelberg and New York, 2nd edition, 1971.
[90] B. J. Cain. Determination of mean elements for Brouwer's satellite theory. *The Astronomical Journal*, 67:391, August 1962.
[91] J. A. Campbell and W. H. Jefferys. Equivalence of the Perturbation Theories of Hori and Deprit. *Celestial Mechanics*, 2(4):467–473, 1970.
[92] É. Cartan. *Leçons sur les invariants intégraux*. A. Hermann & fils, Paris, 1922.
[93] J. R. Cary. Lie transforms and their use in Hamiltonian perturbation theory. Report LBL-6350, Lawrence Berkeley Laboratory, 1 Cyclotron Road, Berkeley CA 94720, June 1978. Also: U.S. Department of Energy Report DOE/ET-0074, January, 1979.
[94] J. R. Cary. Lie transform perturbation theory for Hamiltonian systems. *Physics Reports*, 79(2):129–159, 1981.
[95] D. Casanova, A. Petit, and A. Lemaître. Long-term evolution of space debris under the J_2 effect, the solar radiation pressure and the solar and lunar perturbations. *Celestial Mechanics and Dynamical Astronomy*, 123(2):223–238, October 2015.
[96] S. Casotto, S. Padovan, R. P. Russell, and M. Lara. Detecting a Subsurface Ocean From Periodic Orbits at Enceladus. In *AGU Fall Meeting Abstracts*, volume 2008, pages P23B–1366, December 2008.
[97] M. Ceccaroni and J. D. Biggs. Analytic perturbative theories in highly inhomogeneous gravitational fields. *Icarus*, 224:74–85, May 2013.
[98] P. J. Cefola and R. A. Broucke. On the formulation of the gravitational potential in terms of equinoctial variables. In *13th Aerospace Sciences Meeting*, Pasadena, California, pages 1–25. American Institute of Aeronautics and Astronautics, USA, January 1975. AIAA Paper No. 75-9.
[99] P. J. Cefola and D. J. Fonte. Extension of the Naval Space Command Satellite Theory PPT2 to Include a General Tesseral m-daily Model. In *Paper AIAA 96-3606, Astrodynamics Conference*, San Diego, CA, pages 1–45. American Institute of Aeronautics and Astronautics, USA, July 1996.
[100] A. Celletti and L. Chierchia. Construction of analytic KAM surfaces and effective stability bounds. *Communications in Mathematical Physics*, 118:119–161, March 1988.
[101] A. Celletti, C. Galeş, G. Pucacco, and A. J. Rosengren. Analytical development of the lunisolar disturbing function and the critical inclination secular resonance. *Celestial Mechanics and Dynamical Astronomy*, 127(3):259–283, March 2017.
[102] A. Celletti and P. Negrini. Non-integrability of the problem of motion around an oblate planet. *Celestial Mechanics and Dynamical Astronomy*, 61(3):253–260, March 1995.
[103] A. Celletti, G. Pucacco, and D. Stella. Lissajous and Halo Orbits in the Restricted Three-Body Problem. *Journal of NonLinear Science*, 25:343–370, April 2015.
[104] J. W. Chamberlain. Depletion of satellite atoms in a collisionless exosphere by radiation pressure. *Icarus*, 39(2):286–294, August 1979.
[105] J. W. Chamberlain and J. Bishop. Radiation pressure dynamics in planetary exospheres. II - Closed solutions for the evolution of orbital elements. *Icarus*, 106:419–427, December 1993.
[106] C. C. Chao. *Applied orbit perturbation and maintenance*. The Aerospace Press. American Institute of Aeronautics and Astronautics, El Segundo, California, 1998.
[107] C. C. Chao. Analytical Investigation of GEO Debris with High Area-to-Mass Ratio (AIAA 2006-6514). In *AIAA/AAS Astrodynamics Specialist Conference and Exhibit*, 21–24 August 2006, Keystone, Colorado, AIAA Meeting Paper, page 9, American Institute of Aeronautics and Astronautics, August 2006.
[108] C. C. Chao and R. A. Gick. Long-term evolution of navigation satellite orbits: GPS/GLONASS/GALILEO. *Advances in Space Research*, 34(5):1221–1226, 2004.

[109] J. Chapront and G. Francou. The lunar theory ELP revisited. Introduction of new planetary perturbations. *Astronomy and Astrophysics*, 404:735–742, June 2003.

[110] M. Chapront-Touze and J. Chapront. ELP 2000-85 - A semi-analytical lunar ephemeris adequate for historical times. *Astronomy and Astrophysics*, 190:342–352, January 1988.

[111] R. Cid and J. F. Lahulla. Perturbaciones de corto periodo en el movimiento de un satélite artificial, en función de las variables de Hill. *Publicaciones de la Revista de la Academia de Ciencias de Zaragoza*, 24:159–165, 1969.

[112] C. Circi, E. Condoleo, and E. Ortore. A vectorial approach to determine frozen orbital conditions. *Celestial Mechanics and Dynamical Astronomy*, 128(2–3):361–382, June 2017.

[113] W. H. Clohessy and R. S. Wiltshire. Terminal Guidance System for Satellite Rendezvous. *Journal of the Aerospace Sciences*, 27(9):653–658, September 1960.

[114] S. Coffey and K. T. Alfriend. An analytical orbit prediction program generator. *Journal of Guidance, Control and Dynamics*, 7(5):575–581, 1984.

[115] S. Coffey and A. Deprit. Fast evaluation of Fourier series. *Astronomy and Astrophysics*, 81:310–315, January 1980.

[116] S. Coffey, A. Deprit, E. Deprit, and L. Healy. Painting the Phase Space Portrait of an Integrable Dynamical System. *Science*, 247(4944):833–836, February 1990.

[117] S. L. Coffey and K. T. Alfriend. Short Period Elimination for the Tesseral Harmonics. In A. L. Friedlander, P. J. Cefola, B. Kaufman, W. Williamson and G. T. Tseng, editors, *Advances in the Astronautical Sciences*, P.O. Box 28130, San Diego, California 92198, volume 46 of *AAS/AIAA Astrodynamics Conference 1981*, pages 87–101. American Astronautical Society, Univelt, Inc., 1982.

[118] S. L. Coffey and A. Deprit. Third-Order Solution to the Main Problem in Satellite Theory. *Journal of Guidance, Control and Dynamics*, 5(4):366–371, 1982.

[119] S. L. Coffey, A. Deprit, and E. Deprit. Frozen orbits for satellites close to an earth-like planet. *Celestial Mechanics and Dynamical Astronomy*, 59(1):37–72, May 1994.

[120] S. L. Coffey, A. Deprit, and B. R. Miller. The critical inclination in artificial satellite theory. *Celestial Mechanics*, 39(4):365–406, December 1986.

[121] S. L. Coffey, H. L. Neal, A. M. Segerman, and J. J. Travisano. An Analytic Orbit Propagation Program for Satellite Catalog Maintenance. In K. T. Alfriend, I. M. Ross, A. K. Misra and C. F. Peters, editors, *AAS/AIAA Astrodynamics Conference 1995*, P.O. Box 28130, San Diego, California 92198, volume 90 of *Advances in the Astronautical Sciences*, pages 1869–1892. American Astronautical Society, Univelt, Inc., 1996.

[122] S. K. Collins and P. J. Cefola. Double averaged third body model for prediction of super-synchronous orbits over long time spans. In *Paper AAS 79-135, American Astronautical Society*, June 1979.

[123] C. Colombo. Planetary Orbital Dynamics (PlanODyn) suite for long term propagation in perturbed environment. In *Proceedings of the 6th International Conference on Astrodynamics Tools and Techniques, ICATT*, pages 1–7. ESA, March 2016.

[124] C. Colombo, E. M. Alessi, W. van der Weg, S. Soldini, F. Letizia, M. Vetrisano, M. Vasile, A. Rossi, and M. Landgraf. End-of-life disposal concepts for libration point orbit and highly elliptical orbit missions. *Acta Astronautica*, 110:298–312, 2015. Dynamics and Control of Space Systems.

[125] C. Colombo, C. Lücking, and C. R. McInnes. Orbital dynamics of high area-to-mass ratio spacecraft with J_2 and solar radiation pressure for novel Earth observation and communication services. *Acta Astronautica*, 81(1):137–150, 2012.

[126] C. Colombo and C. R. McInnes. Orbital Dynamics of "Smart-Dust" Devices with Solar Radiation Pressure and Drag. *Journal of Guidance Control Dynamics*, 34:1613–1631, November 2011.

[127] C. C. Conley. Low energy transit orbits in the restricted three-body problem. *SIAM Journal on Applied Mathematics*, 16:732–746, 1968.

[128] G. E. Cook. Luni-Solar Perturbations of the Orbit of an Earth Satellite. *Geophysical Journal*, 6:271–291, April 1962.
[129] G. E. Cook. Satellite drag coefficients. *Planetary and Space Science*, 13(10):929–946, October 1965.
[130] G. E. Cook. Perturbations of near-circular orbits by the Earth's gravitational potential. *Planetary and Space Science*, 14:433–444, May 1966.
[131] G. E. Cook, D. G. King-Hele, and D. M. C. Walker. The Contraction of Satellite Orbits Under the Influence of Air Drag. I. With Spherically Symmetrical Atmosphere. *Proceedings of the Royal Society of London Series A*, 257(1289):224–249, September 1960.
[132] R. A. Cook. The long-term behavior of near-circular orbits in a zonal gravity field (AAS 91-463). In B. Kaufman, K. T. Alfriend, R. L. Roehrich and R. R. Dasenbrock, editors, *Astrodynamics 1991*, P.O. Box 28130, San Diego, California 92198, volume 76 of *Advances in the Astronautical Sciences*, pages 2205–2221. American Astronautical Society, Univelt, Inc., 1992.
[133] V. T. Coppola and J. F. Palacián. Elimination of the latitude in artificial satellite theory. *The Journal of the Astronautical Sciences*, 42:27–34, January 1994.
[134] R. Cushman. Geometry of the Bifurcations of the Normalized Reduced Henon-Heiles Family. *Proceedings of the Royal Society of London Series A*, 382:361–371, August 1982.
[135] R. Cushman. Reduction, Brouwer's Hamiltonian, and the critical inclination. *Celestial Mechanics*, 31(4):401–429, December 1983.
[136] R. Cushman and D. L. Rod. Reduction of the semisimple 1:1 resonance. *Physica D Nonlinear Phenomena*, 6(1):105–112, October 1982.
[137] E. Cutting, J. C. Frautnick, and G. H. Born. Orbit analysis for Seasat-A. *The Journal of the Astronautical Sciences*, 26:315–342, December 1978.
[138] J. M. A. Danby. Motion of a Satellite of a Very Oblate Planet. *The Astronomical Journal*, 73(10):1031–1038, December 1968.
[139] J. M. A. Danby. *Fundamentals of Celestial Mechanics*. Willmann-Bell, Richmond VA, 2nd edition, 1992.
[140] J. M. A. Danby, A. Deprit, and A. R. M. Rom. The Symbolic Manipulation of Poisson Series. Mathematical Note No. 432 D1-82-0481, Boeing Scientific Research Laboratories, Seattle, Washington, 1965.
[141] Z. Dang, J. Luo, P. Shi, and H. Zhang. General Characteristics of the Motion on J_2-Perturbed Equatorial Orbits. *Journal of Guidance Control Dynamics*, 42(10):2319–2324, October 2019.
[142] J. Daquin, I. Gkolias, and A. J. Rosengren. Drift and Its Mediation in Terrestrial Orbits. *Frontiers in Applied Mathematics and Statistics*, 4(35):1–17, 2018.
[143] V. Dehant, O. de Viron, O. Karatekin, and T. van Hoolst. Excitation of Mars polar motion. *Astronomy & Astrophysics*, 446(1):345–355, 2006.
[144] C. E. Delaunay. *La Théorie du Mouvement de la Lune, Premier volume*, volume 28 of *Mémoires de l'Academie des Sciences de l'Institut Impérial de France*. Mallet-Bachellier, Paris, 1860.
[145] F. Deleflie, P. Legendre, P. Exertier, and F. Barlier. Long term evolution of the Galileo constellation due to gravitational forces. *Advances in Space Research*, 36:402–411, 2005.
[146] F. Deleflie, A. Rossi, C. Portmann, G. Métris, and F. Barlier. Semi-analytical investigations of the long term evolution of the eccentricity of Galileo and GPS-like orbits. *Advances in Space Research*, 47(5):811–821, March 2011.
[147] M. Delgado. Classroom Note: The Lagrange–Charpit Method. *SIAM Review*, 39(2):298–304, January 1997.
[148] F. Delhaise. Analytical treatment of air drag and earth oblateness effects upon an artificial satellite. *Celestial Mechanics and Dynamical Astronomy*, 52:85–103, March 1991.
[149] F. Delhaise and J. Henrard. The problem of critical inclination combined with a resonance in mean motion in artificial satellite theory. *Celestial Mechanics and Dynamical Astronomy*, 55:261–280, March 1993.

[150] A. Deprit. Free Rotation of a Rigid Body Studied in the Phase Space. *American Journal of Physics*, 35:424–428, 1967.

[151] A. Deprit. Canonical transformations depending on a small parameter. *Celestial Mechanics*, 1(1):12–30, 1969.

[152] A. Deprit. Ideal elements for perturbed Keplerian motions. *Journal of Research of the National Bureau of Standards*, 79:1–15, 1975.

[153] A. Deprit. Celestial Mechanics: Never Say No To A Computer. *Journal of Guidance Control Dynamics*, 4(5):577–581, September 1981.

[154] A. Deprit. The elimination of the parallax in satellite theory. *Celestial Mechanics*, 24(2):111–153, 1981.

[155] A. Deprit. The Main Problem in the Theory of Artificial Satellites to Order Four. *Journal of Guidance Control Dynamics*, 4(2):201–206, March 1981.

[156] A. Deprit. Delaunay normalisations. *Celestial Mechanics*, 26:9–21, January 1982.

[157] A. Deprit. Elimination of the nodes in problems of N bodies. *Celestial Mechanics*, 30:181–195, June 1983.

[158] A. Deprit. The reduction to the rotation for planar perturbed Keplerian systems. *Celestial Mechanics*, 29:229–247, March 1983.

[159] A. Deprit. Dynamics of orbiting dust under radiation pressure. In A. Berger, editor, *The Big-Bang and Georges Lemaître*, pages 151–180. Springer, Dordrecht, 1984.

[160] A. Deprit. The Lissajous transformation. I - Basics. *Celestial Mechanics and Dynamical Astronomy*, 51:201–225, 1991.

[161] A. Deprit and A. Delie. Trojan orbits. I. d'Alembert Series at L_4. *Icarus*, 4:242–266, July 1965.

[162] A. Deprit and E. Deprit. Massively Parallel Symbolic Computation. In *Proceedings of the ACM-SIGSAM 1989 International Symposium on Symbolic and Algebraic Computation, ISSAC '89*, pages 308–316. ACM, New York, NY, USA, 1989.

[163] A. Deprit and A. Elipe. The Lissajous transformation. II - Normalization. *Celestial Mechanics and Dynamical Astronomy*, 51:227–250, September 1991.

[164] A. Deprit and A. Elipe. Complete reduction of the Euler-Poinsot problem. *The Journal of the Astronautical Sciences*, 41:603–628, October 1993.

[165] A. Deprit and S. Ferrer. Note on Cid's Radial Intermediary and the Method of Averaging. *Celestial Mechanics*, 40(3–4):335–343, 1987.

[166] A. Deprit and S. Ferrer. Simplifications in the theory of artificial satellites. *The Journal of the Astronautical Sciences*, 37(4):451–463, December 1989.

[167] A. Deprit and B. Miller. Simplify or Perish. *Celestial Mechanics*, 45:189–200, 1989.

[168] A. Deprit and B. R. Miller. Normalization in the Face of Integrability. *Annals of the New York Academy of Sciences*, 536(1):101–126, August 1988.

[169] A. Deprit, J. F. Palacián, and E. Deprit. The Relegation Algorithm. *Celestial Mechanics and Dynamical Astronomy*, 79(3):157–182, 2001.

[170] A. Deprit and D. L. Richardson. Comments on Aksnes' intermediary. *Celestial Mechanics*, 28(3):253–273, 1982.

[171] A. Deprit and A. Rom. The Main Problem of Artificial Satellite Theory for Small and Moderate Eccentricities. *Celestial Mechanics*, 2(2):166–206, June 1970.

[172] A. Deprit and A. R. M. Rom. Computerized Expansions in Elliptic Motion. Mathematical Note No. 504 D1-82-0601, Boeing Scientific Research Laboratories, Seattle, Washington, 1965.

[173] A. Deprit and A. R. M. Rom. Lindstedt's Series on a Computer. *The Astronomical Journal*, 73:210–213, April 1968.

[174] E. Deprit and A. Deprit. Poincaré's méthode nouvelle by skew composition. *Celestial Mechanics and Dynamical Astronomy*, 74(3):175–197, July 1999.

[175] NIST Digital Library of Mathematical Functions. http://dlmf.nist.gov/, Release 1.0.27 of 2020-06-15. F. W. J. Olver, A. B. Olde Daalhuis, D. W. Lozier, B. I. Schneider, R. F. Boisvert, C. W. Clark, B. R. Miller, B. V. Saunders, H. S. Cohl, and M. A. McClain, eds.

[176] E. J. Doedel, R. C. Paffenroth, H. B. Keller, D. J. Dichmann, J. Galán-Vioque, and A. Vanderbauwhede. Computation of Periodic Solutions of Conservative Systems with Application to the 3-Body Problem. *International Journal of Bifurcation and Chaos*, 13:1353–1381, June 2003.

[177] A. J. Dragt. Lie Methods for Nonlinear Dynamics with Applications to Accelerator Physics. University of Maryland: http://www.physics.umd.edu/dsat/, 2019. Accessed: 2020-11-21.

[178] A. J. Dragt and J. M. Finn. Lie series and invariant functions for analytic symplectic maps. *Journal of Mathematical Physics*, 17:2215–2227, December 1976.

[179] G. Duffing. *Erzwungene Schwingungen bei veränderlicher Eigenfrequenz und ihre technische Bedeutung*, volume 41/42 of *Sammlung Vieweg*. F. Vieweg & sohn, Braunschweig, 1918. In German.

[180] P. Dunn. Geopotential resonance in a Landsat orbit. *Bulletin Geodesique*, 55(2):143–158, June 1981.

[181] M. C. Eckstein and F. Hechler. A reliable derivation of the perturbations due to any zonal and tesseral harmonics of the geopotential for nearly-circular satellite orbits. Scientific Report ESRO SR-13, European Space Research Organisation, Darmstadt, Federal Republic of Germany, June 1970.

[182] M. C. Eckstein, Y. Y. Shi, and J. Kevorkian. Satellite motion for arbitrary eccentricity and inclination around the smaller primary in the restricted three-body problem. *The Astronomical Journal*, 71:248–263, May 1966.

[183] T. A. Ely. Eccentricity Impact on East-West Stationkeeping for Global Positioning System Class Orbits. *Journal of Guidance Control Dynamics*, 25:352–357, March 2002.

[184] T. A. Ely. Mean Element Propagations Using Numerical Averaging. *The Journal of the Astronautical Sciences*, 61(3):275–304, September 2014.

[185] T. A. Ely. Transforming Mean and Osculating Elements Using Numerical Methods. *The Journal of the Astronautical Sciences*, 62(1):21–43, March 2015.

[186] T. A. Ely and K. C. Howell. Long-term evolution of artificial satellite orbits due to resonant tesseral harmonics. *The Journal of the Astronautical Sciences*, 44:167–190, April 1996.

[187] T. A. Ely and K. C. Howell. Dynamics of artificial satellite orbits with tesseral resonances including the effects of luni-solar perturbations. *Dynamics and Stability of Systems*, 12(4):243–269, 1997.

[188] T. A. Ely and K. C. Howell. East-west stationkeeping of satellite orbits with resonant tesseral harmonics. *Acta Astronautica*, 46:1–15, February 2000.

[189] D. Eui Chang and J. E. Marsden. Geometric Derivation of the Delaunay Variables and Geometric Phases. *Celestial Mechanics and Dynamical Astronomy*, 86(2):185–208, June 2003.

[190] R. W. Farquhar. The control and use of libration-point satellites. Technical Report R-346, Goddard Space Flight Center, Greenbelt, Maryland September 1970.

[191] R. W. Farquhar and A. A. Kamel. Quasi-Periodic Orbits about the Translunar Libration Point. *Celestial Mechanics*, 7:458–473, June 1973.

[192] F. Fassò. Superintegrable hamiltonian systems: Geometry and perturbations. *Acta Applicandae Mathematica*, 87(1):93–121, May 2005.

[193] J. M. Ferrandiz. Linearization in special cases of perturbed Keplerian motions. *Celestial Mechanics*, 39:23–31, May 1986.

[194] J. M. Ferrándiz and L. Floría. Towards a systematic definition of intermediaries in the theory of artificial satellites. *Bulletin of the Astronomical Institutes of Czechoslovakia*, 42:401–407, November 1991.

[195] S. Ferraz Mello. Analytical Study of the Earth's Shadowing Effects on Satellite Orbits. *Celestial Mechanics*, 5:80–101, January 1972.

[196] S. Ferraz-Mello. Elimination of secular terms generated by the coupling of perturbations. *Celestial Mechanics*, 25:293–296, November 1981.

[197] S. Ferraz-Mello. Do Average Hamiltonians Exist? *Celestial Mechanics and Dynamical Astronomy*, 73:243–248, January 1999.

[198] S. Ferraz-Mello. *Canonical Perturbation Theories – Degenerate Systems and Resonance*, volume 345 of *Astrophysics and Space Science Library*. Springer, New York, January 2007.

[199] S. Ferrer and M. Lara. Families of Canonical Transformations by Hamilton-Jacobi-Poincaré Equation. Application to Rotational and Orbital Motion. *Journal of Geometric Mechanics*, 2(3):223–241, 2010.

[200] S. Ferrer and M. Lara. Integration of the Rotation of an Earth-like Body as a Perturbed Spherical Rotor. *The Astronomical Journal*, 139(5):1899–1908, 2010.

[201] S. Ferrer, M. Lara, J. F. Palacián, J. F. San Juan, A. Viartola, and P. Yanguas. The Hénon and Heiles Problem in Three Dimensions: I. Periodic Orbits Near the Origin. *International Journal of Bifurcation and Chaos*, 8:1199–1213, June 1998.

[202] S. Ferrer and B. R. Miller. Coordinates for Perturbed Keplerian Systems with Axial Symmetry. *Celestial Mechanics and Dynamical Astronomy*, 53(1):3–10, March 1992.

[203] M. P. Francis, G. S. Gedeon, and B. C. Douglas. Perturbations of repeating groundtrack satellites by tesseral harmonics in the gravitational potential. *AIAA Journal*, 4:1281–1286, July 1966.

[204] L. Froideval and B. Schutz. Long arc analysis of GPS orbits (AAS 06-142). In S. R. Vadali, L. A. Cangahuala, P. W. Schumacher Jr. and J. J. Guzman, editors, *Spaceflight Mechanics 2006*, P.O. Box 28130, San Diego, California 92198, volume 124 of *Advances in the Astronautical Sciences*, pages 653–664. American Astronautical Society, Univelt, Inc., 2006.

[205] C. de la Fuente Marcos and R. de la Fuente Marcos. Asteroid 2014 OL_{339}: yet another Earth quasi-satellite. *Monthly Notices of the Royal Astronomical Society*, 445:2985–2994, December 2014.

[206] T. Fukushima. New Canonical Variables for Orbital and Rotational Motions. In H. Kinoshita and H. Nakai, editors, *25th Symposium on Celestial Mechanics*, pages 100–122, 1992.

[207] T. Fukushima. New canonical variables for orbital and rotational motions. *Celestial Mechanics and Dynamical Astronomy*, 60:57–68, September 1994.

[208] T. Fukushima. Efficient integration of torque-free rotation by energy scaling method. In A. Brzeziński, N. Capitaine and B. Kolaczek, editors, *Proceedings of the Journées Systèmes de Référence Spatio-Temporels 2005*, pages 101–104. Space Research Centre PAS, Warsaw, Poland, October 2006.

[209] L. P. Fulcher and B. F. Davis. Theoretical and experimental study of the motion of the simple pendulum. *American Journal of Physics*, 44:51–55, January 1976.

[210] F. Gachet, A. Celletti, G. Pucacco, and C. Efthymiopoulos. Geostationary secular dynamics revisited: application to high area-to-mass ratio objects. *Celestial Mechanics and Dynamical Astronomy*, 128:149–181, June 2017.

[211] G. Gaias, J.-S. Ardaens, and O. Montenbruck. Model of J_2 perturbed satellite relative motion with time-varying differential drag. *Celestial Mechanics and Dynamical Astronomy*, 123:411–433, December 2015.

[212] G. Gaias, C. Colombo, and M. Lara. Analytical Framework for Precise Relative Motion in Low Earth Orbits. *Journal of Guidance Control Dynamics*, 43(5):915–927, March 2020.

[213] G. Gaias, M. Lara, and C. Colombo. Accurate Osculating/Mean Orbital Elements Conversions for Spaceborne Formation Flying. In *Proceedings of the 27th International Symposium on Space Flight Dynamics*, Melbourne, Australia, pages 1–14. ISSFD, February 2019.

[214] E. M. Gaposchkin. Calculation of satellite drag coefficients. NASA STI/Recon Technical Report N, MIT Lincoln Laboratory, Lexinton, Massachusetts July 1994.

[215] D. García Yárnoz, D. J. Scheeres, and C. R. McInnes. On the "a" and "g" families of orbits in the Hill problem with solar radiation pressure and their application to asteroid orbiters. *Celestial Mechanics and Dynamical Astronomy*, 121:365–384, April 2015.

[216] B. Garfinkel. On the motion of a satellite of an oblate planet. *The Astronomical Journal*, 63(1257):88–96, March 1958.

[217] B. Garfinkel. The orbit of a satellite of an oblate planet. *The Astronomical Journal*, 64(9):353–367, November 1959.

[218] B. Garfinkel. On the motion of a satellite in the vicinity of the critical inclination. *The Astronomical Journal*, 65:624–627, December 1960.

[219] B. Garfinkel. An improved theory of motion of an artificial satellite. *The Astronomical Journal*, 69(3):223–229, April 1964.

[220] B. Garfinkel. Tesseral harmonic perturbations of an artificial satellite. *The Astronomical Journal*, 70:784–786, December 1965.

[221] B. Garfinkel. The disturbing function for an artificial satellite. *The Astronomical Journal*, 70:699–704, November 1965.

[222] B. Garfinkel. Formal solution in the problem of small divisors. *The Astronomical Journal*, 71(8):657–669, October 1966.

[223] B. Garfinkel. A Theory of Libration. *Celestial Mechanics*, 13(2):229–246, March 1976.

[224] B. Garfinkel. Comments on 'About an unsuspected integrable problem' by F. Mignard and M. Henon. *Celestial Mechanics*, 35(4):343–344, April 1985.

[225] B. Garfinkel and K. Aksnes. Spherical Coordinate Intermediaries for an Artificial Satellite. *The Astronomical Journal*, 75(1):85–91, February 1970.

[226] R. L. Garwin. Solar sailing: a practical method of propulsion within the Solar System. *Jet Propulsion*, 28:188–190, 1958.

[227] C. F. Gauss. *Theoria Motus Corporum Coelestium in Sectionibus Conicis Solem Ambientiun*. Frid. Perthes et I.H. Besser, Hamburg, 1809. Reprint: Cambridge University Press, 2001. English translation by C.H. Davis, Boston, 1857, reprinted by Dover Publications, 1963.

[228] G. S. Gedeon. Tesseral Resonance Effects on Satellite Orbits. *Celestial Mechanics*, 1:167–189, June 1969.

[229] G. E. O. Giacaglia. The influence of high-order zonal harmonics on the motion of an artificial satellite without drag. *The Astronomical Journal*, 69:303–308, May 1964.

[230] G. E. O. Giacaglia. Lunar Perturbations of Artificial Satellites of the Earth. *Celestial Mechanics*, 9:239–267, April 1974.

[231] G. E. O. Giacaglia. A note on Hansen's coefficients in satellite theory. *Celestial Mechanics*, 14:515–523, December 1976.

[232] G. E. O. Giacaglia. A note on the inclination functions of satellite theory. *Celestial Mechanics*, 13:503–509, June 1976.

[233] G. E. O. Giacaglia. Transformations of spherical harmonics and applications to geodesy and satellite theory. *Studia Geophysica et Geodaetica*, 24:1–11, March 1980.

[234] G. E. O. Giacaglia. Hansen Coefficients and Generalized Spherical Harmonics. *Publications of the Astronomical Society of Japan*, 39:171–178, 1987.

[235] P. J. S. Gil and J. Schwartz. Simulations of Quasi-Satellite Orbits Around Phobos. *Journal of Guidance Control Dynamics*, 33:901–914, May 2010.

[236] A. Giorgilli. A computer program for integrals of motion. *Computer Physics Communications*, 16:331–343, 1979.

[237] A. Giorgilli, A. Delshams, E. Fontich, L. Galgani, and C. Simó. Effective stability for a Hamiltonian system near an elliptic equilibrium point, with an application to the restricted three body problem. *Journal of Differential Equations*, 77:167–198, 1989.

[238] A. Giorgilli and L. Galgani. Formal integrals for an autonomous Hamiltonian system near an equilibrium point. *Celestial Mechanics*, 17:267–280, April 1978.

[239] A. Giorgilli and L. Galgani. Rigorous estimates for the series expansions of Hamiltonian perturbation theory. *Celestial Mechanics*, 37:95–112, 1985.

[240] I. Gkolias, J. Daquin, F. Gachet, and A. J. Rosengren. From Order to Chaos in Earth Satellite Orbits. *The Astronomical Journal*, 152(5):119, November 2016.

[241] H. Goldstein. Prehistory of the "Runge-Lenz" vector. *American Journal of Physics*, 43(8):737–738, August 1975.

[242] H. Goldstein. More on the prehistory of the Laplace or Runge-Lenz vector. *American Journal of Physics*, 44(11):1123–1124, November 1976.

[243] H. Goldstein, C. P. Poole, and J. L. Safko. *Classical Mechanics*. Addison-Wesley, New York, 3rd edition, 2001.

[244] A. R. Golikov. THEONA—a numerical-analytical theory of motion of artificial satellites of celestial bodies. *Cosmic Research*, 50(6):449–458, November 2012.

[245] V. V. Golubev. *Lectures on Integration of the Equations of Motion of a Rigid Body about a Fixed Point*. Israel Program for Scientific Translations. S. Monson, Jerusalem, 1960.

[246] G. Gómez, A. Jorba, J. Masdemont, and C. Simó. Study Refinement of Semi-Analytical Halo Orbit Theory. Technical Report Contract 8625/89/D/MD(SC), European Space Operations Center, Robert-Bosch-Strasse 5, 64293 Darmstadt, Germany, 1991.

[247] G. Gómez, M. Lara, and R. Russell. A dynamical systems approach to the design of the science orbit around Europa. In *Paper ISTS 2006-d-02, 19th International Symposium on Space Flight Dynamics*, Kanazawa, Japan, June 4–11, 2006, pages 1–6. ISSFD, February 2006.

[248] G. Gómez, J. Llibre, R. Martínez, and C. Simó. *Dynamics and Mission Design Near Libration Points. Vol. I Fundamentals: The Case of Collinear Libration Points*, volume 2 of *World Scientific Monograph Series in Mathematics*. World Scientific, Singapore, 2001.

[249] G. Gómez, M. W. Lo, and J. J. Masdemont. *Libration Point Orbits and Applications*. World Scientific Publishing Co. Pte. Ltd., 5 Toh Tuck Link, Singapore 596224, 2003.

[250] G. Gómez, M. Marcote, and J. M. Mondelo. The invariant manifold structure of the spatial Hill's problem. *Dynamical Systems*, 20(1):115–147, 2005.

[251] R. H. Gooding and C. A. Wagner. On the inclination functions and a rapid stable procedure for their evaluation together with derivatives. *Celestial Mechanics and Dynamical Astronomy*, 101:247–272, July 2008.

[252] R. A. Gordon, G. D. Mistretta, and J. S. Watson. A Comparison of Classical Analytic Theories for the Motion of Artificial Satellites. *Journal of Guidance Control Dynamics*, 2(3):184–189, May 1979.

[253] A. J. Green. *Orbit determination and prediction processes for low altitude satellites*. PhD thesis, Dept. of Aeronautics and Astronautics, Massachusetts Institute of Technology, 77 Massachusetts Ave, Cambridge, MA February 1980.

[254] W. Gröbner. *Die Lie-Reihen und Ihre Anwendungen*, volume 3 of *Mathematische Monographien*. Deutscher Verlag der Wissenschaften, Berlin, 1960.

[255] J. Guckenheimer and P. J. Holmes. *Nonlinear Oscillations, Dynamical Systems, and Bifurcations of Vector Fields*, volume 42 of *Applied Mathematical Sciences*. Springer-Verlag, New York, 1983.

[256] J. R. Guinn. Periodic gravitational perturbations for conversion between osculating and mean orbit elements (AAS 91-430). In B. Kaufman, K. T. Alfriend, R. L. Roehrich and R. R. Dasenbrock, editors, *AAS/AIAA Astrodynamics Conference 1991*, P.O. Box 28130, San Diego, California 92198, volume 76 of *Advances in the Astronautical Sciences*, pages 1–25. American Astronautical Society, Univelt, Inc., 1991.

[257] P. Gurfil and N. J. Kasdin. Canonical modelling of coorbital motion in Hill's problem using epicyclic orbital elements. *Astronomy and Astrophysics*, 409:1135–1140, October 2003.

[258] P. Gurfil and M. Lara. Motion near frozen orbits as a means for mitigating satellite relative drift. *Celestial Mechanics and Dynamical Astronomy*, 116(3):213–227, July 2013.

[259] P. Gurfil and M. Lara. Satellite onboard orbit propagation using Deprit's radial intermediary. *Celestial Mechanics and Dynamical Astronomy*, 120(2):217–232, October 2014.

[260] M. C. Gutzwiller. Moon-Earth-Sun: The oldest three-body problem. *Reviews of Modern Physics*, 70(2):589–639, April 1998.

[261] Y. Hagihara. Libration of an Earth Satellite with Critical Inclination. *Smithsonian Contributions to Astrophysics*, 5(5):39–51, 1961.

[262] D. P. Hamilton and A. V. Krivov. Circumplanetary Dust Dynamics: Effects of Solar Gravity, Radiation Pressure, Planetary Oblateness, and Electromagnetism. *Icarus*, 123(2):503–523, October 1996.

[263] P. A. Hansen. Expansions of the product of a power of the radius vector with the sinus or cosinus of a multiple of the true anomaly in terms of series containing the sinuses or cosinuses of the multiples of the true, eccentric or mean anomaly. *Abhandlungen der Koniglich Sachsischen Gesellschaft der Wissenschaften*, 2(3):183–281, 1855. English translation by J. C. Van der Ha, ESA/ESOC, Darmstadt, Germany, 1977.

[264] R. S. Harrington. Dynamical evolution of triple stars. *Astronomical Journal*, 73:190–194, April 1968.

[265] I. Harris and W. Priester. Theoretical Models for the Solar-Cycle Variation of the Upper Atmosphere. *Journal of Geophysical Research*, 67(12):4585–4591, November 1962.

[266] I. Harris and W. Priester. Time-Dependent Structure of the Upper Atmosphere. *Journal of Atmospheric Sciences*, 19:286–301, July 1962.

[267] D. Hautesserres. Extrapolation long terme de l'orbite du satellite SimbolX par la methode de Gragg-Bulirsch-Stoer (GBS). Technical Report DCT/SB/OR/2009-2474, Centre National d'Études Spatiales, 18, avenue Edouard Belin – 31401 Toulouse Cedex 9, France, January 2009.

[268] D. Hautesserres and M. Lara. A fast and efficient algorithm for onboard LEO intermediary propagation. In *Proceedings of the 6th International Conference on Astrodynamics Tools and Techniques, ICATT* ESA, 2016.

[269] D. Hautesserres and M. Lara. Intermediary LEO propagation including higher order zonal harmonics. *Celestial Mechanics and Dynamical Astronomy*, 127:505–526, April 2017.

[270] L. Healy and E. Deprit. Paint by number: Uncovering phase flows of an integrable dynamical system. *Computers in Physics*, 5(5):491–496, 1991.

[271] L. M. Healy. The Main Problem in Satellite Theory Revisited. *Celestial Mechanics and Dynamical Astronomy*, 76(2):79–120, 2000.

[272] L. M. Healy and J. J. Travisano. Automatic rendering of astrodynamics expressions for efficient evaluation. *Journal of the Astronautical Sciences*, 46(1):65–81, 1998.

[273] M. Hénon. Exploration numérique du problème restreint. I. Masses égales ; orbites périodiques. *Annales d'Astrophysique*, February 28:499–511, 1965.

[274] M. Hénon. Exploration numérique du problème restreint. II. Masses égales, stabilité des orbites périodiques. *Annales d'Astrophysique*, 28:992–1007, February 1965.

[275] M. Hénon. Numerical Exploration of the Restricted Problem, V. Hill's Case: Periodic Orbits and their Stability. *Astronomy and Astrophysics*, 1:223–238, February 1969.

[276] M. Hénon. Numerical exploration of the restricted problem. VI. Hill's case: Non-periodic orbits. *Astronomy and Astrophysics*, 9:24–36, November 1970.

[277] M. Hénon. Vertical Stability of Periodic Orbits in the Restricted Problem. II. Hill's Case. *Astronomy and Astrophysics*, 30:317–321, January 1974.

[278] M. Hénon. *Generating Families in the Restricted Three-Body Problem*, volume 52 of *Lecture Notes in Physics Monographs*. Springer-Verlag, Berlin, Heidelberg, 1st edition, 1997.

[279] M. Hénon. New Families of Periodic Orbits in Hill's Problem of Three Bodies. *Celestial Mechanics and Dynamical Astronomy*, 85:223–246, March 2003.

[280] M. Hénon and C. Heiles. The applicability of the third integral of motion: Some numerical experiments. *The Astronomical Journal*, 69:73–79, February 1964.
[281] M. Hénon and J.-M. Petit. Series expansion for encounter-type solutions of Hill's problem. *Celestial Mechanics*, 38:67–100, January 1986.
[282] J. Henrard. On a perturbation theory using Lie transforms. *Celestial Mechanics*, 3:107–120, March 1970.
[283] J. Henrard. Periodic Orbits Emanating from a Resonant Equilibrium. *Celestial Mechanics*, 1:437–466, September 1970.
[284] J. Henrard. Virtual singularities in the artificial satellite theory. *Celestial Mechanics*, 10(4):437–449, December 1974.
[285] J. Henrard. Analytical drag theory of an artificial satellite with small eccentricity In K. B. Bhatnagar, editor, *Space Dynamics and Celestial Mechanics. Proceedings of the International Workshop held in Delhi, India*, 14–16 November 1985, volume 127 of *Astrophysics and Space Science Library*, pages 261–272. D. Reidel Publishing Company, P.O. Box 17, 3300 AA Dordrecht, Holland, 1986.
[286] J. Henrard. A survey of Poisson series processors. *Celestial Mechanics*, 45:245–253, March 1988.
[287] J. Henrard and M. Moons. Hamiltonian Theory of the Libration of the Moon. In V. G. Szebehely, editor, *Dynamics of planets and satellites and theories of their motion*, volume 72 of *Astrophysics and Space Science Library*, pages 125–135. Proceedings of the International Astronomical Union colloquium no. 41. D. Reidel Publishing Company, Dordrecht: Holland/Boston: U.S.A., 1978.
[288] J. Henrard and J. Roels. Equivalence for Lie Transforms. *Celestial Mechanics*, 10:497–512, December 1974.
[289] P. Herget and P. Musen. The calculation of literal expansions. *The Astronomical Journal*, 64:11–19, February 1959.
[290] S. Herrick. A modification of the "Variation of Constants" method for special perturbations. *Publications of the Astronomical Society of the Pacific*, 60:321–323, October 1948.
[291] S. Herrick. Icarus and the variation of parameters. *The Astronomical Journal*, 58:156–164, August 1953.
[292] G. W. Hill. Researches in the Lunar Theory. *American Journal of Mathematics*, 1:5–26, 1878.
[293] G. W. Hill. Motion of a system of material points under the action of gravitation. *The Astronomical Journal*, 27:171–182, April 1913.
[294] G. Hintz. Survey of Orbit Element Sets. *Journal of Guidance, Control, and Dynamics*, 31(3):785–790, May-June 2008.
[295] D. L. Hitzl and J. V. Breakwell. Resonant and non-resonant gravity-gradient perturbations of a tumbling tri-axial satellite. *Celestial Mechanics*, 3:346–383, September 1971.
[296] F. R. Hoots. Reformulation of the Brouwer geopotential theory for improved computational efficiency. *Celestial Mechanics*, 24(2):367–375, August 1981.
[297] F. R. Hoots and R. G. France. An analytic satellite theory using gravity and a dynamic atmosphere. *Celestial Mechanics*, 40:1–18, 1987.
[298] F. R. Hoots and R. L. Roehrich. Models for Propagation of the NORAD Element Sets. Project SPACETRACK, Rept. 3, U.S. Air Force Aerospace Defense Command, Colorado Springs, CO, December 1980.
[299] F. R. Hoots, P. W. Schumacher Jr., and R. A. Glover. History of Analytical Orbit Modeling in the U. S. Space Surveillance System. *Journal of Guidance, Control, and Dynamics*, 27(5):174–185, March 2004.
[300] H. Hopf. Über die Abbildungen der dreidimensionalen Sphäre auf die Kugelfläche. *Mathematische Annalen*, 104:637–665, 1931.

[301] G.-I. Hori. The motion of an artificial satellite in the vicinity of the critical inclination. *The Astronomical Journal*, 65:291–300, June 1960.

[302] G.-I. Hori. A new approach to the solution of the main problem of the lunar theory. *The Astronomical Journal*, 68:125–146, April 1963.

[303] G.-I. Hori. Theory of General Perturbation with Unspecified Canonical Variables. *Publications of the Astronomical Society of Japan*, 18(4):287–296, 1966.

[304] G.-I. Hori. Theory of General Perturbations for Non-Canonical Systems. *Publications of the Astronomical Society of Japan*, 23:567–587, 1971.

[305] M. E. Hough. Sun-synchronous orbits near critical inclination. *Celestial Mechanics*, 25(2):137–157, October 1981.

[306] S. Hughes. Satellite orbits perturbed by direct solar radiation pressure - General expansion of the disturbing function. *Planetary and Space Science*, 25:809–815, September 1977.

[307] S. Hughes. Earth satellite orbits with resonant lunisolar perturbations. I. Resonances dependent only on inclination. *Proceedings of the Royal Society of London Series A*, 372:243–264, August 1749. 1980.

[308] S. Hughes. Earth satellite orbits with resonant lunisolar perturbations. II - Some resonances dependent on the semi-major axis, eccentricity and inclination. *Proceedings of the Royal Society of London Series A*, 375:379–396, March 1981.

[309] D. Ineichen, G. Beutler, and U. Hugentobler. Sensitivity of GPS and GLONASS orbits with respect to resonant geopotential parameters. *Journal of Geodesy*, 77(7):478–486, October 2003.

[310] K. A. Innanen and F. C. House. The Existence of the Third Integral for a Family of High-Velocity Stars. *Astrophysics and Space Science*, 7:139–150, April 1970.

[311] K. A. Innanen, J. Q. Zheng, S. Mikkola, and M. J. Valtonen. The Kozai Mechanism and the Stability of Planetary Orbits in Binary Star Systems. *Astronomical Journal*, 113:1915–1919, May 1997.

[312] M. Irigoyen and C. Simó. Non integrability of the J_2 problem. *Celestial Mechanics and Dynamical Astronomy*, 55(3):281–287, March 1993.

[313] T. Ito and K. Ohtsuka. The Lidov-Kozai Oscillation and Hugo von Zeipel. *Monographs on Environment, Earth and Planets*, 7(1):1–113, November 2019.

[314] J. Ivory. On the Figure Requisite to Maintain the Equilibrium of a Homogeneous Fluid Mass That Revolves Upon an Axis. *Philosophical Transactions of the Royal Society of London, Series I*, 114:85–150, January 1824.

[315] I. G. Izsak. A Theory of Satellite Motion about an Oblate Planet. I. A Second-Order Solution of Vinti's Dynamical Problem. *SAO Special Report*, 52, November 1960. Reprinted in: Smithsonian Contributions to Astrophysics, vol. 6, 1963, pp. 81–107.

[316] I. G. Izsak. On the Critical Inclination in Satellite Theory. *SAO Special Report*, 90, March 1962.

[317] I. G. Izsak. A note on perturbation theory. *The Astronomical Journal*, 68(8):559–561, October 1963.

[318] L. G. Jacchia. Revised Static Models of the Thermosphere and Exosphere with Empirical Temperature Profiles. *SAO Special Report*, 332, May 1971.

[319] J. Jackson. Retrograde satellite orbits. *Monthly Notices of the Royal Astronomical Society*, 74:62–82, December 1913.

[320] C. G. J. Jacobi. Ueber eine besondere Gattung algebraischer Functionen, die aus der Entwicklung der Function $(1 - 2xz + z^2)^{1/2}$ entstehen. *Journal für die reine und angewandte Mathematik (Crelles Journal)*, 1827(2):223–226, January 1827.

[321] C. G. J. Jacobi. Sur l'élimination des noeuds dans le problème des trois corps. *Astronomische Nachrichten*, 20:81–98, December 1842.

[322] C. G. J. Jacobi. Sur la rotation d'un corps. *Comptes Rendus de l'Académie des Sciences*, 24:97–106, 1849.

[323] W. H. Jefferys. A FORTRAN-based list processor for Poisson series. *Celestial Mechanics*, 2(4):474–480, December 1970.

[324] W. H. Jefferys. Automated Closed Form Integration of Formulas in Elliptic Motion. *Celestial Mechanics*, 3:390–394, September 1971.

[325] W. H. Jefferys. New treatment of the critical argument in resonance problems. *The Astronomical Journal*, 81:132–134, February 1976.

[326] D. J. Jezewski. A noncanonical analytic solution to the J_2 perturbed two-body problem. *Celestial Mechanics*, 30(4):343–361, August 1983.

[327] D. J. Jezewski. An analytic solution for the J_2 perturbed equatorial orbit. *Celestial Mechanics*, 30(4):363–371, August 1983.

[328] À. Jorba. A methodology for the numerical computation of normal forms, centre manifolds and first integrals of Hamiltonian systems. *Experimental Mathematics*, 8(2):155–195, 1999.

[329] À. Jorba and J. J. Masdemont. Dynamics in the center manifold of the collinear points of the restricted three body problem. *Physica D Nonlinear Phenomena*, 132:189–213, July 1999.

[330] A. H. Jupp. The problem of the critical inclination revisited. *Celestial Mechanics*, 11(3):361–378, May 1975.

[331] A. H. Jupp. The critical inclination problem - 30 years of progress. *Celestial Mechanics*, 43(1–4):127–138, 1988.

[332] W. Kahan. Lecture Notes on the Status of IEEE Standard 754 for Binary Floating-Point Arithmetic. Technical report, Electrical Engineering and Computer Science, University of California, Berkeley, CA, October 1997.

[333] A. A. Kamel. Perturbation Method in the Theory of Nonlinear Oscillations. *Celestial Mechanics*, 3:90–106, March 1970.

[334] A. A. Kamel, D. Ekman, and R. Tibbitts. East-West Stationkeeping Requirements of nearly Synchronous Satellites Due to Earth's Triaxiality and Luni-Solar Effects. *Celestial Mechanics*, 8:129–148, August 1973.

[335] J-P. R. Kaniecki. Short periodic variations in the first-order semianalytical satellite theory. Master thesis, Department of Aeronautics and Astronautics, Massachusetts Institute of Technology, 77 Massachusetts Ave, Cambridge, MA, September 1979.

[336] N. J. Kasdin, P. Gurfil, and E. Kolemen. Canonical Modelling of Relative Spacecraft Motion Via Epicyclic Orbital Elements. *Celestial Mechanics and Dynamical Astronomy*, 92:337–370, August 2005.

[337] B. Kaufman. First order semianalytic satellite theory with recovery of the short period terms due to third body and zonal perturbations. *Acta Astronautica*, 8(5–6):611–623, 1981.

[338] B. Kaufman and R. Dasenbrock. Higher Order Theory for Long-Term Behavior of Earth and Lunar Orbiters. Technical Report AD-754 738, Naval Research Laboratory, Washington, D.C., Mathematics and Information Sciences Division, December 1972.

[339] W. M. Kaula. Analysis of Gravitational and Geometric Aspects of Geodetic Utilization of Satellites. *Geophysical Journal*, 5:104–133, July 1961.

[340] W. M. Kaula. Development of the lunar and solar disturbing functions for a close satellite. *The Astronomical Journal*, 67:300–303, June 1962.

[341] W. M. Kaula. *Theory of satellite geodesy. Applications of satellites to geodesy*. Blaisdell, Waltham, Massachusetts, 1966. Reprint: Dover, Mineola, New York, 2000.

[342] T. S. Kelly. A note on first-order normalizations of perturbed Keplerian systems. *Celestial Mechanics and Dynamical Astronomy*, 46:19–25, March 1989.

[343] J. Kevorkian. Uniformly valid asymptotic representation for all times of the motion of a satellite in the vicinity of the smaller body in the restricted three-body problem. *The Astronomical Journal*, 67:204–211, May 1962.

[344] D. King-Hele. *Theory of Satellite Orbits in an Atmosphere. Butterworths Mathematical Texts*. Butterworths, London, 1964.

[345] H. Kinoshita. First-Order Perturbations of the Two Finite Body Problem. *Publications of the Astronomical Society of Japan*, 24:423–457, July 1972.
[346] H. Kinoshita. Third-Order Solution of an Artificial-Satellite Theory. *SAO Special Report*, 379, July 1977.
[347] H. Kinoshita. Analytical expansions of torque-free motions for short and long axis modes. *Celestial Mechanics and Dynamical Astronomy*, 53(4):365–375, December 1992.
[348] H. Kinoshita, G. Hori, and H. Nakai. Modified Jacobi polynomial and its application to expansions of disturbing functions. *Annals of the Tokyo Astronomical Observatory*, 14:14–35, January 1974.
[349] H. Kinoshita and H. Nakai. Analytical Solution of the Kozai Resonance and its Application. *Celestial Mechanics and Dynamical Astronomy*, 75(2):125–147, October 1999.
[350] H. Kinoshita and H. Nakai. General solution of the Kozai mechanism. *Celestial Mechanics and Dynamical Astronomy*, 98(1):67–74, May 2007.
[351] Z. Knežević and A. Milani. Orbit maintenance of a lunar polar orbiter. *Planetary and Space Science*, 46(11–12):1605–1611, December 1998.
[352] A. Y. Kogan. Distant satellite orbits in the restricted circular three-body problem. *Cosmic Research*, 26:705–710, May 1989.
[353] N. A. Kolmogorov. On the conservation of quasi-periodic motions for a small change in the Hamiltonian function. *Doklady Akademii Nauk SSSR*, 98:527–530, 1954. (in Russian).
[354] A. S. Konopliv. A Perturbation Method and Some Applications. *Celestial Mechanics and Dynamical Astronomy*, 47:305–320, 1990.
[355] A. S. Konopliv. A Third Order of J_2 Solution with a Transformed Time. Interoffice Memorandum IOF 314.3 - 970, Jet Propulsion Laboratory, 4800 Oak Grove Dr, Pasadena, CA 91109, USA, March 1991.
[356] P.-V. Koseleff. Comparison Between Deprit and Dragt-Finn Perturbation Methods. *Celestial Mechanics and Dynamical Astronomy*, 58:17–36, January 1994.
[357] J. Kovalevsky. Sur la théorie du mouvement d'un satellite à fortes inclinaison et excentricité. In G. I. Kontopoulos, editor, *The Theory of Orbits in the Solar System and in Stellar Systems*, volume 25 of *IAU Symposium*, pages 326–344, January 1966.
[358] J. Kovalevsky. *Introduction to celestial mechanics*, volume 7 of *Astrophysics and Space Science Library*. Springer Netherlands, Dordrecht, Holland, 1967.
[359] Y. Kozai. On the Effects of the Sun and the Moon upon the Motion of a Close Earth Satellite. *SAO Special Report*, 22, March 1959.
[360] Y. Kozai. The motion of a close earth satellite. *The Astronomical Journal*, 64:367–377, November 1959.
[361] Y. Kozai. Effects of Solar Radiation Pressure on the Motion of an Artificial Satellite. *SAO Special Report*, 56:25–34, January 1961.
[362] Y. Kozai. Mean values of cosine functions in elliptic motion. *The Astronomical Journal*, 67:311–312, June 1962.
[363] Y. Kozai. Second-order solution of artificial satellite theory without air drag. *The Astronomical Journal*, 67:446–461, September 1962.
[364] Y. Kozai. Secular perturbations of asteroids with high inclination and eccentricity. *The Astronomical Journal*, 67:591–598, November 1962.
[365] Y. Kozai. Motion of a Lunar Orbiter. *Publications of the Astronomical Society of Japan*, 15:301–312, 1963.
[366] Y. Kozai. Lunisolar Perturbations with Short Periods. SAO Special Report, 235, December 1966.
[367] Y. Kozai. A New Method to Compute Lunisolar Perturbations in Satellite Motions. *SAO Special Report*, 349, February 1973.

[368] H. A. Kramers. Über den Einfluß eines elektrischen Feldes auf die Feinstruktur der Wasserstofflinien. *Zeitschrift für Physik*, 3(4):199–223, July 1920.

[369] A. V. Krivov and J. Getino. Orbital evolution of high-altitude balloon satellites. *Astronomy and Astrophysics*, 318:308–314, February 1997.

[370] A. V. Krivov, L. L. Sokolov, and V. V. Dikarev. Dynamics of Mars-Orbitting Dust: Effects of Light Pressure and Planetary Oblateness. *Celestial Mechanics and Dynamical Astronomy*, 63:313–339, January 1996.

[371] V. W. Kudielka. Balanced Earth Satellite Orbits. *Celestial Mechanics and Dynamical Astronomy*, 60(4):455–470, December 1994.

[372] V. W. Kudielka. Equilibria Bifurcations of Satellite Orbits. In R. Dvorak and J. Henrard, editors, *The Dynamical Behaviour of our Planetary System*, page 243–255, 1997.

[373] M. Kummer. On resonant non linearly coupled oscillators with two equal frequencies. *Communications in Mathematical Physics*, 48:53–79, February 1976. Erratum: *Communications in Mathematical Physics* 60:192, 1978.

[374] A. L. Kutuzov. Application of symbolic computer operations for solving the main problem in the theory of satellite motion. *Soviet Astronomy Letters*, 1:42–44, February 1975.

[375] P. Lála and L. Sehnal. The Earth's shadowing effects in the short-periodic perturbations of satellite orbits. *Bulletin of the Astronomical Institutes of Czechoslovakia*, 20:327–330, January 1969.

[376] T. Lam and G. J. Whiffen. Exploration of Distant Retrograde Orbits Around Europa (AAS 05-110). In D. A. Vallado, M. J. Gabor and P. N. Desai, editors, *AAS/AIAA Spaceflight Mechanics Meeting 2005*, P.O. Box 28130, San Diego, California 92198, Jan. 23–27, 2005, volume 120 of *Advances in the Astronautical Sciences*, pages 135–153. American Astronautical Society, Univelt, Inc., 2005.

[377] L. D. Landau and E. M. Lifshitz. *Mechanics*. Butterworth-Heinemann, Oxford, UK, and Burlington, MA, USA, 3rd edition, 1976.

[378] M. H. Lane. The development of an artificial satellite theory using a power-law atmospheric density representation. In *Proceedings of the 2nd Aerospace Sciences Meeting*, pages 1–16. American Institute of Aeronautics and Astronautics, January 1965.

[379] M. H. Lane and K. Cranford. An improved analytical drag theory for the artificial satellite problem. In *AIAA/AAS Astrodynamics Conference*, Princeton, New Jersey, pages 1–12. American Institute of Aeronautics and Astronautics, USA, August 1989. AIAA Paper No. 69-925.

[380] M. H. Lane and F. R. Hoots. General perturbations theories derived from the 1965 Lane drag theory. Special Astrodynamic Report SPACETRACK REPORT NO. 2, Aerospace Defense Command, US Air Force, Peterson AFB, CO, 80914, December 1979.

[381] M. T. Lane. An Analytical Treatment of Resonance Effects on Satellite Orbits. *Celestial Mechanics*, 42:3–38, March 1987.

[382] M. T. Lane. On Analytic Modeling of Lunar Perturbations of Artificial Satellites of the Earth. *Celestial Mechanics and Dynamical Astronomy*, 46(4):287–305, December 1989.

[383] D. Lantukh, R. P. Russell, and S. Broschart. Heliotropic orbits at oblate asteroids: balancing solar radiation pressure and J_2 perturbations. *Celestial Mechanics and Dynamical Astronomy*, 121:171–190, February 2015.

[384] P. S. Laplace. *Traité de Mécanique Céleste*, volume 4. Chez Courcier, Paris, 1805. Livre viii, §37.

[385] M. Lara. On numerical continuation of families of periodic orbits in a parametric potential. *Mechanics Research Communications*, 23(3):291–298, 1996.

[386] M. Lara. SADSaM: A Software Assistant for Designing SAtellite Missions. Technical Report DTS/MPI/MS/MN/99-053, Centre National d'Études Spatiales, 18, avenue Edouard Belin - 31401 Toulouse Cedex 9, France, May 1999.

[387] M. Lara. Searching for Repeating Ground Track Orbits: A Systematic Approach. *The Journal of the Astronautical Sciences*, 47(3–4):177–188, 1999.
[388] M. Lara. Repeat Ground Track Orbits of the Earth Tesseral Problem as Bifurcations of the Equatorial Family of Periodic Orbits. *Celestial Mechanics and Dynamical Astronomy*, 86(2):143–162, June 2003.
[389] M. Lara. Simplified Equations for Computing Science Orbits Around Planetary Satellites. *Journal of Guidance Control Dynamics*, 31(1):172–181, January 2008.
[390] M. Lara. Design of long-lifetime lunar orbits: A hybrid approach. *Acta Astronautica*, 69(3–4):186–199, August 2011.
[391] M. Lara. Short-axis-mode rotation of a free rigid body by perturbation series. *Celestial Mechanics and Dynamical Astronomy*, 118(3):221–234, 2014.
[392] M. Lara. Efficient Formulation of the Periodic Corrections in Brouwer's Gravity Solution. *Mathematical Problems in Engineering*, 2015:980652, 2015.
[393] M. Lara. LEO intermediary propagation as a feasible alternative to Brouwer's gravity solution. *Advances in Space Research*, 56(3):367–376, August 2015.
[394] M. Lara. On inclination resonances in Artificial Satellite Theory. *Acta Astronautica*, 110:239–246, May 2015.
[395] M. Lara. Analytical and Semianalytical Propagation of Space Orbits: The Role of Polar-Nodal Variables. In G. Gómez and J. J. Masdemont, editors, *Astrodynamics Network AstroNet-II: The Final Conference*, volume 44 of *Astrophysics and Space Science Proceedings*, pages 151–166. Springer, Cham, 2016.
[396] M. Lara. A Hopf variables view on the libration points dynamics. *Celestial Mechanics and Dynamical Astronomy*, 129(3):285–306, November 2017.
[397] M. Lara. Note on the ideal frame formulation. *Celestial Mechanics and Dynamical Astronomy*, 129:137–151, September 2017.
[398] M. Lara. A fast an efficient algorithm for the computation of distant retrograde orbits. In G. Ortega et al., editors, *Proceedings of the 7th International Conference on Astrodynamics Tools and Techniques, ICATT*, pages 1–10. ESA, November 2018.
[399] M. Lara. Complex variables approach to the short-axis-mode rotation of a rigid body. *Applied Mathematics and Nonlinear Sciences*, 3(2):537–552, 2018.
[400] M. Lara. Exploring Sensitivity of Orbital Dynamics with Respect to Model Truncation: The Frozen Orbits Approach. In M. Vasile, E. Minisci, L. Summerer and P. McGinty, editors, *Stardust Final Conference*, volume 52 of *Astrophysics and Space Science Proceedings*, pages 69–83. Springer, Cham, 2018.
[401] M. Lara. Nonlinear librations of distant retrograde orbits: a perturbative approach—the Hill problem case. *Nonlinear Dynamics*, 93(4):2019–2038, April 2018.
[402] M. Lara. A new radial, natural, higher-order intermediary of the main problem four decades after the elimination of the parallax. *Celestial Mechanics and Dynamical Astronomy*, 131(9):42, September 2019.
[403] M. Lara. Design of distant retrograde orbits based on a higher order analytical solution. *Acta Astronautica*, 161:562–578, August 2019.
[404] M. Lara. Design of quasi-satellite orbits: Analytical alternatives. In *Proceedings of the 27th International Symposium on Space Flight Dynamics*, Melbourne, Australia, pages 1–12. ISSFD, February 2019.
[405] M. Lara. Review of analytical solutions for low earth orbit propagation and study of the precision improvement in the conversion of osculating to mean elements. Technical Report CM 2019/SER0023, Universidad de La Rioja, Logroño, La Rioja, September 2019.
[406] M. Lara. A perturbation solution to Vinti's dynamical problem. *Romanian Journal of Technical Sciences – Applied Mechanics*, 65(3):163–180, 2020.

[407] M. Lara. Brouwer's satellite solution redux. arXiv, eprint 2009.10665 [math.DS]:1–20, September 2020.

[408] M. Lara. Solution to the main problem of the artificial satellite by reverse normalization. *Nonlinear Dynamics*, 101(2):1501–1524, August 2020.

[409] M. Lara, A. Deprit, and A. Elipe. Numerical Continuation of Families of Frozen Orbits in the Zonal Problem of Artificial Satellite Theory. *Celestial Mechanics and Dynamical Astronomy*, 62(2):167–181, 1995.

[410] M. Lara and S. Ferrer. Computing Long Lifetime Orbits for the Europa Observation Mission. In *International Symposium on Space Technology and Science, ISTS 2006*, P.O. Box 28130, San Diego, California 92198, USA, volume 19 of *International Symposium on Space Flight Dynamics*, pages 1983–2002, 2006. ISTS.

[411] M. Lara and S. Ferrer. Computing long-lifetime science orbits around natural satellites. *Discrete and Continuous Dynamical Systems, Series S*, 1(2):293–302, 2008.

[412] M. Lara and S. Ferrer. Closed form perturbation solution of a fast rotating triaxial satellite under gravity-gradient torque. *Cosmic Research*, 51(4):289–303, July 2013.

[413] M. Lara and S. Ferrer. Expanding the simple pendulum's rotation solution in action-angle variables. *European Journal of Physics*, 36(5):055040, September 2015.

[414] M. Lara and S. Ferrer. Expanding the simple pendulum's rotation solution in action-angle variables. *ArXiv e-prints*, eprint 1503.03358 [nlin.SI], March 2015.

[415] M. Lara, T. Fukushima, and S. Ferrer. Ceres' rotation solution under the gravitational torque of the Sun. *Monthly Notices of the Royal Astronomical Society*, 415(1):461–469, July 2011.

[416] M. Lara and P. Gurfil. Integrable approximation of J_2-perturbed relative orbits. *Celestial Mechanics and Dynamical Astronomy*, 114(3):229–254, November 2012.

[417] M. Lara, R. López, I. Pérez, and J. F. San-Juan. Exploring the long-term dynamics of perturbed Keplerian motion in high degree potential fields. *Communications in Nonlinear Science and Numerical Simulation*, 82:105053, 2020.

[418] M. Lara and J. F. Palacián. Hill Problem Analytical Theory to the Order Four: Application to the Computation of Frozen Orbits around Planetary Satellites. *Mathematical Problems in Engineering, 2009(Article ID*, 753653:1–18, 2009.

[419] M. Lara, J. F. Palacián, and R. P. Russell. Mission design through averaging of perturbed Keplerian systems: the paradigm of an Enceladus orbiter. *Celestial Mechanics and Dynamical Astronomy*, 108(1):1–22, 2010.

[420] M. Lara, J. F. Palacián, P. Yanguas, and C. Corral. Long-Term Behavior of a Mercury Orbiter Perturbed by the Elliptic Motion of the Sun (AAS 08-236). In J. H. Seago, B. Neta, T. J. Eller and F. J. Pelletier, editors, *Spaceflight Mechanics 2008*, P.O. Box 28130, San Diego, California 92198, USA, volume 130 of *Advances in the Astronautical Sciences*, pages 1937–1951. American Astronautical Society, Univelt, Inc., 2008.

[421] M. Lara, J. F. Palacián, P. Yanguas, and C. Corral. Analytical theory for spacecraft motion about Mercury. *Acta Astronautica*, 66(7–8):1022–1038, April 2010.

[422] M. Lara and J. Peláez. On the numerical continuation of periodic orbits. An intrinsic, 3-dimensional, differential, predictor-corrector algorithm. *Astronomy and Astrophysics*, 389:692–701, July 2002.

[423] M. Lara, J. Peláez, and H. Urrutxua. Modifying the atlas of low lunar orbits using inert tethers. *Acta Astronautica*, 79:52–60, October 2012.

[424] M. Lara, I. Pérez, and R. López. Higher Order Approximation to the Hill Problem Dynamics about the Libration Points. *Communications in Nonlinear Science and Numerical Simulation*, 59:612–628, June 2018.

[425] M. Lara, A. J. Rosengren, and E. Fantino. Non-singular recursion formulas for third-body perturbations in mean vectorial elements. *Astronomy and Astrophysics*, 634:A61, February 2020.

[426] M. Lara and R. P. Russell. On the family "g" of the restricted three-body problem. *Monografías de la Real Academia de Ciencias de Zaragoza*, 30:51–66, 2006.
[427] M. Lara and R. P. Russell. Computation of a Science Orbit About Europa. *Journal of Guidance, Control, and Dynamics*, 30(1):259–263, 2007.
[428] M. Lara and R. P. Russell. Fast Design of Repeat Ground Track Orbits in High-Fidelity Geopotentials. *The Journal of the Astronautical Sciences*, 56(3):311–324, 2008.
[429] M. Lara, R. P. Russell, and B. F. Villac. Classification of the Distant Stability Regions at Europa. *Journal of Guidance Control Dynamics*, 30:409–418, March 2007.
[430] M. Lara, R. P. Russell, and B. F. Villac. Fast estimation of stable regions in real models. *Meccanica*, 42(5):511–515, October 2007.
[431] M. Lara, B. de Saedeleer, and S. Ferrer. Preliminary design of low lunar orbits. In *Proceedings of the 21st International Symposium on Space Flight Dynamics*, Toulouse, France, pages 1–15. ISSFD, October 2009.
[432] M. Lara and J. F. San-Juan. Dynamic Behavior of an Orbiter Around Europa. *Journal of Guidance, Control and Dynamics*, 28(2):291–297, March-April 2005.
[433] M. Lara, J. F. San-Juan, and S. Ferrer. Secular motion around synchronously orbiting planetary satellites. *Chaos: An Interdisciplinary Journal of Nonlinear Science*, 15(4):043101, 2005.
[434] M. Lara, J. F. San-Juan, Z. J. Folcik, and P. J. Cefola. Deep Resonant GPS-Dynamics Due to the Geopotential. *The Journal of the Astronautical Sciences*, 58(4):661–676, October 2011.
[435] M. Lara, J. F. San-Juan, and D. Hautesserres. Semi-analytical propagator of high eccentricity orbits. Technical Report R-S15/BS-0005-024, Centre National d'Études Spatiales, 18, avenue Edouard Belin - 31401 Toulouse Cedex 9, France, January 2016.
[436] M. Lara, J. F. San-Juan, and D. Hautesserres. HEOSAT: a mean elements orbit propagator program for highly elliptical orbits. *CEAS Space Journal*, 10:3–23, March 2018.
[437] M. Lara, J. F. San-Juan, L. M. López, and P. J. Cefola. On the third-body perturbations of high-altitude orbits. *Celestial Mechanics and Dynamical Astronomy*, 113:435–452, August 2012.
[438] M. Lara, J. F. San-Juan, and L. M. López-Ochoa. Averaging Tesseral Effects: Closed Form Relegation versus Expansions of Elliptic Motion. *Mathematical Problems in Engineering*, 2013:570127, 2013.
[439] M. Lara, J. F. San-Juan, and L. M. López-Ochoa. Precise analytical computation of frozen-eccentricity, low Earth orbits in a tesseral potential, *Mathematical Problems in Engineering*, 2013:191384, 2013.
[440] M. Lara, J. F. San-Juan, and L. M. López-Ochoa. Delaunay variables approach to the elimination of the perigee in Artificial Satellite Theory. *Celestial Mechanics and Dynamical Astronomy*, 120(1):39–56, September 2014.
[441] M. Lara, J. F. San-Juan, and L. M. López-Ochoa. Efficient semi-analytic integration of GNSS orbits under tesseral effects. *Acta Astronautica*, 102(0):355–366, 2014.
[442] M. Lara, J. F. San-Juan, and L. M. López-Ochoa. Proper Averaging Via Parallax Elimination (AAS 13-722). In S. B. Broschart, J. D. Turner, K. C. Howell and F. R. Hoots, editors, *Astrodynamics 2013*, P.O. Box 28130, San Diego, California 92198, USA, January 2014, volume 150 of *Advances in the Astronautical Sciences*, pages 315–331. American Astronautical Society, Univelt, Inc., 2014.
[443] M. Lara, J. F. San-Juan, L. M. López-Ochoa, and P. J. Cefola. Long-term evolution of Galileo operational orbits by canonical perturbation theory. *Acta Astronautica*, 94(2):646–655, 2014.
[444] M. Lara, R. Vilhena de Moraes, D. M. Sanchez, and A. F. Bertachini de Almeida Prado. Efficient computation of short-period analytical corrections due to third-body effects (AAS 15-295). In R. Furfaro, S. Casotto, A. Trask and S. Zimmer, editors, *AAS/AIAA Spaceflight Mechanics Meeting 2015*, P.O. Box 28130, San Diego, California 92198, USA, volume 155 of *Advances in the Astronautical Sciences*, pages 437–455. American Astronautical Society, Univelt, Inc., 2015.

[445] W. Larson and J. R. Wertz, editors. In *Space Mission Analysis and Design*, volume 2 of *Space Technology Library*, Springer Netherlands, Dordrecht, Holland, 1st edition, 1991.
[446] D. F. Lawden. Fundamentals of space navigation. *Journal of the British Interplanetary Society*, 13(2):87–101, March 1954.
[447] D. F. Lawden. *Elliptic Functions and Applications*, volume 80 of *Applied Mathematical Sciences*. Springer-Verlag, New York, 1st edition, 1989.
[448] C. Leubner. Correcting a widespread error concerning the angular velocity of a rotating rigid body. *American Journal of Physics*, 49:232–234, March 1981.
[449] A. Lichtenberg and M. Lieberman. *Regular and Chaotic Dynamics*, volume 38 of *Applied Mathematical Sciences*. Springer-Verlag, New York, 2nd edition, 1992.
[450] M. L. Lidov. The evolution of orbits of artificial satellites of planets under the action of gravitational perturbations of external bodies. *Planetary and Space Science*, 9:719–759, October 1962.
[451] M. L. Lidov. On the approximated analysis of the orbit evolution of artificial satellites. In M. Roy, editor, *Dynamics of Satellites/Dynamique des Satellites*, volume 62 of *IAU Symposium*, pages 168–179. Springer, Berlin, Heidelberg, 1963.
[452] M. L. Lidov. Integrable cases of satellite problem with the third body and the oblate planet. In Y. Kozai, editor, *Stability of the Solar System and of Small Stellar Systems*, volume 62 of *IAU Symposium*, pages 117–124, 1974.
[453] M. L. Lidov. Semianalytical methods for calculating satellite motion. *TRUDY Institut Teoreticheskoi Astronomii, Akademiia Nauk SSSR*, 17:54–61, 1978. In Russian.
[454] M. L. Lidov and M. A. Vashkov'yak. Perturbation theory and analysis of the evolution of quasi-satellite orbits in the restricted three-body problem. *Cosmic Research*, 31:187–207, September 1993.
[455] M. L. Lidov and M. A. Vashkov'yak. On quasi-satellite orbits for experiments on refinement of the gravitation constant. *Astronomy Letters*, 20:188–198, March 1994.
[456] M. L. Lidov and M. A. Vashkov'yak. On quasi-satellite orbits in a restricted elliptic three-body problem. *Astronomy Letters*, 20:676–690, September 1994.
[457] M. L. Lidov and M. V. Yarskaya. Integrable Cases in the Problem of the Evolution of a Satellite Orbit under the Joint Effect of an Outside Body and of the Noncentrality of the Planetary Field. *Cosmic Research*, 12:139–152, March 1974.
[458] S. Lie. Theorie der Transformationsgruppen I. *Mathematische Annalen*, 16:441–528, 1880.
[459] A. Lindstedt. Über die allgemeine Form der Integrale des Dreikörperproblems. *Astronomische Nachrichten*, 105:97, April 1883.
[460] J. C. Liou and J. K. Weaver. Orbital Dynamics of High Area-To-Mass Ratio Debris and Their Distribution in the Geosynchronous Region. In D. Danesy, editor, *4th European Conference on Space Debris*, volume 587 of *ESA Special Publication*, pages 285–290, August 2005.
[461] J. J. F. Liu. Satellite motion about an oblate earth. *AIAA Journal*, 12:1511–1516, January 1974.
[462] J. J. F. Liu and R. L. Alford. Semianalytic Theory for a Close-Earth Artificial Satellite. *Journal of Guidance Control Dynamics*, 4:576, September 1981.
[463] H. S. London. Some Exact Solutions of the Equations of Motion of a Solar Sail With Constant Sail Setting. *ARS Journal*, 30(2):198–200, 1960.
[464] A. C. Long, J. O. Cappellari, C. E. Velez, and A. J. Fluchs. Goddard Trajectory Determination System (GTDS). Mathematical Theory. Revision I. Technical Report FDD/552-89/001, National Aeronautics and Space Administration, Goddard Space Flight Center, Greenbelt, MD., July 1989.
[465] D. W. Lozier. NIST Digital Library of Mathematical Functions. *Annals of Mathematics and Artificial Intelligence*, 38(1–3):105–119, May 2003.
[466] L. G. Lukyanov, N. V. Emeljanov, and G. I. Shirmin. Generalized Problem of Two Fixed Centers or the Darboux-Gredeaks Problem. *Cosmic Research*, 43(3):186–191, May 2005.

[467] D. Lutzky and C. Uphoff. Short-periodic variations and second-order numerical averaging. In *AIAA, 13th Aerospace Sciences Meeting*, January 20–22, 1975, volume 11 of *Aerospace Sciences Meeting*, pages 31–38, American Institute of Aeronautics and Astronautics, Reston, VA, January 1975. AIAA Paper 75-11.

[468] R. H. Lyddane. Small eccentricities or inclinations in the Brouwer theory of the artificial satellite. *Astronomical Journal*, 68(8):555–558, October 1963.

[469] R. H. Lyddane and C. J. Cohen. Numerical comparison between Brouwer's theory and solution by Cowell's method for the orbit of an artificial satellite. *The Astronomical Journal*, 67:176–177, April 1962.

[470] M. Macdonald and C. R. McInnes. Solar sail science mission applications and advancement. *Advances in Space Research*, 48(11):1702–1716, December 2011.

[471] B. Mahajan, S. R. Vadali, and K. T. Alfriend. Exact Delaunay normalization of the perturbed Keplerian Hamiltonian with tesseral harmonics. *Celestial Mechanics and Dynamical Astronomy*, 130:25, March 2018.

[472] A. Marchesiello and G. Pucacco. Bifurcation Sequences in the Symmetric 1:1 Hamiltonian Resonance. *International Journal of Bifurcation and Chaos*, 26:1630011, 2016.

[473] L. Markley and J. Crassidis. *Fundamentals of Spacecraft Attitude Determination and Control*, volume 33 of *Space Technology Library*. Springer-Verlag, New York, 2014.

[474] J. E. Marsden and T. Ratiu. *Introduction to Mechanics and Symmetry*, volume 17 of *Texts in Applied Mathematics*. Springer-Verlag, New York, 2nd edition, 1999.

[475] V. Martinusi, L. Dell'Elce, and G. Kerschen. Analytic propagation of near-circular satellite orbits in the atmosphere of an oblate planet. *Celestial Mechanics and Dynamical Astronomy*, 123(1):85–103, September 2015.

[476] V. Martinusi, L. Dell'Elce, and G. Kerschen. First-order analytic propagation of satellites in the exponential atmosphere of an oblate planet. *Celestial Mechanics and Dynamical Astronomy*, 127:451–476, April 2017.

[477] J. J. Masdemont. High-order expansions of invariant manifolds of libration point orbits with applications to mission design. *Dynamical Systems*, 20(1):59–113, 2005.

[478] W. D. McClain. *A Recursively Formulated First-Order Semianalytic Artificial Satellite Theory Based on the Generalized Method of Averaging, Volume 1: The Generalized Method of Averaging Applied to the Artificial Satellite Problem*. NASA CR-156782. NASA, Greenbelt, Maryland, 2nd edition, 1977. Provided by the NASA Technical Reports Server (NTRS).

[479] C. R. McInnes. *Solar sailing. Technology, dynamics and mission applications. Astronomy and Planetary Sciences*. Springer, London (UK), 1st edition, 1999.

[480] J-C. van der Meer and R. H. Cushman. Orbiting dust under radiation pressure. In H. B. Doebner and J. D. Hennig, editors, *Proceedings of the XVth International Conference on Differential Geometric Methods in Theoretical Physics*, Clausthal-Zellerfeld, Germany, 1986, pages 403–414. World Scientific, Singapore, 1987.

[481] J. Meeus. *Mathematical astronomy morsels*. Willmann-Bell, Richmond, VA, 1997.

[482] J. Meeus. *Astronomical algorithms*. Willmann-Bell, Richmond, VA, 2nd edition, 1998.

[483] P. M. Mehta, A. Walker, E. Lawrence, R. Linares, D. Higdon, and J. Koller. Modeling satellite drag coefficients with response surfaces. *Advances in Space Research*, 54(8):1590–1607, October 2014.

[484] G. Mengali and A. A. Quarta. Heliocentric trajectory analysis of Sun-pointing smart dust with electrochromic control. *Advances in Space Research*, 57:991–1001, February 2016.

[485] D. Merritt. *Dynamics and Evolution of Galactic Nuclei. Princeton Series in Astrophysics*. Princeton University Press, Princeton, New Jersey, 08540.

[486] W. A. Mersman. Explicit recursive algorithms for the construction of equivalent canonical transformations. *Celestial Mechanics*, 3:384–389, September 1971.

[487] G. Metris. Mean values of particular functions in the elliptic motion. *Celestial Mechanics and Dynamical Astronomy*, 52:79–84, March 1991.

[488] G. Métris and P. Exertier. Semi-analytical theory of the mean orbital motion. *Astronomy and Astrophysics*, 294:278–286, February 1995.

[489] G. Metris, P. Exertier, Y. Boudon, and F. Barlier. Long period variations of the motion of a satellite due to non-resonant tesseral harmonics of a gravity potential. *Celestial Mechanics and Dynamical Astronomy*, 57:175–188, October 1993.

[490] K. R. Meyer. Lie Transform Tutorial – II. In K. R. Meyer and D. S. Schmidt, editors, *Computer Aided Proofs in Analysis*, volume 28 of *The IMA Volumes in Mathematics and Its Applications*. Springer, New York, NY, 1991.

[491] K. R. Meyer, G. R. Hall, and D. Offin. *Introduction to Hamiltonian Dynamical Systems and the N-Body Problem*, volume 90 of *Applied Mathematical Sciences*. Springer, New York, 2nd edition, 2009.

[492] M. Michalodimitrakis. Hill's problem - Families of three-dimensional periodic orbits. I. *Astrophysics and Space Science*, 68:253–268, March 1980.

[493] F. Mignard. Radiation pressure and dust particle dynamics. *Icarus*, 49(3):347–366, March 1982.

[494] F. Mignard. Effects of radiation forces on dust particles in planetary rings. In R. Greenberg and A. Brahic, editors, *IAU Colloq. 75: Planetary Rings*, pages 333–366, 1984.

[495] F. Mignard and M. Hénon. About an Unsuspected Integrable Problem. *Celestial Mechanics*, 33(3):239–250, July 1984.

[496] A. Milani, A. M. Nobili, and P. Farinella. *Non-gravitational perturbations and satellite geodesy*. Adam Hilger Ltd., Bristol, UK, 1987.

[497] M. Milankovitch. *Kanon der Erdbestrahlung und seine Anwendung auf das Eiszeitenproblem. Mechanics of Space Flight*. Königlich Serbische Akademie, Belgrade, 1941. English translation: Canon of Insolation and the Ice-age Problem. Israel Program for Scientific Translations, Jerusalem, 1969.

[498] B. R. Miller. The Lissajous transformation. III - Parametric bifurcations. *Celestial Mechanics and Dynamical Astronomy*, 51:251–270, September 1991.

[499] X. Ming and X. Shijie. Exploration of distant retrograde orbits around Moon. *Acta Astronautica*, 65:853–860, September 2009.

[500] H. D. Mittelmann. A Pseudo-Arclength Continuation Method for Nonlinear Eigenvalue Problems. *SIAM Journal on Numerical Analysis*, 23(5):1007–1016, October 1986.

[501] K. Moe and M. M. Moe. Gas surface interactions and satellite drag coefficients. *Planetary and Space Science*, 53(8):793–801, July 2005.

[502] M. M. Moe. Solar-Lunar Perturbations of the Orbit of an Earth Satellite. *ARS Journal*, 30(5):485–487, 1960.

[503] M. M. Moe, S. D. Wallace, and K. Moe. Recommended drag coefficients for aeronomic satellites. *Geophysical Monograph Series*, 87:349–356, 1995.

[504] N. D. Moiseev. On some simplified models of celestial mechanics based on the restricted circular three-body problem. 1, The plane circular problem. *Publications of the Sternberg State Astronomical Institute*, 15(1):75–99, January 1945. In Russian.

[505] N. D. Moiseev. On some simplified models of celestial mechanics based on the restricted circular three-body problem. 2, Averaged versions of the bounded spatial circular problem. *Publications of the Sternberg State Astronomical Institute*, 15(1):100–117, January 1945. In Russian.

[506] O. Montenbruck and E. Gill. *Satellite Orbits. Models, Methods and Applications. Physics and Astronomy*. Springer-Verlag, Berlin, Heidelberg, New York, 2001.

[507] P. Moore. A resonance problem of two degrees of freedom. *Celestial Mechanics*, 30:31–47, May 1983.

[508] P. Moore. A problem of libration with two degrees of freedom. *Celestial Mechanics*, 33:49–69, May 1984.

[509] B. Morando. Orbites de Résonance des Satellites de 24 H. *Bulletin Astronomique*, 24:47–67, 1963.

[510] A. Morbidelli. *Modern Celestial Mechanics: Aspects of Solar System Dynamics. Advances in Astronomy and Astrophysics*. Taylor & Francis, London, 2002.

[511] J. A. Morrison. Generalized Method of Averaging and the Von Zeipel Method. In R. L. Duncombe and V. G. Szebehely, editors, *Methods in Astrodynamics and Celestial Mechanics*, volume 17 of *Progress in Astronautics and Rocketry*, pages 117–138. Elsevier, 1966.

[512] J. Moser. Regularization of Kepler's problem and the averaging method on a manifold. *Communications in Pure Applied Mathematics*, 23:609–636, January 1970.

[513] D. Mostaza Prieto, B. P. Graziano, and P. C. E. Roberts. Spacecraft drag modelling. *Progress in Aerospace Sciences*, 64:56–65, January 2014.

[514] F. R. Moulton. *An Introduction to Celestial Mechanics*. The MacMillan Company, New York, 2nd edition, 1914. Also, Dover Books on Astronomy, Dover Publications Inc., 1984.

[515] F. R. Moulton, D. Buchanan, T. Buck, F. L. Griffin, W. R. Longley, and W. D. MacMillan. *Periodic orbits*, volume 161. Publications of the Carnegie Institution of Washington, Washington, D. C., 1920.

[516] J. A. Murdock. *Perturbations: Theory and Methods*, volume 27 of *Classics in Applied Mathematics*. SIAM–Society for Industrial and Applied Mathematics, Philadelphia, 1999.

[517] C. D. Murray and S. F. Dermott. *Solar system dynamics*. Cambridge University Press, New York, 1999.

[518] P. Musen. Special perturbations of the vectorial elements. *The Astronomical Journal*, 59:262–267, August 1954.

[519] P. Musen. Contributions to the theory of satellite orbits. In H. K. Bijl, editor, *Space Research*, pages 434–447. North-Holland, New York, 1960.

[520] P. Musen. The Influence of the Solar Radiation Pressure on the Motion of an Artificial Satellite. *Journal of Geophysical Research*, 65:1391–1396, May 1960.

[521] P. Musen, A. Bailie, and E. Upton. Development of the Lunar and Solar Perturbations in the Motion of an Artificial Satellite. *NASA Special Publication*, 54:24, 1965. First published: NASA-TN-D-494, 1961.

[522] P. Musen and A. E. Bailie. On the Motion of a 24-Hour Satellite. *Journal of Geophysical Research*, 67(3):1123–1132, March 1962.

[523] P. Musen, R. Bryant, and A. Bailie. Perturbations in Perigee Height of Vanguard I. *Science*, 131:935–936, March 1960.

[524] F. Namouni. Secular Interactions of Coorbiting Objects. *Icarus*, 137:293–314, February 1999.

[525] S. Naoz. The Eccentric Kozai-Lidov Effect and Its Applications. *Annual Reviews of Astronomy and Astrophysics*, 54:441–489, September 2016.

[526] A. H. Nayfeh. *Perturbation Methods*. Wiley-VCH Verlag GmbH & Co. KGaA, Weinheim, Germany, 2004.

[527] X. X. Newhall and J. G. Williams. Estimation of the Lunar Physical Librations. *Celestial Mechanics and Dynamical Astronomy*, 66(1):21–30, March 1996.

[528] L. Niccolai, M. Bassetto, A. A. Quarta, and G. Mengali. A review of Smart Dust architecture, dynamics, and mission applications. *Progress in Aerospace Sciences*, 106:1–14, April 2019.

[529] M. Nicoli. First Order Analytical Solution for Distant Retrograde Orbits in the Circular Restricted Three-Body Problem. Master thesis, School of Industrial and Information Engineering, Department of Aerospace Sciences and Technologies, Politecnico di Milano, Via La Masa 34, 20156 Milano, Italy, April 2019.

[530] J. Oberst, K. Willner, and K. Wickhusen. DePhine – The Deimos and Phobos Interior Explorer – A Proposal to ESA's Cosmic Vision Program. In *European Planetary Science Congress, EPSC2017–539*, September 2017.

[531] P. Oberti. The main problem of geosynchronous satellite theory around an equilibrium position. *Astronomy and Astrophysics*, 284:281–284, April 1994.

[532] P. Oberti. A simple intermediary orbit for the J_2 problem. *Astronomy & Astrophysics*, 437:333–338, July 2005.

[533] K. Ochs. A comprehensive analytical solution of the nonlinear pendulum. *European Journal of Physics*, 32:479–490, March 2011.

[534] E. J. Öpik. *Interplanetary encounters: close-range gravitational interactions. Developments in solar system and space science*. Elsevier, Amsterdam, 1976.

[535] A. A. Orlov. Second-order short-period solar perturbations in the motion of the satellites of planets. *Byulleten' Instituta Teoreticheskoj Astronomii (Leningrad)*, 12:302–309, January 1970.

[536] C. Osácar and J. F. Palacián. Decomposition of functions for elliptical orbits. *Celestial Mechanics and Dynamical Astronomy*, 60(2):207–223, October 1994.

[537] C. Osácar, J. F. Palacián, and M. Palacios. Numerical Evaluation of the Dilogarithm of Complex Argument. *Celestial Mechanics and Dynamical Astronomy*, 62:93–98, May 1995.

[538] J. F. Palacián. *Teoría del satélite artificial: armónicos teserales y su relegación mediante simplificaciones algebraicas*. PhD thesis, Dept. Física de la Tierra y el Cosmos, Universidad de Zaragoza, Pedro Cerbuna 12, Zaragoza, Spain, May 1992.

[539] J. F. Palacián. An Analytical Solution for Artificial Satellites at Low Altitudes. In K. Kurzyńska, F. Barlier, P. K. Seidelmann and I. Wytrzyszczak, editors, *Dynamics and Astrometry of Natural and Artificial Celestial Bodies, Proceedings of the Conference on Astrometry and Celestial Mechanics held in Poznań, Poland*, September 13–17, 1993, pages 365–370. Astronomical Observatory of A. Mickiewicz University, Poznan, Poland, 1994.

[540] J. F. Palacián. Normal forms for perturbed Keplerian systems. *Journal of Differential Equations*, 180(2):471–519, 2002.

[541] J. F. Palacián, J. F. San-Juan, and P. Yanguas. Analytical Theory for the Spot Satellite. In K. C. Howell, D. A. Cicci, J. E. Cochran Jr. and T. S. Kelso, editors, *AAS/AIAA Spaceflight Mechanics Meeting 1997*, P.O. Box 28130, San Diego, California 92198, USA, volume 95 of *Advances in the Astronautical Sciences*, pages 375–382. American Astronautical Society, Univelt, Inc., 1997.

[542] J. F. Palacián, J. Vanegas, and P. Yanguas. Compact normalisations in the elliptic restricted three body problem. *Astrophysics and Space Science*, 362:215, November 2017.

[543] M. E. Parke, R. H. Stewart, D. L. Farless, and D. E. Cartwright. On the choice of orbits for an altimetric satellite to study ocean circulation and tides. *Journal of Geophysical Research: Oceans*, 92(C11):11693–11707, 1987.

[544] D. L. Parker. An Analytic Derivation of the Relationship between the Angular Velocity Vector and the Euler Angles and their Time Derivatives. *American Journal of Physics*, 37(9):925–927, 1969.

[545] R. W. Parkinson, H. M. Jones, and I. I. Shapiro. Effects of Solar Radiation Pressure on Earth Satellite Orbits. *Science*, 131(3404):920–921, 1960.

[546] P. Pascua, J. L. Rubio, A. Viartola, and S. Ferrer. Visualizing relative equilibria and bifurcations by painting Hamiltonians on personal computers. *International Journal of Bifurcation and Chaos*, 06(08):1411–1424, 1996.

[547] M. E. Paskowitz and D. J. Scheeres. Design of Science Orbits About Planetary Satellites: Application to Europa. *Journal of Guidance Control Dynamics*, 29:1147–1158, September 2006.

[548] M. E. Paskowitz and D. J. Scheeres. Control of Science Orbits About Planetary Satellites. *Journal of Guidance Control Dynamics*, 32(1):223–231, 2009.

[549] S. J. Peale. Dust belt of the Earth. *Journal of Geophysical Research*, 71(3):911–933, February 1966.

[550] J. Peláez and M. Lara. Dynamics of fast-rotating tethered satellites. *Monografías de la Real Academia de Ciencias de Zaragoza*, 32:75–83, 2009.

[551] J. Peláez, M. Lara, C. Bombardelli, F. R. Lucas, M. Sanjurjo-Rivo, D. Curreli, E. C. Lorenzini, and D. J. Scheeres. Periodic Orbits of a Hill-Tether Problem Originated from Collinear Points. *Journal of guidance, control, and dynamics*, 35(1):222–233, 2012.

[552] I. C. Percival and D. Richards. *Introduction to Dynamics*. Cambridge University Press, Cambridge, UK, January 1983.

[553] S. Persson, B. Jacobsson, and E. Gill. PRISMA – Demonstration Mission for Advanced Rendezvous and Formation Flying Technologies and Sensors (paper IAC-05-B56B07). In *Proceedings of the 56th International Astronautical Congress (IAC)*, October 17–21 2005, Fukuoka, Japan, pages 1–10. International Astronautical Federation, IAF, October 2005.

[554] J.-M. Petit and M. Hénon. Satellite encounters. *Icarus*, 66:536–555, June 1986.

[555] C. M. Petty and J. V. Breakwell. Satellite orbits about a planet with rotational symmetry. *Journal of the Franklin Institute*, 270(4):259–282, 1960.

[556] H. C. Plummer. On the Possible Effects of Radiation on the Motion of Comets, with special reference to Encke's Comet. *Monthly Notices of the Royal Astronomical Society*, 65:229–238, January 1905.

[557] H. C. K. Plummer. *An introductory treatise on dynamical astronomy*. Cambridge, University press, 1918.

[558] H. Poincaré. *Les méthodes nouvelles de la mécanique céleste*. Paris, Gauthier-Villars et fils, 1892-1899.

[559] H. Poincaré. Sur les planètes du type d'Hécube. *Bulletin Astronomique, Serie I*, 19:289–310, January 1902.

[560] A. Pousse, P. Robutel, and A. Vienne. On the co-orbital motion in the planar restricted three-body problem: the quasi-satellite motion revisited. *Celestial Mechanics and Dynamical Astronomy*, 128:383–407, August 2017.

[561] J. H. Poynting. Radiation in the Solar System: its Effect on Temperature and its Pressure on Small Bodies. *Philosophical Transactions A*, 202:525–552, June 1903. Also, Collected Scientific Papers, Cambridge, 1920, pp. 304–330.

[562] R. J. Proulx and W. D. McClain. Series representations and rational approximations for Hansen coefficients. *Journal of Guidance Control Dynamics*, 11:313–319, July 1988.

[563] R. J. Proulx, W. D. McClain, L. W. Early, and P. J. Cefola. A theory for the short-periodic motion due to the tesseral harmonic gravity field. In *AAS/AIAA, Astrodynamics Specialist Conference*. American Astronautical Society, August 1981.

[564] G. Pucacco. Structure of the centre manifold of the L_1, L_2 collinear libration points in the restricted three-body problem. *Celestial Mechanics and Dynamical Astronomy*, 131(10):44, September 2019.

[565] P. Puig Adam. *Curso teórico prãctico de ecuaciones diferenciales aplicado a la Física y Técnica*. Nuevas Gráficas, S.A., Madrid, 10th edition, 1967. In Spanish.

[566] D. L. Richardson. Analytic construction of periodic orbits about the collinear points. *Celestial Mechanics*, 22:241–253, October 1980.

[567] C. E. Roberts Jr.. An Analytic Model for Upper Atmosphere Densities Based Upon Jacchia's 1970 Models. *Celestial Mechanics*, 4(3–4):368–377, December 1971.

[568] H. P. Robertson. Dynamical Effects of Radiation in the Solar System. *Monthly Notices of the Royal Astronomical Society*, 97:423–437, April 1937.

[569] O. Rodrigues. Mémoire sur l'attraction des sphéroïdes. *Correspondence Sur l'École Impériale Polytechnique*, 3(3):361–385, January 1816. Thesis for the Faculty of Science of the University of Paris.

[570] A. Rom. Mechanized Algebraic Operations (MAO). *Celestial Mechanics*, 1:301–319, September 1970.

[571] M. Romano. Concise form of the dynamic and kinematic solutions of the Euler-Poinsot problem (IAA-AAS-DyCoSS1-01-05 – AAS 12-304). In A. D. Guerman, P. M. Bainum and J.-M. Contant, editors, *Dynamics and Control of Space Systems DyCoSS'2012*, P.O. Box 28130, San Diego, California 92198, volume 145 of *Advances in the Astronautical Sciences*, pages 55–68. American Astronautical Society, Univelt, Inc., 2012.

[572] B. A. Romanowicz. On The Tesseral-Harmonics Resonance Problem in Artificial-Satellite Theory. *SAO Special Report*, 365, March 1975.

[573] B. A. Romanowicz. On the Tesseral-Harmonics Resonance Problem in Artificial-Satellite Theory. Part II. *SAO Special Report*, 373, March 1976.

[574] G. W. Rosborough and C. Ocampo. Influence of Higher Degree Zonals on the Frozen Orbit Geometry (AAS 91-428). In B. Kaufman, K. T. Alfriend, R. L. Roehrich and R. R. Dasenbrock, editors, *Astrodynamics 1991*, P.O. Box 28130, San Diego, California 92198, volume 76 of *Advances in the Astronautical Sciences*, pages 1291–1304. American Astronautical Society, Univelt, Inc., 1992.

[575] A. J. Rosengren, H. Namazyfard, and G. E. O. Giacaglia. Effects of higher-order multipoles of the lunar disturbing potential on elongated orbits in cislunar space. *European Physical Journal Special Topics*, 229(8):1545–1555, May 2020.

[576] A. J. Rosengren and D. J. Scheeres. Long-term dynamics of high area-to-mass ratio objects in high-Earth orbit. *Advances in Space Research*, 52(8):1545–1560, October 2013.

[577] A. J. Rosengren and D. J. Scheeres. On the Milankovitch orbital elements for perturbed Keplerian motion. *Celestial Mechanics and Dynamical Astronomy*, 118:197–220, March 2014.

[578] A. J. Rosengren, D. J. Scheeres, and J. W. McMahon. The classical Laplace plane as a stable disposal orbit for geostationary satellites. *Advances in Space Research*, 53(8):1219–1228, April 2014.

[579] S. D. Ross and D. J. Scheeres. Multiple Gravity Assists, Capture, and Escape in the Restricted Three-Body Problem. *SIAM Journal on Applied Dynamical Systems*, 6(3):576–596, January 2007.

[580] A. Rossi. Resonant dynamics of Medium Earth Orbits: space debris issues. *Celestial Mechanics and Dynamical Astronomy*, 100(4):267–286, April 2008.

[581] A. E. Roy. Luni-Solar Perturbations of an Earth Satellite. *Astrophysics and Space Science*, 4:375–386, August 1969.

[582] A. E. Roy. *Orbital motion*. Institute of Physics Publishing, Bristol (UK), 4th edition, 2005.

[583] A. E. Roy and P. E. Moran. Studies in the Application of Recurrence Relations to Special Perturbation Methods. III. Non-Singular Differential Equations for Special Perturbations. *Celestial Mechanics*, 7:236–255, February 1973.

[584] D. P. Rubincam and N. S. Weiss. Earth Albedo and the Orbit of LAGEOS. *Celestial Mechanics*, 38(3):233–296, March 1986.

[585] R. P. Russell and M. Lara. On the design of an Enceladus science orbit. *Acta Astronautica*, 65(1–2):27–39, 2009.

[586] R. P. Russell and M. Lara. Long-Lifetime Lunar Repeat Ground Track Orbits. *Journal of Guidance, Control, and Dynamics*, 30(4):982–993, July-August 2007.

[587] Yu. A. Sadov. The Action-Angles Variables in the Euler-Poinsot Problem. *PMM-Journal of Applied Mathematics and Mechanics*, 34(5):922–925, 1970.

[588] Yu. A. Sadov. The Action-Angles Variables in the Euler-Poinsot Problem. Preprint No. 22 KIAM Russian Academy of Sciences Moscow (in Russian), 1970.

[589] B. de Saedeleer. Complete Zonal Problem of the Artificial Satellite: Generic Compact Analytic First Order in Closed Form. *Celestial Mechanics and Dynamical Astronomy*, 91:239–268, March 2005.

[590] J. F. San-Juan. ATESAT: Automatization of theories and ephemeris in the artificial satellite problem. Technical Report CT/TI/MS/MN/94-250, Centre National d'Études Spatiales, 18, avenue Edouard Belin - 31401 Toulouse Cedex 9, France, May 1994.

[591] J. F. San-Juan, A. Abad, M. Lara, and D. J. Scheeres. First-Order Analytical Solution for Spacecraft Motion About (433) Eros. *Journal of Guidance Control Dynamics*, 27:290–293, March 2004.
[592] J. F. San-Juan, M. Lara, and S. Ferrer. Phase Space Structure Around Oblate Planetary Satellites. *Journal of Guidance Control Dynamics*, 29:113–120, January 2006.
[593] J. F. San-Juan, D. Ortigosa, L. M. López-Ochoa, and R. López. Deprit's Elimination of the Parallax Revisited. *The Journal of the Astronautical Sciences*, 60:137–148, June 2015.
[594] J. A. Sanders, F. Verhulst, and J. Murdock. *Averaging Methods in Nonlinear Dynamical Systems*, volume 59 of *Applied Mathematical Sciences*. Springer, New York, 2nd edition, 2007.
[595] M. Sansottera and M. Ceccaroni. Rigorous estimates for the relegation algorithm. *Celestial Mechanics and Dynamical Astronomy*, 127:1–18, January 2017.
[596] D. J. Scheeres. Satellite Dynamics about small bodies: Averaged Solar Radiation Pressure Effects. *The Journal of the Astronautical Sciences*, 41(1):25–46, 1999.
[597] D. J. Scheeres. *Orbital Motion in Strongly Perturbed Environments. Astronautical Engineering*. Springer-Verlag, Berlin, 2012.
[598] D. J. Scheeres, M. D. Guman, and B. F. Villac. Stability Analysis of Planetary Satellite Orbiters: Application to the Europa Orbiter. *Journal of Guidance Control Dynamics*, 24(4):778–787, July 2001.
[599] G. Scheifele and O. Graf. Analytical satellite theories based on a new set of canonical elements. In *Mechanics and Control of Flight Conference*, pages 1–20. American Institute of Aeronautics and Astronautics, Reston, Virigina, February 1974.
[600] J. Schubart. Long-Period Effects in Nearly Commensurable Cases of the Restricted Three-Body Problem. *SAO Special Report*, 149, April 1964.
[601] B. Schutz and D. Craig. GPS orbit evolution – 1998-2000 (Paper AIAA-2000-4237). In *Astrodynamics Specialist Conference*, pages 1–11, 2000.
[602] A. M. Segerman and S. L. Coffey. An analytical theory for tesseral gravitational harmonics. *Celestial Mechanics and Dynamical Astronomy*, 76(3):139–156, 2001.
[603] H. Sellamuthu and R. K. Sharma. Hybrid Orbit Propagator for Small Spacecraft Using Kustaanheimo–Stiefel Elements. *Journal of Spacecraft and Rockets*, 55:1282–1288, September 2018.
[604] B. E. Shapiro. Phase Plane Analysis and Observed Frozen Orbit for the Topex/ Poseidon Mission. In P. M. Bainum, G. L. May, Y. Ohkami, K. Uesugi, Q. Faren and L. Furong, editors, *6th AAS/JRS/CSA Symposium, International Space Conference of Pacific-Basin Societies, Strengthening Cooperation in the 21st Century*, P.O. Box 28130, San Diego, California 92198, volume 91 of *Advances in the Astronautical Sciences*, pages 853–872. American Astronautical Society, Univelt, Inc., 1996.
[605] I. I. Shapiro. The prediction of satellite orbits. In M. Roy, editor, *Dynamics of Satellites/Dynamique des Satellites*, pages 257–312. Springer, Berlin, Heidelberg, 1963.
[606] R. K. Sharma. Analytical integration of K-S element equations with J_2 for short-term orbit predictions. *Planetary and Space Science*, 45(11):1481–1486, November 1997.
[607] I. I. Shevchenko. *The Lidov-Kozai Effect – Applications in Exoplanet Research and Dynamical Astronomy*, volume 441 of *Astrophysics and Space Science Library*. Springer, Cham, Switzerland, 2017.
[608] V. V. Sidorenko, A. I. Neishtadt, A. V. Artemyev, and L. M. Zelenyi. Quasi-satellite orbits in the general context of dynamics in the 1:1 mean motion resonance: perturbative treatment. *Celestial Mechanics and Dynamical Astronomy*, 120:131–162, October 2014.
[609] C. L. Siegel and J. Moser. *Lectures on celestial mechanics. Classics in Mathematics*. Springer-Verlag, Berlin Heidelberg, 1st edition, 1971. Reprint of the 1971 Edition (Grundlehren der mathematischen Wissenschaften, Vol. 187), translated by C. I. Kalme.

[610] C. Simo. On the Analytical and Numerical Approximation of Invariant Manifolds. In D. Benest and C. Froeschle, editors, *Modern Methods in Celestial Mechanics*, page 285, January 1990.

[611] C. Simó and T. J. Stuchi. Central stable/unstable manifolds and the destruction of KAM tori in the planar Hill problem. *Physica D Nonlinear Phenomena*, 140:1–32, June 2000.

[612] A. S. Sochilina. On the motion of a satellite in resonance with its rotating planet. *Celestial Mechanics*, 26:337–352, April 1982.

[613] E. M. Soop. *Handbook of Geostationary Orbits*, volume 3 of *Space Technology Library*. Springer, Netherlands, 1st edition, 1994.

[614] J. Souchay, M. Folgueira, and S. Bouquillon. Effects of the Triaxiality on the Rotation of Celestial Bodies: Application to the Earth, Mars and Eros. *Earth Moon and Planets*, 93(2):107–144, October 2003.

[615] J. Souchay, H. Kinoshita, H. Nakai, and S. Roux. A precise modeling of Eros 433 rotation. *Icarus*, 166(2):285–296, October 2003.

[616] F. Spirig and J. Waldvogel. The Three-Body Problem With Two Small Masses: A Singular-Perturbation Approach to the Problem of Saturn's Coorbiting Satellites. In V. G. Szebehely, editor, *Stability of the Solar System and Its Minor Natural and Artificial Bodies*, volume 154 of *NATO Advanced Science Institutes (ASI) Series C*, pages 53–63. Springer Netherlands, Dordrecht, January 1985.

[617] R. Sridharan and W. P. Seniw. An intermediate-averaged theory for high altitude orbits. Technical Note 1979-25, MIT Lincoln Laboratory, Lexington, MA 02173, June 1979.

[618] D. Steichen. An averaging method to study the motion of lunar artificial satellites II: Averaging and Applications. *Celestial Mechanics and Dynamical Astronomy*, 68(3):225–247, July 1998.

[619] S. Steinberg. Lie series, Lie transformations, and their applications. In J. Sánchez Mondragón and K. B. Wolf, editors, *Lie Methods in Optics*, volume 250 of *Lecture Notes in Physics*, page 45. Springer-Verlag, Berlin, 1986.

[620] T. E. Sterne. The gravitational orbit of a satellite of an oblate planet. *The Astronomical Journal*, 63:28–40, January 1958.

[621] E. L. Stiefel and G. Scheifele. *Linear and Regular Celestial Mechanics*, volume 174 of *Grundlehren der mathematischen Wissenschaften*. Springer-Verlag, Berlin Heidelberg, 1st edition, 1971.

[622] M. Stramacchia, C. Colombo, and F. Bernelli-Zazzera. Distant Retrograde Orbits for space-based Near Earth Objects detection. *Advances in Space Research*, 58:967–988, September 2016.

[623] B. Strömgren. Formeln zur genaeherten Stoerungsrechnung in Bahnelementen. Angewandt auf die Planeten 633 Zelima, 956 [1921 IW], 979 Ilsewa, 1035 Amata und 1049 [1925RB]. *Publikationer og mindre Meddeler fra Kobenhavns Observatorium*, 65:1–26, January 1929.

[624] E. Strömgren. Connaisance actuelle des orbites dans le problème des trois corps. *Bulletin Astronomique*, 9:87–130, January 1933.

[625] R. A. Struble. The geometry of the orbits of artificial satellites. *Archive for Rational Mechanics and Analysis*, 7(1):87–104, January 1961.

[626] K. F. Sundman. Mémoire sur le problème des trois corps. *Acta Mathematica*, 36(1):105–179, December 1913.

[627] G. J. Sussman and J. Wisdom. *Structure and interpretation of classical mechanics*. MIT Press, Cambridge, Massachusetts, 1st edition, 2001.

[628] V. Szebehely. *Theory of Orbits. The Restricted Problem of Three Bodies*. Academic Press Inc., New York and London, 1967.

[629] J. L. Tabor and J. D. Vedder. Long-Term Evolution of Uncontrolled Geosynchronous Orbits: Orbital Debris Implications. *The Journal of the Astronautical Sciences*, 40(3):407–418, July-September 1992.

[630] B. Tapley, J. Ries, S. Bettadpur, D. Chambers, M. Cheng, F. Condi, B. Gunter, Z. Kang, P. Nagel, R. Pastor, T. Pekker, S. Poole, and F. Wang. GGM02 An improved Earth gravity field model from GRACE. *Journal of Geodesy*, 79:467–478, November 2005.
[631] F. Tisserand. *Traité de mécanique céleste. Tome I: Perturbations des planètes d'aprés la méthode de la variation des constantes arbitraries*. Gauthier-Villars et fils, Quai des Grands-Augustins, 55, Paris, 1889.
[632] F. Tisserand. *Traité de mécanique céleste. Tome IV: Théories des satellites de Jupiter et de Saturne. Perturbations des petit planetes*. Gauthier-Villars et fils, Quai des Grands-Augustins, 55, Paris, 1896. Chapter vi.
[633] F. Tong and L-d. Wu. Perturbation equations of the elements of Vinti's intermediate orbit. *Chinese Astronomy and Astrophysics*, 5:282–294, September 1981.
[634] T. C. Tsu. Interplanetary Travel by Solar Sail. *ARS Journal*, 29(6):422–427, 1959.
[635] C. Uphoff. Numerical averaging in orbit prediction. *AIAA Journal*, 11:1512–1516, 1973.
[636] H. Urrutxua, M. Sanjurjo-Rivo, and J. Peláez. DROMO propagator revisited. *Celestial Mechanics and Dynamical Astronomy*, 124:1–31, January 2016.
[637] B. A. Ustinov. Motion of Satellites in Small-Eccentricity Orbits in the Noncentral Gravitational Field of the Earth. *Cosmic Research*, 5:159–168, March 1967.
[638] J. Vagners. Modified long-period behavior due to tesseral harmonics. *AIAA Journal*, 6(7):1229–1234, July 1968.
[639] S. Valk, A. Lemaître, and L. Anselmo. Analytical and semi-analytical investigations of geosynchronous space debris with high area-to-mass ratios. *Advances in Space Research*, 41:1077–1090, 2008.
[640] S. Valk, A. Lemaître, and F. Deleflie. Semi-analytical theory of mean orbital motion for geosynchronous space debris under gravitational influence. *Advances in Space Research*, 43(7):1070–1082, April 2009.
[641] D. A. Vallado, P. Crawford, R. Hujsak, and T. S. Kelso. Revisiting Spacetrack Report #3 (AIAA 2006-6753). In *AIAA/AAS Astrodynamics Specialist Conference and Exhibit*, USA, August 2006, *Guidance, Navigation, and Control and Co-located Conferences*, pages 1–88. American Institute of Aeronautics and Astronautics, 2006.
[642] G. B. Valsecchi. In *Geometric Conditions for Quasi-Collisions in Öpik's Theory*, volume 682, pages 145–158. Springer, Berlin, Heidelberg, 2006.
[643] G. B. Valsecchi, E. M. Alessi, and A. Rossi. An analytical solution for the swing-by problem. *Celestial Mechanics and Dynamical Astronomy*, 123(2):151–166, October 2015.
[644] G. B. Valsecchi, T. J. Jopek, and C. Froeschlé. Meteoroid stream identification: a new approach - I. Theory. *Monthly Notices of the Royal Astronomical Society*, 304:743–750, April 1999.
[645] G. B. Valsecchi, A. Milani, G. F. Gronchi, and S. R. Chesley. Resonant returns to close approaches: Analytical theory. *Astronomy and Astrophysics*, 408:1179–1196, September 2003.
[646] M. Valtonen and H. Karttunen. *The Three-Body Problem*. Cambridge University Press, Cambridge, UK, 2006.
[647] J. C. van der Ha. Non-singular and non-conventional orbit perturbation equations. *Zeitschrift für Flugwissenschaften und Weltraumforschung*, 9:217–224, August 1985.
[648] J. C. van der Ha and V. J. Modi. Long-term evaluation of three-dimensional heliocentric solar sail trajectories with arbitrary fixed sail setting. *Celestial Mechanics*, 19:113–138, February 1979.
[649] M. A. Vashkov'yak. Stability of Circular Satellite Orbits for Combined Action of Perturbations from an External Body and from the Noncentrality of the Planetary Gravitational Field. *Cosmic Research*, 12:757–769, November 1974.
[650] M. A. Vashkov'yak. On the special particular solutions of a double-averaged Hill's problem with allowance for flattening of the central planet. *Astronomy Letters*, 22:207–216, March 1996.

[651] M. A. Vashkov'Yak. On the development of M.L. Lidov's techniques on the evolution of satellite orbits. In *Celestial Mechanics*, volume 8 of *IAA Transactions*, pages 174–177, 2002.

[652] M. A. Vashkov'Yak. A Numerical-Analytical Method for Studying the Orbital Evolution of Distant Planetary Satellites. *Astronomy Letters*, 31(1):64–72, January 2005.

[653] M. A. Vashkov'Yak. Particular Solutions of the Singly Averaged Hill Problem. *Astronomy Letters*, 31(7):487–493, July 2005.

[654] M. A. Vashkov'Yak. Periodic solutions of the singly averaged Hill problem. *Astronomy Letters*, 32(10):707–715, October 2006.

[655] M. A. Vashkov'yak and N. M. Teslenko. On the Stability of Particular Solutions of the Singly Averaged Hill Problem. *Astronomy Letters*, 31(12):844–852, December 2005.

[656] M. A. Vashkov'yak and N. M. Teslenko. Refined model for the evolution of distant satellite orbits. *Astronomy Letters*, 35(12):850–865, December 2009. Erratum: *Astronomy Letters*, 36(4):307, April 2010.

[657] F. Verhulst. Discrete Symmetric Dynamical Systems at the Main Resonances with Applications to Axi-Symmetric Galaxies. *Philosophical Transactions of the Royal Society of London Series A*, 290(1375):435–465, January 1979.

[658] R. Vilhena de Moraes. Combined solar radiation pressure and drag effects on the orbits of artificial satellites. *Celestial Mechanics*, 25:281–292, May November 1981.

[659] R. Vilhena de Moraes. Non-gravitational disturbing forces. *Advances in Space Research*, 14, May 1994.

[660] B. F. Villac and D. J. Scheeres. Escaping Trajectories in the Hill Three-Body Problem and Applications. *Journal of Guidance Control Dynamics*, 26:224–232, March 2003.

[661] N. X. Vinh. Recurrence Formulae for the Hansen's Developments. *Celestial Mechanics*, 2:64–76, March 1970.

[662] J. P. Vinti. New Method of Solution for Unretarded Satellite Orbits. *Journal of Research of the National Bureau of Standards*, 62B(2):105–162, October–December 1959.

[663] J. P. Vinti. Theory of an Accurate Intermediary Orbit for Satellite Astronomy. *Journal of Research of the National Bureau of Standards*, 65B(3):169–201, July–September 1961.

[664] J. P. Vinti. Inclusion of the third zonal harmonic in an accurate reference orbit of an artificial satellite. *Journal of Research of the National Bureau of Standards*, 70B(1):17–46, January–March 1966.

[665] J. P. Vinti. Invariant Properties of the Spheroidal Potential of an Oblate Planet. *Journal of Research of the National Bureau of Standards*, 70B(1):1–16, January–March 1966.

[666] J. P. Vinti. Improvement of the Spheroidal Method for Artificial Satellites. *Astronomical Journal*, 74:25–34, February 1969.

[667] J. P. Vinti. In G. J. Der and N. L. Bonavito, editors, *Orbital and Celestial Mechanics*, volume 177 of *Progress in Astronautics and Aeronautics*. American Institute of Aeronautics and Astronautics, Reston, Virginia, 1998.

[668] D. Vokrouhlicky, P. Farinella, and F. Mignard. Solar radiation pressure perturbations for Earth satellites. 1: A complete theory including penumbra transitions. *Astronomy and Astrophysics*, 280:295–312, December 1993.

[669] D. Vokrouhlicky, P. Farinella, and F. Mignard. Solar radiation pressure perturbations for Earth satellites II. An approximate method to model penumbra transitions and their long-term orbital effects on LAGEOS. *Astronomy and Astrophysics*, 285:333–343, May 1994.

[670] D. Vokrouhlicky, P. Farinella, and F. Mignard. Solar radiation pressure perturbations for Earth satellites. III. Global atmospheric phenomena and the albedo effect. *Astronomy and Astrophysics*, 290:324–334, October 1994.

[671] D. Vokrouhlicky and L. Sehnal. On some aspects of the albedo effect application to the ERS-1 satellite. *Celestial Mechanics and Dynamical Astronomy*, 56(3):471–490, July 1993.

[672] V. Volterra. Sur la théorie des variations des latitudes. *Acta Mathematica*, 22:201–357, 1899.

[673] G. Voyatzis and K. I. Antoniadou. On quasi-satellite periodic motion in asteroid and planetary dynamics. *Celestial Mechanics and Dynamical Astronomy*, 130(9):59, September 2018.
[674] C. A. Wagner, B. C. Douglas, and R. G. Williamson. The ROAD program. Technical Report NASA-TM-X-70676, Goddard Space Flight Center, Greenbelt, Maryland, January 1974.
[675] H. G. Walter. Conversion of osculating orbital elements into mean elements. *The Astronomical Journal*, 72:994–997, October 1967.
[676] Y. Wang and P. Gurfil. Dynamical modeling and lifetime analysis of geostationary transfer orbits. *Acta Astronautica*, 128:262–276, November 2016.
[677] J. R. Wertz, editor. In *Spacecraft Attitude Determination and Control*, volume 73 of *Astrophysics and Space Science Library*. Springer Netherlands, Dordrecht, Holland, 1st edition, 1978.
[678] A. D. Wheelon. Free Flight of a Ballistic Missile. *ARS Journal*, 29(12):915–926, December 1959.
[679] E. T. Whittaker. *A Treatise on the Analytical Dynamics of Particles and Rigid Bodies*. Cambridge University Press, Cambridge, UK, 1st edition, February 1904.
[680] E. T. Whittaker. *A Treatise on the Analytical Dynamics of Particles and Rigid Bodies*. Cambridge University Press, Cambridge, UK, 2nd edition, February 1917.
[681] W. E. Wiesel. Numerical Solution to Vinti's Problem. *Journal of Guidance Control Dynamics*, 38(9):1757–1764, September 2015.
[682] J. M. Wilkes. Rotations as solutions of a matrix differential equation. *American Journal of Physics*, 46(6):685–687, 1978.
[683] R. R. Williams and J. Lorell. The Theory of Long-Term Behavior of Artificial Satellite Orbits Due to Third-Body Perturbations. Report NASA TR-32-916, Jet Propulsion Laboratory, Pasadena, CA, February 1966.
[684] A. Wintner. *The analytical foundations of celestial mechanics*, volume 5 of *Princeton Mathematical Series*. Princeton University Press, Princeton, N. J., 1947.
[685] E. Wnuk. Tesseral Harmonic Perturbations for High Order and Degree Harmonics. *Celestial Mechanics*, 44:179–191, March 1988.
[686] E. Wnuk and T. Jopek. Satellite orbit calculations using geopotential coefficients up to high degree and order. *Advances in Space Research*, 14(5):35–42, May 1994.
[687] E. Wnuk and S. Ł. Breiter. The motion of natural and artificial satellites in Mars gravity field. *Advances in Space Research*, 11(6):183–188, 1991.
[688] S. Wright. *Orbit Determination Using Vinti's Solution*. PhD thesis, Air Force Institute of Technology, Wright-Patterson Air Force Base, Ohio, September 2016.
[689] L-d. Wu and F. Tong. A third-order solution of Vinti's problem with explicit expressions for the poisson brackets. *Chinese Astronomy and Astrophysics*, 5:192–201, June 1981.
[690] S. P. Wyatt and F. L. Whipple. The Poynting-Robertson effect on meteor orbits. *The Astrophysical Journal*, 111:134–141, January 1950.
[691] H. Yan, S. R. Vadali, and K. T. Alfriend. A Recursive Formulation of the Satellite Perturbed Relative Motion Problem. In *AIAA/AAS Astrodynamics Specialist Conference*, San Diego, CA, pages 1–14, August 2014.
[692] V. Yurasov. Universal Semianalytic Satellite Motion Propagation Method. In *Proceedings of the U.S. – Russian Second Space Surveillance Workshop*, pages 198–211. Adam Mickiewicz University, Poznan, July 1996.
[693] C. Zagouras and V. V. Markellos. Three-dimensional periodic solutions around equilibrium points in Hill's problem. *Celestial Mechanics*, 35:257–267, March 1985.
[694] M. Zamaro and J. D. Biggs. Natural motion around the Martian moon Phobos: the dynamical substitutes of the Libration Point Orbits in an elliptic three-body problem with gravity harmonics. *Celestial Mechanics and Dynamical Astronomy*, 122(3):263–302, July 2015.
[695] M. Zamaro and J. D. Biggs. Identification of new orbits to enable future mission opportunities for the human exploration of the Martian moon Phobos. *Acta Astronautica*, 119:160–182, February 2016.

[696] T. de Zeeuw and D. Merritt. Stellar orbits in a triaxial galaxy. I - Orbits in the plane of rotation. *The Astrophysical Journal*, 267:571–595, April 1983.

[697] H. von Zeipel. Sur l'application des séries de M. Lindstedt à l'étude du mouvement des comètes périodiques. *Astronomische Nachrichten*, 183(22):345, March 1910.

[698] H. von Zeipel. Research on the Motion of Minor Planets. NASA TT F-9445, 1965. (NASA Translation of: Recherches sur le mouvement des petites planètes, Arkiv för matematik, astronomi och fysik, vol. 11, 1916, vol. 12, 1917, vol. 13, 1918).

[699] Y. Zhao, P. Gurfil, and S. Zhang. Long-Term Orbital Dynamics of Smart Dust. *Journal of Spacecraft and Rockets*, 55:125–142, January 2018.

[700] N. Y. Zhukovsky. On the motion of a rigid body having cavities filled with a homogeneous liquid drop. *Russian Journal of the Physical Chemical Society*, 17:31–152, 1885. Also, Collected Works, vol. 1, Moscow-Leningrad: Gostekhisdat, 1949 (in Russian).

Index

De Gruyter Studies in Mathematical Physics

Volume 14 – 2nd Edition
Oleg N. Kirillov
Nonconservative Stability Problems of Modern Physics, 2021
ISBN: 978-3-11-065377-9, e-ISBN (PDF): 978-3-11-065540-7,
e-ISBN (EPUB) 978-3-11-065386-1

Volume 5 – 2nd Edition
Victor K. Andreev, Yuri A. Gaponenko, Olga N. Goncharova, Vladislav V. Pukhnachev
Mathematical Models of Convection, 2020
ISBN: 978-3-11-065378-6, e-ISBN (PDF): 978-3-11-065546-9,
e-ISBN (EPUB) 978-3-11-065394-6

Volume 24 – 2nd Edition
Igor Olegovich Cherednikov, Tom Mertens, Frederik Van der Veken
Wilson Lines in Quantum Field Theory, 2020
ISBN: 978-3-11-065092-1, e-ISBN (PDF): 978-3-11-065169-0,
e-ISBN (EPUB) 978-3-11-065103-4

Volume 3 – 2nd Edition
Sergei Yu. Pilyugin
Spaces of Dynamical Systems, 2019
ISBN: 978-3-11-064446-3, e-ISBN (PDF): 978-3-11-065716,
e-ISBN (EPUB) 978-3-11-065399-1

Volume 53
Vladimir K. Dobrev
Invariant Differential Operators: Volume 4: AdS/CFT, (Super-)Virasoro, Affine (Super-)Algebras, 2019
ISBN 978-3-11-060968-4, e-ISBN (PDF) 978-3-11-061140-3,
e-ISBN (EPUB) 978-3-11-060971-4

Volume 52
Alexey V. Borisov, Ivan S. Mamaev
Rigid Body Dynamics, 2018
ISBN 978-3-11-054279-0, e-ISBN (PDF) 978-3-11-054444-2,
e-ISBN (EPUB) 978-3-11-054297-4

www.degruyter.com

Bei Fragen zur Produktsicherheit wenden Sie sich bitte an:
If you have any questions regarding product safety,
please contact:

Walter de Gruyter GmbH
Genthiner Straße 13
10785 Berlin
productsafety@degruyterbrill.com